# ALMOST FREE MODULES

## Set-theoretic Methods

# North-Holland Mathematical Library

*Board of Advisory Editors:*

VOLUME 46

NORTH-HOLLAND
AMSTERDAM • NEW YORK • OXFORD • TOKYO

# Almost Free Modules

## Set-theoretic Methods

Paul C. EKLOF
*Department of Mathematics*
*University of California, Irvine*
*CA, U.S.A.*

Alan H. MEKLER
*Department of Mathematics and Statistics*
*Simon Fraser University*
*Burnaby, B.C., Canada*

1990

NORTH-HOLLAND
AMSTERDAM • NEW YORK • OXFORD • TOKYO

MATH
seplae

ELSEVIER SCIENCE PUBLISHERS B.V.
Sara Burgerhartstraat 25
P.O. Box 211, 1000 AE Amsterdam, The Netherlands

*Distributors for the U.S.A. and Canada:*

ELSEVIER SCIENCE PUBLISHING COMPANY, INC.
655 Avenue of the Americas
New York, N.Y. 10010, U.S.A.

**Library of Congress Cataloging-in-Publication Data**

Eklof, Paul C.
    Almost free modules : set-theoretic methods / Paul C. Eklof, Alan
H. Mekler.
        p.    cm. -- (North-Holland mathematical library ; v. 46)
    Includes bibliographical references.
    ISBN 0-444-88502-1 (U.S.)
    1. Modules (Algebra)  2. Set theory.  3. Abelian groups.
I. Mekler, Alan H.  II. Title.  III. Series.
QA247.E385  1990
512'.4--dc20                                                    90-6864
                                                               CIP

ISBN: 0 444 88502 1

PRINTED IN THE NETHERLANDS

To Sherry
and Barbara

# PREFACE

The modern era in set-theoretic methods in algebra can be said to have begun on July 11, 1973 when Saharon Shelah borrowed László Fuchs' *Infinite Abelian Groups* from the Hebrew University library. Soon thereafter, he showed that Whitehead's Problem — to which many talented mathematicians had devoted much creative energy — was not solvable in ordinary set theory (ZFC). In the years since, Shelah and others have made a good deal of progress on other natural and important problems in algebra, in some cases showing that the problem is not solvable in ZFC, and in others, proving results in ZFC using powerful techniques from set theory (combined, of course, with algebraic methods). One purpose of this book is to make these set-theoretic methods available to the algebraist through an exposition of their use in solving a few major problems. In addition, the logician will find here non-trivial applications of set-theoretic techniques, a few topics such as λ-systems and the Black Box which are less well-known, and also the construction of many structures of interest to model theorists.

Actually, there has long been an affinity between set theory and abelian group theory, going back at least as far as Łoś' discovery that measurable cardinals arise naturally in the study of slender groups; also, constructions by transfinite induction have been common, exemplified by Hill's work on almost free groups. Modern set-theoretic techniques were used in work of Gregory, Eklof and Mekler which came immediately before Shelah's work. The modern developments take advantage of the great progress that has been made in set theory in the last two decades. The major methods used can be summarized as follows: (1) stationary sets and their generalization, λ-systems; (2) prediction principles, including the diamond principles and the Black Box(es); (3) combinatorial consequences of the Axiom of Constructibility in addition to the diamond principles; (4) internal forcing axioms, especially Martin's Axiom and the Proper Forcing Axiom; and (5) large cardinal axioms. These are explained in the text, principally in Chapters II,

VI and XIII; little is presumed of the algebraist besides a nodding acquaintance with cardinal and ordinal numbers. (One major set theoretic technique which is not discussed is that of forcing; for this the non-logician might want to consult Dales-Woodin 1987 or Kunen 1980.) The algebraic prerequisites are no more than would be covered in an introductory course on groups, rings and fields.

It is no longer possible to give, in a reasonable amount of space, an account of all the *results* in abelian group theory and homological algebra which use set-theoretic methods. So, as an organizing principle, we have chosen to restrict attention to the torsion-free case, in particular, to the general area described by the title of the book. This means that we have been forced to leave out such interesting work as, for example, that on Crawley's Problem, the socles of *p*-groups, or uncountable Butler groups; however, the set-theoretic *methods* used there are largely represented here. (We have included some of this work in the Bibliography.) Our title and focus are inspired by the notes, *Almost Free Abelian Groups*, of a seminar given by George A. Reid at Tulane in 1966-67; this seminar was "concerned with giving an account of the information known to date concerning Whitehead's problem ... The investigations of this question naturally lead to the consideration of other classes of groups, and other conditions, which, while not perhaps implying freeness, do in one sense or another get fairly close" (from the Introduction).

We concentrate here on four major problems, the first three of which are explicitly represented in the Reid notes. Our philosophy has been to broaden our treatment to modules over rings other than $\mathbb{Z}$ where this can be done without many additional complications (though we have generally avoided the false generality of replacing $\mathbb{Z}$ by a (countable) p.i.d.).

**1. Almost free modules:** *for which cardinals $\kappa$ are there non-free abelian groups of cardinality $\kappa$ which are $\kappa$-free, that is, every subgroup of cardinality $< \kappa$ is free?* This question is implicit in Fuchs' 1960 text and explicit in his 1970 volume. (Higman's 1951 paper solves a case of the problem for non-commutative groups and is, perhaps, the first use of the term "almost free.") It turns out that the answer to this question varies with the set-theoretic assumptions used, although Shelah's Singular Compactness Theorem (IV.3.5) shows, in ZFC, that a cardinal which has the property stated above must be regular. Very complete information can be obtained if we assume $V = L$, the Axiom of Constructibility. All of this is discussed in Chapters IV and VII. We broaden the discussion there to modules over general rings because we are able to give a very satisfactory solution, with the major dividing line being between rings (such as fields) which are left perfect and those (such as $\mathbb{Z}$) which are not. The existence of almost free objects makes sense as a question in universal algebra and our treatment also deals with this wider setting, although we do not aim for total generality. Chapter VIII deals with a related question, that of the structure and classification of abelian groups which are $\aleph_1$-separable, that is, every countable subset is contained in a countable free direct summand.

**2. The structure of Ext:** *solve Whitehead's Problem — does* $\text{Ext}(A, \mathbb{Z}) = 0$ *imply A is free?; determine the structure of the divisible group* $\text{Ext}(A, \mathbb{Z})$ *for any torsion-free A.* Whitehead's Problem appears in a 1955 paper of Ehrenfeucht and is attributed to J. H. C. Whitehead. Call a group $A$ a *W-group* if $\text{Ext}(A, \mathbb{Z}) = 0$; an alternative characterization is that if $F \supseteq K$ are free groups such that $F/K \cong A$, then every homomorphism from $K$ to $\mathbb{Z}$ lifts to one on $F$. It was early observed, independently by several people, that W-groups are $\aleph_1$-free; later Rotman showed that they are even separable (that is, every finite subset is contained in a free direct summand) and slender and Chase showed that the Continuum Hypothesis implies that W-groups are what we call strongly $\aleph_1$-free. As mentioned above, the complete solution to the problem cannot be given in ZFC; this is discussed in Chapter XII. New difficulties arise over other rings, even for uncountable p.i.d.'s, so we restrict ourselves largely to abelian groups.

For these first two problems, the solution depends very strongly on the set theory; in fact, an explicit reduction of the problem to a purely set-theoretic form is given, respectively, in sections VII.3 and XII.3 (although the latter is only for groups of cardinality $\aleph_1$). For the next two problems, the situation is different: though there are some results that are not provable in ZFC, the major results are theorems of ZFC whose proofs use powerful set-theoretic tools.

**3. The structure of Hom:** *characterize abelian groups of the form* $A^*$ $(= \text{Hom}(A, \mathbb{Z}))$ *in some group-theoretic way; is every such group reflexive, that is, canonically isomorphic to its double dual* $(A^*)^{**}$? The fact that $\mathbb{Z}$ is slender is very important here, since it implies that free groups (of non-measurable cardinality) are reflexive (cf. III.3.8); we discuss slenderness in Chapter III, including Eda's extension of Los' theorem to the measurable case. Call a group a *dual group* if it is of the form $A^*$. Another question which occurs in the Reid notes is whether every dual group belongs to the *Reid class*, that is the smallest class of groups containing $\mathbb{Z}$ and closed under direct sums and products. The first negative answers (in ZFC) to the latter question and to the second question above were given by Eda and Ohta. This is discussed in Chapters X and XI. One may also ask if a dual group is *strongly non-reflexive*, that is, not isomorphic, in any way, to its double dual. A great diversity of constructions of dual groups, reflexive, non-reflexive, and strongly non-reflexive is presented in Chapter XIV, almost all of which is new; other questions about dual groups are raised there, and some are answered.

**4. Endomorphism rings:** *characterize the rings which can appear as the endomorphism rings of certain classes of groups.* Baer in 1943 characterized the endomorphism rings of bounded groups. Fuchs asked about this problem in his 1960 text. Important work was done by Corner in the early 1960's when, among other results, he characterized the countable rings which can be realized as endomorphism rings of countable torsion-free reduced abelian groups. Results like his have implications for the structure of groups: obviously, a group whose endomorphism ring is isomorphic to $Z$ is indecomposable; and Corner showed that the realization of certain rings can lead to the existence of groups with "weird" decomposition properties, e.g., a group $A$ such that $A$ is not isomorphic to $A \oplus A$ but $A \oplus A$ is isomorphic to $A \oplus A \oplus A \oplus A$. There has been much work by Corner, Dugas, Göbel, Shelah and others which uses the diamond and Black Box prediction principles to solve cases of this problem. Due to limitations in space and in our expertise, our discussion of this problem, in Chapter XIII, is less comprehensive than that of the other three and serves mainly as an illustration of the use of a new version of the Black Box, due to Shelah, introduced here (and also applied in section 5 of XIV).

Detailed historical information is given in the Notes at the end of each chapter. We have left most of the attribution of results to those Notes; theorems are, naturally, attributed to their original authors, though the proofs here may be different in some cases. We make personal claim only to those results specifically so claimed in the Notes. (However, all mistakes are due to us!) The Exercises at the end of the chapters are, in many cases, guides to further results (often highly interesting and non-trivial but which we have not had room to include in full); we have attempted to point the reader to their sources, whenever possible. We have tried to be accurate in our attributions, but wish to apologize for the inevitable mistakes and oversights.

The Table of Contents contains a guide to the major dependencies between chapters. We have tried to design the book so that the reader can enter it at many different points, and then easily refer back to earlier results as needed.

We are indebted to Saharon Shelah for his help and for allowing some of his work to appear first in this book. We would also like to thank: Bernhard Thomé for his careful reading of most of the manuscript; Aboulmotalab Ihwil and Ali Sagar for taking the lecture notes at SFU which were the genesis of this book; Paul Cohn, Mark Davis, Mark DeBonis, Manfred Dugas, Katsuya Eda,

Martin Gilchrist, Rudiger Göbel, Menachem Magidor and Martin Ziegler for their comments on parts of the book; and Rob Ballantyne, Katy Eklof, Sherry Eklof, Mike Fried, Martin Gilchrist, Roger Hunter and Barbara Kukan for their help in the production of the final camera-ready copy.

This work was partially supported by NSF Grant No. DMS-8400451 and NSERC Grant No. A8948.

University of California, Irvine                          PAUL C. EKLOF
Simon Fraser University                                ALAN H. MEKLER
December, 1989

# TABLE OF CONTENTS

The numbers after each chapter indicate the chapters on which it depends (where dependence is a transitive relation). Everything depends on I and II.

# CHAPTER I
# ALGEBRAIC PRELIMINARIES

In the first two sections of this chapter we review the algebraic background which is assumed in the rest of the book; this also gives us the opportunity to fix notation and conventions. In the last section, we discuss linear topologies on modules. We assume the reader is already familiar with most of the material in this chapter, so it is presented informally and largely without proofs; for more on the topics covered we refer the reader to such texts as Fuchs 1970/1973, Anderson-Fuller 1974, and Rotman 1979, as well as any standard graduate text in algebra.

All rings in this book will have a multiplicative identity and all modules will be unitary modules. Unless otherwise specified, "module" will mean left $R$-module. Much of the time we will focus on abelian groups, that is $\mathbb{Z}$-modules, and we will often refer to these simply as groups.

If $\varphi\colon A \to B$ is a function, and $X \subseteq A$, $\varphi[X] \stackrel{\text{def}}{=} \{\varphi(a)\colon a \in X\}$; $\varphi[A]$ will also be denoted im$(\varphi)$ or rge$(\varphi)$. If $Y \subseteq B$, $\varphi^{-1}[Y]$, $= \{a \in A\colon \varphi(a) \in Y\}$; if $\varphi$ is a homomorphism, ker$(\varphi) = \varphi^{-1}[\{0\}]$. The restriction of $\varphi$ to $X$ is denoted $\varphi{\restriction}X$, i.e. $\varphi{\restriction}X = \{(x, \varphi(x))\colon x \in X\}$. In an abuse of notation, sometimes we will write $M = 0$ instead of $M = \{0\}$ and $\varphi^{-1}[x]$ instead of $\varphi^{-1}[\{x\}]$. If $\psi\colon B \to C$, $\psi \circ \varphi$ is the composition of $\psi$ with $\varphi$, a function from $A$ to $C$.

If $M$ is a module and $Y \subseteq M$, then $\langle Y \rangle$ denotes the submodule generated by $Y$. The notation $Y \subset X$ means $Y \subseteq X$ and $Y \neq X$.

## §1.Homomorphisms and extensions

If $M$ and $H$ are left $R$-modules, $\text{Hom}_R(M, H)$ denotes the group of $R$-homomorphisms from $M$ to $H$, which is an abelian group under the operation defined by: $(f + g)(x) = f(x) + g(x)$. We will sometimes refer to $\text{Hom}_R(M, H)$ as the $H$-*dual* of $M$. Often we will write $\text{Hom}(M, H)$, if $R$ is clear from context. If $x \in M$ and $y \in \text{Hom}(M, H)$, we denote by $\langle y, x \rangle$ or $\langle x, y \rangle$, interchangeably, the element $y(x)$ of $H$, i.e., the result of applying $y$ to $x$.

If $H$ is an $R$-$S$ bimodule (that is, a left $R$-module and a right $S$-module such that $(ra)s = r(as)$ for all $r \in R$, $a \in H$, $s \in S$), then $\mathrm{Hom}_R(M, H)$ has a right $S$-module structure defined by: $(fs)(x) = f(x)s$. The bimodule structure that we will be interested in arises as follows. If $H$ is an $R$-module, let $\mathrm{End}_R(H) = \mathrm{Hom}_R(H, H)$; then $\mathrm{End}_R(H)$ is a ring under composition of homomorphisms, where we define the product $f \cdot g$ to be $g \circ f$; $H$ has a right $\mathrm{End}_R(H)$-module structure defined by: $af = f(a)$ for all $a \in H$ and $f \in \mathrm{End}_R(H)$. This makes $H$ into an $R$-$\mathrm{End}_R(H)$-bimodule. Note that $\mathrm{End}_R(R) \cong R$.

If $M$ is an abelian group, then $M^*$, without further explanation, will denote $\mathrm{Hom}_{\mathbb{Z}}(M, \mathbb{Z})$, called the *dual group of $M$*. A group is called a *dual group* if and only if it is of the form $\mathrm{Hom}(M, \mathbb{Z})$ for some group $M$. The structure and properties of dual groups will be one of the principal subjects of this book. In particular, we will be interested in when a group is (canonically) isomorphic to its double dual. Sometimes it will be convenient to consider this question in the more general context of $H$-duals.

Let us temporarily fix an $R$-module $H$, and let $S$ denote $\mathrm{End}_R(H)$, so that $H$ is an $R$-$S$-bimodule. For convenience denote $\mathrm{Hom}_R(M, H)$ by $M^*$. Then $M^*$ is a right $S$-module, and $\mathrm{Hom}_S(M^*, H)$ is a left $R$-module in the obvious fashion; we denote the latter by $M^{**}$. There is a canonical homomorphism

$$\sigma_M \colon M \to M^{**}$$

defined by: $\langle \sigma_M(x), y \rangle = \langle x, y \rangle$ for all $x \in M$ and $y \in M^*$. We say that $M$ is *$H$-torsionless* if $\sigma_M$ is one-one, and that $M$ is *$H$-reflexive* if $\sigma_M$ is one-one and onto $M$, i.e., an isomorphism. If $H = R$, we say *torsionless* or *reflexive* instead of $R$-torsionless or $R$-reflexive, respectively.

For every $R$-homomorphism $\varphi \colon M \to N$ there is an induced $S$-homomorphism $\varphi^* \colon \mathrm{Hom}(N, H) \to \mathrm{Hom}(M, H)$ defined by: $\varphi^*(f) = f \circ \varphi$ for all $f \in \mathrm{Hom}(N, H)$. There is also an induced $S$-homomorphism $\varphi_* \colon \mathrm{Hom}(H, M) \to \mathrm{Hom}(H, N)$ defined by: $\varphi_*(g) = \varphi \circ g$

for all $g \in \mathrm{Hom}(H, M)$. If $\varphi$ is an isomorphism, then so are $\varphi_*$ and $\varphi^*$. If $\varphi$ is surjective, then $\varphi^*$ is injective; if $\varphi$ is injective, then $\varphi_*$ is injective. But $\varphi^*$ may not be surjective when $\varphi$ is injective; and $\varphi_*$ may not be surjective when $\varphi$ is surjective. In fact, we have the following situation. A sequence of homomorphisms

$$\ldots \rightarrow M_{n-1} \xrightarrow{\varphi_{n-1}} M_n \xrightarrow{\varphi_n} M_{n+1} \rightarrow \ldots$$

is called *exact* if $\ker(\varphi_n) = \mathrm{im}(\varphi_{n-1})$ for all $n$. A *short exact sequence* (or *s.e.s.*) is an exact sequence of the form

$$0 \rightarrow L \xrightarrow{\psi} M \xrightarrow{\varphi} N \rightarrow 0.$$

Given a short exact sequence of $R$-homomorphisms as above and given an $R$-module $H$, the sequences

$$0 \rightarrow \mathrm{Hom}(N, H) \xrightarrow{\varphi^*} \mathrm{Hom}(M, H) \xrightarrow{\psi^*} \mathrm{Hom}(L, H)$$

$$0 \rightarrow \mathrm{Hom}(H, L) \xrightarrow{\psi_*} \mathrm{Hom}(H, M) \xrightarrow{\varphi_*} \mathrm{Hom}(H, N)$$

are exact. We have the following fundamental theorem of Cartan-Eilenberg 1956. (In its statement we will ignore the complication that the domain of the function is a proper class.)

**1.1 Theorem.** *For all $n \geq 1$ and $R$ there is a binary function* $\mathrm{Ext}_R^n(\_,\_)$ *from the class of $R$-modules to the class of abelian groups so that for any short exact sequence*

$$0 \rightarrow L \xrightarrow{\psi} M \xrightarrow{\varphi} N \rightarrow 0$$

*there are exact sequences*

$$0 \rightarrow \mathrm{Hom}(N, H) \xrightarrow{\varphi^*} \mathrm{Hom}(M, H) \xrightarrow{\psi^*} \mathrm{Hom}(L, H) \rightarrow \mathrm{Ext}_R^1(N, H) \rightarrow$$
$$\mathrm{Ext}_R^1(M, H) \rightarrow \mathrm{Ext}_R^1(L, H) \rightarrow \mathrm{Ext}_R^2(N, H) \rightarrow \mathrm{Ext}_R^2(M, H) \rightarrow \ldots$$

*and*

$$0 \rightarrow \mathrm{Hom}(H, L) \xrightarrow{\psi_*} \mathrm{Hom}(H, M) \xrightarrow{\varphi_*} \mathrm{Hom}(H, N) \rightarrow \mathrm{Ext}_R^1(H, L) \rightarrow$$
$$\mathrm{Ext}_R^1(H, M) \rightarrow \mathrm{Ext}_R^1(H, N) \rightarrow \mathrm{Ext}_R^2(H, L) \rightarrow \mathrm{Ext}_R^2(H, M) \rightarrow \ldots$$

These sequences are called the *long exact* (or Cartan-Eilenberg) sequences induced by

$$0 \to L \xrightarrow{\psi} M \xrightarrow{\varphi} N \to 0$$

We will be concerned only with $\mathrm{Ext}_R^1$ which we will often write as $\mathrm{Ext}_R$ or even as Ext, especially when $R$ is a p.i.d. (for in that case $\mathrm{Ext}_R^n(M, H) = 0$ for all $M$, $H$ and all $n \geq 2$). (See page 8 for an explicit definition of Ext when $R$ is a p.i.d.)

## §2. Direct sums and products

An *indexed family* of modules is a function from a set $I$, the index set, to a set of modules. We will write the indexed family as $(M_i : i \in I)$, or, more often, abuse notation and write it as $\{M_i : i \in I\}$, keeping in mind that we allow the possibility that $M_i = M_j$ for $i \neq j$. The *direct product* of the indexed family $\{M_i : i \in I\}$, is defined to be the set of all functions $x : I \to \bigcup_{i \in I} M_i$ such that $x(i) \in M_i$ for all $i \in I$; it is given a module structure via coordinate-wise operations: $(x_1 + x_2)(i) = x_1(i) + x_2(i)$ and $(rx)(i) = rx(i)$. This direct product is denoted $\prod_{i \in I} M_i$ or $\prod_I M_i$, and if $M_i = M$ for all $i$, it is denoted $M^I$, or $M^\kappa$ if $I$ has cardinality $\kappa$. Sometimes we will denote an element $x$ of $\prod_I M_i$ by $(a_i)_I$ or $(a_i)_{i \in I}$ if $x(i) = a_i$ for all $i$. For any $x \in \prod_{i \in I} M_i$ let

$$\mathrm{supp}(x) = \{i \in I : x(i) \neq 0\}.$$

The *direct sum* of the indexed family $\{M_i : i \in I\}$ is defined to be the submodule of $\prod_I M_i$ consisting of all $x$ such that $\mathrm{supp}(x)$ is finite. It is denoted $\bigoplus_{i \in I} M_i$ or $\bigoplus_I M_i$; if $M_i = M$ for all $i$, it is denoted $M^{(I)}$, or $M^{(\kappa)}$ if $I$ has cardinality $\kappa$.

Associated with the direct product or sum we have a couple of canonical homomorphisms. For each $j \in I$ we have the *canonical surjection* $\rho_j : \prod_I M_i \to M_j$ which takes $x \in \prod_I M_i$ to $x(j)$; we sometimes also denote by $\rho_j$ the restriction of $\rho_j$ to $\bigoplus_I M_i$. For each $j \in I$ we also have the *canonical injection* $\lambda_j : M_j \to \bigoplus_I M_i$

which takes $a \in M_j$ to $x$ defined by:

$$x(i) = \begin{cases} a & \text{if } i = j \\ 0 & \text{otherwise} \end{cases}$$

Obviously, $\lambda_j$ is an isomorphism of $M_j$ with a submodule of $\bigoplus_I M_i$, and we sometimes identify $M_j$ with this submodule; also, we sometimes regard $\lambda_j$ as a map into $\prod_I M_i$.

For any indexed family $\{M_i : i \in I\}$ and any module $H$, the map:

$$\text{Hom}(\bigoplus_I M_i, H) \to \prod_{i \in I} \text{Hom}(M_i, H)$$

which takes $f : \bigoplus_I M_i \to H$ to $(f \circ \lambda_i)_I$ is an isomorphism of groups (and of $S$-modules, if $H$ has a right $S$-structure). Moreover, the map:

$$\text{Hom}(H, \prod_I M_i) \to \prod_{i \in I} \text{Hom}(H, M_i)$$

which takes $g : H \to \prod_I M_i$ to $(\rho_i \circ g)_I$ is an isomorphism. (These facts express the universal mapping properties of the direct sum and direct product, respectively.) We also have natural isomorphisms

$$\text{Ext}(\bigoplus_I M_i, H) \cong \prod_{i \in I} \text{Ext}(M_i, H)$$

and

$$\text{Ext}(H, \prod_I M_i) \cong \prod_{i \in I} \text{Ext}(H, M_i).$$

Since $\text{Hom}_R(R, H) \cong H$, $\text{Hom}_R(R^{(\kappa)}, H) \cong H^\kappa$ for any cardinal $\kappa$. The question of when $\text{Hom}_R(R^\kappa, H) \cong H^{(\kappa)}$ is the subject of Chapter III.

**2.1 Lemma.** *For any $R$-modules $M$ and $H$, $M$ is $H$-torsionless if and only if $M$ is isomorphic to a submodule of $H^I$ for some $I$.*

PROOF. Here we are using the same convention as in section 1 and regarding $H$ as an $R$-$S$-bimodule, where $S = \text{End}_R(H)$; $M^*$ and $M^{**}$ are also defined as in section 1. Now suppose first that $M$

is $H$-torsionless. Let $I = M^*$ and define $\theta: M \to H^I$ as follows: for all $x \in M$, $\theta(x)(y) = \langle y, x \rangle$ for all $y \in I$. Then $\theta$ is clearly a homomorphism, and because $M$ is $H$-torsionless, $\theta$ is injective. So $\theta$ embeds $M$ as a submodule of $H^I$. Conversely, suppose $M$ is a submodule of $H^I$; to see that $\sigma_M$ is one-one, consider a nonzero element $x$ of $M$. There exists $j \in I$ such that $x(j) \neq 0$; then if $\rho_j$ is the canonical surjection, $\rho_j \restriction M$ belongs to $M^*$ and $\sigma_M(x)(\rho_j \restriction M) \neq 0$. $\square$

It follows easily from this lemma, or from the definition, that a submodule of an $H$-torsionless module is $H$-torsionless.

If $M_0$ and $M_1$ are submodules of $M$ such that $M = M_0 + M_1$ (i.e. $M$ is the smallest submodule of $M$ containing $M_0$ and $M_1$), and $M_0 \cap M_1 = \{0\}$, $M$ is said to be the (internal) *direct sum* of $M_0$ and $M_1$, written $M = M_0 \oplus M_1$. In that case, $M$ is isomorphic to $\prod_I M_i$ and to $\bigoplus_I M_i$ where $I = \{0, 1\}$. For any submodule $M_0$ of $M$, there is another submodule $M_1$ of $M$ such that $M = M_0 \oplus M_1$ if and only if there is a *projection* of $M$ onto $M_0$, i.e. a homomorphism $\pi: M \to M_0$ such that $\pi \restriction M_0 = $ the identity on $M_0$. (Let $M_1 = \ker(\pi)$.) In this case we say that $M_0$ is a (*direct*) *summand* of $M$.

Note that if $M = M_0 \oplus M_1$ and $A$ is a submodule of $M$ containing $M_0$, then $A = M_0 \oplus (A \cap M_1)$. So if $M_0$ is a direct summand of $M$ and $M_0 \subseteq A \subseteq M$, then $M_0$ is a direct summand of $A$.

We say that a short exact sequence

$$0 \to L \xrightarrow{\psi} M \xrightarrow{\varphi} N \to 0$$

*splits* if there is a homomorphism $p: N \to M$ such that $\varphi \circ p = 1_N$, the identity on $N$. In this case, $p$ is called a *splitting* of $\varphi$, and there is a homomorphism $q: M \to L$, called a *splitting* of $\psi$, such that $q \circ \psi = 1_L$; then the short exact sequence is called *split exact*. Moreover, $M$ is the direct sum of $\ker(\varphi)$ $(= \psi[L])$ and $\operatorname{im}(p)$. Conversely, if $\ker(\varphi)$ is a direct summand of $M$, then the short exact sequence splits. The group $\operatorname{Ext}^1_R(A, H)$ equals $0$ if and only if every short exact sequence

$$0 \to H \to M \to A \to 0$$

is a split exact sequence.

Let $H$ and $S$ be as in section 1, and denote $\mathrm{Hom}_R(M, H)$ by $M^*$, $\mathrm{Hom}_S(M^*, H)$ by $M^{**}$, and $\mathrm{Hom}_R(M^{**}, H)$ by $M^{***}$. Define a map

$$\rho \colon M^{***} \to M^*$$

by: $\langle \rho(z), x \rangle = \langle z, \sigma_M(x) \rangle$ for all $z \in M^{***}$ and $x \in M$.

**2.2 Lemma.** $\rho \circ \sigma_{M^*}$ *is the identity on* $M^*$. *Hence* $\sigma_{M^*}$ *is one-one, so* $M^*$ *is H-torsionless.*

PROOF. Let $y \in M^*$. Then $\langle \rho(\sigma_{M^*}(y)), x \rangle = \langle \sigma_{M^*}(y), \sigma_M(x) \rangle = \langle y, \sigma_M(x) \rangle = \langle y, x \rangle$ for all $x \in M$. Hence $\rho(\sigma_{M^*}(y)) = y$. $\square$

Notice also that if

$$0 \to L \xrightarrow{\psi} M \xrightarrow{\varphi} N \to 0$$

splits, then

$$0 \to N^* \xrightarrow{\varphi^*} M^* \xrightarrow{\psi^*} L^* \to 0$$

is a split short exact sequence because $q \circ \psi = 1_L$ implies $\psi^* \circ q^* = 1_{L^*}$.

An $R$-module $N$ is said to be *free* in case there is a subset, $B$, of $N$ such that $B$ generates $N$ and every set map $f \colon B \to M$, into an arbitrary $R$-module $M$, extends to a homomorphism: $N \to M$; in this case, $B$ is called a *basis* of $N$, and every element of $N$ is uniquely a linear combination of elements of $B$. An $R$-module is free if and only if it is isomorphic to $R^{(\kappa)}$ for some cardinal $\kappa$ $(= |B|)$. For any index set $I$ and $i \in I$, define $e_i \in R^I$ by $e_i(j) = 1$, if $i = j$ and $e_i(j) = 0$, otherwise (in other words, $e_i = \lambda_i(1)$). Then $\{e_i \colon i \in I\}$ is a basis for $R^{(I)}$.

If $N$ is free, then every short exact sequence

$$0 \to L \xrightarrow{\psi} M \xrightarrow{\varphi} N \to 0$$

splits, for arbitrary $L$ and $M$; indeed, if $B$ is a basis of $N$, the set map $f \colon B \to M$ which takes each element, $b$, of $B$ to a pre-image in $M$ under $\varphi$, extends to a homomorphism $p: N \to M$, which is a splitting of $\varphi$. Thus $\text{Ext}_R^1(N, H) = 0$ whenever $N$ is free.

If $R$ is a p.i.d., every submodule of a free $R$-module is free. (A p.i.d. is a *principal ideal domain*, that is, an integral domain such that every ideal is principal.)

For any module $M$, there is a free module $F$ and a surjective homomorphism $\varphi: F \to M$; if we let $K = \ker(\varphi)$ and $\psi$ be the inclusion of $K$ into $F$, then we obtain a short exact sequence

$$0 \to K \xrightarrow{\psi} F \xrightarrow{\varphi} M \to 0.$$

If $R$ is a p.i.d., then $K$ is also free, since it is a submodule of a free module, and the short exact sequence above is called a *free resolution* of $M$. In any case, for every $R$ and every $R$-module $H$, by Theorem 1.1, this short exact sequence induces the exact sequence

$$\text{Hom}(F, H) \xrightarrow{\psi^*} \text{Hom}(K, H) \to \text{Ext}(M, H) \to \text{Ext}(F, H) = 0.$$

(The last term is 0 because $F$ is free.) So $\text{Ext}(M, H)$ is isomorphic to the group of homomorphisms from $K$ to $H$ modulo those which extend to a homomorphism from $F$ to $H$. In particular, $\text{Ext}(M, H) = 0$ if and only if every homomorphism from $K$ to $H$ extends to a homomorphism from $F$ to $H$; this is a criterion for the vanishing of Ext that we will make use of in Chapters VII and XII.

Now we turn to projective and injective modules. An $R$-module $M$ is called *projective* (respectively, *injective*) if and only if for every short exact sequence

$$0 \to A \xrightarrow{\alpha} B \xrightarrow{\beta} C \to 0,$$

the induced map $\beta_*: \text{Hom}(M, B) \to \text{Hom}(M, C)$ (respectively, $\alpha^*: \text{Hom}(B, M) \to \text{Hom}(A, M)$) is surjective. It is easy to see that a free module is projective, a direct sum of projectives is projective, and a direct summand of a projective is projective. Moreover, $M$ is projective if and only if every short exact sequence of the form

$$0 \to H \to A \to M \to 0$$

is a split sequence, i.e., if and only if $\text{Ext}^1_R(M, H) = 0$ for all $H$. Since there is such a short exact sequence with $A$ free, $M$ is isomorphic to a direct summand of a free module if $M$ is projective; thus $M$ is projective if and only if $M$ is isomorphic to a summand of a free module. If $R$ is a p.i.d., $M$ is projective if and only if $M$ is free.

The following are two useful facts about projectives, the first due to Eilenberg, the second to Kaplansky .

**2.3 Lemma.** *If $M$ is projective, then there is a free module $F$ such that $M \oplus F$ is free.*

PROOF. Since $M$ is projective, there is a module $P$ such that $P \oplus M \cong K$ where $K$ is free. Now let $F = K^{(\omega)}$. Then $M \oplus F \cong M \oplus (P \oplus M)^{(\omega)} \cong M \oplus P \oplus M \oplus P \oplus M \oplus \ldots \cong (M \oplus P)^{(\omega)} \cong F$ which is free. $\square$

**2.4 Theorem.** *Every projective $R$-module is a direct sum of countably-generated projective modules.*

PROOF. If $P$ is projective, then $P \oplus M$ is free for some module $M$; say $P \oplus M = \bigoplus_{i \in I} R_i$ where each $R_i$ is isomorphic to $R$. We claim that there is a chain $\{F_\nu : \nu \in \alpha\}$ of submodules of $P \oplus M$ such that $F_0 = 0$; $P \oplus M = \bigcup_{\nu < \alpha} F_\nu$; and for all $\nu$
  (a) $F_\nu \subseteq F_{\nu+1}$;
  (b) if $\nu$ is a limit ordinal, $F_\nu = \bigcup_{\mu < \nu} F_\mu$;
  (c) $F_\nu = (F_\nu \cap P) \oplus (F_\nu \cap M)$;
  (d) $F_\nu = \bigoplus_{i \in J_\nu} R_i$ for some $J_\nu \subseteq I$;
  (e) $F_{\nu+1}/F_\nu$ is countably-generated.
Suppose for the moment that the claim is true. Then $P = \bigcup_{\nu < \alpha} (F_\nu \cap P)$. For all $\nu < \alpha$, $F_{\nu+1} \cap P = (F_\nu \cap P) \oplus C_\nu$ for some $C_\nu$ because, by (c), $F_\nu \cap P$ is a direct summand of $F_\nu$ and, by (d), $F_\nu$ is a direct summand of $P \oplus M$, so $F_\nu \cap P$ is a direct summand of $F_{\nu+1} \cap P$. Moreover, $C_\nu$ is countably-generated by (e) since $(F_{\nu+1} \cap P)/(F_\nu \cap P)$ is a homomorphic image of

$$F_{\nu+1}/F_\nu = ((F_{\nu+1} \cap P)/(F_\nu \cap P)) \oplus ((F_{\nu+1} \cap M)/(F_\nu \cap M)).$$

Then $P = \bigoplus_{\nu < \alpha} C_\nu$, so we are done once the claim is proved.

The chain is constructed by transfinite induction. We can suppose that $I$ is well-ordered. Suppose that $F_\mu$ has been constructed for all $\mu < \nu$. If $\nu$ is a limit ordinal, let $F_\nu = \bigcup_{\mu < \nu} F_\mu$ to satisfy (b). If $\nu = \mu + 1$ for some $\mu$, and $F_\mu \neq P \oplus M$, let $i \in I$ be minimal such that $R_i$ is not contained in $F_\mu$. Let $G_0 = F_\mu$, $G_1 = F_\mu \oplus R_i$ and define by induction on $n \in \omega$ submodules $G_n$ of $P \oplus M$ such that $G_n \subseteq (G_{n+1} \cap P) \oplus (G_{n+1} \cap M)$, $G_n = \bigoplus_{j \in J_n} R_j$ for some $J_n \subseteq I$, and $G_{n+1}/G_n$ is countably-generated. Then $F_\nu \overset{\text{def}}{=} \bigcup_{n \in \omega} G_n$ will have the desired properties. $\square$

An $R$-module $M$ is called *flat* if $\_ \otimes_R M$ preserves the exactness of all sequences $0 \to A \to B$ (of right $R$-modules). Every projective module is flat.

A direct product of injective modules is always injective, and a direct summand of an injective module is injective. A module $M$ is injective if and only if every short exact sequence of the form

$$0 \to M \to A \to H \to 0$$

is a split sequence, i.e., if and only if $\text{Ext}_R^1(H, M) = 0$ for all $H$.

For every module $M$ there is an injective module $Q$ and a one-one homomorphism $\psi \colon M \to Q$.

Let $R$ be an integral domain. We say that an $R$-module $M$ is *torsion-free* if for all $a \in M \setminus \{0\}$ and $r \in R \setminus \{0\}$, $ra \neq 0$; and we call $M$ *divisible* if $M = rM$ for all $r \in R \setminus \{0\}$. We call $M$ *reduced* if $\{0\}$ is the only divisible submodule of $M$. Every injective $R$-module is divisible, and if $R$ is a p.i.d. (or a Dedekind domain), the converse holds. Moreover, for any integral domain $R$, every torsion-free divisible $R$-module is injective (cf. Cartan-Eilenberg 1956, p.128).

If $R$ is a p.i.d., we say that a submodule $B$ of an $R$-module $A$ is *relatively divisible*, or *pure*, in $A$ if $rB = rA \cap B$ for every $r \in R$. If $A$ is torsion-free, then $B$ is pure in $A$ if and only if $A/B$ is torsion-free. We will have more to say about pure submodules in Chapters IV and V.

**2.5.** We conclude this section with a discussion of some particular **Z**-modules, i.e, abelian groups. Of course, every projective group is free, since a subgroup of a free group is free. Also, the injective groups are just the divisible groups. There is a structure theorem for divisible groups. Let **Q** be the group of rationals, under addition. Let $Z(p^\infty)$ denote the $p$-primary part of **Q**/**Z**; $Z(p^\infty)$ can also be described as the subgroup of $\mathbf{C}^*$ (the non-zero complex numbers under multiplication) consisting of the $p^n$th roots of unity for all $n \in \omega$. Then every divisible group is isomorphic to

$$\bigoplus_p Z(p^\infty)^{(\gamma_p)} \oplus \mathbf{Q}^{(\delta)}$$

where $p$ ranges over the primes and the $\gamma_p$ and $\delta$ are cardinals $\geq 0$; moreover, these cardinals are uniquely determined by the divisible group.

Every group can be embedded in a divisible group. Furthermore, if $A$ is torsion-free, $A$ is a subgroup of a divisible group $D \cong \mathbf{Q}^{(\delta)}$ such that $D/A$ is a torsion group; indeed, $D$ can be taken to be (isomorphic to) $\mathbf{Q} \otimes A$. $D$ is called the *injective hull* of $A$.

If $A$ is a group, we let $tA$ denote the torsion subgroup of $A$, i.e., $\{a \in A : na = 0$ for some $n \neq 0\}$; then $A/tA$ is a torsion-freegroup. For any $n \geq 2$, we let $Z(n)$ denote the cyclic group of order $n$, i.e., $\mathbf{Z}/n\mathbf{Z}$. A fundamental theorem says that every finitely-generated abelian group is isomorphic to a direct sum of copies of $\mathbf{Z}$ and of the $Z(n)$. Hence, every finitely-generated torsion-free group is free.

A group is said be of *bounded order* if there exists $n \neq 0$ such that $nA = \{0\}$. Every group of bounded order is a direct sum of cyclic groups.

## §3. Linear topologies

A *linear topology* on a module $M$ is one which is given by a base of neighborhoods, $\mathcal{U}$, about 0 such that each element, $U$, of $\mathcal{U}$ is a submodule of $M$; then for every point $a \in M$, $\{a + U : U \in \mathcal{U}\}$ is a base of neighborhoods about $a$. In that case, a subset $O$ of $M$ is open if and only if for every $a \in O$, there exists $U \in \mathcal{U}$ such that $a + U \subseteq O$. Moreover, addition $+ \colon M \times M \to M$ is continuous, and for every $r \in R$, scalar multiplication by $r \colon M \to M$ is continuous; that is, $M$ is a topological $R$-module. We shall use $M$ to denote a topological module equipped with a given linear topology, and $\mathcal{U}$ to denote the associated neighborhood base of 0; in context there will be no ambiguity.

**3.1 Examples.** The $R$-*topology* on $M$ is the linear topology which has a base of neighborhoods of 0 consisting of all finite intersections of submodules of the form $rM$ for $r \in R \setminus \{0\}$. If $M^1$ is defined to be $\cap\{rM \colon r \in R \setminus \{0\}\}$, then $M$ is Hausdorff in its $R$-topology if and only if $M^1 = 0$. Thus, if $R$ is an integral domain and $M$ is torsion-free, $M$ is Hausdorff in its $R$-topology if and only if $M$ is reduced. We shall also refer to the $\mathbb{Z}$-topology on a group as the $\mathbb{Z}$-adic topology.

If $S$ is a subset of $R$, we can similarly define the $S$-*topology* to be the linear topology with base of neighborhoods of 0 consisting of all finite intersections of submodules of the form $sM$ for $s \in S \setminus \{0\}$. For any $S$ and homomorphism $\varphi \colon M \to N$, $\varphi$ is continuous with respect to the $S$-topology on $M$ and $N$ since $\varphi[sM] \subseteq sN$.

If $R$ is a p.i.d. and $p$ is a prime of $R$, the $p$-*adic topology* on $M$ is defined to be the $S$-topology on $R$ where $S = \{p^n \colon n \in \omega\}$; it has a base of neighborhoods of 0 consisting of the submodules of the form $p^n M$. This topology is a metrizable topology with metric given by: $d(a, b) = p^{-n}$ if and only if $a - b \in p^n M \setminus p^{n+1} M$. $M$ is Hausdorff in the $p$-adic topology if and only if $\cap\{p^n M \colon n \in \omega\} = 0$.

If a linear topology on $M$ has a countable base of neighborhoods of 0, then the topology is metrizable, and therefore describable in terms of convergent sequences, but in general we must use nets to describe the topology.

A (neat) *Cauchy net* in $M$ is defined to be an indexed family $\{a_U \colon U \in \mathcal{U}\}$ of elements of $M$ with the property that for all $U \subseteq V$ in $\mathcal{U}$, $a_U - a_V \in V$. We say that $a$ is the *limit* of the Cauchy net $\{a_U \colon U \in \mathcal{U}\}$ if for all $U \in \mathcal{U}$, $a - a_U \in U$. We say that $M$ is *complete* if $M$ is Hausdorff and every Cauchy net has a (unique) limit in $M$. Two Cauchy nets $\{a_U : U \in \mathcal{U}\}$ and $\{b_U : U \in \mathcal{U}\}$ are *equivalent* if and only if for every $U$, $a_U - b_U \in U$. It is easy to see that if $M$ is complete, then two Cauchy nets have the same limit if and only if they are equivalent.

It is fairly routine to check the following:

**3.2 Proposition.** *A direct product $\prod_{i \in I} M_i$ of modules is complete in the R-topology (respectively, the p-adic topology) if and only if each $M_i$ is complete in the R-topology (resp., the p-adic topology).* □

Given modules $M$ and $\hat{M}$ with linear topologies, we say that $\theta : M \to \hat{M}$ is the *completion* of $M$ if $\hat{M}$ is complete and $\theta$ is an algebraic and topological embedding such that $\theta[M]$ is dense in $\hat{M}$. The completion of $M$ is unique up to isomorphism over $M$.

We can construct the completion of $M$ either by working directly with equivalence classes of Cauchy nets or as an inverse limit; we shall describe the latter. (For a discussion of inverse limits, see section XI.1.) If we define $V \leq U$ to mean $U \subseteq V$ for all $U$, $V$ in $\mathcal{U}$, then $\mathcal{U}$ becomes a directed set, i.e., for all $U$, $V \in \mathcal{U}$ there exists $W (= U \cap V)$ such that $W \geq U$, $V$. For any $U \geq V$ in $\mathcal{U}$ we have a canonical map $\pi_{U,V} : A/U \to A/V$.

**3.3 Proposition.** *Suppose $M$ is a Hausdorff module. Let $\hat{M} = \varprojlim(M/U, \pi_{U,V} : U \geq V \in \mathcal{U})$ where $\hat{M}$ has the topology induced by the product topology on $\prod_{U \in \mathcal{U}} M/U$, and $M/U$ has the discrete topology. If $\theta : M \to \hat{M}$ is the map induced by the canonical maps: $M \to M/U$, then $\theta$ is the completion of $M$.*

PROOF. The inverse limit, $\varprojlim(M/U, \pi_{U,V} : U \geq V \in \mathcal{U})$, is (as a module) the submodule of the direct product $\prod_{U \in \mathcal{U}} M/U$ consisting of those elements $(a_U + U)_{U \in \mathcal{U}}$ such that for all $U \geq V$, $\pi_{U,V}(a_U + U) = a_V + V$. So $(a_U + U)_{U \in \mathcal{U}}$ is in the inverse limit if and only if $(a_U)_{U \in \mathcal{U}}$ is a Cauchy net. As well, two Cauchy nets give rise to the same element of the inverse limit if and only if they are equivalent. The rest of the proof is routine. □

**3.4 Example.** Let $R$ be a p.i.d. equipped with the $p$-adic topology for some prime $p$. Then $R$ is Hausdorff since $\cap_n p^n R = 0$ by unique factorization. The completion, $\hat{R}_p$, is a discrete valuation ring, that is, a local p.i.d., with maximal ideal $p\hat{R}_p$, whose topology is the $p$-adic topology. Every element of $\hat{R}_p$ can be represented uniquely in the form $\sum_{n \in \omega} r_n p^n$ where $r_n$ belongs to a fixed set of representatives of $R/pR$; that is, $\sum_{n \in \omega} r_n p^n$ is the limit in the $p$-adic topology of the sequence $\{\sum_{n<m} r_n p^n : m \in \omega\}$. Indeed, given $x = (\ldots, x_n + p^n R, \ldots) \in \hat{R}_p$, we define $r_n$ inductively so that for all $m$, $x_m - \sum_{n<m} r_n p^n \in p^m R$.

$\hat{R}_p$ is also the completion of $R_{(p)}$, the localization of $R$ at $pR$, equipped with the $p$-adic topology. If $R = \mathbf{Z}$, the ring $\hat{\mathbf{Z}}_p$ is called the ring of *p-adic integers*; its additive group is denoted $J_p$.

Now we turn to the $R$-topology on a p.i.d. $R$. The completion of $R$, $\hat{R}$, is isomorphic to $\varprojlim(R/rR, \pi_{r,t} : r, t \in R)$ where $\pi_{r,t}: R/rR \to R/tR$ is the obvious surjection if $rR \subseteq tR$. Then $\hat{R}$ inherits a ring structure (as well as a topological structure) from the direct product structure on $\prod_{r \in R} R/rR$.

**3.5 Proposition.** *Let $R$ be a p.i.d. equipped with the $R$-topology and let $\hat{R}$ be its completion.*

(i) *The topology on $\hat{R}$ is the $R$-topology;*

(ii) *$\hat{R}$ is a torsion-free reduced $R$-module and $R$ is a pure submodule of $\hat{R}$;*

(iii) *$\hat{R}$ is isomorphic (as ring) to $\prod_p \hat{R}_p$ where $p$ ranges over a representative set of generators of prime ideals of $R$.*

PROOF. (i) Since $\hat{R}$ has a topology induced from the product topology on $\prod_{r \in R} R/rR$, a basic neighborhood of $0$ is of the form $\hat{R} \cap U$, where $U$ consists of all elements of $\prod_{r \in R} R/rR$ which are zero in a fixed finite set of coordinates, say $r_1, \ldots, r_k$. Clearly $r_1\hat{R} \cap \ldots \cap r_k\hat{R} \subseteq \hat{R} \cap U$. Conversely, if $x = (\ldots, x_r + rR, \ldots) \in \hat{R} \cap U$, then for any $i = 1, \ldots, k$, $x_{r_i} \in r_i R$; so for any $r \in R$, $x_{r_i r} \in r_i R$ and hence (for a fixed $i$) there exists $y_r \in R$ such that $r_i y_r = x_{r_i r}$. Thus if $y = (\ldots, y_r + rR, \ldots)$, $y$ belongs to $\hat{R}$ and satisfies $r_i y = x$.

Therefore $\hat{R} \cap U \subseteq r_1 \hat{R} \cap \ldots \cap r_k \hat{R}$.

(ii) $\hat{R}$ is reduced since it is Hausdorff in the $R$-topology. $R$ is embedded in $\hat{R}$ by the diagonal mapping which is a pure embedding because if $a \in R$ is divisible by $r$ in $\hat{R}$, then $a + rR = 0 + rR$ so $a \in rR$. To see that $\hat{R}$ is torsion-free, it is enough to show that $\hat{R}/R$ is torsion-free. Suppose that $x = (\ldots, x_r + rR, \ldots) \in \hat{R}$ and $sx = a \in R$ for some $s \in R \setminus \{0\}$. Then $sx_s + sR = a + sR$, so $a = sb$ for some $b \in R$. Hence, for all $r \in R$, $x_{sr} - b \in rR$; so since $x_r - x_{sr} \in rR$, we have $x_r - b \in rR$. Therefore $x = b \in R$.

(iii) Define $\psi$ to be the map from $\hat{R}$ to $\prod_p \hat{R}_p$ which takes $x = (\ldots, x_r + rR, \ldots)$ to the element of $\prod_p \hat{R}_p$ whose $p$th coordinate is $(x_p + pR, \ldots, x_{p^n} + p^n R, \ldots)$. It is clear that $\psi$ is a ring homomorphism. Moreover, $\psi$ is one-one because if $\psi(x) = 0$, then for any $r$, if $p^n$ divides $r$, then $p^n$ divides $x_r$ since $x_r \equiv x_{p^n}$ (mod $p^n$) and $p^n$ divides $x_{p^n}$; hence $x_r \in rR$. Finally, using the Chinese Remainder Theorem we can show that $\psi$ is surjective. $\square$

**3.6 Examples.** (i) If $\mathbb{Z}$ is equipped with the $\mathbb{Z}$-adic topology, $\mathbb{Z}$ has a base of neighborhoods of 0 consisting of the subgroups $n!\mathbb{Z}(n \in \omega \setminus \{0\})$. Then every element of $\hat{\mathbb{Z}}$ can be represented in the form $\sum_{n \in \omega} n! a_n$ where $a_n \in \mathbb{Z}$. Indeed, given $x = (\ldots, x_n + n\mathbb{Z}, \ldots) \in \hat{\mathbb{Z}}$, we define $a_n$ inductively so that for all $m, x_{m!} - \sum_{n < m} n! a_n \in m!\mathbb{Z}$. (This representation is not unique because it depends on the choice of the $x_n$ in the coset $x_n + n\mathbb{Z}$.)

(ii) Let $M$ be the $\mathbb{Z}$-adic closure of $\mathbb{Z}^{(\kappa)}$ in $\mathbb{Z}^\kappa$. Suppose $H$ is a reduced torsion-free group. If $f \in \mathrm{Hom}(M, H)$, then $f$ is completely determined by $f{\restriction}\mathbb{Z}^{(\kappa)}$, because $f$ is continuous when $M$ and $H$ are given the $\mathbb{Z}$-adic topology. Thus $\mathrm{Hom}(M, H)$ is naturally isomorphic to a subgroup of $\mathrm{Hom}(\mathbb{Z}^{(\kappa)}, H) \cong H^\kappa$. Suppose $f : \mathbb{Z}^{(\kappa)} \to \hat{\mathbb{Z}}$. If $f(e_\alpha) \neq 0$, for infinitely many $\alpha$, then $f[M] \not\subseteq \mathbb{Z}$. So $\mathrm{Hom}(M, \mathbb{Z})$ is naturally isomorphic to $\mathbb{Z}^{(\kappa)}$.

# CHAPTER II
# SET THEORY

In this chapter we work within ordinary set theory, ZFC (defined in section 1). Although we presume some familiarity with ordinal and cardinal numbers and other elementary concepts from axiomatic set theory, we briefly review these notions, without proofs, in the first section. Then we turn to some notions which are, perhaps, less familiar. In section 2, we discuss filters and ultrafilters and use these to define some large cardinals, including measurable cardinals. In section 3, we define ultraproducts, and encounter for the first time the key notion of a $\kappa$-free module, which will play a major role in the rest of the book. In section 4, we introduce a couple of other definitions which will be very important, namely that of a stationary set and of a $\kappa$-filtration. Finally, we conclude, in section 5, with a short discussion of game theory and of trees.

## §1. Ordinary set theory

Our basic set theory will be Zermelo-Fraenkel set theory with the Axiom of Choice, or ZFC. For the record we state the axioms of this theory. (More precisely, these axioms define the primitive concepts of "set" and "member of". The latter is also referred to as "element of" and denoted with the symbol "$\in$"; all variable symbols stand for sets.)

*Extensionality*: if two sets $x$ and $y$ have the same members, then $x = y$;

*Pairing*: for any sets $x$ and $y$, there is a set, $\{x, y\}$, whose sole members are $x$ and $y$;

*Union*: for any set $x$ there is a set $y$, denoted $\cup x$, such that $z \in y$ if and only if $z$ is a member of a member of $x$;

*Power Set*: for any set $x$ there is a set $y$, denoted $\mathcal{P}(x)$, whose members are precisely the subsets of $x$;

*Separation*: for every formula $\varphi(u, v)$ and sets $x$ and $p$, there is a set $y$ such that $z \in y$ if and only if $z \in x$ and $\varphi(z, p)$ holds;

*Infinity*: there is a set $x$ such that $\emptyset$, the empty set, belongs to $x$, and for every $y \in x$, $y \cup \{y\} \in x$;

*Replacement*: for every formula $\varphi(u, v, w)$ and sets $X$ and $p$ such that $\varphi(\cdot, \cdot, p)$ defines a function, there is a set $Y$ such that for all $x \in X$, if $\varphi(x, y, p)$ holds, then $y \in Y$;

*Regularity*: every set $x$ has a member $y$ such that no member of $y$ is a member of $x$;

*Choice*: for every set $X$ of pairwise-disjoint non-empty sets, there is a set $Y$ such that for all $x \in X$, $Y \cap x$ has exactly one member.

Separation and Replacement are actually axiom *schemes*, one axiom for each formula $\varphi$. Here *formula* means a formula of the first-order predicate calculus built up from the atomic formulas

$$x \in y \quad \text{and} \quad x = y$$

using the connectives $\wedge, \vee$ and $\neg$, and the quantifiers $\forall$ and $\exists$. (See section VI.2.)

As is well known, almost all of "every-day" mathematics can be carried out within the context of ZFC, so it is not necessary, in ordinary practice, to refer to these axioms. However, one of the main themes of this book, is that there are mathematical — even algebraic — problems which cannot be solved on the basis of ZFC.

An axiom system is *consistent* if no contradiction can be derived from it. Gödel's second incompleteness theorem implies that ZFC cannot be proved (by finitistic means) to be consistent. If ZFC is inconsistent, then *every* sentence $\varphi$ (of the first-order predicate calculus) is provable from ZFC. On the other hand, if ZFC is consistent, Gödel's first incompleteness theorem implies that ZFC is *incomplete*, i.e., there are sentences $\varphi$ such that neither $\varphi$ nor its negation, $\neg\varphi$, is provable from ZFC. The best known example is the Continuum Hypothesis (CH) defined below. These incompleteness properties are not peculiar defects of ZFC but properties of every axiom system which is powerful enough to be a basis of "every-day" mathematics. Thus, for example, there are sentences $\varphi$ such that neither $\varphi$ nor $\neg\varphi$ is provable from ZFC + CH. (But of course CH is not one of them!)

We shall not concern ourselves with the unlikely possibility that ZFC is inconsistent: there is good empirical and philosophical evidence for its consistency. But the reader should keep in mind that when we make a statement such as "CH is consistent with ZFC" we really mean that *if ZFC is consistent, then* ZFC + CH is consistent. Similarly, by "CH is independent of ZFC" we mean that *if ZFC is consistent, then* neither CH nor $\neg$CH is provable from ZFC, which is the same as saying that CH is consistent with ZFC and $\neg$CH is consistent with ZFC.

Strictly speaking, ZFC does not speak of "classes", but we shall use this informal language to refer to collections which may be too large to be sets. Given any property $\varphi(u, v)$ and set $p$, we can speak of the class defined by $\varphi$ and $p$, denoted $\{x : \varphi(x, p)\}$. For example, $V$, the class of all sets, equals $\{x : x = x\}$. Russell's Paradox implies that $V$ is a *proper class*, that is, a class which is not a set. (If it were a set, then, by Separation, $R \overset{\text{def}}{=} \{x \in V : x \notin x\}$ would be a set; but then considering whether or not $R \in R$ leads to a contradiction.)

A set $X$ is called *transitive* if every $x \in X$ is a subset of $X$, i.e., $y \in x \in X$ implies $y \in X$. Given any set $S$, define by induction

$$S_0 = S \quad \text{and} \quad S_{n+1} = \cup S_n.$$

Then $\cup\{S_n : n \in \omega\}$ is transitive and is the smallest transitive set containing $S$; it is called the *transitive closure* of $S$, and is denoted $TC(S)$.

The class, *Ord*, of ordinal numbers (or ordinals) is defined to be the class of all sets which are transitive and well-ordered by $\in$. It may be proved that every member of an ordinal is an ordinal (that is, *Ord* is a transitive class) and for any two different ordinals $\alpha$ and $\beta$, either $\alpha \in \beta$ or $\beta \in \alpha$. Moreover, every well-ordered set, $S$, is isomorphic, as a linear ordering, with one and only one ordinal number, $\alpha$; we say that $S$ has *order type* $\alpha$. *Ord* is well-ordered by $\in$.

If $\alpha$ and $\beta$ are ordinals, we shall sometimes write $\alpha < \beta$ instead of $\alpha \in \beta$. Notice that the supremum of a set of ordinals is equal to the union of the set. The least ordinal, 0, is the empty set, $\emptyset$. Every ordinal $\alpha$ has an immediate successor, denoted $\alpha + 1$, which equals $\alpha \cup \{\alpha\}$. An ordinal which does not have an immediate predecessor (i.e., is not of the form $\alpha + 1$) is called a *limit ordinal*; otherwise it is a *successor ordinal*. Thus 0 is a limit ordinal. We let

$$\lim(\gamma) = \{\alpha \in \gamma : \alpha \text{ is a limit ordinal}\} \qquad \text{and}$$
$$\mathrm{succ}(\gamma) = \{\alpha \in \gamma : \alpha \text{ is a successor ordinal}\}.$$

We identify the finite ordinals with the natural numbers: $0 = \emptyset$, $1 = 0 + 1$, $2 = 1 + 1$, etc. The set of finite ordinals is denoted $\omega$; it is the first infinite ordinal.

By transfinite induction on ordinals we can define: $V_0 = \emptyset$; $V_{\alpha+1} = \mathcal{P}(V_\alpha)$; and for limit ordinals $\alpha$, $V_\alpha = \bigcup_{\beta < \alpha} V_\beta$. Then each $V_\alpha$ is transitive and $V_\alpha \subseteq V_\beta$ if $\alpha \leq \beta$. The Axiom of Regularity implies that $V = \bigcup\{V_\alpha : \alpha \in Ord\}$, i.e., every set belongs to some $V_\alpha$.

An ordinal number is called a *cardinal* (*number*) if it is not equinumerous with any of its members. The class of all cardinals is denoted *Card*. All of the finite ordinals are cardinals, and so is $\omega$, which is also denoted $\aleph_0$. For every cardinal $\kappa$, there is a next largest cardinal number, denoted $\kappa^+$. The alephs are defined as follows: $\aleph_0 = \omega$, the first infinite cardinal; $\aleph_{\alpha+1} = (\aleph_\alpha)^+$; and if $\alpha$ is a limit ordinal, $\aleph_\alpha = \bigcup\{\aleph_\beta : \beta < \alpha\}$. A cardinal of the form $\kappa^+$ (or, equivalently, of the form $\aleph_{\alpha+1}$ if it is infinite) is called a *successor cardinal*; otherwise it is a *limit cardinal*. (Note that every infinite cardinal is a limit *ordinal*.) We shall sometimes denote $\aleph_\alpha$ by $\omega_\alpha$: generally we write $\aleph_\alpha$ when we think of it as a cardinal, and $\omega_\alpha$ when we regard it as an ordinal. Every set $X$ is equinumerous with one and only one cardinal number. If $\kappa$ is the unique cardinal equinumerous with $X$, we write $|X| = \kappa$ and say that $X$ has cardinality $\kappa$. We say that $X$ is *countable* if it has cardinality $\leq \aleph_0$.

If $\kappa$ and $\lambda$ are cardinals, $\kappa + \lambda$ is defined to be the cardinality of $(\kappa \times \{0\}) \cup (\lambda \times \{1\})$ and $\kappa \cdot \lambda$ is defined to be the cardinality of $\kappa \times \lambda$. Addition and multiplication of infinite cardinals is easy:

$$\aleph_\alpha + \aleph_\beta = \aleph_\alpha \cdot \aleph_\beta = \max\{\aleph_\alpha, \aleph_\beta\}.$$

Exponentiation is another matter. For any sets $A$ and $B$, let $^A B$ denote the set of all functions from $A$ to $B$. Then define $\kappa^\lambda$ to be the cardinality of $^\lambda \kappa$. Define $\kappa^{<\lambda}$ to be the sup of $\{\kappa^\mu : \mu < \lambda\}$. Then for example, $2^\kappa$ equals the cardinality of $\mathcal{P}(\kappa)$, for the latter is equinumerous with the set of characteristic functions of subsets of $\kappa$. If $\kappa$ is infinite, $2^\kappa$ equals $\kappa^\kappa$. But which aleph is $2^{\aleph_\alpha}$ equal to? The Continuum Hypothesis (CH) says that

$$2^{\aleph_0} = \aleph_1$$

and the Generalized Continuum Hypothesis (GCH) says that for all ordinals $\alpha$,

$$2^{\aleph_\alpha} = \aleph_{\alpha+1}.$$

It was proved by Gödel that GCH is consistent with ZFC, and by Cohen that $\neg$ CH is consistent with ZFC; thus GCH (and CH) are independent of ZFC.

Define $\kappa$ to be a *strong limit cardinal* if it is a cardinal and for every $\lambda < \kappa$, $2^\lambda < \kappa$.

If $\alpha$ is a limit ordinal $> 0$, a subset $X$ of $\alpha$ is said to be *cofinal* in $\alpha$ if sup $X = \alpha$, i.e., for every $\beta < \alpha$ there exists $\gamma \in X$ such that $\gamma > \beta$. The *cofinality* of $\alpha$ is defined to be the least cardinal $\lambda$ such that there exists a subset $X$ of $\alpha$ of cardinality $\lambda$ which is cofinal in $\alpha$; we write $\mathrm{cf}(\alpha) = \lambda$. A cardinal $\kappa$ is said to be *regular* if $\mathrm{cf}(\kappa) = \kappa$; otherwise, $\kappa$ is *singular*.

Every successor cardinal, $\aleph_{\alpha+1}$, is regular because $\aleph_\alpha \cdot \aleph_\alpha = \aleph_\alpha$. (If $X$ is a subset of $\aleph_{\alpha+1}$ of cardinality at most $\aleph_\alpha$, then since every member of $X$ has cardinality $\leq \aleph_\alpha$, $\cup X$ has cardinality $\leq \aleph_\alpha \cdot \aleph_\alpha = \aleph_\alpha$, and so sup $X$ cannot equal $\aleph_{\alpha+1}$.) If $\aleph_\alpha$ is a limit cardinal and $\alpha > 0$, then $\mathrm{cf}(\aleph_\alpha) = \mathrm{cf}(\alpha)$. The first infinite limit cardinal, $\aleph_0$, is regular, but, generally, limit cardinals are singular; for example $\aleph_\omega$, $\aleph_{\omega+\omega}$, $\aleph_{\omega^2}$ have cofinality $\aleph_0$ and $\aleph_{\omega_1}$ has cofinality $\aleph_1$. A regular limit cardinal $> \aleph_0$ is called a *weakly inaccessible* cardinal. If $\aleph_\alpha$ is weakly inaccessible, then $\mathrm{cf}(\aleph_\alpha) = \mathrm{cf}(\alpha) \leq \alpha \leq \aleph_\alpha$ so $\aleph_\alpha = \alpha$. A cardinal $\kappa$ is called *strongly inaccessible*, or just *inaccessible*, if it is regular and uncountable and a strong limit cardinal. A strongly inaccessible cardinal is weakly inaccessible; if GCH holds, then the converse is true.

If $\kappa$ is strongly inaccessible, then $V_\kappa$ is a model of ZFC, i.e., all the ZFC axioms hold in $V_\kappa$. Thus if Inac is the statement "there is an inaccessible cardinal", Inac cannot be proved from ZFC (if ZFC is consistent) because if it could, then we could prove that there is a model of ZFC and hence ZFC is consistent; but this violates Gödel's second incompleteness theorem. The same applies for the statement "there is a weakly inaccessible cardinal" (see VI.3).

Moreover, we cannot even prove that ZFC + Inac is consistent assuming ZFC is consistent. For we have already observed that in ZFC + Inac we can prove "ZFC is consistent"; thus if we could prove that "ZFC is consistent implies ZFC + Inac is consistent", then in ZFC + Inac we could prove "ZFC + Inac is consistent", which violates Gödel's second incompleteness theorem (for the axiom system ZFC + Inac). Similarly, we cannot prove that the existence of a weakly inaccessible cardinal is relatively consistent. The large cardinals we encounter in the next sections will all be inaccessible, so their existence cannot be shown to be consistent. On the other hand, no one has (yet) proved that their existence is inconsistent.

For any infinite cardinal $\kappa$, let

$$\mathrm{H}(\kappa) = \{x\colon |TC(x)| < \kappa\}.$$

$\mathrm{H}(\aleph_0)$ is called the class of *hereditarily finite* sets, and equals $V_\omega$. $\mathrm{H}(\aleph_1)$ is called the class of *hereditarily countable* sets. In general, $\mathrm{H}(\kappa)$ is called the class of sets *hereditarily of cardinality* $< \kappa$; it is a transitive set and contains every transitive set of cardinality

$< \kappa$; thus, in particular, an ordinal $\alpha$ belongs to H($\kappa$) if and only if $\alpha < \kappa$.

We will also have occasion to refer to ordinal arithmetic. If $\alpha$, $\beta$ are ordinals then $\alpha + \beta$ denotes the (unique) ordinal which is isomorphic to a copy of $\alpha$ followed by a copy of $\beta$. As well, $\alpha \cdot \beta$ is the ordinal isomorphic to $\beta$ copies of $\alpha$; i.e., $\alpha \cdot \beta$ is isomorphic to $\beta \times \alpha$ under the lexicographic order defined by $(\gamma, \delta) < (\tau, \sigma)$ if $\gamma < \tau$ or $\gamma = \tau$ and $\delta < \sigma$. The notation we are using is ambiguous in that the operations of cardinal arithmetic and ordinal arithmetic are denoted by the same symbols. But it is usually clear from the context which operation is intended.

## §2. Filters and large cardinals

Throughout this section $I$ will denote an arbitrary (infinite) set.

**2.1 Definition.** A *filter* on $I$ is a subset $D$ of $\mathcal{P}(I)$ satisfying:
   (1) $\emptyset \notin D$, $I \in D$;
   (2) if $X, Y \in D$, then $X \cap Y \in D$;
   (3) if $X \in D$ and $X \subseteq Y \subseteq I$, then $Y \in D$.
If $Y$ is a subset of $I$, the *principal filter generated by* $Y$ is the filter $D \overset{\text{def}}{=} \{X \in \mathcal{P}(I): Y \subseteq X\}$. If $D$ is not equal to $D_Y$ for any $Y \subseteq I$, $D$ is called *non-principal*. An *ultrafilter* on $I$ is a filter $D$ on $I$ such that for every $X \subseteq I$, either $X \in D$ or $I \setminus X \in D$. A set $N \subseteq \mathcal{P}(I)$, is called an *ideal* if $\{I \setminus X : X \in N\}$ is a filter.

**2.2 Lemma.** *If $S$ is a subset of $\mathcal{P}(I)$, $S$ is contained in a filter on $I$ if and only if $S$ has the finite intersection property (FIP), i.e., the intersection of any finite subset of $S$ is non-empty.*

PROOF. Clearly a filter, and therefore any subset of, it has the finite intersection property. Conversely if $S$ has FIP, let $D$ be the set of all subsets of $I$ which contain the intersection of a finite subset of $S$; it is easy to check that $D$ is a filter. □

**2.3 Lemma.** *A filter $D$ on $I$ is an ultrafilter if and only if it is a maximal filter, i.e., there is no filter on $I$ which properly contains $D$.*

PROOF. If $D$ is an ultrafilter and $D'$ is a filter on $I$ containing $D$, then every $X$ in $D'$ must belong to $D$, because otherwise, $I \setminus X$ would belong to $D$ and then we would have $\emptyset = X \cap (I \setminus X) \in D'$, which contradicts the definition of a filter. Conversely, suppose that $D$ is a maximal filter on $I$. For any $Y \subseteq I$, if $\{Y\} \cup D$ has FIP, then $\{Y\} \cup D$ is, by 2.2, contained in a filter $D'$. But then the maximality of $D$ implies that $D' = D$, so $Y$ belongs to $D$. For any $X \subseteq I$, it is easy to verify that either $\{X\} \cup D$ or $\{I \setminus X\} \cup D$ has FIP. $\square$

**2.4 Definition.** For any infinite cardinal $\kappa$, a filter $D$ on $I$ is called $\kappa$-*complete* if for every subset $\mathcal{S}$ of $D$ of cardinality $< \kappa$, $\bigcap \mathcal{S}$ belongs to $D$. Otherwise, it is $\kappa$-*incomplete*. An $\omega_1$-complete filter is also called *countably complete*.

Generally, having defined a property of filters, we shall say that an ideal $N$ has that property if the filter $\{I \setminus X : X \in N\}$ does. For example, we say that an ideal $N$ is $\kappa$-complete if $\{I \setminus X : X \in N\}$ is $\kappa$-complete; thus $N$ is $\kappa$-complete if and only if for every subset $\mathcal{S}$ of $N$ of cardinality $< \kappa$, $\bigcup \mathcal{S}$ belongs to $N$.

**2.5 Examples.** Every filter is $\omega$-complete. A principal filter is $\kappa$-complete for every $\kappa$.

If $|I| \geq \kappa$, the co-$\kappa$ *filter*, $C_\kappa$, on $I$ is $\{X \subseteq I : |I \setminus X| < \kappa\}$; $C_\omega$ is called the *cofinite filter*. If $\kappa$ is regular, $C_\kappa$ is $\kappa$-complete but not $\kappa^+$-complete. However, if $|I| \geq \kappa^+$, $C_\kappa$ has the property that every intersection of $\kappa$ members of $C_\kappa$ is non-empty.

**2.6 Lemma.** *Let $D$ be an ultrafilter on $I$ and $\kappa$ an infinite cardinal. The following are equivalent:*

(1) *$D$ is $\kappa$-complete;*

(2) *for every subset $\mathcal{D}$ of $D$ of cardinality $< \kappa$, $\bigcap \mathcal{D} \neq \emptyset$;*

(3) *for every partition $\Pi$ of $I$ into fewer than $\kappa$ sets, there exists a unique $Z \in \Pi$ which belongs to $D$;*

(4) *for every subset $\mathcal{S}$ of $\mathcal{P}(I)$ of cardinality $< \kappa$, if $\bigcup \mathcal{S} \in D$, then $D \cap \mathcal{S} \neq \emptyset$.*

PROOF. (1) $\Rightarrow$ (2) is immediate. Assume (2). Suppose $\Pi$ is a partition of $I$ i.e., a family of pairwise disjoint subsets of $I$ whose union is $I$) of size $< \kappa$. Let $\mathcal{D} = \{(I \setminus Z): Z \in \Pi\}$. Then $\mathcal{D}$ is a subset of $\mathcal{P}(I)$ of size $< \kappa$, and $\bigcap \mathcal{D} = \emptyset$; so by (2), there exists $Z \in \Pi$ such that $I \setminus Z \notin D$; but then $Z \in D$. Notice also that a partition of $I$ can have at most one member belonging to $D$ because a filter is closed under intersections but does not contain the empty set. Thus (3) holds.

To prove (3) $\Rightarrow$ (4), suppose we are given $\mathcal{S} = \{Y_\nu : \nu < \lambda\} \subseteq \mathcal{P}(I)$, such that $\bigcup \mathcal{S} \in D$ and $\lambda < \kappa$. Inductively define $Y'_\nu = Y_\nu \setminus \bigcup_{\mu < \nu} Y'_\mu$ where $Y'_0 = Y_0$. Then $\Pi \overset{\text{def}}{=} (\{Y'_\nu : \nu < \lambda\} \cup \{I \setminus (\bigcup \mathcal{S})\}) \setminus \{\emptyset\}$ is a partition of $I$, so by (3), one of its members belongs to $D$. This member is certainly not $I \setminus (\bigcup \mathcal{S})$, so it must be some $Y'_\nu$; but $Y'_\nu \subseteq Y_\nu$, so $Y_\nu$ also belongs to $D$.

To see that (4) $\Rightarrow$ (1), suppose that $\mathcal{D}$ is a subset of $D$ of cardinality $< \kappa$. Let $\mathcal{S} = \{I \setminus Y : Y \in \mathcal{D}\}$. Then no element of $\mathcal{S}$ belongs to $D$, so by (4), $\bigcup \mathcal{S} \notin D$. Hence $\bigcap \mathcal{D} = I \setminus (\bigcup \mathcal{S})$ does belong to $D$. $\square$

**2.7 Proposition.** *For any ultrafilter $D$ on $I$, the following are equivalent*:

(1) *$D$ is non-principal*;
(2) *$C_\omega \subseteq D$*;
(3) *for all $i \in I$, $\{i\} \notin D$.*

PROOF. (1) $\Rightarrow$ (2): If some finite set belongs to $D$, then by 2.6, since $D$ is $\omega$-complete, some $\{i\}$ belongs to $D$ and then obviously $D$ is principal (generated by $\{i\}$).

(2) $\Rightarrow$ (3) is clear.

(3) $\Rightarrow$ (1): If $D$ is a principal ultrafilter, generated by $Y$, then $Y$ must be a singleton, because otherwise, if $a \in Y$, $\{a\} \notin D$ and $I \setminus \{a\} \notin D$. $\square$

**2.8 Theorem.** *Every filter on $I$ is contained in an ultrafilter. Hence, for any infinite set, $I$, there exists a non-principal ultrafilter on $I$.*

PROOF. For any filter $F$, by Zorn's Lemma there is a maximal filter, $D$, containing $F$, which, by 2.3, is an ultrafilter. If $F$ contains the cofinite filter, then $D$ is non-principal. $\square$

**2.9 Corollary.** *A non-principal ultrafilter on $I$ is not $|I|^+$-complete.*

PROOF. If $D$ is a non-principal ultrafilter on $I$, then by 2.7, $I \setminus \{i\} \in D$ for all $i \in I$. But $\cap\{I \setminus \{i\}: i \in I\} = \emptyset$ does not belong to $D$. $\square$

Given an ultrafilter $D$ on $I$ we can define a two-valued measure on $I$; more precisely, we define $\mu_D: \mathcal{P}(I) \to \{0, 1\}$ by: $\mu_D(X) = 1$ if $X \in D$ and $\mu_D(X) = 0$ if $X \notin D$. If $D$ is non-principal, points (singletons) have measure 0; and if $D$ is $\kappa$-complete then $\mu_D$ is $\kappa$-additive, i.e., whenever $\{X_\nu: \nu < \lambda\}$ is a partition of $Y$ and $\lambda < \kappa$, then $\mu_D(Y) = \sum_{\nu<\lambda} \mu_D(X_\nu)$. Thus $\mu_D$ is always finitely additive, i.e., $\omega$-additive; it is natural to ask whether there is a countably additive, two-valued non-trivial measure on a set $I$. Obviously the only relevant property of $I$ is its cardinality.

**2.10 Definition.** A cardinal $\kappa$ is called $\omega$-*measurable* if there is a non-principal $\omega_1$-complete ultrafilter on $\kappa$ (or equivalently on any set $I$ of size $\kappa$). $\kappa$ is called *measurable* if it is uncountable and there is a non-principal $\kappa$-complete ultrafilter on $\kappa$ (or equivalently on any set $I$ of size $\kappa$).

We say that $I$ is measurable ($\omega$-measurable) if its cardinality is measurable ($\omega$-measurable).

Notice that although there is a non-principal $\omega$-complete ultrafilter on $\omega$, we have defined measurable so that $\omega$ is not measurable. This avoids making some special exceptions for $\omega$ when discussing the properties of measurable and $\omega$-measurable cardinals. For example it is then clear that every measurable cardinal is $\omega$-measurable. The converse is not true, as we shall see, but the corollary to the following theorem says that the existence questions are equivalent.

**2.11 Theorem.** *Assume that there is an $\omega$-measurable cardinal and let $\kappa$ be the first one. Then*

(i) *every cardinal* $\geq \kappa$ *is $\omega$-measurable; and*

(ii) *for every set I, every $\omega_1$-complete ultrafilter on I is $\kappa$-complete.*

PROOF. (i) If $D$ is a non-principal $\omega_1$-complete ultrafilter on $\kappa$ and $\lambda$ is a cardinal $\geq \kappa$, we can define a non-principal $\omega_1$-complete ultrafilter $D'$ on $\lambda$, as follows. For any subset $Y$ of $\lambda$, let $Y \in D'$ if and only if $Y \cap \kappa \in D$. It is routine to check that $D'$ has the desired properties.

(ii) Suppose, to obtain a contradiction, that there is an ultrafilter $D$ on $I$ which is $\omega_1$-complete but not $\kappa$-complete. Then $D$ is non-principal and there is a partition $\Pi = \{Y_j : j \in J\}$ of $I$ such that $J$ is a set of cardinality $\lambda < \kappa$ and no member of $\Pi$ belongs to $D$. We will show that there is a non-principal $\omega_1$-complete ultrafilter on $J$, thereby contradicting the minimality of $\kappa$. Define $U \subseteq \mathcal{P}(J)$ as follows: for any $X \subseteq J$, $X \in U$ if and only if $\cup\{Y_j : j \in X\} \in D$. It is easy to see that $U$ is a non-principal ultrafilter on $J$. To see that $U$ is countably complete, suppose that $X_n \in U$ for $n \in \omega$. Let $X = \bigcap_{n \in \omega} X_n$. Then $\cup\{Y_j : j \in X\} = \bigcap_{n \in \omega}(\cup\{Y_j : j \in X_n\})$ which belongs to $D$ since $D$ is $\omega_1$-complete and each $X_n$ belongs to $U$. Therefore $X \in U$. $\square$

**2.12 Corollary.** *If there is an $\omega$-measurable cardinal, then there is a measurable cardinal. In fact, the first $\omega$-measurable cardinal is measurable.* $\square$

An ultrafilter $D$ on $\kappa$ is called *uniform* if every member of $D$ has cardinality $\kappa$. Any non-principal $\kappa$-complete ultrafilter on $\kappa$ is uniform because any subset of size $< \kappa$ is the union of fewer than $\kappa$ singletons, none of which are in the ultrafilter.

**2.13 Theorem.** *Every measurable cardinal is strongly inaccessible.*

PROOF. Let $\kappa$ be such that there is a $\kappa$-complete non-principal ultrafilter $D$ on $\kappa$. To show that $\kappa$ is regular, consider a subset $S$ of $\kappa$ of cardinality $< \kappa$. For each $\alpha \in S$, $\alpha \notin D$ since $D$ is uniform and $\alpha$ has cardinality $< \kappa$. Thus $\cup S$ does not equal $\kappa$ since $D$ is $\kappa$-complete; so $\sup S \neq \kappa$.

Suppose, to obtain a contradiction, that there is a cardinal $\lambda <$ $\kappa$ such that $2^\lambda \geq \kappa$. Then there is a set $F$ of functions: $\lambda \to \{0,1\}$ such that $|F| = \kappa$. We can regard $D$ as a filter on $F$. For each $\nu < \lambda$, let $Y_\nu$ be the one of the two sets $\{f \in F: f(\nu) = \ell\}(\ell = 0,$ 1) which belongs to $D$. Since $D$ is $\kappa$-complete, $\bigcap_{\nu<\lambda} Y_\nu$ belongs to $D$, which is a contradiction since it has only one element. $\square$

It follows that we cannot prove from ZFC the existence of measurable cardinals, or even prove the relative consistency with ZFC of the existence of measurable cardinals (see section 1). (Of course, it is consistent that there are no measurable cardinals.) We should also note that the first measurable cardinal is much larger than the first strongly inaccessible cardinal; indeed, it can be proved that there are $\kappa$ strongly inaccessible cardinals below a measurable cardinal $\kappa$.

Notice also that 2.13 implies that if there are $\omega$-measurable cardinals, then there are $\omega$-measurable cardinals which are not measurable; indeed every cardinal above the first $\omega$-measurable is $\omega$-measurable, but most of them (for example all the successor and singular cardinals) are not strongly inaccessible, much less measurable.

**2.14 Definition.** Given an indexed family of subsets $\{X_\nu: \nu < \kappa\}$ of $\kappa$, the *diagonal intersection* of the family, denoted $\triangle\{X_\nu: \nu < \kappa\}$, is defined to be

$$\{\alpha < \kappa: \alpha \in \bigcap_{\nu<\alpha} X_\nu\}.$$

We say that a filter on $\kappa$ is *normal* if it closed under diagonal intersections.

We state without proof the following (see for example Jech 1978).

**2.15 Theorem.** *If $\kappa$ is a measurable cardinal, then there is a normal non-principal $\kappa$-complete ultrafilter $D$ on $\kappa$. Moreover, $\{\alpha \in \kappa: \alpha$ is a regular cardinal$\} \in D$.* $\square$

A function $\theta$ on a subset $S$ of $\kappa$ is called *regressive* if $\theta(\alpha) < \alpha$, for all $\alpha \in S$. The following proposition connects normality and regressive functions.

**2.16 Proposition.** *Suppose $F$ is a normal filter on a cardinal $\kappa$, $S \subseteq \kappa$, $\kappa \setminus S \notin F$, and $\theta$ is a regressive function on $S$. Then there is a set $X \subseteq S$ so that $\kappa \setminus X \notin F$ and $\theta {\restriction} X$ is constant.*

PROOF. Assume no such $X$ exists. Then for every $\alpha$, $X_\alpha \overset{\text{def}}{=} \{\beta \in S : \theta(\beta) = \alpha\}$ is such that $\kappa \setminus X_\alpha$ is in $F$. Let $Y = \triangle \{\kappa \setminus X_\alpha : \alpha < \kappa\}$. Then $Y \in F$ since $F$ is normal. Moreover, $Y \subseteq \kappa \setminus S$ since $\theta$ is regressive; so $\kappa \setminus S \in F$, a contradiction. $\square$

The converse of this result is also true for non-principal $\kappa$-complete filters.

We will briefly discuss other large cardinals which can be defined in terms of the existence of ultrafilters.

**2.17 Definition.** A cardinal $\kappa$ is called *strongly compact* if for every set $I$, every $\kappa$-complete filter on $I$ is contained in a $\kappa$-complete ultrafilter on $I$. We say that $\kappa$ is $L_{\lambda\omega}$-*compact* if for every set $I$, every $\kappa$-complete filter on $I$ is contained in a $\lambda$-complete ultrafilter on $I$. (The terminology comes from infinitary logic.)

It is clear that $L_{\lambda\omega}$-compact implies $L_{\mu\omega}$-compact if $\lambda \geq \mu$, and that $\kappa$ is strongly compact if and only if $\kappa$ is $L_{\kappa\omega}$-compact. If $\kappa$ is $L_{\lambda\omega}$-compact, then every cardinal $\geq \kappa$ is also $L_{\lambda\omega}$-compact. Moreover, every $L_{\omega_1\omega}$-compact cardinal is $\omega$-measurable. Magidor has shown (assuming the existence of certain large cardinals) that it is consistent that the first measurable cardinal is strongly compact, and also that it is consistent that there is an $L_{\omega_1\omega}$-compact cardinal but no strongly compact cardinals.

There are also cardinals called *weakly compact*, which we want to define. We say that $\kappa$ is weakly compact if $\kappa$ is strongly inaccessible and for every $\kappa$-complete Boolean subalgebra $\mathcal{B}$ of $\mathcal{P}(\kappa)$ such that $\mathcal{B}$ has cardinality $\kappa$, every $\kappa$-complete filter on $\mathcal{B}$ is contained in a $\kappa$-complete ultrafilter on $\mathcal{B}$. (Here, the notions of filter and $\kappa$-complete are extended in a natural way to this setting.)

Measurable cardinals are weakly compact, and there are $\kappa$ weakly compact cardinals below a measurable cardinal $\kappa$. There are $\kappa$ strongly inaccessible cardinals below a weakly compact cardinal $\kappa$.

**2.18.** We conclude with some facts about the number of $\kappa$-complete filters on $\kappa$, which will be relevant in section III.3. (See Comfort and Negrepontis 1974, sections 7 and 8 for most of these.) Obviously there are at most $2^{2^\kappa}$ different ultrafilters on any cardinal $\kappa$ (and in fact if $\kappa$ is infinite, there *are* that many ultrafilters.). If $\kappa$ is measurable, then there are at least $2^\kappa$ $\kappa$-complete ultrafilters on $\kappa$: see Exercise 12. Kunen 1970 has shown that it is consistent that there is a measurable $\kappa$ with exactly $2^\kappa$ $\kappa$-complete ultrafilters on $\kappa$. On the other hand, if there is a measurable cardinal, then there is an $\omega$-measurable cardinal $\kappa$ such that there are only $\kappa$ $\omega_1$-complete ultrafilters on $\kappa$: see Exercise 11. Comfort and Negrepontis 1972 and Kunen 1970 have shown that if $\kappa$ is strongly compact and $\lambda \geq \kappa$ such that $\lambda^{<\kappa} = \lambda$, then there are $2^{2^\lambda}$ $\kappa$-complete ultrafilters on $\lambda$.

## §3. Ultraproducts

The ultraproduct construction is an important one in model theory and as such can be applied to arbitrary (first-order) structures. Since our concern is with modules, we shall confine our discussion to that setting; also we will not deal specifically with results, such as the Fundamental Theorem of Ultraproducts, which involve formal languages in their statement (cf. Exercise 6).

Again, $I$ will always denote an infinite set.

**3.1 Definition.** Let $\{M_i : i \in I\}$ be an indexed family of $R$-modules. Recall that if $a \in \prod_{i \in I} M_i$ , $\operatorname{supp}(a) = \{i \in I : a(i) \neq 0\}$. Given a filter $D$ on $I$, let $K_D = \{x \in \prod_{i \in I} M_i : I \setminus \operatorname{supp}(x) \in D\}$; i.e., $x \in K_D$ if and only if $\{i \in I : x(i) = 0\} \in D$. It follows immediately from the properties of a filter that $K_D$ is a submodule of $\prod_{i \in I} M_i$. The quotient module $\prod_{i \in I} M_i / K_D$ is denoted $\prod_{i \in I} M_i / D$ and called the *reduced product* of $\{M_i : i \in I\}$ with respect to $D$; if $D$ is an ultrafilter it is called the *ultraproduct* with respect to $D$.

If $a \in \prod_{i \in I} M_i$, the coset of $a$ in $\prod_{i \in I} M_i/D$ is denoted $a_D$. Notice that $a_D = b_D$ if and only if $\{i \in I : a(i) = b(i)\}$ belongs to $D$.

If $M_i = M$ for all $i \in I$, we denote $\prod_{i \in I} M_i/D$ by $M^I/D$ and call it the *reduced power* or *ultrapower* (if $D$ is an ultrafilter) of $M$ with respect to $D$. In this case there is a canonical map $\delta_M : M \to M^I/D$ which takes $m \in M$ to $\bar{m}_D$, where $\bar{m} : I \to M$ is the constant function with value $m$.

It is easy to see that if $D$ is a principal ultrafilter, generated by $\{j\}$, then $\prod_{i \in I} M_i/D$ is canonically isomorphic to $M_j$.

**3.2 Proposition.** *For any $M$ and $D$, $\delta_M$ is an embedding. If $D$ is a $\kappa$-complete ultrafilter on $I$ and $|M| < \kappa$, then $\delta_M$ is an isomorphism.*

PROOF. If $\delta_M(m) = 0$ $(= \bar{0}_D)$, then $\{i \in I : \bar{m}(i) = 0\} \in D$; since this set is either $\emptyset$ or $I$, it must be $I$, so $m = 0$. Similarly $\delta_M(m+n) = \delta_m(m) + \delta_M(n)$ because $I = \{i \in I : \overline{m+n}(i) = \bar{m}(i) + \bar{n}(i)\}$; and $\delta_M$ preserves scalar multiplication for the same reason.

For the second part of the proposition we must show, under the hypotheses, that $\delta_M$ is surjective. For any $x \in M^I/D$, and any $m \in M$, let $Y_m = \{i \in I : x(i) = m\}$. Then $\{Y_m : m \in M\}$ is a partition of $I$ of cardinality $< \kappa$, so by 2.6, there exists a unique $m \in M$ such that $Y_m \in D$. But then $x_D = \delta_M(m)$. $\square$

**3.3 Corollary.** *If the cardinality of $M$ is not $\omega$-measurable and $D$ is an $\omega_1$-complete ultrafilter on $I$, then $M^I/D$ is canonically isomorphic to $M$.*

PROOF. This follows immediately from 3.2 and 2.11. (If there is no measurable cardinal then every $\omega_1$-complete ultrafilter is principal.) $\square$

We shall make only limited use of ultraproducts, so we shall discuss here only those results that we need. The following proposition is called the Wald-Los Lemma.

**3.4 Proposition.** *Let $\{M_i : i \in I\}$ be an indexed family of $R$-modules. Let $D$ be a $\kappa$-complete filter on $I$ such that $|R| < \kappa$. If*

*A is a subgroup of* $\prod_{i \in I} M_i / D$ *of cardinality* $< \kappa$, *then there is an embedding* $\gamma: A \to \prod_{i \in I} M_i$ *such that* $\pi \circ \gamma$ *is the identity on A.* (*Here* $\pi$ *is the canonical projection of* $\prod_{i \in I} M_i$ *onto* $\prod_{i \in I} M_i / D$.)

PROOF. For each element $a$ of $A$, choose a representative $x_a$ in $\prod_{i \in I} M_i$. Then for $a, b \in A$ and $r \in R$, the sets

$$Y_{a,b} = \{i \in I : x_a(i) + x_b(i) = x_{a+b}(i)\}$$

and

$$Y_{r,a} = \{i \in I : r x_a(i) = x_{ra}(i)\}$$

belong to $D$. Since $D$ is $\kappa$-complete, the set

$$Y = \cap\{Y_{a,b} : a, b \in A\} \cap \bigcap\{Y_{r,a} : r \in R, a \in A\}$$

also belongs to $D$. Then we define $\gamma$ by:

$$\gamma(a)(i) = \begin{cases} x_\alpha(i) & \text{if } i \in Y \\ 0 & \text{otherwise} \end{cases}$$

and easily check that $\gamma$ is the desired map. $\square$

We want to discuss the cardinality of ultraproducts, but first we need a lemma.

**3.5 Lemma.** *Let* $\kappa$ *be a cardinal such that* $2^{<\kappa} = \kappa$. *Then there is a family of* $2^\kappa$ *functions* $f_\nu: \kappa \to \kappa$ ($\nu < 2^\kappa$) *such that for* $\mu \neq \nu$, $\{\alpha \in \kappa : f_\nu(\alpha) = f_\mu(\alpha)\}$ *has cardinality* $< \kappa$.

PROOF. Let $S = \{X \subseteq \kappa : \sup(X) < \kappa\}$. By hypothesis $S$ has cardinality $\kappa$, so we can identify it with $\kappa$ and define functions from $\kappa$ to $S$. In fact we define a function $f_A: \kappa \to S$ for every $A \subseteq \kappa$: let $f_A(\alpha) = A \cap \alpha$. Now if $A \neq B$, then there exists $\beta \in (A \setminus B) \cup (B \setminus A)$; for all $\alpha > \beta$, $f_A(\alpha) \neq f_B(\alpha)$. $\square$

**3.6 Proposition.** *Let* $\kappa$ *be a cardinal such that* $2^{<\kappa} = \kappa$. *Let* $D$ *be a uniform ultrafilter on a set* $I$ *of cardinality* $\kappa$. *Let* $M_i$ *be a module of cardinality* $\kappa$ *for all* $i \in I$. *Then* $\prod_{i \in I} M_i / D$ *has cardinality* $2^\kappa$.

PROOF. We must show that $\prod_{i \in I} M_i / D$ has cardinality $\geq 2^\kappa$. Without loss of generality we can assume that the universe of each $M_i$ is $\kappa$, and that $I = \kappa$. Then the functions $f_\nu$ of the previous lemma are elements of $\prod_{i \in I} M_i$, and by the uniformity of $D$ and the definition of the ultraproduct, different functions represent different elements of $\prod_{i \in I} M_i / D$. $\square$

There is an extensive literature on cardinalities of ultraproducts to which we refer the interested reader. See, for example, Chang-Keisler 1973 or Keisler 1965.

Our final topic is ultraproducts of free modules. We will be able to prove some results about our main topic of "almost free" modules. Here we shall make use of some of the large cardinals defined in the previous section.

If $B_i$ is a subset of the module $M_i$ for each $i \in I$, then $\prod_{i \in I} B_i$ is a subset of $\prod_{i \in I} M_i$; we denote by $\prod_{i \in I} B_i / D$ the image of $\prod_{i \in I} B_i$ under the canonical projection of $\prod_{i \in I} M_i$ onto $\prod_{i \in I} M_i / D$, i.e., $\prod_{i \in I} B_i / D$ is the set of elements of $\prod_{i \in I} M_i / D$ represented by an element of $\prod_{i \in I} B_i$.

**3.7 Lemma.** *If $B_i$ is a linearly independent subset of $M_i$ for each $i \in I$, then for any ultrafilter $D$ on $I$, $\prod_{i \in I} B_i / D$ is a linearly independent subset of $\prod_{i \in I} M_i / D$.*

PROOF. Suppose not: let $(x_0)_D, \ldots, (x_n)_D$ be distinct elements of $\prod_{i \in I} B_i / D$ such that there are non-zero elements $r_0, \ldots, r_n$ of $R$ such that $\sum_{j \leq n} r_j (x_j)_D = 0$ in $\prod_{i \in I} M_i / D$. Then

$$Y \stackrel{\text{def}}{=} \{i \in I : \sum_{j \leq n} r_j x_j(i) = 0 \text{ in } M_i\}$$

belongs to $D$, as does

$$Z_{j,k} \stackrel{\text{def}}{=} \{i \in I : x_j(i) \neq x_k(i)\}$$

for $j \neq k$ (since $(x_j)_D \neq (x_k)_D$). Therefore

$$W \stackrel{\text{def}}{=} Y \cap \bigcap_{j \neq k} Z_{j,k}$$

belongs to $D$, so $W$ is non-empty. For any $i \in W$, since $B_i$ is linearly independent and the $x_j(i)$'s are distinct, and $\sum_{j \le n} r_j x_j(i) = 0$, we conclude that $r_j = 0$ for all $j$. □

**3.8 Theorem.** *Let $D$ be a $\kappa$-complete ultrafilter on $I$ where $\kappa > |R| + \aleph_0$. If $M_i$ is a free module for each $i \in I$, then $\prod_{i \in I} M_i / D$ is a free module.*

PROOF. Let $B_i$ be a basis of $M_i$. By Lemma 3.7, $\prod_{i \in I} B_i / D$ is a linearly independent subset of $\prod_{i \in I} M_i / D$ so we need only prove that it generates $\prod_{i \in I} M_i / D$. Let $y \in \prod_{i \in I} M_i$. For any $(n{+}1)$-tuple $\vec{r} = (r_0, \ldots, r_n) \in R^{n+1}$, let

$$Y(\vec{r}) \overset{\text{def}}{=} \{i \in I : \exists x_0(i) \ldots, x_n(i) \in B_i(y(i) = \sum_{j \le n} r_j x_j(i))\}.$$

Since each $B_i$ generates $M_i$, $\cup \{Y(\vec{r}) : \vec{r} \in R^{n+1},\ n \in \omega\} = I$. By the hypothesis on $|R|$, there *are* $< \kappa$ sequences. Therefore there exists $\vec{r} = (r_0, \ldots, r_n)$ such that $Y(\vec{r})$ belongs to $D$. Then $y_D = \sum_{j \le n} r_j (x_j)_D$ where $x_j(i)$ is defined so that $y(i) = \sum_{j \le n} r_j x_j(i)$ for $i \in Y(\vec{r})$. □

Let us say that an $R$-module $M$ is $\kappa$-*free* if every submodule of $M$ of cardinality $< \kappa$ is free. This notion will be vacuous unless every submodule of a free $R$-module is free, so let us assume that $R$ is a principal ideal domain; we will also assume that $\kappa > |R|$. We are interested in the question of when a $\kappa$-free module is free. (See Chapters IV and VII for much more on this question, including a generalization of the notion of $\kappa$-free to modules over arbitrary rings.)

We will be dealing with those cardinals $\kappa$, defined at the end of section 2, which have the property that $\kappa$-complete filters are contained in ultrafilters which are $\lambda$-complete (for certain $\lambda$). We begin by defining a particular $\kappa$-complete filter.

**3.9 Definition.** Given a set $J$ and a regular cardinal $\kappa \le |J|$, let $\mathcal{P}_\kappa(J) = \{Y \in \mathcal{P}(J) : |Y| < \kappa\}$. For each $Y \in \mathcal{P}_\kappa(J)$, let $U_Y = \{X \in \mathcal{P}_\kappa(J) : Y \subseteq X\}$. Let $F_\kappa(J)$ be the filter on $\mathcal{P}_\kappa(J)$ generated by $\{U_Y : Y \in \mathcal{P}_\kappa(J)\}$, i.e., $F_\kappa(J) = \{Z \subseteq \mathcal{P}_\kappa(J) : Z \supseteq U_Y$ for some $Y \in \mathcal{P}(J)\}$. Then $F_\kappa(J)$ is a $\kappa$-complete filter, since for any cardinal $\lambda < \kappa$ and sets $Y_\nu (\nu < \lambda)$ in $\mathcal{P}_\kappa(J)$, $\bigcap_{\nu < \lambda} U_{Y_\nu} \supseteq U_W$ where $W = \bigcup_{\nu < \lambda} Y_\nu$ which has cardinality $< \kappa$.

**3.10 Theorem.** *Let $\lambda$ be an uncountable cardinal and $\kappa$ an $L_{\lambda\omega}$-compact cardinal. Let $R$ be a p.i.d. of cardinality $< \lambda$. If $M$ is a $\kappa$-free $R$-module (of arbitrary cardinality), then $M$ is free.*

PROOF. Note that $\kappa \geq \lambda$. Let $J$ be the underlying set of $M$, and let $I = \mathcal{P}_\kappa(J)$ and $F = F_\kappa(J)$ be defined as in Definition 3.9. For every $Y \in I$, $\langle Y \rangle$ is a submodule of $M$ of cardinality $< \kappa$, so it is free. By hypothesis on $\kappa$, $F$ is contained in a $\lambda$-complete ultrafilter $D$. Form the ultraproduct $\prod_{Y \in I} \langle Y \rangle / D$ which we will denote by $M^*$. Then by 3.8, $M^*$ is free. Define a map $\varphi : M \to M^*$ as follows: for any $a \in M$, let

$$\bar{a}(Y) = \begin{cases} a & \text{if } a \in Y \\ 0 & \text{otherwise} \end{cases}$$

and let $\varphi(a) = \bar{a}_D$. Now $a \in Y$ for some $Y \in I$, so $\bar{a}(Z) = a$ for all $Z \in U_Y$. Therefore it is clear that $\varphi$ is an embedding. Since $M^*$ is free, and $M$ is isomorphic to a submodule of $M^*$, and $R$ is a p.i.d., it follows that $M$ is free. $\square$

**3.11 Corollary.** *If $\kappa$ is an $L_{\omega_1\omega}$-compact cardinal, then every $\kappa$-free abelian group (of arbitrary cardinality) is free.* $\square$

Corollary 3.11 is an example of a "compactness result", related to those we discuss in section IV.3.

## §4. Cubs and stationary sets

Our theme in this section will be the description of "relatively large" subsets of a limit ordinal $\gamma$; we will have analogs of "sets of measure 1" (the closed unbounded sets, or cubs) and of "sets of non-zero measure" (the stationary sets).

**4.1 Definition.** A subset $C$ of $\gamma$ is called a *cub* (in $\gamma$) if

(1) $C$ is *closed* in $\gamma$, i.e., for all $Y \subseteq C$, if sup $Y \in \gamma$, then sup $Y \in C$; and

(2) $C$ is *unbounded* in $\gamma$, i.e., sup $C = \gamma$.

In some sources, a cub is called a *club*. The definition above makes sense even for ordinals of countable cofinality, but most of the good properties of cubs occur for ordinals of uncountable cofinality. Throughout the rest of this section, $\gamma$ will denote a limit ordinal of cofinality $> \aleph_0$, and $\kappa$ will denote a regular uncountable cardinal.

Clearly $\gamma$ itself and $\lim(\gamma)$ are cubs in $\gamma$. More generally, if $C$ is a cub, then

$$C^* \stackrel{\text{def}}{=} \{\alpha \in C : \alpha \text{ is a limit point of } C\}$$

is also a cub. (We say that $\alpha$ is a *limit point* of $C$ if for every $\beta < \alpha$ there exists $\gamma \in C$ such that $\beta < \gamma < \alpha$.)

**4.2 Example.** Let $\kappa$ be a regular uncountable cardinal and let $f: \kappa^n \to \kappa$ for some $n \geq 1$. Let $C = \{\sigma \in \kappa : f(x) \in \sigma$ for every $x \in \sigma^n\}$. We claim that $C$ is a cub in $\kappa$. It is clear that $C$ is closed in $\kappa$. To prove that it is unbounded we must show that for every $\beta \in \kappa$ there exists $\alpha > \beta$ such that $\alpha \in C$. We define an increasing sequence of $\beta_m$'s as follows: let $\beta_0 = \beta$; if $\beta_m$ has been defined then since the cardinality of $\beta_m^n$ is $< \kappa$, there exists $\beta_{m+1} > \beta_m$ such that $f(x) \in \beta_{m+1}$ for all $x \in \beta_m^n$. Then if we let $\alpha = \sup\{\beta_m : m \in \omega\}$, it is clear that $\alpha \in C$. We say, in short, that there is a cub of subsets of $\kappa$ which is closed under $f$. (This argument that $C$ is unbounded is a typical one.)

It is clear that the intersection of any number of cubs in $\gamma$ is again closed in $\gamma$, but it may not be unbounded. For example, if $\lambda = \text{cf}(\gamma)$ and $f: \lambda \to \gamma$ has range cofinal in $\gamma$, let $C_\alpha \stackrel{\text{def}}{=} \{\nu \in \gamma : \nu > f(\alpha)\}$ for each $\alpha \in \lambda$; then each $C_\alpha$ is clearly a cub, but $\cap\{C_\alpha : \alpha \in \lambda\}$ is empty. However, we have the following important positive result:

**4.3 Proposition.** *The intersection of fewer than* $\mathrm{cf}(\gamma)$ *cubs in* $\gamma$ *is again a cub in* $\gamma$.

PROOF. Suppose $\{C_\nu : \nu < \mu\}$ are cubs in $\gamma$, where $\mu < \mathrm{cf}(\gamma)$. We must show that for any $\beta \in \gamma$ there is an element *of* $\bigcap_{\nu < \mu} C_\nu$ which is greater than $\beta$. Since $\mathrm{cf}(\gamma)$ is uncountable we can assume that $\mu$ is an infinite cardinal; then there is a function $f: \mu \to \mu$ such that for all $\alpha \in \mu$, $f^{-1}[\alpha]$ is cofinal in $\mu$ (because $\mu \cdot \mu = \mu$). Define $\beta_\nu (\nu < \mu)$ by induction: $\beta_\nu$ is the first member of $C_{f(\nu)}$ which is $> \beta + \sup\{\beta_\tau : \tau < \nu\}$; this is possible since $C_{f(\nu)}$ is unbounded and since $\nu < \mathrm{cf}(\gamma)$. Then let $\delta = \sup\{\beta_\nu : \nu < \mu\}$; $\delta$ belongs to $\gamma$ since $\mu < \mathrm{cf}(\gamma)$; and $\delta \in C_\alpha$ for all $\alpha < \mu$ because $C_\alpha$ is closed and by choice of $f$, $\delta = \sup\{\beta_\nu : f(\nu) = \alpha\}$. $\square$

Now we can define the *closed unbounded filter* on $\gamma$ to be the set of all subsets $X$ of $\gamma$ which contain a cub on $\gamma$. By the Proposition, this is in fact a filter in the sense of Definition 2.1. If we consider the quotient of $\mathcal{P}(\gamma)$ by this filter we are led to the following definition.

**4.4 Definition.** For $X$ and $Y$ in $\mathcal{P}(\gamma)$, define $X \sim Y$ if and only if there is a cub $C$ in $\gamma$ such that $X \cap C = Y \cap C$. It is easy to see that this is an equivalence relation on $\mathcal{P}(\gamma)$. Denote the equivalence class of $X$ by $\tilde{X}$. The set of equivalence classes is the quotient Boolean algebra $\mathcal{P}(\gamma)/F$, where $F$ is the closed unbounded filter on $\gamma$: it has a Boolean algebra structure induced by the natural structure on $\mathcal{P}(\gamma)$. We will denote this quotient algebra by $D(\gamma)$. The greatest element of $D(\gamma)$ is denoted by 1 and equals $\tilde{\gamma}$ (or $\tilde{C}$ for any cub). The least element of $D(\gamma)$ is denoted by 0 and equals $\tilde{\emptyset}$ (or the equivalence class of $\gamma \setminus C$ for any cub).

We say that $X \subseteq \gamma$ is *stationary* in $\gamma$ if $\tilde{X} \neq 0$; otherwise said, $X$ is stationary in $\gamma$ if and only if for all cubs $C$, $X \cap C \neq \emptyset$. A set $X \subseteq \gamma$ is called *co-stationary* if $\gamma \setminus X$ is stationary; $X$ is called *thin* if it is not stationary.

We can now make precise the analogy mentioned at the beginning of the section. We can define a "$D(\gamma)$-valued measure" $\mu: \mathcal{P}(\gamma) \to D(\gamma)$ by $\mu(X) = \tilde{X}$. By Proposition 4.3, $\mu$ is $< \mathrm{cf}(\gamma)$-additive. A set $X$ has $\mu$-measure 1 if and only if it contains a cub, and it has $\mu$-measure 0 if and only if it is thin. As a special case of the additivity we have:

**4.5 Corollary.** *If $\lambda < \text{cf}(\gamma)$ and $\bigcup\{X_\nu : \nu < \lambda\}$ is stationary in $\gamma$, then there exists $\nu < \lambda$ such that $X_\nu$ is stationary in $\gamma$.*

PROOF. If not, then for each $\nu$ there exists a cub $C_\nu$ such that $C_\nu \cap X_\nu = \emptyset$. But then $(\bigcap_{\nu<\lambda} C_\nu) \cap (\bigcup_{\nu<\lambda} X_\nu) = \emptyset$ which is a contradiction of the assumption that $\bigcup_{\nu<\lambda} X_\nu$ is stationary because, by 4.3, $\bigcap_{\nu<\lambda} C_\nu$ is a cub. $\square$

**4.6 Definition.** Let $\kappa$ be a regular uncountable cardinal. If $E$ is a subset of $\kappa$, let $E' = \{\gamma \in \kappa : \text{cf}(\gamma) > \aleph_0 \text{ and } E \cap \gamma \text{ is stationary in } \gamma\}$. We say that $E$ is *non-reflecting* (or *sparse*) if $E' = \emptyset$.

**4.7 Example.** For any infinite regular cardinal $\rho < \kappa$, let $E_\rho = \{\alpha \in \kappa : \text{cf}(\alpha) = \rho\}$. Then $E_\rho$ is a stationary subset of $\kappa$. Indeed, given a cub $C$ in $\kappa$, one can easily define by transfinite induction a continuous strictly increasing $f : \kappa \to C$. (Here *continuous* means that for every limit ordinal $\sigma$, $f(\sigma) = \sup\{f(\tau) : \tau < \sigma\}$.) Then $f(\rho)$ is a member of $C$ which has cofinality $\rho$ and hence belongs to $E_\rho$, so $C \cap E_\rho \neq \emptyset$. If $\kappa = \aleph_1$, then $E_{\aleph_0}$ is a cub; in fact, $E_{\aleph_0} = \lim(\aleph_1)$. For $\kappa > \aleph_1$, $E_\rho$ does not contain a cub since, by the same argument as above, any cub contains elements of all possible cofinalities $< \kappa$. If $\kappa = \lambda^+$, where $\lambda$ is regular, then $E_\lambda$ is a non-reflecting subset of $\kappa$, since for any $\gamma \in \kappa$ of cofinality $> \aleph_0$, there is a cub in $\gamma$ whose elements all have cofinality $< \text{cf}(\gamma) \leq \lambda$.

Thus we know that for successor cardinals $\kappa \geq \aleph_2$, $D(\kappa)$ has cardinality $\geq 3$; that is, there is a stationary set which is not a cub. This result cannot be proved for $\kappa = \aleph_1$ without use of the Axiom of Choice. Our next results will give us more information about the size of $D(\kappa)$. For later use, we state our next result in a very general form. (Note that if $F$ is the closed unbounded filter on $\kappa$, then $S \subseteq \kappa$ is stationary if and only if $\kappa \setminus S \notin F$.)

**4.8 Proposition.** *Suppose $F$ is a $\kappa$-complete filter on $\kappa = \lambda^+$ which contains the cofinite filter. Let $S \subseteq \kappa$ such that $\kappa \setminus S \notin F$.*

*Then there is a decomposition of $S$, $S = \amalg_{\beta \in \kappa} S_\beta$, into $\kappa$ pairwise disjoint subsets $S_\beta$ such that for all $\beta$, $\kappa \setminus S_\beta \notin F$.*

PROOF. For each $\alpha < \kappa$, choose a surjection $g_\alpha: \lambda \to \alpha$. (Here we use the Axiom of Choice.) For all $\nu \in \lambda$, $\beta \in \kappa$, let

$$S_\beta^\nu = \{\alpha \in S: g_\alpha(\nu) = \beta\}.$$

We claim that there is a $\nu$ such that $\{S_\beta^\nu : \beta \in \kappa,\ \kappa \setminus S_\beta^\nu \notin F\}$ has cardinality $\kappa$. First fix $\beta$ and let $Y_\beta = \cup\{S_\beta^\nu : \nu \in \lambda\}$; then $Y_\beta = \{\alpha \in \kappa : \alpha > \beta\} \cap S$, since the $g_\alpha$'s are surjective. Since $F$ is $\kappa$-complete and contains the cofinite filter, the hypothesis implies that $\{\alpha \in \kappa : \alpha > \beta\} \in F$; so $\kappa \setminus Y_\beta$ does not belong to $F$. Since $F$ is $\kappa$-complete, there exists $\nu(\beta)$ such that $\kappa \setminus S_\beta^{\nu(\beta)} \notin F$ (because otherwise $\kappa \setminus Y_\beta = \bigcap_\nu \kappa \setminus S_\beta^\nu$ belongs to $F$). Now $\beta$ ranges over $\kappa$ and $\nu(\beta)$ ranges over $\lambda$ which is less than $\kappa$, so there exists $\nu \in \lambda$ such that $\nu = \nu(\beta)$ for $\kappa$ many $\beta$'s — which is exactly the claim.

Let $\nu$ be as in the claim, let $I = \{\beta \in \kappa : \kappa \setminus S_\beta^\nu \notin F\}$, and let $W = S \setminus \cup \{S_\beta^\nu : \beta \in I\}$. Notice that, for fixed $\nu$, the $S_\beta^\nu$ are pairwise disjoint. Let $\mu$ be the first element of $I$; let $S_\mu = S_\mu^\nu \cup W$; and for $\beta \in I \setminus \{\mu\}$, let $S_\beta = S_\beta^\nu$. Then $S = \amalg_{\beta \in I} S_\beta$ is the desired decomposition. $\square$

**4.9 Corollary.** *If $\kappa$ is a regular uncountable cardinal which is a successor cardinal, then every stationary subset of $\kappa$ can be partitioned into $\kappa$ disjoint stationary subsets. Moreover, $D(\kappa)$ has cardinality $2^\kappa$.*

PROOF. Apply 4.8 to the closed unbounded filter $F$, and remember that $S$ is stationary precisely when $\kappa \setminus S \notin F$. To see that $D(\kappa)$ has cardinality $2^\kappa$, partition $\kappa$ into $\kappa$ disjoint stationary subsets: $\kappa = \amalg_{\beta < \kappa} S_\beta$. For each $X \subseteq \kappa$, let $E_X = \cup \{S_\beta : \beta \in X\}$. Then for all $X \neq Y$, $E_X$ and $E_Y$ represent different elements of $D(\kappa)$: indeed, if $\alpha \in X \setminus Y$, say, then for all cubs $C$ there exists an element of $S_\alpha \cap C$ — and hence of $E_X \cap C$ — which does not belong to $E_Y$. $\square$

Corollary 4.9 — which was proved by Ulam in this manner — holds as well for regular limit cardinals $\kappa$ — as was proved by

Solovay — but we shall not give that proof here. (See Jech 1978, Theorem 85, *p.* 433; see also Exercise 16.)

Proposition 4.3 says that the closed unbounded filter on $\kappa$ is a $\kappa$-complete filter. Now we shall prove that it is a normal filter (cf. Definition 2.14).

**4.10 Proposition.** *For any regular uncountable $\kappa$, the diagonal intersection of a $\kappa$-sequence of cubs in $\kappa$ is a cub in $\kappa$.*

PROOF. Let $\{C_\nu : \nu < \kappa\}$ be a $\kappa$-sequence of cubs in $\kappa$. We must show that $\{\alpha < \kappa : \alpha \in C_\nu$ for all $\nu < \alpha\}$ is a cub in $\kappa$. If we let $C'_\nu = \bigcap_{\mu < \nu} C_\mu$ for all $\nu$, then $\triangle\{C_\nu : \nu < \kappa\} = \triangle\{C'_\nu : \nu < \kappa\}$, so without loss of generality, we can assume that $C_\mu \subseteq C_\nu$ whenever $\mu \geq \nu$. Let $C$ be $\triangle\{C_\nu : \nu < \kappa\}$.

First we show that $C$ is closed in $\kappa$. Suppose that $Y \subseteq C$ such that $\alpha = \sup Y \in \kappa$. If $\nu < \alpha$, let $Y_\nu = \{\beta \in Y : \beta > \nu\}$. Then $Y_\nu \subseteq C_\nu$ by definition of the diagonal intersection. Moreover $\alpha = \sup Y_\nu$, so $\alpha$ belongs to $C_\nu$ since $C_\nu$ is closed. Thus $\alpha \in C$ since $\alpha \in C_\nu$ for all $\nu < \alpha$.

Secondly, we show that $C$ is unbounded in $\kappa$. Given $\beta \in \kappa$ we must find $\alpha \in C$ such that $\alpha > \beta$. Define $\beta_n$ by induction on $n \in \omega$: $\beta_0 = \beta$; if $\beta_n$ has been defined, chose $\beta_{n+1} > \beta_n$ such that $\beta_{n+1} \in C_{\beta_n}$. Let $\alpha = \sup\{\beta_n : n \in \omega\}$. Then for any $\nu < \alpha$, there exists $m$ so that $\beta_m \geq \nu$; for any $n \geq m$, $\beta_{n+1} \in C_{\beta_n} \subseteq C_{\beta_m} \subseteq C_\nu$; hence $\alpha = \sup\{\beta_n : n > m\}$ belongs to $C_\nu$. $\square$

As an immediate application of 2.16 we have:

**4.11 Corollary.** (**Fodor's Lemma**) *Suppose $\theta : S \to \kappa$ is a regressive function where $S$ is a stationary subset of $\kappa$. Then there is a stationary $S' \subseteq S$ such that $\theta \restriction S'$ is constant.* $\square$

We conclude this section by illustrating the preceding ideas and results in two contexts. In doing so we will introduce two important notions that will play a role in the following chapters: that of a $\kappa$-filtration and that of a coloring of a ladder system.

**4.12.** Let $A$ be a set of cardinality $\leq \kappa$. A $\kappa$-*filtration* of $A$ is an indexed sequence $\{A_\nu : \nu < \kappa\}$ such that for all $\mu$, $\nu$ in $\kappa$:

(1) the cardinality of $A_\nu$ is $< \kappa$;

(2) $\mu < \nu$ implies $A_\mu \subseteq A_\nu$;

(3) (*continuity*) $\nu \in \lim(\kappa)$ implies $A_\nu = \bigcup_{\tau < \nu} A_\tau$;

(4) $A = \bigcup_{\nu < \kappa} A_\nu$.

Suppose that $\{A'_\nu : \nu < \kappa\}$ is another $\kappa$-filtration of $A$. We claim that

$$(4.12.1) \qquad\qquad C \overset{\text{def}}{=} \{\nu \in \kappa : A_\nu = A'_\nu\}$$

is a cub in $\kappa$.

It is clear that $C$ is closed in $\kappa$ (by continuity of the filtrations). To see that $C$ is unbounded, let $\beta \in \kappa$; define $\beta_0 = \beta$; if $\beta_n$ has been defined and $n$ is odd (resp. even), let $\beta_{n+1} > \beta_n$ be chosen so that $A_{\beta_n} \subseteq A'_{\beta_{n+1}}$ (resp. $A'_{\beta_n} \subseteq A_{\beta_{n+1}}$); this is possible by conditions (1) and (4) and the regularity of $\kappa$. Then if $\alpha = \sup\{\beta_n : n \text{ is odd}\} = \sup\{\beta_n : n \text{ is even}\}$, we have that $A_\alpha = A'_\alpha$ so $\alpha \in C$.

More generally, if $\{B_\nu : \nu < \kappa\}$ is a $\kappa$-filtration of another set $B$ of cardinality $\kappa$, and $\theta : A \to B$ is onto, then there is a cub $C$ so that for $\nu \in C$, $\theta[A_\nu] = B_\nu$.

If $f : A^n \to A$, then just as in Example 4.2, one can show that $\{\nu < \kappa : A_\nu \text{ is closed under } f\}$ is a cub in $\kappa$. If $\theta : \kappa \to A$ such that for all $\nu$, $\theta(\nu) \in A_\nu$, then, as in 4.11, there is a stationary set $S$ so that $\theta$ is constant on $S$ (cf. Exercise 19).

**4.13.** Let $E$ be a subset of $\lim(\omega_1)$. If $\delta \in E$, a *ladder on* $\delta$ is a function $\eta_\delta : \omega \to \delta$ which is strictly increasing and has range cofinal in $\delta$. A *ladder system on* $E$ is an indexed family $\eta = \{\eta_\delta : \delta \in E\}$ such that each $\eta_\delta$ is a ladder on $\delta$. Fodor's Lemma implies that if $E$ is stationary in $\omega_1$, then the ladders in a given ladder system cannot be disjoint; indeed, given a ladder system $\eta$, for a fixed $n$, define $f_n : E \to \omega_1$ by $f_n(\delta) = \eta_\delta(n)$; this function is regressive, so by 4.11, there is an $\alpha$ and a stationary subset $E'$ of $E$ such that for all $\delta \in E'$, $\eta_\delta(n) = \alpha$.

For a cardinal $\lambda \geq 2$, a $\lambda$-*coloring* of a ladder system $\eta$ on $E$ is a family $c = \{c_\delta : \delta \in E\}$ such that $c_\delta : \omega \to \lambda$. (We think of $\lambda$ as a set of "colors", and of $c_\delta$ as "coloring" $\eta_\delta(n)$ with color $c_\delta(n)$.) We would like to "uniformize" this coloring in the sense that we regard $c$ as assigning a fixed "color" to each ordinal $\alpha$ independently of the ladder(s) on which $\alpha$ appears. By the application of Fodor's Lemma given above, we cannot hope to do this in general, because we will have $\eta_\delta(n) = \eta_\gamma(m) = \alpha$ with $\delta \neq \gamma$ and we could have $c_\delta(n) \neq c_\gamma(m)$. So we use a modified notion: a *uniformization* of a coloring $c$ of a ladder system $\eta$ on $E$ is a pair $(f, f^*)$ where $f: \omega_1 \rightarrow \lambda$, $f^*: E \rightarrow \omega$ and for all $\delta \in E$ and all $n \geq f^*(\delta)$, if $\alpha = \eta_\delta(n)$, then $f(\alpha) = c_\delta(n)$. If such a pair exists, we say that $c$ can be uniformized. In order for the pair to exist it is enough to have either member of the pair; i.e., either $f$ so that for all $\delta \in E$, $f(\eta_\delta(n)) = c_\delta(n)$, for all but finitely many $n$, or $f^*$ so that for all $\delta$, $\alpha \in E$, if $n \geq f^*(\delta)$, $m \geq f^*(\alpha)$ and $\eta_\delta(n) = \eta_\alpha(m)$, then $c_\delta(n) = c_\alpha(m)$. If $f$ is given, then $f^*(\delta)$ can be defined to be the least $n$ so that $f(\eta_\delta(m)) = c_\delta(m)$, for all $m \geq n$. On the other hand, if $f^*$ is given, choose any $f$ so that $f(\eta_\delta(n)) = c_\delta(n)$ for all $n \geq f^*(\delta)$. We say that $(\eta, \lambda)$-*uniformization holds* if every $\lambda$-coloring of $\eta$ can be uniformized. We shall see in Chapter VI that whether $(\eta, \lambda)$-uniformization holds cannot be settled in ZFC. We shall make use of ladder systems and colorings in Chapters VIII and XII.

## §5. Games and trees

In this section we will introduce some useful notions from set theory which did not find a home in earlier sections.

We begin by studying (two-person) games. These games can be thought of intuitively as two players alternately choosing moves (numbered by ordinals) according to some rules. The two players will usually be called Player I and Player II. Player I moves at the even ordinals and Player II moves at the odd ordinals. Thus Player I starts with move 0, and moves at limit stages, with move $\omega$, $\omega+\omega$, etc. After some (ordinal) number of moves, the play of the game ends and one player or the other is declared the winner according to the rules. Clearly games such as chess fit into this structure. A more formal approach to games would be to describe a game as a pair $(X, W)$ where $X$ is a set of sequences (with the property that no sequence is a proper initial segment of another) and $W$ is a subset of $X$. Here $X$ is supposed to be the set of legal plays of the game and $W$ is the plays that Player I wins. We will adopt the more informal description of games in most of our applications.

Another notion which is quite important is the notion of a *winning strategy* for one player or the other. To simplify the description, let us assume that $G$ is a game of length $\lambda$, where $\lambda$ is a limit ordinal. A game has *length* $\lambda$ if every play of the game has length $\lambda$. Player I has a winning strategy for $G$ if there is a sequence $\{f_{2\alpha} : \alpha < \lambda\}$ of functions so that

$$f_0, a_1, f_2(a_1), a_3, f_4(a_1, a_3), \ldots, f_{2\alpha}(\langle a_{2\beta+1} : \beta < \alpha \rangle), \ldots$$

is a winning play for I whenever $a_1, a_3, \ldots$ are legal moves for II. Similarly Player II has a winning strategy for $G$ if there is a sequence $\{f_{2\alpha+1} : \alpha < \lambda\}$ of functions so that

$$a_0, f_1(a_0), a_2, f_3(a_0, a_2), \ldots, f_{2\alpha+1}(\langle a_{2\beta} : \beta \leq \alpha \rangle), \ldots$$

is a winning play for II whenever $a_0, a_2, \ldots$ are legal moves for I.

**5.1 Example.** Let $W$ be any set of countable subsets of $\omega_1$. Define $G_W$ as follows. $G_W$ is a game of length $\omega$. Players I and II alternately choose elements $\alpha_0, \alpha_1, \ldots$ of $\omega_1$. Player I wins $G_W$ if $\{\alpha_n : n \in \omega\} \in W$. We first claim that Player I has a winning strategy if and only if $W$ contains a cub. (Since a countable ordinal is the set of ordinals less than it, a cub is a set of countable subsets of $\omega_1$.) Suppose that $W$ contains a cub $C$. Then Player I can win as follows. First choose $g : \omega \setminus \{0\} \to \omega \times \omega$ so that $g$ is onto and if $g(n) = (k, m)$, then $k < n$. If Player II plays $\alpha_{2n+1}$ then Player I chooses $\beta_n \in C$ so that $\alpha_{2n+1} < \beta_n$ and enumerates $\beta_n$ as $\{\gamma_{(n, m)} : m \in \omega\}$. At stage 0, I plays 0, then for $n > 0$, at stage $2n$, I plays $\gamma_{g(n)}$. At the end of the play of the game, $\{\alpha_n : n \in \omega\} = \sup \beta_n \in C$.

Suppose now that the sequence of functions $\{f_{2n} : n \in \omega\}$ forms a winning strategy for Player I. There is a cub $D$ so that for all

$\delta \in D$, $\delta$ is closed under each of the functions $f_{2n}$. We will show that $D \subseteq W$. If $\delta \in D$, enumerate $\delta$ as $\{\alpha_{2n+1} : n \in \omega\}$. Consider the game where at stage $2n + 1$, Player II plays $\alpha_{2n+1}$. Since $\delta$ is closed under the functions, $\{f_0, \alpha_1, f_2(\alpha_1), \alpha_3, f_4(\alpha_1, \alpha_3), \ldots\} = \delta$. Thus $\delta \in W$ since Player I wins.

Similarly we can show that Player II has a winning strategy if and only if the complement of $W$ contains a cub.

Since there are stationary co-stationary sets, there are games in which neither player has a winning strategy. A game is called *determined* if some player has a winning strategy. There are important classes of games which are determined; one such class is the class of closed or open games. A game is said to be *closed* if any play $(a_\alpha : \alpha < \lambda)$ of the game is a winning play for I if for all $\delta < \lambda$, $(a_\alpha : \alpha < \delta)$ can be extended to a winning play for I. (A topology can be put on the set of legal plays of the game so that $W$ becomes a closed set.) Similarly we define a game to be *open* if any play $(a_\alpha : \alpha < \lambda)$ of the game is a winning play for II if for all $\delta < \lambda$, $(a_\alpha : \alpha < \delta)$ can be extended to a winning play for II. The following theorem, known as the Gale-Stewart theorem, is almost an immediate consequence of the definitions.

**5.2 Theorem.** *Every closed or open game is determined.*

PROOF. We will only consider the case where the game is closed. Suppose that II does not have a winning strategy. Then the following strategy is a winning strategy for Player I. At every even stage, I moves so that II does not have a winning strategy for the game after I makes that move. The assumption that II does not have a winning strategy means that such a move exists for I at every stage. Since the game is closed, Player I will win such a game. $\square$

**5.3** We will often have occasion to talk about trees. For set-theorists, a tree is a partially ordered set $(T, <)$ with a least element so that the predecessors of any element are well ordered. An element of $T$ is sometimes referred to as a *node*. A *final node* of $T$ is one which has no successors. A node $x$ of $T$ is said to have *height*

$\alpha$, denoted ht$(x) = \alpha$, if the order-type of $\{y \in T : y < x\}$ is $\alpha$. The *height* of $T$ is defined to be sup$\{$ht$(x) + 1 : x \in T\}$.

If $(T, <)$ is a tree, then $(T_1, < \upharpoonright T_1)$ is said to be a *subtree* of $T$ if whenever $t \in T_1$ and $s < t$, then $s \in T_1$. If $A$ is any set and $\lambda$ is an ordinal then

$$^{<\lambda}A \overset{\text{def}}{=} \{g : \alpha \to A : \alpha < \lambda\}$$

forms a *tree of sequences* ordered by $\subseteq$ . There is little loss of generality (see Exercise 22) in assuming that every tree is a subtree of a tree of sequences. It is common to abuse notation and refer to $T$ as a tree if $<$ is understood.

If $T$ is a tree, a *branch* $B$ of $T$ is a maximal linearly ordered initial subset of $T$. The *length* of a branch is its order type. (If $B$ is a branch then $(B, <)$ is easily seen to isomorphic to an ordinal.) The height of $T$ is then the supremum of the lengths of its branches. A tree is said to be $\kappa$-*branching* if every element has $< \kappa$ immediate successors; it is *finitely branching* if it is $\aleph_0$-branching, i.e., every element of the tree has only finitely many immediate successors.

One of the simplest, but most useful, results in set theory is the following.

**5.4 Theorem. (König's Lemma)** *An infinite finitely branching tree has an infinite branch.*

PROOF. Let $T$ be an infinite finitely branching tree. We claim that if $t \in T$ is such that $\{s \in T : t < s\}$ is infinite, then there is $t_1 > t$ so that $\{s \in T : t_1 < s\}$ is infinite. Given this claim, we can inductively choose an increasing sequence $t_0 < t_1 < t_2 \ldots$ so that for all $n$, $\{s \in T : t_n < s\}$ is infinite: we can take $t_0$ to be the minimum element of the tree and the claim allows us to pick $t_{n+1}$ if $t_n$ has been chosen. Then $\{t_n : n \in \omega\}$ is contained in a branch.

As for the claim, suppose $t$ is as in the claim. There is an $m \in \omega$ such that $t$ has exactly $m$ immediate successors $\{s_k : k < m\}$. Since

$$\{s \in T : t < s\} \setminus \{s_k : k < m\} = \bigcup_{k<m} \{s \in T : s_k < s\},$$

one of the sets on the right-hand side must be infinite. $\square$

One use of the tree $^{<\omega}2$ is to construct a family of almost disjoint sets. A family $\mathcal{F}$ of countable sets is said to be *almost disjoint* if for all $X \neq Y \in \mathcal{F}$, $X \cap Y$ is finite.

**5.5 Proposition.** *There is a family of $2^{\aleph_0}$ almost disjoint subsets of $\omega$.*

PROOF. Since $^{<\omega}2$ has cardinality $\aleph_0$, it is enough to find a family of $2^{\aleph_0}$ almost disjoint subsets of $^{<\omega}2$. In fact, the branches of this tree form such a family. Here are all the details. For each $\eta \in {}^{\omega}2$ (= the set of functions from $\omega$ to 2), let $X_\eta = \{\eta \restriction n : n \in \omega\}$. If $\eta \neq \nu$, then $X_\eta \cap X_\nu = \{\eta \restriction n : \eta \restriction n = \nu \restriction n\}$, a finite set. $\square$

The last topic we want to discuss in this section is Ramsey's theorem, which we will use in section XIV.3. First we fix the notation. If $X$ is a set then by $[X]^n$ we mean the set of subsets of $X$ of cardinality exactly $n$. If $g : [X]^n \to k$, then $Y \subseteq X$ is said to be *homogeneous for $g$* if there exists $m \in k$ such that $g(S) = m$ for all $S \in [Y]^n$.

**5.6 Theorem. (Ramsey's Theorem)** *For every infinite set $X$, every $n$, $k \in \omega$, and $g : [X]^n \to k$, there is an infinite set $Y \subseteq X$, so that $Y$ is homogeneous for $g$.* $\square$

A proof of Ramsey's theorem can be found in many books. (See also Exercise 25.)

## EXERCISES

1. If $\kappa$ is weakly compact and $M$ is a $\kappa$-free abelian group of cardinality $\kappa$, then $M$ is free. [Hint: in the notation of the proof of 3.10, choose an appropriate $\kappa$-complete subalgebra $\mathcal{B}$ of $I$ and form an "ultraproduct" with respect to an ultrafilter $D$ on $\mathcal{B}$. See also Theorem IV.3.2.]

2. If $D$ is a principal ultrafilter, generated by $\{j\}$, then $\prod_{i \in I} M_i / D$ is canonically isomorphic to $M_j$ via the map which takes $a_D$ to $a(j)$.

3. Given a finitely additive measure: $\mu : \mathcal{P}(I) \to \{0,1\}$ there is an ultrafilter $D$ such that $\mu = \mu_D$.

4. $D$ is called a *regular ultrafilter* on $I$ if there exists a subset $X$ of $D$ of cardinality $|I|$ such that each element of $I$ belongs to only finitely many members of $X$. Prove:

(i) every non-principal ultrafilter on $\omega$ is regular;

(ii) if $D$ is a regular ultrafilter on $I$ then $D$ is $\omega_1$-incomplete; and

(iii) if $D$ is regular and $M$ is an infinite module, then the cardinality of $M^I/D$ is $|M|^{|I|}$. [Hint: define a one-one function from $^IM$ into $\tilde{M}^I/D$ where $\tilde{M} = M^{(\omega)}$.]

5. If $D$ is an ultrafilter on $I$ which is $\omega_1$-incomplete, then for any modules $M_i(i \in I)$, any countable set of linear equations with co-efficients from $\prod_{i \in I} M_i/D$ that is finitely solvable in $\prod_{i \in I} M_i/D$ is solvable in $\prod_{i \in I} M_i/D$.

6. Let $D$ be an ultrafilter on $I$, $M_i(i \in I)$ a family of $R$-modules and

$$\sum_{k \leq n_j} r_{kj} x_{kj} = y_j \quad (j \in J)$$

a system of linear equations (where $J$ is any finite set).

(i) For any $a_j \in \prod_{i \in I} M_i$ ,

$$\sum_{k \leq n_j} r_{kj} x_{kj} = (a_j)_D \quad (j \in J)$$

has a solution in $\prod_{i \in I} M_i/D$ if and only if the set of $i \in I$ such that

$$\sum_{k \leq n_j} r_{kj} x_{kj} = a_j(i) \quad (j \in J)$$

has a solution in $M_i$ is a member of $D$. (This is a special case of the Fundamental Theorem of Ultraproducts referred to at the beginning of section 3.)

(ii) If $D$ is $\kappa$-complete, (i) holds for any set $J$ of cardinality $< \kappa$.

7. If $\lambda > |R|$, an ultraproduct of torsionless $R$-modules with respect to a $\lambda$-complete ultrafilter is torsionless. [Hint: if $f_i: M_i \to R(i \in I)$ are homomorphisms, then — in the obvious notation — $\prod_I f_i/D$ is a homomorphism from $\prod_I M_i/D$ to $R$.]

8. Say that an $R$-module $M$ is $\kappa$-*torsionless* if every submodule of cardinality $< \kappa$ is torsionless. Prove that $\mathbf{Z}^I/F$ is $\kappa$-torsionless if $F$ is a $\kappa$-complete filter on $I$.

9. If $\kappa$ is a $L_{\lambda\omega}$-compact cardinal, where $\lambda > |R| + \aleph_0$, and $M$ is $\kappa$-torsionless, then $M$ is torsionless. [Hint: cf. proof of 3.10.]

10. Prove that an abelian group $A$ is $\aleph_1$-free if and only if $A$ can be embedded in a reduced power $\mathbf{Z}^I/F$ where $F$ is $\omega_1$-complete. [Hint: We will prove that for any $I$, $\mathbf{Z}^I$ is $\aleph_1$-free (see IV.2.8). For ($\Leftarrow$), use 3.4; for ($\Rightarrow$), construct an embedding of $A$ into a reduced product of countable free modules as in 3.10.]

11. If there is a measurable cardinal, then there is an $\omega$-measurable cardinal $\kappa$ such that there are exactly $\kappa$ $\omega_1$-complete ultrafilters on $\kappa$. [Hint: Define inductively: $\kappa_0$ = the first measurable, $\kappa_{n+1} = 2^{\kappa_n}$; $\kappa = \bigcup_{n \in \omega} \kappa_n$; show that if $D$ is an $\omega_1$-complete ultrafilter on $\kappa$ then some $\kappa_n$ belongs to $D$.]

12. Suppose $\kappa$ is a measurable cardinal.
(i) For any subset $X$ of $\kappa$ of cardinality $\kappa$ there is a $\kappa$-complete ultrafilter on $\kappa$ which contains $X$.
(ii) There is a subset $S$ of $\mathcal{P}(\kappa)$ such that $S$ has cardinality $2^\kappa$ and for every $X \neq Y$ in $S$, $|X \cap Y| < \kappa$ [cf. II.5.5].
(iii) There are $2^\kappa$ different $\kappa$-complete ultrafilters on $\kappa$.

13. If $\kappa$ is $L_{\omega_1\omega}$-compact, then there are at least $2^\kappa$ different $\omega_1$-complete ultrafilters on $\kappa$.

14. Let $D$ be a non-principal ultrafilter on $\omega$. Suppose that for each $k \in \omega$, $\{\alpha_{n,k} : n \in \omega\}$ is a non-increasing sequence of cardinals. For each $n \in \omega$ let $\beta_n = \prod_{k \in \omega} \alpha_{n,k}/D$; and let $\beta = \lim_{n \to \infty} \beta_n$. Prove that there exists $h : \omega \to \omega$ such that $h$ is not constant mod $D$ (i.e., $h^{-1}[\{n\}] \notin D$ for any $n \in \omega$) and $\prod_{k \in \omega} \alpha_{h(k),k}/D = \beta$.

15. Let $\kappa$ be a regular uncountable cardinal. A *cub* on $\mathcal{P}_\kappa(A)$ is a set $C \subseteq \mathcal{P}_\kappa(A)$ such that $C$ is closed under unions of chains of length $< \kappa$ and for every $Y \in \mathcal{P}_\kappa(A)$, there exists $X \in C$ such that $Y \subseteq X$. $S \subseteq \mathcal{P}_\kappa(A)$ is said to be *stationary* if for all cubs $C$ in $\mathcal{P}_\kappa(A)$, $C \cap S \neq \emptyset$.

(i) Prove analogs of 4.3 and 4.10. For $\kappa = \aleph_1$, prove the analog of 5.1.

(ii) (Fodor's Lemma) Suppose $S$ is a stationary subset of $\mathcal{P}_\kappa(A)$ and $\theta: S \to \mathcal{P}_\kappa(A)$ such that for all $Y \in S$, $\theta(Y) \subset Y$. Then there is a stationary $S' \subseteq S$ such that $\theta{\restriction}S'$ is constant.

16. If $\kappa$ is a regular limit cardinal, then $\kappa$ is the disjoint union of $\kappa$ stationary subsets. [Hint: let $E_\lambda$ be defined as in 4.7 for each regular $\lambda < \kappa$.]

17. (i) A subset $C$ of a regular uncountable cardinal $\kappa$ is a cub in $\kappa$ if and only if it is the range of a continuous strictly-increasing function $f: \kappa \to \kappa$.

(ii) If $\gamma$ is a singular cardinal, then there is a cub $C$ in $\gamma$ such that no member of $C$ is a regular cardinal. [Hint: if $\lambda = \mathrm{cf}(\gamma)$, choose $C$ to be the range of a continuous function $f: \lambda \to \{\alpha \in \gamma : \alpha > \lambda\}$.]

18. If $\{A_\nu : \nu < \kappa\}$ is a $\kappa$-filtration of a set $A$ of cardinality $\kappa$ ($a$ regular cardinal), there is a cub $C$ in $\kappa$ such that for all $\nu \in C$, $|A_{\nu^+} \setminus A_\nu| = |\nu^+ \setminus \nu|$ (where $\nu^+$ denotes $\inf\{\alpha \in C : \alpha > \nu\}$).

19. (Fodor's Lemma) Let $\kappa$ be a regular cardinal and $\{A_\nu : \nu < \kappa\}$ a $\kappa$-filtration of a set $A$ of cardinality $\kappa$. Prove: if $S$ is a stationary subset of $\kappa$ and $\theta: S \to A$ such that for all $\alpha$, $\theta(\alpha) \in A_\alpha$, then there is a stationary $S' \subseteq S$ such that $\theta{\restriction}S'$ is constant.

20. For any ladder system $\eta$ on a stationary set $E \subseteq \omega_1$, if $c$ is the coloring on $\eta$ defined by $c_\delta(n) = \eta_\delta(n + 1)$, then $c$ cannot be uniformized. [Hint: if $(f, f^*)$ is a uniformization, apply Fodor's Lemma (Exercise 19) to $\theta(\delta) \overset{\text{def}}{=} (f^*(\delta), \eta_\delta(f^*(\delta)))$.] (Compare this with VI.4.6.)

21. If $E$ is a stationary subset of a regular cardinal $\kappa$, and $f$ and $g$ are functions from $E$ into $\kappa$ such that for every $\nu \in E$, $f(\nu) \neq g(\nu)$, then there is a stationary subset $E'$ of $E$ such that $\{f(\nu): \nu \in E'\} \cap \{g(\nu): \nu \in E'\} = \emptyset$. [Hint: there is a stationary $E_1 \subseteq E$ such that for all $\mu, \nu \in E_1$, $f(\mu) < \mu \iff f(\nu) < \nu$ and $g(\mu) < \mu \iff g(\nu) < \nu$; by Fodor, there is a stationary $E_2 \subseteq E_1$ such that

if $f(\nu) < \nu$ for all $\nu \in E_1$ then $f \restriction E_2$ is constant, and similarly for $g$; there is a cub $C$ such that for every $\alpha \in C$ and every $\nu < \alpha$, $f(\nu) < \alpha$ and $g(\nu) < \alpha$; let $E' = E_2 \cap C$.]

22. Suppose $(T, <)$ is a tree. If $x \in T$, let $[x] = \{s \in T : s < x\}$. Let $T_1 = T \cup \{[x] : ht(x) \text{ is a limit ordinal} > 0\}$. Let $<_1$ be the transitive closure of the relation extending $<$ so that if $ht(x)$ is a limit ordinal $> 0$, then for all $s < x$, $s <_1 [x] < x$. Show that $T_1$ is isomorphic to a tree of sequences.

23. For any two trees $T_1$ and $T_2$, define the game $G(T_1, T_2)$ as follows: Player I chooses elements of $T_1$ forming a strictly increasing sequence in $T_1$ and Player II chooses similarly in $T_2$. The first player to be unable to move loses. Let $T$ be any tree and let $T'$ be the tree of all initial segments of branches of $T$, partially-ordered by inclusion. Prove that Player II has a winning strategy in the game $G(T, T')$.

24. (i) Show that Ramsey's Theorem is equivalent to the statement that for all $n, k \in \omega$ if $g$ is a function from increasing $n$-tuples of elements of $\omega$ to $k$ then there is an infinite $Y$ so that $g$ is constant on increasing $n$-tuples from $Y$. Call such a $Y$ *homogeneous* for $g$.
(ii) Prove the equivalent version of Ramsey's theorem for $n = 1$. (This is called the pigeon hole principle.)

25. Prove Ramsey's theorem. [Hint: Prove it by induction on $n$. If $g(x_1, \ldots, x_{n+1})$ is a function, let $g_a = g(a, x_2, \ldots, x_n)$. Let $X_0 = \omega$. Inductively choose $a_m$ and $X_m$ so that $X_m$ is homogeneous for $g_{a_m}$, $a_{m+1}$ is the least element of $X_m$, and $X_{m+1} \subseteq X_m$. Partition $\{a_m : m \geq 1\}$ according to the value of $g(a_m, b_2, \ldots, b_{n+1})$, where $b_2 < \ldots < b_{n+1} \in X_m$. Show that any infinite block of the partition is homogeneous for $g$.]

26. (i) For all cardinals $\lambda$, $\mu$, $(\mu^\lambda)^\lambda = \mu^\lambda$.
(ii) If $\mu$ is an infinite cardinal and $\mu^\lambda = \mu$, then $(\mu^+)^\lambda = \mu^+$. [Hint: $\lambda < \mu$ and $^\lambda(\mu^+) = \bigcup_{\alpha < \mu^+} {}^\lambda\alpha$.]

27. If $E \subseteq \lim(\kappa)$ is a non-reflecting stationary subset of $\kappa$ and $X$ is a non-stationary subset of $E$, we can choose for every $\gamma \in X$ a

cub $C_\gamma$ in $\gamma$ such that for all $\beta \neq \gamma$ in $X$, $C_\beta \cap C_\gamma = \emptyset$. [Hint: prove by induction on sup $X$ that we can choose the $C_\gamma$ such that inf $C_\gamma$ is greater than any fixed $\nu < $ inf $X$.]

## NOTES

We refer to Jech 1978 for the history of, and more information about, the topics of sections 1 and 2. The terminology "$L_{\lambda\omega}$-compact" is taken from Bell 1974. The ultraproduct construction goes back to work of Skolem; the reduced product construction and its use in model theory appears first in Łoś 1955, where the Fundamental Theorem of Ultra-products is proved. For a general introduction to ultraproducts for al-gebraists, see Eklof 1977b.

Some sources do not include the condition of strong inaccessibility in the definition of weakly compact. For a discussion of equivalent defini-tions of weakly compact, see Drake 1974, Chapter 10. The consistency of the existence of an $L_{\omega_1\omega}$-compact cardinal with the non-existence of strongly compact cardinals will appear in Magidor-Solovay 19??.

Proposition 3.4 appears in Wald 1983a. (See also Dugas-Göbel 1985a, *p.* 86.) Proposition 3.6 is from Keisler 1964a. We discuss $\kappa$-free groups in the Notes to IV.

4.8 and 4.9 are due to Ulam 1930; see also Solovay 1971. Fodor's Lemma (4.11) appears in Fodor 1956. Ladders and colorings are dis-cussed in Devlin-Shelah 1978.

Theorem 5.2 is proved in Gale-Stewart 1953. Theorem 5.4 appears in König 1926. Theorem 5.6 is due to Ramsey 1930.

Exercises. 1:see Notes for IV.3.2; 4:Keisler 1964; 6:Łoś 1955; 9:Berg-man-Solovay 1987/Eda-Abe 1987; 14:Eklof 1973 (for more on this so-called "Eklof Property", see Comfort-Negrepontis 1974, p.304); 21:She-lah (cf. Eklof-Mekler 1981); 25: Ramsey 1930 (this proof has been at-tributed to Rado); 27:Hyttinen-Väänänen.

# CHAPTER III
# SLENDER MODULES

In section I.2 we noted that $\text{Hom}(\bigoplus_{i\in I} M_i, H)$ is canonically isomorphic to $\prod_{i\in I} \text{Hom}(M_i, H)$ for any modules $M_i$ and $H$. There is in general no corresponding isomorphism for $\text{Hom}(\prod_{i\in I} M_i, H)$, but there is an important class of modules, the *slender* modules, for which we have such an isomorphism: a module $H$ is slender precisely when $\text{Hom}(\prod_{i\in I} M_i, H)$ is canonically isomorphic to $\bigoplus_{i\in I} \text{Hom}(M_i, H)$ for any *countable* set $I$ (see Corollary 1.5). We shall study this in more detail in section 1, and also consider briefly some related notions. In section 2 we shall look at examples of slender rings and modules and prove, in particular, that $\mathbb{Z}$ is slender. Finally, in section 3 we shall consider what happens for uncountable index sets, both measurable and non-measurable.

## §1. Introduction to slenderness

The notion of slenderness is due to J. Łoś. An $R$-module $H$ is said to be *slender* if for every homomorphism $\theta: R^\omega \to H$, $\theta(e_n) = 0$ for all but finitely many $n \in \omega$; recall that $e_n$ is the element of $R^\omega$ such that $\text{supp}(e_n) = \{n\}$ and $e_n(n) = 1$.

Recall that in section I.2 we defined, for any family $\{M_i : i \in I\}$ of modules, the canonical injection

$$\lambda_j : M_j \to \bigoplus_{i\in I} M_i \subseteq \prod_{i\in I} M_i.$$

**1.1 Lemma.** *Suppose $H$ is slender. For any family $\{M_i : i \in I\}$ of $R$-modules, and any homomorphism $\varphi: \prod_{i\in I} M_i \to H$, $\varphi \circ \lambda_i \equiv 0$ for all but finitely many $i \in I$.*

PROOF. Suppose this is false; let $\varphi: \prod_{i\in I} M_i \to H$ be a counterexample. Then there is an infinite subset $\{i_n : n \in \omega\}$ of $I$ such that for every $n \in \omega$, there exists $a_n \in M_{i_n}$ such that $\varphi \circ \lambda_{i_n}(a_n) \neq 0$.

Define $\psi: R^\omega \to \prod_{i \in I} M_i$ by

$$\psi((r_n)_{n \in \omega})(i) = \begin{cases} r_m a_m & \text{if } i = i_m \text{ for some } m \\ 0 & \text{otherwise} \end{cases}$$

and let $\theta = f \circ \psi$. Then $\theta(e_n) = \varphi \circ \lambda_{i_n}(a_n) \neq 0$ for every $n \in \omega$, which contradicts the fact that $H$ is slender. $\square$

While the lemma above holds for families of modules of arbitrary size, the following theorem requires a restriction in the size of the index set. In section 3 we shall see how to generalize this result to index sets of larger size (cf. 3.3).

For any countable family $\{M_n: n \in \omega\}$, let $\prod_{n \geq m} M_n$ denote the subgroup of $\prod_{n \in \omega} M_n$ consisting of all $x$ such that $\text{supp}(x) \subseteq \{n: n \geq m\}$, i.e., $x(n) = 0$ for $n < m$.

**1.2 Theorem.** *$H$ is a slender module if and only if for any countable family $\{M_n: n \in \omega\}$ of modules and any homomorphism $\varphi: \prod_{n \in \omega} M_n \to H$, there exists $m \in \omega$ such that $\varphi[\prod_{n \geq m} M_n] = \{0\}$.*

PROOF. Sufficiency is clear; to prove necessity suppose that $H$ is slender, but there is some $\varphi: \prod_{n \in \omega} M_n \to H$ which is a counterexample. Then for every $m \in \omega$ there exists $x^{(m)} \in \prod_{n \geq m} M_n$ such that $\varphi(x^{(m)}) \neq 0$. For any sequence of elements $r_m \in R$ ($m \in \omega$), let $\sum_{m \in \omega} r_m x^{(m)}$ be the element, $z$, of $\prod_{n \in \omega} M_n$ such that for all $n$,

$$z(n) = \sum_{m \in \omega} r_m x^{(m)}(n) = \sum_{m \leq n} r_m x^{(m)}(n)$$

(a finite sum). Define $\theta: R^\omega \to H$ by the rule

$$\theta((r_n)_{n \in \omega}) = \varphi(\sum_{n \in \omega} r_n x^{(n)}).$$

Then $\theta$ is clearly a homomorphism such that for all $n \in \omega$, $\theta(e_n) \neq 0$. Since this contradicts the assumption that $H$ is slender, we have completed the proof. $\square$

**1.3 Corollary.** *Let $H$ be a slender module. For any countable family $\{M_n: n \in \omega\}$ and any homomorphism $\varphi: \prod_{n \in \omega} M_n \to H$, if $\varphi \upharpoonright \bigoplus_{n \in \omega} M_n \equiv 0$, then $\varphi \equiv 0$.*

PROOF. By 1.2 there is an $m$ such that $\varphi[\prod_{n \geq m} M_n] = \{0\}$. Now for any $x \in \prod_{n \in \omega} M_n$, let $b \in \bigoplus_{n \in \omega} M_n$ such that $x - b \in \prod_{n \geq m} M_n$. Then $\varphi(x - b) = 0$, so $\varphi(x) = \varphi(b) = 0$. Hence $\varphi$ is identically zero. □

**1.4 Definition.** Given any indexed family $\{M_i : i \in I\}$ of $R$-modules and any module $H$, define a homomorphism

$$\Phi = \Phi(\{M_i : i \in I\}, H) : \bigoplus_{i \in I} \text{Hom}(M_i, H) \to \text{Hom}(\prod_{i \in I} M_i, H)$$

by the rule:

$$\Phi((g_i)_{i \in I})((m_i)_{i \in I}) = \sum_{i \in I} g_i(m_i)$$

for any family $\{g_i : i \in I\}$ of homomorphisms $g_i : M_i \to H$ such that $g_i \equiv 0$ for almost all $i$.

(*Warning*: in the case where $I$ is $\omega$-measurable, we shall change the definition of $\Phi$ in section 3.)

It is easy to see that $\Phi$ is a monomorphism. If $H$ is slender, then by Lemma 1.1 we can define

$$\Phi' : \text{Hom}(\prod_{i \in I} M_i, H) \to \bigoplus_{i \in I} \text{Hom}(M_i, H)$$

by $\Phi'(\varphi) = ((\varphi \circ \lambda_i)_I)$. Then $\Phi' \circ \Phi$ is clearly the identity. The following result implies that $\Phi \circ \Phi'$ is the identity if $I$ is countable.

**1.5 Corollary.** *$H$ is slender if and only if for any countable family $\{M_n : n \in \omega\}$,*

$$\Phi(\{M_n : n \in \omega\}, H) : \bigoplus_{n \in \omega} \text{Hom}(M_n, H) \to \text{Hom}(\prod_{n \in \omega} M_n, H)$$

*is an isomorphism*

PROOF. Suppose first that $H$ is slender. Given a family $\{M_n : n \in \omega\}$, we already know $\Phi$ is a monomorphism, so it remains to prove that $\Phi$ is surjective. Given $\varphi \in \text{Hom}(\prod_{n \in \omega} M_n, H)$, let $m$ be as in Theorem 1.2, so $\varphi[\prod_{n \geq m} M_n] = \{0\}$. For each $n \in \omega$, let $g_n = \varphi \circ \lambda_n$.

Then $g_n \equiv 0$ for $n \geq m$. Now $\prod_{n\in\omega} M_n = (\bigoplus_{n<m} M_n)\oplus(\prod_{n\geq m} M_n)$, and by construction $\Phi((g_n)_\omega)$ agrees with $\varphi$ on $(\bigoplus_{n<m} M_n)$; moreover both functions are identically zero on $\prod_{n\geq m} M_n$; so we conclude that $\varphi = \Phi((g_n)_{n\in\omega})$.

For the converse, if $\theta: R^\omega \to H$, then $\theta$ belongs to the image of $\Phi$, so $\theta \circ \lambda_n \equiv 0$ for almost all $n$. $\square$

Now we shall consider some basic properties of slender modules.

**1.6 Lemma.** *Let $H$ be a slender $R$-module. Then*
  (i) *every submodule of $H$ is slender; and*
  (ii) *$H$ is not injective unless it is zero.*

PROOF. The first part is clear from the definition of slender. For the second part, if $H$ is a non-zero injective, we can define a homomorphism $\theta$ from $R^{(\omega)}$ into $H$ such that for all $n$, $\theta(e_n) \neq 0$, and then extend it to $R^\omega$ by the injectivity of $H$. $\square$

Recall that a module over a p.i.d. is *reduced* if the maximal divisible submodule is zero.

**1.7 Proposition.** *If $R$ is a p.i.d. and $H$ is a slender $R$-module, then $H$ is reduced and torsion-free.*

PROOF. $H$ is reduced by Lemma 1.6 since a divisible $R$-module is injective. To show that $H$ is torsion-free, it suffices by 1.6(i) to prove that for any prime $p$ of $R$, $R/pR$ is not slender. Since $R/pR$ is injective as $R/pR$-module we can, as in 1.6, define $\theta: (R/pR)^\omega \to R/pR$ such that for all $n$, $\theta(e_n) \neq 0$; composing $\theta$ with the canonical map $\varphi: R^\omega \to R^\omega/p(R^\omega) \cong (R/pR)^\omega$, we obtain a map that shows that $R/pR$ is not slender as $R$-module. $\square$

For more on when slender modules are torsion-free, see 2.7. Proposition 1.7 implies that a slender module over $\mathbf{Z}$ does not contain a copy of $\mathbf{Q}$, $Z(p^\infty)$, or $\mathbf{Z}/p\mathbf{Z}$ for any prime $p$. In Chapter IX, we shall prove Nunke's characterization of slender abelian groups by the subgroups they do not contain.

The next result, of Chase, will enable us to conclude immediately that a direct sum of slender modules over a p.i.d. is slender; it will be an important tool for us in other ways as well. (See Chapter X.) Given a module $C$ over an arbitrary ring $R$, define

$$C^1 = \{x \in C : x \in rC \text{ for all } r \in R \setminus \{0\}\}.$$

Notice that if $R$ is a p.i.d. and $C$ is torsion-free, then $C^1$ is the maximal divisible submodule of $C$.

**1.8 Theorem.** *Let $R$ be any ring, and let $\{M_n : n \in \omega\}$ and $\{C_j : j \in J\}$ be indexed families of $R$-modules. Then for any homomorphism*

$$\varphi \colon \prod_{n \in \omega} M_n \to \bigoplus_{j \in J} C_j$$

*there exists $m \in \omega$, $r \in R \setminus \{0\}$, and a finite subset $S$ of $J$ such that*

$$\varphi[r \prod_{n \geq m} M_n] \subseteq \bigoplus_{j \in S} C_j + \bigoplus_{j \in J} C_j^1.$$

PROOF. Suppose that the theorem is false and let $\varphi \colon \prod_{n \in \omega} M_n \to \bigoplus_{j \in J} C_j$ be a counterexample. Let $\pi_i \colon \bigoplus_{j \in J} C_j \to C_i$ be the canonical projection. We will define by induction on $m \in \omega$, elements $x^{(m)} \in \prod_{n \in \omega} M_n$, $r_m \in R$, and $j_m \in J$ such that:

(a) $r_{m+1} \in r_m R$;
(b) $x^{(m)} \in r_m \prod_{n \geq m} M_n$;
(c) $j_m \notin \bigcup_{n < m} \operatorname{supp}(\varphi(x^{(n)}))$; and
(d) $\pi_{j_m} \varphi(x^{(m)}) \notin r_{m+1} C_{j_m}$.

Let $r_0 = 1$. Suppose now that for some $m \geq 0$, $r_n$ has been defined for $n \leq m$ and that $x^{(n)}$ and $j_n$ have been defined for $n < m$. Then there must exist $x^{(m)} \in r_m \prod_{n \geq m} M_n$ and $j_m \in J \setminus \bigcup_{n < m} \operatorname{supp}(\varphi(x^{(n)}))$ such that $\pi_{j_m} \varphi(x^{(m)}) \notin C_{j_m}^1$, because otherwise $r = r_m$ and $S = \bigcup_{n < m} \operatorname{supp}(\varphi(x^{(n)}))$ would show that $\varphi$ is not a counterexample to the theorem. Hence there exists $r_{m+1} \in R$ such that $\pi_{j_m} \varphi(x^{(m)}) \notin r_{m+1} C_{j_m}$; without loss of generality we may suppose that $r_{m+1} \in r_m R$.

Let $x = \sum_{n \in \omega} x^{(n)}$ denote the obvious element of $\prod_{n \in \omega} M_n$ (cf. the proof of 1.2). We claim that for all $m$, $\pi_{j_m} \varphi(x) \neq 0$; this will be a contradiction of the fact that $\varphi(x)$ belongs to the direct sum $\bigoplus_{j \in J} C_j$, since the $j_m$ are distinct because $j_m \in \text{supp}(\varphi(x^{(m)}) \setminus \text{supp}(\varphi(x^{(n)})$ for all $n < m$. So it remains to prove the claim. For any $m$, (a) and (b) imply that

$$x = x^{(1)} + \ldots + x^{(m)} + r_{m+1} y$$

for some $y \in \prod_{n \geq m+1} M_n$. Thus

$$
\begin{aligned}
\pi_{j_m}\varphi(x) &= \pi_{j_m}\varphi(x^{(1)}) + \ldots + \pi_{j_m}\varphi(x^{(m)}) + r_{m+1}\varphi(y) \\
&= \pi_{j_m}\varphi(x^{(m)}) + r_{m+1}\varphi(y)
\end{aligned}
$$

(the last because $j_m \notin \text{supp}(\varphi(x^{(n)})$ for $n < m$). But $\pi_{j_m}\varphi(x^{(m)}) \notin r_{m+1}C_{j_m}$ by (d), so $\pi_{j_m}\varphi(x) \neq 0$. $\square$

In section 3 we shall generalize this result to homomorphisms on a product over an arbitrary index set.

**1.9 Corollary.** *If $R$ is a p.i.d. and $\varphi: \prod_{n \in \omega} M_n \to \bigoplus_{j \in J} C_j$ is a homomorphism where each $C_j$ is torsion-free and reduced, then there exists $m \in \omega$ and a finite subset $S \subseteq J$ such that $\varphi[\prod_{n \geq m} M_n] \subseteq \bigoplus_{j \in S} C_j$.* $\square$

**1.10 Corollary.** *A direct sum of slender modules over a p.i.d. $R$ is also slender.*

PROOF. Suppose that $\{H_j : j \in J\}$ is an indexed family of slender modules. Consider a homomorphism $\varphi: \prod_{n \in \omega} M_n \to \bigoplus_{j \in J} H_j$. Since each $H_j$ is torsion-free and reduced by 1.7, Corollary 1.9 implies that there exists a finite subset $S$ of $J$ such that $\varphi[\prod_{n \geq m} M_n] \subseteq \bigoplus_{j \in S} H_j$. Now a finite sum of slender modules is clearly slender, so it follows that $\varphi(e_n) = 0$ for almost all $n$. $\square$

In fact, a direct sum of slender modules over any ring is slender: see Exercise 1.

We conclude this section by briefly considering some notions related to slenderness. Since these will not be needed later we shall leave some of the development to the Exercises; here we shall just give the definitions and some examples of applications of Theorem 1.8.

For the rest of this section $R$ is a p.i.d. An $R$-module $H$ is called *almost slender* if for every $\varphi\colon R^\omega \to H$, there exists $r \neq 0$ in $R$ such that for almost all $n$, $r\varphi(e_n) = 0$. It follows, as in 1.5, that for any countable family $\{M_n\colon n \in \omega\}$ of modules and almost slender module $H$,

$$\mathrm{Hom}(\prod_{n\in\omega} M_n, H) = \mathrm{im}(\Phi(\{M_n\colon n \in \omega\}, H)) + T$$

where $T$ is a torsion module.

Obviously $H$ is slender if and only if it is almost slender and torsion-free. But there are almost slender torsion modules:

**1.11 Proposition.** *If $T$ is a reduced torsion module, then $T$ is almost slender.*

PROOF. It suffices to show that the image of $\varphi$ is of bounded order for every homomorphism $\varphi\colon R^\omega \to T$. So replacing $T$ by $\mathrm{im}(\varphi)$, we can assume that $\varphi$ is an epimorphism. Now $T = \bigoplus_p T_p$, the direct sum of its $p$-torsion components, and each $T_p$ has a basic submodule $B_p$ (cf. V.1.5). By a fundamental result (Fuchs 1970, Theorem 36.1), there is an epimorphism $h_p$ of $T_p$ onto $B_p$. Composing $\varphi$ with these epimorphisms, we obtain an epimorphism $g\colon R^\omega \to \bigoplus_p B_p$. Now $\bigoplus_p B_p = \bigoplus_{j\in J} C_j$ where each $C_j$ is of bounded order. Thus it follows from Theorem 1.8 that there exists $m \in \omega$ and $r \neq 0$ such that $rg[\prod_{n\geq m} R] = \{0\}$. But then since $R^m$ is finitely-generated and $T$ is torsion we can conclude that the image, $\bigoplus_p B_p$, of $R^\omega$ under $g$ is of bounded order. Since $T$ is reduced and its basic submodule is of bounded order, it follows that $T = \bigoplus_p B_p$, and hence has bounded order. $\square$

In fact, any module of countable torsion-free rank is almost slender: see Exercise 12 and also IX.2.6.

Given any indexed family of modules $\{C_j\colon j \in J\}$ and any module $M$ we can define a monomorphism

$$\Psi = \Psi(M, \{C_j\colon j \in J\})\colon \bigoplus_{j\in J} \mathrm{Hom}(M, C_j) \to \mathrm{Hom}(M, \bigoplus_{j\in J} C_j)$$

by the rule:

$$\Psi((g_j)_{j\in J})(a) = (g_j(a))_{j\in J}.$$

Then $\varphi: M \to \bigoplus_{j\in J} C_j$ belongs to the image of $\Psi$ if and only if $\pi_j \circ \varphi \equiv 0$ for almost all $j$. We say that $M$ is *dually slender* if $\Psi$ is surjective for any family $\{C_j : j \in J\}$ of *reduced* modules. Equivalently, $M$ is dually slender if and only if for any $\varphi: M \to \bigoplus_{j\in J} C_j$ where the $C_j$'s are reduced, there is a finite set $S$ such that $\varphi[M] \subseteq \bigoplus_{j\in S} C_j$. We say that $M$ is *almost dually slender* if for any family $\{C_j : j \in J\}$ of reduced modules and any $\varphi \in \mathrm{Hom}(M, \bigoplus_{j\in J} C_j)$, there exists $r \neq 0$ such that $r\varphi$ belongs to the image of $\Psi$. In this case $M$ is also called a *Fuchs-44* module (after Problem 44 of Fuchs 1970). Obviously, any finitely-generated $R$-module is Fuchs-44 (even dually slender). The following implies that, for example, $R^\omega$ is Fuchs-44.

**1.12 Proposition.** *If $\{M_n : n \in \omega\}$ is a sequence of Fuchs-44 modules, then $\prod_{n\in\omega} M_n$ is also Fuchs-44.*

PROOF. Let $\varphi: \prod_{n\in\omega} M_n \to \bigoplus_{j\in J} C_j$ be a homomorphism, where the $C_j$ are reduced modules. Without loss of generality, $C_j = \pi_j \varphi[\prod_{n\in\omega} M_n]$ for all $j$. Theorem 1.8 implies that there exists a non-zero $r$ and a finite set $S$ such that for $j \notin S$, $r\pi_j\varphi[\prod_{n\in\omega} M_n] \subseteq C_j^1$. So for $j \notin S$ we have $rC_j \subseteq C_j^1 \subseteq rC_j$, and hence $C_j^1 = rC_j$. But then for all $s \in R$, $C_j^1 \subseteq srC_j = sC_j^1$, so $C_j^1$ is divisible. Since $C_j$ is reduced, this implies that $C_j^1 = 0$. Thus for all $j \notin S$, $\pi_j \circ r\varphi \equiv 0$. □

In the Exercises we shall see that $R^\omega$ is not dually slender, and in section 3 we shall see that $R^\lambda$ is almost dually slender for *any* cardinal $\lambda$ — provided that $|R|$ is non-$\omega$-measurable.

## §2. Examples of slender modules and rings

A ring $R$ is said to be *slender* if it is a slender module when considered as a left module over itself. One of our goals in this section is to prove that $\mathbf{Z}$ is a slender ring; we shall derive this — twice — as a consequence of more general results. In I.3.6(ii), we have already sketched a proof that $\text{Hom}(M, \mathbf{Z})$ is naturally isomorphic to $\mathbf{Z}^{(\omega)}$, where $M$ is the $\mathbf{Z}$-adic completion of $\mathbf{Z}^{(\omega)}$ inside $\mathbf{Z}^\omega$; since every homomorphism from $\mathbf{Z}^\omega$ to $\mathbf{Z}$ restricts to $M$, we get a third proof that $\mathbf{Z}$ is slender.

We begin with a general lemma which is formulated to cover more than one application. (There are obvious affinities with the Baire category theorem.)

**2.1 Lemma.** *Let* $\varphi: \prod_{n\in\omega} M_n \to B$ *be an* $R$-*module homomorphism. Suppose* $B = \bigcup_{k\in\omega} B_k$ *such that there is a subset* $X$ *of* $R$ *consisting of commuting elements and with the property that for all* $k \in \omega$ *and all* $b \in B \setminus B_k$, *there exists* $r \in X$ *such that* $(b + rB) \cap B_k = \emptyset$. *Then there exist* $m$, $t$ *and* $k$ *in* $\omega$, $y \in \prod_{n\in\omega} M_n$, *and* $r_0, \ldots, r_t$ *in* $X$ *such that*

$$\varphi[y + r_0 \cdots r_t \prod_{n\geq m} M_n] \subseteq B_k.$$

PROOF. Let $U_k = \prod_{n>k} M_n$. Supposing the result is false, we proceed to define by induction elements $a_k \in \prod_{n\in\omega} M_n$ and $r_k \in X$ such that $a_{k+1} \in a_k + r_0 \cdots r_k U_{k+1}$ and

$$(*) \qquad \varphi[a_{k+1} + r_{k+1} \prod_{n\in\omega} M_n] \cap B_k = \emptyset.$$

Let $a_0 = 0$ and $r_0 = 1$. Suppose $a_i$ and $r_i$ have been chosen for $i \leq k$. By the assumption that the result is false, there exists $a_{k+1} \in a_k + r_0 \cdots r_k U_{k+1}$ such that $\varphi(a_{k+1}) \notin B_k$. By the hypothesis of the lemma, there exists $r_{k+1} \in X$ such that $(\varphi(a_{k+1}) + r_{k+1}B) \cap B_k = \emptyset$. This implies $(*)$ since $\varphi$ is a homomorphism and thus completes the inductive step. Notice that we have:

(1) $a_m(i) = a_k(i)$ for all $i \leq k \leq m$; and
(2) $r_j$ divides $a_m(i) - a_k(i)$ for all $i$ and all $j \leq k \leq m$.

Now define $a \in \prod_{n\in\omega} M_n$ by: $a(n) = a_n(n)$ for all $n \in \omega$, and consider $\varphi(a)$. By hypothesis there must exist $k$ such that $\varphi(a) \in B_k$.

But (1) and (2) imply that $a \in a_{k+1} + r_0 \cdots r_{k+1} U_{k+2}$; so $(*)$ implies that $\varphi(a) \notin B_k$, a contradiction. $\square$

**2.2 Theorem.** *Let $H$ be a countable $R$-module, such that there is a subset $X$ of $R$ with the following properties:*

(1) *the elements of $X$ commute;*

(2) *for all $r \in X$ and $a \in H$, $ra = 0$ implies $a = 0$;*

(3) $\cap\{rH : r \in X\} = \{0\}$.

*Then $H$ is a slender module.*

PROOF. Let $\{M_n : n \in \omega\}$ be any countable family of $R$-modules, and let $\varphi : \prod_{n \in \omega} M_n \to H$ be a homomorphism. It suffices to prove that there exists $m \in \omega$ such that $\varphi[\prod_{n \geq m} M_n] = \{0\}$.

As before, let $U_m = \prod_{n \geq m} M_n$. It suffices to prove that for some $r_0, \ldots, r_t$ in $X$, and $m \in \omega$, $\varphi[r_0 \cdots r_t U_m] = \{0\}$, since then (2) implies that $\varphi[U_m] = \{0\}$. Let $\{b_n : n \in \omega\}$ be an enumeration of $H$ and let $B_k = \{b_k\}$. We want to apply Lemma 2.1 with $B = H$, so suppose that $b \in H \setminus B_k$, i.e., $b \neq b_k$; then by (3), there exists $r \in X$ such that $b - b_k \notin rH$, so $(b + rH) \cap B_k = \emptyset$. Hence by 2.1 we have

$$\varphi[y + r_0 r_1 \cdots r_t U_m] \subseteq \{b_k\}$$

for some $r_0, \ldots, r_t \in X$, $y \in \prod_{n \in \omega} M_n$, and $m, t, k \in \omega$. But then $\varphi(y) = b_k$, so we must have $\varphi[r_0 \cdots r_t U_m] = \{0\}$. $\square$

**2.3 Corollary.** *Let $R$ be an integral domain and $H$ a countable torsion-free reduced $R$-module. Then $H$ is slender.*

PROOF. We apply Theorem 2.2 with $X = R \setminus \{0\}$: conditions (1) and (2) are clear, and (3) follows from the hypotheses that $H$ is reduced and torsion-free, since $\cap\{rH : r \in X\}$ is a divisible submodule of $H$. $\square$

The countability hypothesis is necessary in 2.3 since, for example, $\mathbf{Z}^\omega$ is reduced and torsion-free but not slender.

**2.4 Corollary.** *If $R$ is a countable integral domain which is not a field, then $R$ is a slender ring. In particular, $\mathbf{Z}$ is a slender ring.*

PROOF. Since $R$ is obviously torsion-free as an $R$-module, to apply 2.3 it remains to show that $R$ is reduced. Let $x$ be any non-zero element of $R$. If $x$ is a unit, then $x \notin rR$, where $r$ is any non-unit of $R$. If $x$ is a non-unit then $x \notin x^2 R$, since $x = x^2 s$ implies $1 = xs$, or $x$ is a unit. Hence the maximal divisible submodule of $R$ is $\{0\}$. □

Recall that $M^*$ denotes $\mathrm{Hom}(M, \mathbf{Z})$ if $M$ is an abelian group. We know that
$$(\mathbf{Z}^{(\omega)})^* \cong \mathrm{Hom}(\mathbf{Z}, \mathbf{Z})^\omega \cong \mathbf{Z}^\omega$$
(cf. section I.2). Now, since $\mathbf{Z}$ is slender, we also have that
$$(\mathbf{Z}^\omega)^* \cong \mathrm{Hom}(\mathbf{Z}, \mathbf{Z})^{(\omega)} \cong \mathbf{Z}^{(\omega)}$$
(cf. Corollary 1.5; in fact, this isomorphism takes the canonical surjection $\rho_i \in (\mathbf{Z}^\omega)^*$ to $e_i \in \mathbf{Z}^{(\omega)}$). Hence $(\mathbf{Z}^{(\omega)})^{**} \cong \mathbf{Z}^{(\omega)}$ and $(\mathbf{Z}^\omega)^{**} \cong \mathbf{Z}^\omega$. But in fact, more is true:

**2.5 Corollary.** $\mathbf{Z}^{(\omega)}$ *and* $\mathbf{Z}^\omega$ *are* $(\mathbf{Z}\text{-})$*reflexive.*

PROOF. We will give the proof for $\mathbf{Z}^\omega$ and leave the other for the reader. (A proof of a generalization is given in 3.8; see also Exercise 9.) Since $\mathbf{Z}^\omega$ is torsionless by I.2.1, it suffices to prove that $\sigma: \mathbf{Z}^\omega \to (\mathbf{Z}^\omega)^{**}$ is surjective. So let $z \in (\mathbf{Z}^\omega)^{**}$; then for each $n \in \omega$, let $a_n = \langle z, \rho_n \rangle$ where $\rho_n: \mathbf{Z}^\omega \to \mathbf{Z}$ is the canonical surjection on the $n$th factor. Let $a = (a_n)_{n \in \omega}$. It suffices to prove that $\sigma(a) = z$. For any $y \in (\mathbf{Z}^\omega)^*$, $y = \sum_n y(e_n)\rho_n$ by the proof of 1.5. Therefore,
$$\langle z, y \rangle = \sum_n y(e_n)\langle z, \rho_n \rangle = \sum_n y(e_n)a_n = \langle y, a \rangle = \langle \sigma(a), y \rangle$$
and hence $\sigma(a) = z$. □

**2.6 Lemma.** *Given a ring homomorphism* $\theta: R \to S$, *and an $S$-module $H$, if $H$ is slender as a $R$-module (where the $R$-module structure on $H$ is defined by:* $ra = \theta(r)a$ *for any $r \in R$, $a \in H$), then $H$ is slender as an $S$-module.*

PROOF. This is clear since any $S$-module homomorphism $\varphi: S^\omega \to H$ is also an $R$-module homomorphism. □

**2.7 Proposition.** *If $R$ is a Noetherian integral domain in which every prime ideal is maximal, then every slender $R$-module is torsion-free.*

PROOF. Suppose $H$ is a slender $R$-module which is not torsion-free. Let $a$ be a non-zero element of $H$ such that $\text{Ann}(a)$ is non-zero and maximal among annihilators of non-zero elements. Then $\text{Ann}(a)$ is easily seen to be prime, and hence maximal by hypothesis. Now $Ra$ may be regarded as a module over $R/\text{Ann}(a)$, which is a field; by 1.6(i) and 2.6, $Ra$ is slender as an $R/\text{Ann}(a)$-module, which contradicts 1.6(ii) — since every module over a field is injective. $\square$

There are integral domains $R$ for which there are *torsion* slender $R$-modules: see Exercise 3. However, Proposition 2.7 says that this can't happen if, for example, $R$ is a Dedekind domain.

**2.8 Lemma.** *If $H$ is a module which is complete in a non-discrete metrizable linear topology, then $H$ is not slender.*

PROOF. Since the linear topology on $H$ (see §I.3) is given by a metric, there is a neighborhood system of 0, $\{U_n : n \in \omega\}$, consisting of submodules of $H$ such that $0 \neq U_{n+1} \subseteq U_n$. For each $n \in \omega$ choose $a_n \in H$ such that $0 \neq a_n \in U_n$. Define $\tilde{\varphi} : R^{(\omega)} \to H$ by: $\tilde{\varphi}(e_n) = a_n$. Then for any $r_n \in R$, $\{\sum_{n \leq m} r_n a_n : m \in \omega\}$ is a Cauchy sequence, which thus has a limit, denoted $\sum_{n \in \omega} r_n a_n$, in $H$. We extend $\tilde{\varphi}$ to a homomorphism $\varphi : R^\omega \to H$ such that $\varphi((r_n)_{n \in \omega}) = \sum_{n \in \omega} r_n a_n$. $\square$

The lemma proves one direction of the following characterization of slender p.i.d.'s. Recall that a *discrete valuation ring* is a local p.i.d.; it is called *complete* if it is complete in the $p$-adic topology, where $p$ is the (unique) prime in the ring (cf. I.3.1). Notice that the following yields another proof that $\mathbf{Z}$ is slender.

**2.9 Theorem.** *If $R$ is a p.i.d., $R$ is slender if and only if $R$ is not a complete discrete valuation ring.*

PROOF. Lemma 2.8 implies that a complete discrete valuation ring is not slender. Towards proving the converse, we shall show first that if $R$ is a discrete valuation ring which is not slender, then $R$ is complete. Let $p$ be a generator of the unique prime ideal of $R$ and assume $\varphi: R^\omega \to R$ is such that $\varphi(e_n) \neq 0$ for all $n$. Write each $\varphi(e_n)$ as $p^{i_n} a_n$ where $p$ does not divide $a_n$; so $a_n$ is a unit. By composing $\varphi$ with a homomorphism: $R^\omega \to R^\omega$, we can assume that $i_{n+1} > i_n > i_0 = 0$. Any element, $b$, of $\hat{R}$ can be written as $\sum_{n \in \omega} b_n p^n$ where $b_n$ is a either a unit of $R$ or zero (cf. I.3.4). Given $b$, we can define $x(k) \in R$ by induction so that for all $n$,

$$\sum_{k<n} \varphi(1_k) x(k) = \sum_{j<i_{n+1}} b_j p^j.$$

Then $\varphi(x) = b$, so $b \in R$. Hence $R = \hat{R}$.

It remains to show that if $R$ is not slender, then $R$ is local. Suppose $p$ and $q$ are distinct primes of $R$, and $\varphi: R^\omega \to R$ is such that $\varphi(e_n) \neq 0$ for all $n$. For each $n$, write $\varphi(e_n)$ as $p^{i_n} q^{j_n} a_n$ where neither $p$ nor $q$ divides $a_n$. Choose an element $x \in R^\omega$ so that: $x(0) \neq 0$; for all $n \geq 1$, $p^{n-1} | x(n)$ and $p^{n+i_n+1} | \sum_{\ell \leq n} x(\ell) \varphi(e_\ell)$; and if $k$ is the maximum power of $q$ which divides $\varphi(e_0) x(0)$, then for all $n > 0$, $q^{k+1} | x(n)$. (The values $x(n)$ can be defined by induction on $n$, using the fact that $p$ and $q$ are relatively prime.) Let $x^{(n)} = x - \sum_{i \leq n} x(i) e_i$. By construction, $p^n | x^{(n)}$, and $p^n | \sum_{i \leq n} x(i) \varphi(e_i)$. Hence for all $n$, $p^n | \varphi(x)$, so $\varphi(x) = 0$ (by unique factorization, since $R$ is a p.i.d.). On the other hand, $q^{k+1} | x^{(0)}$, but $x(0) \varphi(e_0)$ is not divisible by $q^{k+1}$. Hence $\varphi(x)$ is not divisible by $q^{k+1}$, which contradicts $\varphi(x) = 0$. $\square$

Theorem 2.9 is true for Dedekind domains as well, i.e., the only non-slender Dedekind domains are complete discrete valuations domains: see Exercise 5.

Next we present some results about modules which are slender over their endomorphism rings. We will then use these to produce some reflexive modules.

**2.10 Theorem.** *Let $H$ be an $R$-module and $S = \operatorname{End}_R(H)$. Suppose that $H = \bigoplus_{n \in \omega} H_n$ such that for all $k \in \omega$, $\bigoplus_{n<k} H_n$ contains*

*no non-zero S-submodule of H. Then H is slender as a right S-module.*

PROOF. Let $\pi_i$ be the projection of $H$ onto $\bigoplus_{n \geq i} H_n$. Given an $S$-module homomorphism $\varphi : \prod_{n \in \omega} M_n \to H$, we are going to apply (a right-hand version of) Lemma 2.1 with $R = S$, $B = H$, $B_k = \bigoplus_{n < k} H_n$, and $X = \{\pi_i : i \in \omega\}$ (so note that all our modules are *right* $S$-modules). The elements of $X$ commute since

$$(*) \qquad\qquad \pi_i \pi_j = \pi_j \pi_i = \pi_j \text{ if } j \geq i.$$

Moreover, for any $b \in H \setminus B_k$, there exists $i \geq k$ such that the $i$th-coordinate of $b$ is non-zero; hence $(b + H\pi_{i+1}) \cap B_k = \emptyset$. Therefore 2.1 implies that for some $m$, $k$ and $i$ in $\omega$ and $y \in \prod_{n \in \omega} M_n$,

$$\varphi[y + (\prod_{n \geq m} M_n)\pi_i] \subseteq B_k$$

(Notice that the product of any number of elements of $X$ is equal to one of them by $(*)$). But then $\varphi[(\prod_{n \geq m} M_n)\pi_i] \subseteq B_k$ since $B_k$ is a submodule. Hence $\varphi[\prod_{n \geq m} M_n] \subseteq B_d$, where $d = \max\{i, k\}$. Now $\varphi[\prod_{n \geq m} M_n]$ is an $S$-submodule of $B_d$ since $\varphi$ is an $S$-module homomorphism, and hence equals zero by hypothesis. $\square$

**2.11 Corollary.** *If H is isomorphic (as left R-module) to $N^{(\omega)}$ for some R-module N, then H is slender as a right $\mathrm{End}_R(H)$-module.*

PROOF. This follows easily from the theorem, since for any $k$ we can define an $R$-endomorphism $\sigma$ of $H$ which permutes the copies of $N$, and takes the $(k-1)$st copy onto the $k$th copy of $N$. $\square$

It is convenient to place here the following consequence of 2.11; in order to do so, we will anticipate some results from the next section. Let us fix a ring $R$ and an $R$-module $H$. Recall the following definitions from section I.1. Let $S = \mathrm{End}_R(H)$. If $M$ is any (left) $R$-module, let $M^*$ denote $\mathrm{Hom}_R(M, H)$, a right $S$-module. Then $M^{**} = \mathrm{Hom}_S(M^*, H)$ is a left $R$-module, and $M$ is called $H$-reflexive if the canonical map $\sigma_M : M \to M^{**}$ is an isomorphism.

**2.12 Theorem.** *Suppose $H \cong N^{(\omega)}$ and $M \cong N^{(\kappa)}$ for some R-module N and some infinite non-$\omega$-measurable cardinal $\kappa$. Then M is H-reflexive.*

PROOF. We can identify $M$ with $H^{(\kappa)}$. Then $M^* \cong \operatorname{Hom}_R(H, H)^\kappa \cong S^\kappa$. Since $H$ is slender as right $S$-module by 2.11 and $\kappa$ is non-$\omega$-measurable, $M^{**} \cong H^{(\kappa)}(= M)$ by Corollary 3.6. In fact, $\sigma_M$ is an isomorphism. It is clearly injective by I.2.1. To see that $\sigma_M$ is surjective, consider $z \in M^{**}$; for each $i < \kappa$, let $a_i = \langle z, \rho_i \rangle$ where $\rho_i \colon H^{(\kappa)} \to H$ is the canonical surjection. Then by Corollary 3.3, $a \stackrel{\text{def}}{=} (a_i)_{i<\kappa}$ belongs to $H^{(\kappa)}$ and one can check that $\sigma_M(a) = z$. □

## §3. The Loś-Eda theorem

In this section we shall see how the results of section 1, such as Theorem 1.2 and its corollaries, can be generalized to apply to homomorphisms on uncountable products. Let us begin with an example which shows that the most naïve generalization does not work when the index set is $\omega$-measurable.

**3.1 Example.** Let $R = H = \mathbb{Z}$, and let $I = \kappa$ be a measurable cardinal. For each $i \in I$ let $M_i = \mathbb{Z}$. Let $D$ be a non-principal $\kappa$-complete ultrafilter on $\kappa$. We are going to define a function $\varphi$ from $\prod_{i \in I} M_i = \mathbb{Z}^\kappa$ into $\mathbb{Z}$ as follows. For any $x \in \mathbb{Z}^\kappa$ and any $n \in \mathbb{Z}$, let

$$Y_n(x) = \{\nu \in \kappa \colon x(\nu) = n\}.$$

Then $\{Y_n(x) \colon n \in \mathbb{Z}\}$ is a partition of $\kappa$, so there exists a unique $m \in \mathbb{Z}$ such that $Y_m(x) \in D$ (cf. II.2.6). Define $\varphi(x) = m$. If $x_1, x_2 \in \mathbb{Z}^\kappa$, and $\varphi(x_i) = m_i$, then $Y_{m_1+m_2}(x_1 + x_2) \supseteq Y_{m_1}(x_1) \cap Y_{m_2}(x_2)$, so $\varphi(x_1 + x_2) = m_1 + m_2 = \varphi(x_1) + \varphi(x_2)$. Hence $\varphi$ is a homomorphism. Moreover $\varphi$ is non-zero, since, for example, $\varphi(\bar{k}) = k$, where $\bar{k}(n) = k$ for all $n \in \mathbb{Z}$. However $\varphi \circ \lambda_i \equiv 0$ for all $i$, so $\Phi$ (see 1.4) is not onto. Also, $\varphi \restriction \mathbb{Z}^{(\kappa)} \equiv 0$, since for all $x \in \mathbb{Z}^{(\kappa)}$, $Y_0(x)$ is cofinite; so Corollary 1.3 fails when $\omega$ is replaced by $\kappa$. Finally, notice that for all $\mu < \kappa$, $\varphi[\prod_{i \geq \mu} M_i] \neq \{0\}$ because — for example — the element $z$ of $\prod_{i \geq \mu} M_i$ which is 1 for all $i \geq \mu$ is not sent to 0: $Y_1(z) \in D$

since $|\kappa \setminus Y_1(z)| < \kappa$ and $D$ is non-principal and $\kappa$-complete. Thus the analog of Theorem 1.2 fails.

Notice that for the $\varphi$ of Example 3.1, $\varphi(x) = 0$ if and only if $\mathrm{supp}(x) \notin D$. This is an example of the following theorem of Eda, which generalizes a theorem of Łoś.

**3.2 Theorem.** *Let $H$ be a slender module. Then for any set $I$ and any family $\{M_i : i \in I\}$ of $R$-modules and any homomorphism*

$$\varphi \colon \prod_{i \in I} M_i \to H$$

*of $R$-modules, there are $\omega_1$-complete ultrafilters $D_1, \ldots, D_n$ on $I$ such that for all $a \in \prod_{i \in I} M_i$, if for all $k = 1, \ldots, n$ $\mathrm{supp}(a) \notin D_k$, then $\varphi(a) = 0$.*

Before proving the theorem, let us derive some consequences. First, let us consider what happens when $I$ is a non-$\omega$-measurable index set; then every $\omega_1$-complete ultrafilter, $D$, is principal, i.e., $D = \{X \subseteq I : j \in X\}$ for some $j \in I$. In this case we obtain the following consequence of Theorem 3.2, which shows that 3.2 is indeed a generalization of 1.2:

**3.3 Corollary.** *If $H$ is slender and $I$ is not $\omega$-measurable then for any $\varphi \colon \prod_{i \in I} M_i \to H$ there exists a finite subset $J$ of $I$ such that for any $a \in \prod_{i \in I} M_i$, if $a(j) = 0$ for all $j \in J$, then $\varphi(a) = 0$.*

PROOF. Since $I$ is not $\omega$-measurable, every $\omega_1$-complete ultrafilter on $I$ is principal. Thus given $\omega_1$-complete ultrafilters $D_1, \ldots, D_n$ on $I$, there are elements $j_1, \ldots, j_n$ such that for all $a \in \prod_{i \in I} M_i$ and all $k = 1, \ldots, n$, $\mathrm{supp}(a) \notin D_k$ if and only if $a(j_k) = 0$. $\square$

As a consequence we obtain, by the same proof, the following generalization of Corollary 1.3.

**3.4 Corollary.** *Let $H$ be a slender module and $I$ a set which is not $\omega$-measurable. For any family $\{M_i : i \in I\}$ and any $\varphi \colon \prod_{i \in I} M_i \to H$, if $\varphi \restriction \bigoplus_I M_i \equiv 0$, then $\varphi \equiv 0$.* $\square$

For arbitrary $I$ we have a generalization of 1.5. But first we need to revise the definition of $\Phi$.

Let $\mathcal{D} = \mathcal{D}_I$ be the set of all $\omega_1$-complete ultrafilters on $I$ (principal and non-principal). Given a family of modules $\{M_i : i \in I\}$ and a module $H$ define

$$\Phi = \Phi(\{M_i : i \in I\}, H): \bigoplus_{D \in \mathcal{D}} \mathrm{Hom}(\prod_{i \in I} M_i / D, H) \to \mathrm{Hom}(\prod_{i \in I} M_i, H)$$

by the rule:

$$\Phi((\varphi_D)_{D \in \mathcal{D}}))(x) = \sum_{D \in \mathcal{D}} \varphi_D(x_D)$$

for any $x \in \prod_I M_i$. (Here, as in II.3.1, $x_D$ denotes the equivalence class of $x$ modulo $D$.) It is routine to check that $\Phi$ is a well-defined homomorphism with domain and codomain as stated.

Next let us observe that if $I$ is not $\omega$-measurable then $\Phi$ is exactly the function defined in 1.4. (Recall that we warned there that the definition would change for measurable $I$!) Indeed, if $I$ is not $\omega$-measurable, then $\mathcal{D}$ consists of principal ultrafilters; more precisely

$$\mathcal{D} = \{D_j : j \in I\}$$

where $D_j = \{X \subseteq I : j \in X\}$, the principal ultrafilter generated by $\{j\}$. Moreover, $\prod_I M_i / D_j$ is canonically isomorphic to $M_j$ (cf. Exercise II.2). Under these isomorphisms our new $\Phi$ can be canonically identified with our old $\Phi$.

For all $I$ (measurable or non-measurable) we have:

**3.5 Lemma.** $\Phi$ *is one-one.*

PROOF. Consider a non-zero element $(\varphi_D)_{D \in \mathcal{D}}$ of the domain of $\Phi$, and let $D_1, \ldots, D_n$ be all the elements, $D$, of $\mathcal{D}$ such that $\varphi_D$ is non-zero. For convenience, denote $\varphi_{D_k}$ by $\varphi_k$. Let $x \in \prod_{i \in I} M_i$ such that $\varphi_1(x_{D_1}) \neq 0$. For each $j = 2, \ldots, n$ choose $S_j \in D_1 \setminus D_j$, and let $Y = \cap\{S_j : j = 2, \ldots, n\}$. Then $Y \in D_1 \setminus D_j$ for all $j = 2, \ldots, n$. Without loss of generality we can assume that $x(i) = 0$ for $i \in I \setminus Y$. But then $\varphi(x_{D_j}) = \varphi(\bar{0}_{D_j}) = 0$ for $j \geq 1$. Hence

$$\Phi((\varphi_D)_{D \in \mathcal{D}})(x) = \sum_{j=1}^{n} \varphi_j(x_{D_j}) = \varphi_1(x_{D_1}) \neq 0. \quad \square$$

Before proving the next result we need a piece of notation: for any subset $Y$ of $I$ and any $x \in \prod_{i \in I} M_i$, let $x{\restriction}Y$ denote the element $y$ of $\prod_{i \in I} M_i$ such that $\mathrm{supp}(y) = Y$ and $y(i) = x(i)$ for all $i \in Y$.

**3.6 Corollary.** *If $H$ is a slender $R$-module, then for any set $I$ and any family $\{M_i : i \in I\}$ of $R$-modules,*

$$\Phi \colon \bigoplus_{D \in \mathcal{D}} \mathrm{Hom}(\prod_I M_i/D,\ H) \to \mathrm{Hom}(\prod_I M_i,\ H)$$

*is an isomorphism.*

PROOF. Given $\varphi \in \mathrm{Hom}(\prod_{i \in I} M_i, H)$, we must show that $\varphi$ is in the image of $\Phi$. Let $D_1, \ldots, D_n$ be as in Theorem 3.2. As in the proof of 3.5, we can choose $Y_1 \in D_1$ such that $Y_1 \notin D_j$ for $j \geq 2$. Choose $Y_2' \in D_2$ such that $Y_2' \notin D_j$ for $j \geq 3$ and let $Y_2 = Y_2' \setminus Y_1$. Continuing in this way we can choose $Y_k \in D_k$ for each $k = 1, \ldots, n$, such that $Y_k \cap Y_j = \emptyset$ for $k \neq j$. Define

$$g_{D_k} \colon \prod_{i \in I} M_i/D_k \to H$$

by: $g_{D_k}(x_{D_k}) = \varphi(x{\restriction}Y_k)$. We must check that $g_{D_k}$ is well-defined. If $x_{D_k} = 0$ then $\{i \in I : x(i) = 0\} \in D_k$, so $\{i \in I : (x{\restriction}Y_k)(i) = 0\}$ belongs to $D_j$ for all $j = 1, \ldots, n$ since $I \setminus Y_k \in D_j$ for $j \neq k$. Hence Theorem 3.2 implies that $\varphi(x{\restriction}Y_k) = 0$, so $g_{D_k}$ is well-defined. For $D \notin \{D_1, \ldots, D_n\}$, let $g_D$ be the zero element in $\mathrm{Hom}(\prod_{i \in I} M_i/D, H)$.

Now we claim that $\Phi((g_D)_{D \in \mathcal{D}}) = \varphi$. For any $x \in \prod_{i \in I} M_i$, write $x = x_0 + x_1 + \ldots + x_n$, where $x_k = x{\restriction}Y_k$ for $k = 1, \ldots, n$, and $x_0 = x{\restriction}W$, where $W = I \setminus (\cup_{k=1}^n Y_k)$. Then by Theorem 3.2, $\varphi(x_0) = 0$. So

$$\Phi((g_D)_{D \in \mathcal{D}})(x) = \sum_{k=1}^n g_{D_k}(x_{D_k}) = \sum_{k=1}^n \varphi(x_k) = \varphi(x)$$

and we are done. $\square$

In the next two results, $M^*$ denotes $\mathrm{Hom}_R(M, R)$.

**3.7 Corollary.** (i) *Let $H$ be a slender module, and let $M$ be any module whose cardinality is not $\omega$-measurable. Then for any cardinal $\kappa$ (measurable or non-measurable), there is a cardinal $\lambda \geq \kappa$ such that* $\operatorname{Hom}(M^\kappa, H) \cong \operatorname{Hom}(M, H)^{(\lambda)}$. *If $\kappa$ is not $\omega$-measurable, then $\lambda = \kappa$.*

(ii) *If $R$ is a slender ring whose cardinality is not $\omega$-measurable, then $(R^\kappa)^*$ is free, and if $\kappa$ is not $\omega$-measurable, $(R^\kappa)^* \cong R^{(\kappa)}$.*

PROOF. Let $\lambda$ be the cardinality of $\mathcal{D}_\kappa$, the set of all $\omega_1$-complete ultrafilters on $\kappa$. Since $|M|$ is not $\omega$-measurable, $M^\kappa/D$ is isomorphic to $M$ for any $D \in \mathcal{D}_\kappa$ (cf. II.3.3). Hence the domain of $\Phi$ is isomorphic to $\operatorname{Hom}(M, H)^{(\lambda)}$ and the first conclusion follows immediately from Corollary 3.6. Notice that if $\kappa$ is not $\omega$-measurable, then the elements of $\mathcal{D}$ are in one-one correspondence with the elements of $\kappa$, so $\lambda = \kappa$. Finally, part (ii) follows from (i) because $\operatorname{Hom}_R(R, R) \cong R$. $\square$

The following may be regarded as a generalization of Corollary 2.5.

**3.8 Corollary.** *Let $R$ be a slender ring of non-$\omega$-measurable cardinality. For any cardinal $\kappa$, let $F_\kappa$ denote the free module $R^{(\kappa)}$.*

(i) *$F_\kappa$ is $R$-reflexive if and only if $\kappa$ is not $\omega$-measurable;*

(ii) *for any cardinal $\kappa$, $F^{**}$ is free;*

(iii) *if there is a measurable cardinal, then there are $\omega$-measurable cardinals, $\kappa_0$ and $\kappa_1$, such that $F^{**}_{\kappa_0}$ is isomorphic to $F_{\kappa_0}$, but $F^{**}_{\kappa_1}$ is not isomorphic to $F_{\kappa_1}$.*

PROOF. For any $\kappa$, $F^*_\kappa$ is isomorphic to $R^\kappa$ since $\operatorname{Hom}_R(R, R) \cong R$ (cf. section I.2). Now let $\kappa$ be non-$\omega$-measurable and consider $\varphi \in F^{**}_\kappa$, which we identify with $(R^\kappa)^*$; since $R$ is slender, $(\varphi \circ \lambda_\nu)_{\nu \in \kappa}$ belongs to $\bigoplus_{\nu \in \kappa} \operatorname{Hom}(R, R)$, which is the domain of $\Phi$. By Corollary 3.6, $\Phi((\varphi \circ \lambda_\nu)_\kappa) = \varphi$. Let $x$ be the element of $R^{(\kappa)}$ defined by: $x(\nu) = \varphi(\lambda_\nu(1))$ for all $\nu \in \kappa$. Then $\sigma_{F_\kappa}(x) = \varphi$ because

$$\langle \varphi, y \rangle = \sum_\nu (\varphi \circ \lambda_\nu)(y(\nu)) = \sum_\nu y(\nu) \cdot \varphi(\lambda_\nu(1)) = \langle y, x \rangle = \langle \sigma_{F_\kappa}(x), y \rangle$$

for all $y \in R^\kappa = F_\kappa^*$. (Here, under the identification of $R^\kappa$ with $F_\kappa^*$, $y$ acts on $x$ by $\langle y, x \rangle = \sum_{\nu \in \kappa} y(\nu)x(\nu)$.)

To prove the other direction of (i), suppose that $\kappa$ is $\omega$-measurable; then by II.2.11, there is a non-principal $|R|^+$-complete ultrafilter $D$ on $\kappa$. We then define a function $\varphi \colon R^\kappa \to R$ as in Example 3.1; namely $\varphi(x) = r$ if and only if $\{\nu \in \kappa \colon x(\nu) = r\} \in D$. We claim that $\varphi$ does not belong to the range of $\sigma_{F_\kappa}$. Indeed, for any $x \in F_\kappa \setminus \{0\}$, $\langle \sigma_{F_\kappa}(x), e_\nu \rangle = x(\nu) \neq 0$ for some $\nu$, but $\langle \varphi, e_\nu \rangle = 0$. (Here, under the identification of $F_\kappa^*$ with $R^\kappa$, $e_\nu$ corresponds to the canonical surjection $\rho_\nu$.)

As for (ii), by 3.7, $F_\kappa^{**}$ is isomorphic to $\mathrm{Hom}(R, R)^{(\lambda)} \cong R^{(\lambda)}$, where $\lambda = |\mathcal{D}_\kappa|$. Then for (iii) we need only observe that there are $\omega$-measurable cardinals $\kappa_0$ and $\kappa_1$ such that $|\mathcal{D}_{\kappa_0}| = \kappa_0$ and $|\mathcal{D}_{\kappa_1}| > \kappa_1$; for this we refer to Exercises II.11 and II.12. $\square$

In a similar manner one can prove that $R^\kappa$ is reflexive if and only if $\kappa$ is not $\omega$-measurable. (See Exercise 9.)

Now without further delay we will give the

**Proof of 3.2.** For convenience, let us introduce the following notation: for any subset $Y$ of $I$ let $\prod(Y)$ denote $\prod_{i \in Y} M_i$, i.e., the set of all $x \in \prod_{i \in I} M_i$ such that $\mathrm{supp}(x) \subseteq Y$. Recall that for any $a \in \prod_{i \in I} M_i$, we defined $a{\restriction}Y$ to be the element of $\prod(Y)$ such that $(a{\restriction}Y)(i) = a(i)$ for all $i \in Y$. Now given $\varphi \colon \prod_{i \in I} M_i \to H$, define

$$ \mathcal{S}' = \{Y \subseteq I \colon \varphi[\prod(Y)] \neq \{0\}\} $$

We assert that

(3.2.1)    Any set of pairwise disjoint elements of $\mathcal{S}'$ is finite.

Assuming for the moment that this is true, let us continue the proof. Let $\mathcal{S} = \{Y \in \mathcal{S}' \colon$ for all $Z \subseteq Y$, exactly one of $Z$ and $Y \setminus Z$ belongs to $\mathcal{S}'\} = \{Y \in \mathcal{S}' \colon$ for all $Z \subseteq Y$, either $\varphi{\restriction}\prod(Z) \equiv 0$ or $\varphi{\restriction}\prod(Y \setminus Z) \equiv 0\}$.

Let $\{Y_1, \ldots, Y_n\}$ be a maximal pairwise disjoint subset of $\mathcal{S}$. For $k = 1, \ldots, n$ define

$$D_k = \{X \subseteq I : \varphi[\textstyle\prod(X \cap Y_k)] \neq \{0\}\}.$$

We claim that

(3.2.2)     $D_k$ is an $\omega_1$-complete ultrafilter.

Again, let us skip the proof for the moment and continue on. Next we claim that

(3.2.3)     If $W \overset{\text{def}}{=} I \setminus \cup \{Y_k : k = 1, \ldots, n\}$, then $\varphi[\prod(W)] = \{0\}$.

Assuming this we can quickly complete the proof. Given any $a \in \prod_{i \in I} M_i$ we can write $a = a{\upharpoonright}W + a{\upharpoonright}Y_1 + \ldots + a{\upharpoonright}Y_n$. By (3.2.3), $\varphi(a{\upharpoonright}W) = 0$. Now let $X = \text{supp}(a)$ and suppose that for all $k = 1, \ldots, n$, $X \notin D_k$. Then $\varphi[\prod(X \cap Y_k)] = \{0\}$, so $\varphi(a{\upharpoonright}Y_k) = 0$ since $a{\upharpoonright}Y_k = a{\upharpoonright}(X \cap Y_k)$. Hence $\varphi(a) = 0$, and we are done except for the proofs of the three claims; it is in these that we shall make use of the slenderness of $H$.

First of all, a piece of notation: suppose $\{X_n : n \in \omega\}$ is a family of subsets of $I$ such that each $i \in I$ belongs to only finitely many $X_n$; if $a_n \in \prod(X_n)$ for each $n$, then $\sum_n a_n$ denotes an element of $\prod_{i \in I} M_i$, viz., the element $x$ such that for each $i$, $x(i) = \sum_{n \in \omega} a_n(i)$, which is a finite sum by hypothesis.

*Proof of* (3.2.1): Suppose, to the contrary, that $\{Y_n : n \in \omega\}$ is an infinite set of pairwise disjoint elements of $\mathcal{S}'$. For each $n$, let $a_n \in \prod(Y_n)$ such that $\varphi(a_n) \neq 0$. Define $\theta : R^\omega \to H$ by:

(3.2.4)     $$\theta((r_n)_{n \in \omega}) = \varphi\left(\sum_n r_n a_n\right)$$

Then $\theta(e_n) \neq 0$ for all $n$, which contradicts the slenderness of $H$.

*Proof of* (3.2.2): First we must check that $D_k$ is a filter. It is easy to see that $D_k$ does not contain $\emptyset$ and is closed under upward inclusion. Suppose $X_0$ and $X_1$ belong to $D_k$, but $X_0 \cap X_1 \notin D_k$. Without loss of generality we may suppose that $X_j \subseteq Y_k$ for $j = 0, 1$. Now for some $a \in \prod_{i \in I} M_i$,

$$0 \neq \varphi(a {\restriction} X_0) = \varphi(a {\restriction} (X_0 \cap X_1)) + \varphi(a {\restriction} (X_0 \setminus X_1))$$

and by hypothesis the first term of the latter sum is zero. Thus $\varphi(a {\restriction} (X_0 \setminus X_1)) \neq 0$, and hence $\varphi {\restriction} \prod(X_0 \setminus X_1) \neq 0$. Similarly $\varphi {\restriction} \prod(X_1 \setminus X_0) \neq 0$. But this contradicts the fact that $Y_k$ belongs to $\mathcal{S}$ (let $Z = X_1 \setminus X_0$). Hence $D_k$ is a filter. It is an ultrafilter because if $X \notin D_k$ and $a \in \prod(Y_k)$ such that $\varphi(a) \neq 0$, then

$$0 \neq \varphi(a) = \varphi(a {\restriction} X \cap Y_k) + \varphi(a {\restriction} Y_k \setminus X) = \varphi(a {\restriction} Y_k \setminus X),$$

which shows that $I \setminus X$ belongs to $D_k$.

Finally, to show that $D_k$ is $\omega_1$-complete, suppose, to the contrary, that there is a subset $\{X_n : n \in \omega\}$ of $D_k$ such that $\bigcap_{n \in \omega} X_n = \emptyset$. Without loss of generality we may suppose that, for all $n$, $X_{n+1} \subseteq X_n \subseteq Y_k$. Since $X_n \in D_k$, there exists $a_n \in \prod(X_n)$ such that $\varphi(a_n) \neq 0$. Define $\theta : R^\omega \to H$ by equation (3.2.4) above. (Notice that since the intersection of the $X_n$'s is empty, $\sum_n r_n a_n$ is well-defined.) We obtain a contradiction of the slenderness of $H$ as before, and therefore have completed the proof of (3.2.2).

*Proof of* (3.2.3): Suppose (3.2.3) is false, i.e., that $W \in \mathcal{S}'$. We will define by induction on $n$ a sequence of pairwise disjoint subsets $X_n$ of $W$ as follows. By the maximality of $\{Y_1, \ldots, Y_n\}$, $W$ does not belong to $\mathcal{S}$ so there exists $X_0 \subseteq W$ such that $\varphi[\prod(X_0)] \neq \{0\}$ and $\varphi[\prod(W \setminus X_0)] \neq \{0\}$. Suppose $X_n$ has been chosen so that $\varphi[\prod(X_n)] \neq \{0\}$ and $\varphi[\prod(W \setminus (X_0 \cup \ldots \cup X_n))] \neq \{0\}$; then since $W \setminus (X_0 \cup \ldots \cup X_n) \notin \mathcal{S}$ there must exist $X_{n+1} \subseteq W \setminus (X_0 \cup \ldots \cup X_n)$ such that $\varphi[\prod(X_{n+1})] \neq \{0\}$ and $\varphi[\prod(W \setminus (X_0 \cup \ldots \cup X_n \cup X_{n+1}))] \neq \{0\}$. Now, for each $n$ choose $a_n \in X_n$ such that $\varphi(a_n) \neq 0$; using (3.2.4) we obtain a contradiction of the slenderness of $H$ as before.

This completes the proof of the claims, and hence the proof of Theorem 3.2. □

In a similar manner we can generalize Theorem 1.8, provided $R$ is countable. (For further generalizations see Exercise 10.)

**3.9 Theorem.** *Let $R$ be a countable ring and let $\{M_i : i \in I\}$ and $\{C_j : j \in J\}$ be indexed families of $R$-modules. Then for any homomorphism*

$$\varphi : \prod_{i \in I} M_i \to \bigoplus_{j \in J} C_j$$

*there exist $\omega_1$-complete ultrafilters $D_1, \ldots, D_n$ on $I$, an element $r \neq 0$ of $R$, and a finite subset $S$ of $J$ such that for any $a \in \prod_{i \in I} M_i$, if for all $k = 1, \ldots, n$, supp$(a) \notin D_k$, then $\varphi(ra) \in \bigoplus_{j \in S} C_j + \bigoplus_{j \in J} C_j^1$.*

PROOF. The proof proceeds along the lines of the proof of 3.2, with Theorem 1.8 substituting for the slenderness of the range. Let $\{r_n : n \in \omega\}$ be an enumeration of $R \setminus \{0\}$ such that each element occurs infinitely often. Given a subset $S$ of $J$, let $\pi_S$ denote the canonical projection of $\bigoplus_{j \in J} C_j$ onto $\bigoplus_{j \notin S} C_j$; we will write $\pi_S \varphi$ for $\pi_S \circ \varphi$. For any subset $Y$ of $I$, we will say that "$Y$ satisfies (†)" if:

(†)      for all $r \neq 0$ in $R$, for all finite $S \subseteq J$, $\pi_S \varphi[r \prod(Y)]$ is not contained in $\bigoplus_{j \notin S} C_j^1$.

Otherwise "$Y$ fails (†)". Notice that if $Y$ fails (†), then any subset of $Y$ fails (†). Define $\mathcal{S}'$ to be the set of all subsets $Y$ of $I$ such that $Y$ satisfies (†). We claim that

(3.9.1)      Any set of pairwise disjoint elements of $\mathcal{S}'$ is finite.

Indeed, suppose to the contrary that $\{Y_n : n \in \omega\}$ is an infinite set of pairwise disjoint elements of $\mathcal{S}'$. Define elements $a_n$ of $\prod(Y_n)$ and $j_n$ of $J$ by induction on $n$ such that $j_n \notin \{j_0, \ldots, j_{n-1}\}$ and

$\pi_{j_n}\varphi(r_n a_n) \notin C^1_{j_n}$; this is possible because $Y_n$ satisfies (†). Now define $\theta : R^\omega \to \bigoplus_{j\in J} C_j$ by equation (3.2.4). For any $m \in \omega$, any $r \in R \setminus \{0\}$, and any finite subset $S$ of $J$, there exists $n \geq m$ such that $j_n \notin S$ and $r_n = r$; since

$$\theta(re_n) = \varphi(r_n a_n) \notin \bigoplus_{j\in S} C_j + \bigoplus_{j\in J} C^1_j \ ,$$

we have a contradiction of Theorem 1.8.

Let $\mathcal{S}$ be the set of all $Y \in \mathcal{S}'$ such that for all $Z \subseteq Y$, exactly one of $Z$ and $Y \setminus Z$ satisifes (†). Let $\{Y_1, \ldots, Y_n\}$ be a maximal pairwise disjoint subset of $\mathcal{S}$. For $k = 1, \ldots, n$ define

$$D_k = \{X \subseteq I : X \cap Y_k \text{ satisfies } (†)\}.$$

We claim that

(3.9.2)      $D_k$ is an $\omega_1$-complete ultrafilter

Suppose first that $X_0$ and $X_1$ belong to $D_k$ but $X_0 \cap X_1$ does not. Without loss of generality we can suppose that $X_0 \subseteq Y_k$ and $X_1 \subseteq Y_k$. Thus there exists $r \neq 0$ and a finite set $S$ such that $\pi_S \varphi[r \prod(X_0 \cap X_1)] \subseteq \bigoplus_{j\notin S} C^1_j$. We shall obtain a contradiction by proving that both $X_0 \setminus X_1$ and $X_1 \setminus X_0$ satisfy (†). Given any $r' \neq 0$ and any finite subset $S'$ of $J$, since $X_0$ satisfies (†) there exists $a \in \prod(X_0)$ and $j \notin S \cup S'$ such that $\pi_j\varphi(rr'a) \notin C^1_j$. Since $a = a{\restriction}(X_0 \cap X_1) + a{\restriction}(X_0 \setminus X_1)$, and since $\pi_j\varphi(rr'a{\restriction}(X_0 \cap X_1)) \in C^1_j$ (because $j \notin S$), we can conclude that $\pi_j\varphi(rr'a{\restriction}(X_0 \setminus X_1)) \notin C^1_j$. This shows that $X_0 \setminus X_1$ satisfies (†); of course, the proof that $X_1 \setminus X_0$ satisfies (†) is the same.

We shall omit the rest of the proof that $D_k$ is an ultrafilter, since it proceeds along similar lines, and turn to the proof that $D_k$ is $\omega_1$-complete. Suppose, to obtain a contradiction, that there is a subset $\{X_n : n \in \omega\}$ of $D_k$ such that for all $n$, $X_{n+1} \subseteq X_n \subseteq Y_k$, and $\bigcap_{n\in\omega} X_n = \emptyset$. Define, by induction on $n$, elements $a_n \in \prod(X_n)$ and $j_n \notin \{j_0, \ldots, j_{n-1}\}$ such that $\pi_{j_n}\varphi(r_n a_n) \notin C^1_{j_n}$. This is possible because $X_n$ satisfies (†). Then defining $\theta$ as in (3.2.4), we obtain a contradiction of Theorem 1.8.

Now for our last claim we assert

(3.9.3)    If $W = I \setminus \cup\{Y_k : k = 1, \ldots, n\}$, then $W$ fails (†).

The proof of this is very similar to that of (3.2.3) except that we use Theorem 1.8 to obtain a contradiction.

Now let $a \in \prod_{i \in I} M_i$ such that for all $k$ supp$(a) \notin D_k$. Since $a = a{\upharpoonright}W + a{\upharpoonright}Y_1 + \ldots + a{\upharpoonright}Y_n$ and since $W$ fails (†), it suffices to prove that for each $k$ there exists $r_k \neq 0$ and a finite set $S_k$ such that for all $X \subseteq Y_k$, if $X \notin D_k$, then $\pi_{S_k}\varphi[r_k \prod(X)] \subseteq \bigoplus_{j \notin S_k} C_j^1$. (For then, 3.9 is true with $r = r_1 \cdots r_n$ and $S = S_1 \cup \ldots \cup S_n$.)

Suppose that this is false for some $k$. Then, by methods which are by now becoming familiar, we can inductively define finite subsets $S_m$ of $J$, and elements $j_m \in J$, $d_m \in \omega$, and $X_m \subseteq I$ such that:

$$X_m \notin D_k;$$
$$S_m \subseteq S_{m+1};$$
$$X_m \cap X_\ell = \emptyset \text{ if } \ell \neq m;$$
$$\pi_{S_m}\varphi[r_0 r_1 \cdots r_{d_m} \prod(X_m)] \subseteq \bigoplus_{j \notin S_m} C_j^1;$$
$$j_m \in S_{m+1} \setminus S_m; \text{ and}$$
$$\pi_{j_m}\varphi[r_0 r_1 \cdots r_{d_m} \prod(X_{m+1})] \text{ is not contained in } C_{j_m}^1.$$

Finally, we obtain a contradiction of 1.8 by defining $\theta$ as in (3.2.4). $\square$

**3.10 Corollary.** *Let $R$ be a countable p.i.d. and $\{M_i : i \in I\}$ an indexed family of Fuchs-44 modules. Then $\prod_{i \in I} M_i$ is Fuchs-44 if either:*

(1) *$I$ is not $\omega$-measurable; or*

(2) *$\sup\{|M_i| : i \in I\}$ is not $\omega$-measurable.*

PROOF. Let $\varphi : \prod_{i \in I} M_i \to \bigoplus_{j \in J} C_j$ where the $C_j$ are reduced. By composing $\varphi$ with the canonical surjection: $\bigoplus_{j \in J} C_j \to \bigoplus_{j \in J} C_j / C_j^1$, we can assume that $C_j^1 = 0$ for all $j$ (cf. the proof of 1.12). Let $D_1$, $\ldots$, $D_n$, $Y_1$, $\ldots$, $Y_n$, $r$ and $S$ be as in Theorem 3.9 and its proof. In case (1) each $D_k$ is principal, so there is a finite subset $I'$ of $I$ such that $\varphi[r \prod(I \setminus I')] \subseteq \bigoplus_{j \in S} C_j$. Since each $M_i$ $(i \in I')$ is Fuchs-44 and $I'$ is finite, it follows easily that $\prod_{i \in I} M_i$ is Fuchs-44.

Now in case (2), for each $k = 1, \ldots, n$ define

$$\varphi_k : \prod_{i \in I} M_i / D_k \to \bigoplus_{j \notin S} C_j$$

by: $\varphi_k(x_{D_k}) = \pi_S \varphi(rx \upharpoonright Y_k)$. Then $\varphi_k$ is well-defined because if $x_{D_k} = 0$, then $\text{supp}(x \upharpoonright Y_k) \notin D_\nu$ for all $\nu = 1, \ldots, n$, so $\varphi(rx \upharpoonright Y_k) \in \bigoplus_{j \in S} C_j$ (cf. proof of 3.6). By (2), II.2.13, and II.2.11(ii), we can assume that $M_i = M$ for some fixed $M$ and all $i \in X$ for some $X \in D_k$; hence $\prod_{i \in I} M_i / D_k \cong M$ (by II.3.2). Thus $\prod_{i \in I} M_i / D_k$ is Fuchs-44 and so there exists $r_k \neq 0$ and a finite set $S_k$ such that the image of $r_k \varphi_k$ is contained in $\bigoplus_{j \in S_k} C_j$. Therefore since $x = x \upharpoonright W + x \upharpoonright Y_1 + \ldots + x \upharpoonright Y_n$, if we let $\tilde{r} = r \cdot r_1 \cdots r_n$ and $\tilde{S} = S \cup S_1 \cup \ldots \cup S_n$, then the image of $\tilde{r}\varphi$ is contained in $\bigoplus_{j \in \tilde{S}} C_j$. □

From part (2) we can conclude, for example, that $\mathbf{Z}^\lambda$ is Fuchs-44 for any cardinal $\lambda$ (cf. Proposition 1.12).

The hypothesis on $F$ in the following result is satisfied if, for example, $I$ is not $\omega$-measurable and $F$ contains the cofinite filter. (See II.2.7.)

**3.11 Corollary.** *If $F$ is a filter on $I$ which is not contained in any $\omega_1$-complete ultrafilter on $I$, then for any family of groups $\{M_i : i \in I\}$, the reduced product $\prod_{i \in I} M_i / F$ is a Fuchs-44 group.*

PROOF. Let $\psi : \prod_{i \in I} M_i / F \to \bigoplus_{j \in J} C_j$ where the $C_j$ are reduced. As in 3.10, we can assume that $C_j^1 = 0$ for all $j$. Apply 3.9 to the composition $\varphi \overset{\text{def}}{=} \psi \circ \pi$ where $\pi$ is the canonical surjection: $\prod_{i \in I} M_i \to \prod_{i \in I} M_i / F$. Let $r$, $D_1, \ldots, D_n$, and $S$ be as in the conclusion of 3.9. We will be done if we show that $\psi[r \prod_{i \in I} M_i / F] \subseteq \bigoplus_{j \in S} C_j$. But given $a_F \in \prod_{i \in I} M_i / F$, we can assume that $\text{supp}(a) \notin D_k$ for all $k = 1, \ldots, n$ since $F \not\subseteq D_k$. Therefore $\psi(ra_F) = \varphi(ra) \in \bigoplus_{j \in S} C_j$. □

**3.12 Example.** In Eda-Abe 1987 it is shown that there is a product of Fuchs-44 groups which is not Fuchs-44. For the purposes of this example, which is not used elsewhere, we shall refer ahead to section IX.4, particularly to the proof of Theorem IX.4.13. Let $\kappa$ be the first measurable cardinal and let $S = \{\alpha < \kappa : \alpha$ is a regular uncountable cardinal$\}$. For each $\alpha \in S$, let $A_\alpha$ be the reduced product $(\mathbf{Z}^{(\omega)})^\alpha / C_\alpha$ where $C_\alpha$ is the co-$\alpha$ filter on $\alpha$ (cf. II.2.5). By 3.11, $A_\alpha$ is a Fuchs-44 group. Exactly as in the proof of IX.4.13, there is a homomorphism $\bar\varphi \colon \prod_{\alpha \in S} A_\alpha \to \mathbf{Z}^{(\omega)}$ which is, in fact, an epimorphism. But then $\prod_{\alpha \in S} A_\alpha$ is certainly not Fuchs-44.

## EXERCISES

1. (i) $H$ is slender if and only if every homomorphism $\varphi \colon \prod_{n \in \omega} M_n \to H$ is continuous, where $H$ is given the discrete topology and $\prod_{n \in \omega} M_n$ the product topology where each factor $M_n$ has the discrete topology.
(ii) A direct sum of slender modules (over any ring) is slender. [Hint: use (i) and the Baire Category Theorem.]

2. (i) $R$ is slender if and only if every projective $R$-module is slender. [Hint: use 1(ii).]
(ii) If $R$ is slender, then for any infinite $\kappa$, $R^\kappa$ is not projective.

3. If $R = \mathbf{Z}[x]$ and $H = R/2R$, then $H$ is a torsion slender $R$-module. [Hint: let $X = \{x^n : n \in \omega\}$ and use 2.2.]

4. If $R$ is a slender ring, then $R[x]$ and $M_n(R)$ (the ring of $n \times n$ matrices over $R$) are slender rings. [Hint: use 2 and 2.6.]

5. A non-slender Dedekind domain is a p.i.d. [Hint: use the facts: If $P$ is a prime ideal of a Dedekind domain $R$, then $P$ is a maximal ideal and $R_P$ is a p.i.d.; also, a Dedekind domain with only finitely many primes is a p.i.d.]

6. Let $R$ be an integral domain such that there is a countable subset $X = \{r_n : n \in \omega\}$ of $R \setminus \{0\}$ such that $\bigcap_{n \in \omega} r_n R = \{0\}$. Prove that if $|R| < 2^{\aleph_0}$, then $R$ is slender. [Hint: Suppose that $\varphi \colon R^\omega \to R$ is such that $\varphi(e_n) \neq 0$ for all $n$; define inductively $m_n \in \omega$ such that for all $\epsilon \neq \delta$ in $\{0,1\}^\omega$, $\varphi((r_{m_n} \epsilon_n)_{n \in \omega}) \neq \varphi((r_{m_n} \delta_n)_{n \in \omega})$.]

7. If $H$ is a slender group and $\kappa$ and $\lambda$ are infinite cardinals such that $H^\kappa \cong H^\lambda$, then $\kappa = \lambda$.

8. If $G$ is the subgroup of $\mathbf{Z}^\lambda$ consisting of all elements which are 0 in all but countably many components, then $G^* \cong \mathbf{Z}^{(\lambda)}$. [Hint: every countable subset of $G$ is contained in a subgroup $\cong \mathbf{Z}^\omega$.]

9. If $R$ is a slender ring of non-$\omega$-measurable cardinality, then $R^\kappa$ is reflexive if and only if $\kappa$ is not $\omega$-measurable. [Hint: imitate the proof of 3.8. This can also be proved by using the statement of 3.8 and IX.1.9.]

10. Given a descending chain $N_0 \supseteq N_1 \supseteq \ldots \supseteq N_m \supseteq \ldots$ of finitely generated right ideals of $R$ and a homomorphism $\varphi \colon \prod_{i \in I} M_i \to \bigoplus_{j \in J} C_j$ of left $R$-modules, then there exist finitely many $\omega_1$-complete ultrafilters $D_1, \ldots, D_n$ and a finite subset $S$ of $J$ such that for any $a \in \prod_{i \in I} M_i$, if $\mathrm{supp}(a) \notin D_k$ for all $k$, then $\varphi(a) \in \bigoplus_{j \in S} C_j + \bigcap_{n \in \omega}(N_n \bigoplus_{j \in J} C_j)$.

*In the remaining exercises, all modules are modules over a fixed p.i.d. $R$.*

11. If $0 \to A \to B \to C \to 0$ is exact, and $A$ and $C$ are slender (resp. almost slender), then $B$ is slender (resp. almost slender).

12. Any reduced module of countable torsion-free rank is almost slender. [Hint: use 11, 1.11, and 2.3.]

13. Assume $R$ is countable. Let $H$ be almost slender and $\varphi \colon \prod_{i \in I} M_i \to H$. Then there exists $r \neq 0$ and $\omega_1$-complete ultrafilters $D_1, \ldots, D_n$ on $I$ such that for all $a \in \prod_{i \in I} M_i$, if $\mathrm{supp}(a) \in D_k$ for $k = 1, \ldots, n$, then $r\varphi(a) = 0$. [Hint: let $\mathcal{S} = \{Y \subseteq I \colon$ for all $r \neq 0$, $r\varphi(\prod(Y)) \neq \{0\}$ but for all $Z \subseteq Y$, there exists $r \neq 0$ such that $r\varphi(\prod(Z)) = \{0\}$ or $r\varphi(\prod(Y \setminus Z)) = \{0\}$. Then proceed as in the proof of 3.2.]

14. (i) For any family $\{C_j \colon j \in J\}$ of torsion-free reduced modules, $\Psi(R^\omega, \{C_j \colon j \in J\})$ is an isomorphism. [Hint: use 1.8.]
(ii) If $R$ is not a field, then $R^\omega$ is not dually slender. [Hint: start with the canonical map: $R^\omega \to (R/pR)^\omega$ and use the fact that the range is a vector space over $R/pR$.]

15. Suppose $R$ has only countably many primes.
(i) $H$ is Fuchs-44 if and only if for every $\sum$-cyclic torsion module $T$ and every $\varphi: H \to T$, $\varphi[H]$ has bounded order. [Hint: suppose $\varphi: H \to \bigoplus_{j \in J} C_j$ is a counterexample; define inductively $r_n \in R \setminus \{0\}$, $j_n \in J$, $a_n \in H$ s.t.: $r_{n+1} \in r_n R$; for every $s \in R$ $\exists n$ s.t. $r_n \in sR$; $a_n \in r_n H$; $j_n \notin \bigcup_{k<n} \text{supp}(\varphi(a_n))$; and $\pi_{j_n} \varphi(a_n) \notin r_{n+1} C_{j_n}$. Let $C'_j = r_{n+1} C_j$ if $j = j_n$. Consider the induced map: $H \to \bigoplus_{j \in J} C_j / C'_j$.]
(ii) $H$ is dually slender if and only if for every $\sum$-cyclic torsion $T$ and every $\varphi: H \to T$, $\varphi[H]$ is finitely-generated.

16. If $p$ is a prime of $R$, then $\hat{R}_{(p)}$, the completion of $R$ localized at $p$ in the $p$-adic topology, is dually slender. [Hint: By 15(ii) it suffices to consider $\varphi: \hat{R}_{(p)} \to T$, where $T$ is torsion; then notice that $\varphi[T]$ is a $\hat{R}_{(p)}$-module.]

17. We consider what happens if the definitions of dually slender and almost dually slender are changed to remove the requirement that the $C_j$'s are reduced. For simplicity let $R = \mathbf{Z}$. Let $M$ be an abelian group.
(i) Prove that if $\Psi(M, \{C_j : j \in J\})$ is surjective for *every* family $\{C_j : j \in J\}$, then $M$ is finitely-generated. [Hint: show first that $M$ has finite rank by considering the embedding of $M$ into its divisible hull; now let $F \subseteq M$ be the free module, of rank $n$, generated by a maximal independent subset of $M$ and show that if $T = M/F$ (a torsion group) is not finitely-generated, then there is a homomorphism

$$g: M \to \mathbf{Q}^{(n)} \oplus \bigoplus_p Z(p^\infty)^{(\omega)}$$

which is not in the range of $\Psi$: consider separately the cases (1) $T$ contains a copy of $Z(p^\infty)$ for some $p$; or (2) $T$ contains an infinite direct sum of cyclic groups.]
(ii) Suppose that $M$ is an abelian group such that for *every* family $\{C_j : j \in J\}$, and every $\varphi \in \text{Hom}(M, \bigoplus_{j \in J} C_j)$, there exists $r \neq 0$ such that $r\varphi$ belongs to the range of $\Psi$. Show that $M$ is the direct sum of a bounded group and a finite rank free group. [Hint: show

first, as in (i), that the torsion part of $M$ must be of bounded order, hence a summand of $M$. Then we can suppose that $M$ is torsion-free. Let $F$ be as in (i) and show that if $F$ has rank $n$, then for any prime $p$, $M/F$ has at most $n$ summands of order a power of $p$; so if $M/F$ is not finitely-generated it must be of unbounded order; use this to construct a counterexample.]

## NOTES

Specker 1950 first proved that $(Z^\omega)^* \cong Z^{(\omega)}$ and generalizations were proved by Ehrenfeucht-Łoś 1954 and Zeeman 1955. The notion of slenderness and the main results for the non-measurable case, including 1.1, 1.2, 1.3, 1.5, 3.3 and 3.4 are due to J. Łoś and were published in Fuchs 1960. The generalization to the measurable case appears in Eda 1982 and 1983d. 1.10 is due to Fuchs 1960. The special case of Eda's result characterizing $(Z^\kappa)^*$ for arbitrary $\kappa$ is due to Balcerzyk 1962. Mrówka 1972 generalized the result on $(Z^\omega)^* \cong Z^{(\omega)}$ to a topological setting; see section XI.4 for the group-theoretic consequences of Mrówka's work.

Theorem 1.8 is from Chase 1962b; the generalization to non-measurable $I$ is due to Ivanov 1979 and Dugas-Zimmerman-Husigen 1981; its generalization to the measurable case is from Eda 1982 and 1983b; 1.11 is due to Rychkov 1980.

The Fuchs-44 groups are characterized in Ivanov 1978 and 1.12 and 3.10(1) are proved. Corollaries 3.10(2) and 3.11 are from Eda 1983b; 3.12 is from Eda-Abe 1987. For more on the notions of almost slender, almost dually slender, etc. see also Göbel-Richkov-Wald 1981 and Rychkov-Thomé 1986.

2.1 is adapted from Mader 1984, which corrects Lady 1973; 2.2 to 2.7 (except 2.5) are based on Heinlein 1971 and Lady 1973; 2.9 is due to Heinlein 1971; 2.10 is from Mader 1984; 2.11 and 2.12 are from Huber 1983a. For more on slender modules over arbitrary domains, see Fuchs-Salce 1985 and Dimitric 1983

If we replace $Z^\omega$ in the definition of a slender group by an appropriate (monotone) subgroup $M$ of $Z^\omega$, we obtain the notion of an $M$-slender group: see Göbel-Wald 1979. Göbel-Wald 1980 show that MA (see §VI.4) implies that there are the maximum possible number of different notions of $M$-slender. Blass-Laflamme 1989 show that it is consistent with ZFC that there are exactly the minimal possible number, four, of different notions.

Exercises. 1:Heinlein 1971/Lady 1973; 3,4:Lady 1973; 5:Heinlein 1971; 7:Eda 1983d; 10:Dugas-Zimmermann-Huisgen 1981 (for the non-measurable case); 12:Rychkov-Thomé 1986; 13:Eda 1983b; 14:Rychkov-Thomé 1986; 15:(i)Ivanov 1978, (ii)Rychkov-Thomé 1986; 16:Rychkov-Thomé 1986.

# CHAPTER IV
# ALMOST FREE MODULES

This chapter, and Chapter VII, deal with modules which have the property that "most" submodules generated by fewer than $\kappa$ elements are free; these will be called $\kappa$-free modules. One of the concerns of the first section is giving a precise meaning to "most". In the case of $\mathbf{Z}$-modules, this is easy — "most" means "all" — but for modules over arbitrary rings $R$, the analysis is more subtle. One of the reasons that we deal with modules over arbitrary rings in sections 1 and 3 of this chapter and section 1 of Chapter VII is that we are able to give a satisfactory answer, for all rings, to the question of the existence of $\kappa$-free modules which are are not free; the crucial dichotomy in our analysis occurs between rings (like fields) which are left perfect and those (like $\mathbf{Z}$) which are not. All the results of this chapter are provable in ZFC; many of those in Chapter VII require additional set-theoretic hypotheses, which are introduced in Chapter VI.

Section 1 also introduces some important tools for the analysis and construction of $\kappa$-free modules, namely the $\kappa$-filtration and the $\Gamma$-invariant of a module. Section 2 deals specifically with $\aleph_1$-free $\mathbf{Z}$-modules. Section 3 returns to the general setting and deals with "compactness" results, that is, results about when a $\kappa$-free module of cardinality $\kappa$ is free. ("Incompactness" results — that is, those which say that there is a $\kappa$-free module of cardinality $\kappa$ which is *not* free — are the principal subject of Chapter VII.)

## §1. $\kappa$-free modules

If $\kappa$ is an uncountable cardinal, an abelian group is called $\kappa$-free if every subgroup of cardinality $< \kappa$ is free (cf. II.3). If we try to extend this to modules over arbitrary rings, we encounter two difficulties. First of all, $\kappa$ may be smaller than the cardinality of the ring, $R$, so that there may not even be any non-zero submodules of cardinality $< \kappa$; this is relatively easily disposed of by considering the size of generating sets for submodules, instead of the size of the submodules themselves. A second and more serious problem is the possible absence of the Schreier property, which would mean that not even a free module would be $\kappa$-free if we required *every* submodule of size $< \kappa$ to be free. Here a number of alternatives are possible, and have been considered in the literature, leading to a variety of definitions of $\kappa$-free. The stronger the definition, the stronger the assertion that there exist non-free $\kappa$-free modules. We shall choose one of the strongest for our primary definition.

We say that a submodule $N$ of $M$ is $< \kappa$-generated (resp. $\le \kappa$-generated) if it equals $\langle X \rangle$ for some $X$ of cardinality $< \kappa$ (resp. $\le \kappa$).

Throughout this section $\kappa$ will denote an uncountable cardinal.

**1.1 Definition.** If $\kappa$ is a regular uncountable cardinal, we shall say that a module $M$ is *$\kappa$-free* if there is a subset $\mathcal{C}$ of $\mathcal{P}(M)$ satisfying:

(1) every element of $\mathcal{C}$ is a $< \kappa$-generated free submodule of $M$;

(2) every subset of $M$ of cardinality $< \kappa$ is contained in an element of $\mathcal{C}$; and

(3) $\mathcal{C}$ is closed under unions of well-ordered chains of length $< \kappa$.

Note that we need $\kappa$ to be regular for this to be a reasonable definition: for if $\kappa$ is singular, we may have a submodule which is not $< \kappa$-generated but is the union of a chain of length $< \kappa$ of $< \kappa$-generated submodules.

If $\lambda$ is a singular cardinal, we define $M$ to be $\lambda$-free if $M$ is $\kappa$-free for every regular cardinal $\kappa < \lambda$. (It is also possible to define $\kappa$-free as in Definition 1.1 for successor cardinals $\kappa$ and for limit cardinals $\lambda$, define $M$ to be $\lambda$-free if it is $\kappa$-free for all $\kappa < \lambda$. See Exercise 13 for an application of this definition.)

First, we should observe that free modules are $\kappa$-free:

**1.2 Lemma.** *If $F$ is a free module, then for any infinite cardinal $\lambda$, $F$ is $\lambda$-free.*

PROOF. Let $B$ be a basis of $F$. For any regular $\kappa$, let $\mathcal{C}$ consist of all submodules generated by a subset of $B$ of cardinality $< \kappa$.

Clearly $\mathcal{C}$ satisfies (1) – (3). $\square$

We can loosely describe properties (1) – (3) by saying that $\mathcal{C}$ is a closed unbounded set (or cub) of $< \kappa$-generated free submodules of $M$ (cf. Exercise II.15). A weaker notion of $\kappa$-free is simply to require the existence of a $\mathcal{C}$ with properties (1) and (2); in this case we say that $M$ is $\kappa$-*free in the weak sense*. There are intermediate notions which involve game-theoretic language; for these see Hodges 1981, Eklof-Mekler 1988, and/or Shelah 1975a. All of these definitions of $\kappa$-free will agree when $R$ is a p.i.d. because in that case, a submodule of a free module is free. In fact, if $R$ is a p.i.d. an $R$-module is $\kappa$-free if and only if *every* $< \kappa$-generated submodule is free.

Obviously, a module which is $< \kappa$-generated and $\kappa$-free (even in the weak sense) is free. The question of whether there exist $\leq \kappa$-generated $\kappa$-free modules which are not free is more difficult and more interesting, and will receive a lot of attention from us. We begin by introducing a tool which is important in studying $\leq \kappa$-generated modules. (Compare with Definition II.4.12.)

**1.3 Definition.** If $M$ is a $\leq \kappa$-generated module, a $\kappa$-*filtration* of $M$ is a sequence $\{M_\nu : \nu \in \kappa\}$ of submodules of $M$ whose union is $M$ and which satisfies for all $\mu$, $\nu < \kappa$:

   (a) $M_\nu$ is a $< \kappa$-generated submodule of $M$;
   (b) if $\mu \leq \nu$, then $M_\mu \subseteq M_\nu$;
   (c) if $\nu$ is a limit ordinal, then $M_\nu = \bigcup_{\mu < \nu} M_\mu$.
(A chain of sets with property (c) is said to be *continuous*.)

Next we have an existence result and a kind of uniqueness result. (For the definition of a cub, see II.4.1.)

**1.4 Lemma.** *Let $\kappa$ be a regular uncountable cardinal and let $M$ be a $\leq \kappa$-generated module. Then*

   (i) *$M$ has a $\kappa$-filtration;*
   (ii) *if $\{M_\nu : \nu \in \kappa\}$ and $\{M'_\nu : \nu \in \kappa\}$ are two $\kappa$-filtrations of $M$, there is a cub $C$ in $\kappa$ such that for all $\nu \in C$, $M_\nu = M'_\nu$.*

PROOF. (i) Let $X = \{x_\mu : \mu < \kappa\}$ be a generating set for $M$. Let $M_\nu = \langle x_\mu : \mu < \nu \rangle$.

The proof of (ii) is essentially the same as in II.4.12. □

**1.5 Lemma.** *Let* $\kappa$ *be a regular uncountable cardinal.* $A \leq \kappa$-*generated module* $M$ *is* $\kappa$-*free if and only if it has a* $\kappa$-*filtration* $\{M_\nu : \nu < \kappa\}$ *consisting of free modules.*

PROOF. If $M$ has such a filtration, then $\mathcal{C} \overset{\text{def}}{=} \{M_\nu : \nu < \kappa\}$ satisfies properties (1) – (3) of Definition 1.1. Conversely, given $\mathcal{C}$ as in Definition 1.1, let $X$ be as in the proof of 1.4(i); then choose for each successor ordinal $\nu$, an element $M_\nu$ of $\mathcal{C}$ containing $\{x_\mu : \mu < \nu\}$; it exists by 1.1(2). For limit $\nu$, let $M_\nu = \bigcup_{\mu < \nu} M_\mu$, which belongs to $\mathcal{C}$ by 1.1(3). Since every element of $\mathcal{C}$ is free, we are done. □

Now we want to define an invariant of a $\leq \kappa$-generated $\kappa$-free module $M$, which we denote $\Gamma_\kappa(M)$, or simply $\Gamma(M)$. To say that this is an invariant means that if $M \cong N$, then $\Gamma(M) = \Gamma(N)$; it does *not* however mean the converse, i.e. $\Gamma(M)$ does not necessarily determine $M$ up to isomorphism (cf. VIII.1.2). (However, if $\Gamma(M) = \Gamma(F)$ where $F$ is free, then $M$ is also free: see 1.7). This invariant takes its values in $D(\kappa)$, the set of equivalence classes of $\mathcal{P}(\kappa)$ modulo the equivalence relation of identity on a cub: see II.4.4.

**1.6 Definition.** Let $\kappa$ be a regular cardinal and $M$ a $\kappa$-free $\leq \kappa$-generated module. Let $\{M_\nu : \nu < \kappa\}$ be a $\kappa$-filtration of $M$. Let

(1.6.1) $\quad$ $E = \{\nu < \kappa : \{\mu > \nu : M_\mu/M_\nu \text{ is not free}\}$ is stationary in $\kappa\}$.

Equivalently, $E = \{\nu < \kappa : M/M_\nu \text{ is not } \kappa\text{-free}\}$. Define $\Gamma_\kappa(M)$ to be $\tilde{E}$, the equivalence class of $E$, i.e.

$$\Gamma_\kappa(M) = \{X \subseteq \kappa : \exists \text{ cub } C \subseteq \kappa \text{ s.t. } X \cap C = E \cap C\}.$$

In order to see that $\Gamma_\kappa(M)$ is an invariant of $M$ we must show that its definition does not depend on the choice of the $\kappa$-filtration. So suppose that $\{M'_\nu : \nu < \kappa\}$ is another $\kappa$-filtration, and let

$$E' = \{\nu < \kappa : \{\mu > \nu : M'_\mu/M'_\nu \text{ is not free}\} \text{ is stationary in } \kappa\}.$$

Then by Lemma 1.4(ii), $C \overset{\text{def}}{=} \{\nu < \kappa: M_\nu = M'_\nu\}$ is a cub; moreover, if $\nu \in C$, then $\nu \in E$ if and only if $\nu \in E'$ (Exercise 1). Hence $E \cap C = E' \cap C$, so $\tilde{E} = \tilde{E}'$.

Since $\kappa$ will always be clear from context, we will usually write $\Gamma(M)$ instead of $\Gamma_\kappa(M)$.

Let us remark also that the definition of $\Gamma(M)$ can be simplified in the case when $R$ is a p.i.d. because a submodule of a free module is free. Namely, in this case $E = \{\nu < \kappa: \exists \mu > \nu$ s.t. $M_\mu/M_\nu$ is not free$\}$. This is because if $M_\mu/M_\nu$ is not free then for all $\tau \geq \mu$, $M_\tau/M_\nu$ is not free, and $\{\tau < \kappa: \tau \geq \mu\}$ is a stationary subset of $\kappa$.

**1.7 Proposition.** *Let $\kappa$ be a regular uncountable cardinal. If $M$ is a $\leq \kappa$-generated $\kappa$-free module, then $M$ is free if and only if $\Gamma(M) = 0 (= \tilde{\emptyset}$, the equivalence class of the empty set).*

PROOF. Suppose first that $M$ is free. Let $B = \{b_\nu: \nu < \alpha\}$ be a basis (where $\alpha \leq \kappa$), and let $M_\nu = \langle b_\mu: \mu < \nu \rangle$ for each $\nu < \kappa$. Then $\{M_\nu: \nu < \kappa\}$ is a $\kappa$-filtration such that each $M_\nu$ is a free summand of $M$. For this filtration, $E$ (defined as in (1.6.1)) is clearly empty.

Conversely, suppose that $\Gamma(M) = 0$. Fix a $\kappa$-filtration $\{M_\nu: \nu < \kappa\}$ of $M$ consisting of free modules, and let $E$ be as in (1.6.1). Then $E$ is not stationary, i.e., there is a cub $C$ such that $E \cap C = \emptyset$. So for each $\nu$ in $C$, $\{\mu > \nu: M_\mu/M_\nu$ is not free$\}$ is not stationary, i.e., there is a cub $C_\nu$ which has empty intersection with the latter set. Now let $D$ be the intersection of $C$ with the diagonal intersection of the $C_\nu$ (cf. II.2.14):

$$D = C \cap \{\mu < \kappa: \mu \in \cap\{C_\nu: \nu < \mu\}\}.$$

Notice that

(†)       for any $\nu < \mu$, if $\mu, \nu \in D$, then $M_\mu/M_\nu$ is free

since $\mu \in C_\nu$. Now $D$ is a cub (cf. II.4.10), so $\{M_\nu: \nu \in D\}$ is a continuous chain. Using (†) we can inductively construct for each

$\nu \in D$ a basis $B_\nu$ of $M_\nu$ so that for $\nu \leq \mu$, $B_\nu \subseteq B_\mu$. Then $\bigcup_{\nu < \kappa} B_\nu$ will be a basis of $M$. $\square$

We will use repeatedly the fact, proved by the argument at the end of the preceding proof, that if $\{M_\nu : \nu < \alpha\}$ is a continuous chain of free modules such that $M_{\nu+1}/M_\nu$ is free for all $\nu$, then $\bigcup_{\nu < \alpha} M_\nu$ is free.

When we construct (in Chapter VII) $\kappa$-free modules which are not free, our method will be to choose a stationary subset $E$ of $\kappa$ and then to construct inductively a $\kappa$-filtration so that $E$ is the set defined by (1.6.1), and so that all members of the filtration are free. Then by 1.7 the union of the filtration will not be free. In fact, our construction will yield modules which will satisfy another property, which we want to define now.

**1.8 Definition.** Let $\kappa$ be regular. An $R$-module $M$ is said to be *strongly $\kappa$-free* if there is a set $\mathcal{S}$ of $< \kappa$-generated free submodules of $M$ containing $\{0\}$ and such that: for any subset $S$ of $M$ of cardinality $< \kappa$, and any $N \in \mathcal{S}$, there exists $N' \in \mathcal{S}$ such that $N' \supseteq N \cup S$ and $N'/N$ is free.

This definition is motivated by infinitary logic: if $\kappa > |R|$ the strongly $\kappa$-free $R$-modules are those which are equivalent to a free $R$-module in the language $L_{\infty\kappa}$ (cf. Eklof 1974). The terminology "strongly $\kappa$-free" is misleading here, because although it is clear that a strongly $\kappa$-free module is $\kappa$-free in the weak sense, we do not know if it is always $\kappa$-free in the sense of Definition 1.1. The historical reason for the terminology is that it originated in the case of $R = \mathbb{Z}$; in that case, and for p.i.d.'s in general, we have the following characterization. Here, we say that $N$ is a $\kappa$-*pure* submodule of $M$ if $L/N$ is free whenever $L$ is a submodule of $M$ containing $N$ such that $L/N$ is $< \kappa$-generated, or, equivalently, $M/N$ is $\kappa$-free.

**1.9 Proposition.** *Let $R$ be a p.i.d., and $M$ an $R$-module. Then $M$ is strongly $\kappa$-free if and only if $M$ is $\kappa$-free and every subset of $M$ of cardinality $< \kappa$ is contained in a $< \kappa$-generated $\kappa$-pure submodule of $M$.*

PROOF. The sufficiency of the condition is clear since we can then take $S$ to consist of all $< \kappa$-generated $\kappa$-pure submodules of $M$. For the converse, let $S$ be as in Definition 1.8. Since every $< \kappa$-generated submodule is contained in a member of $S$, and every member of $S$ is free, it follows that every $< \kappa$-generated submodule is free. To complete the argument it suffices to show that every element of $S$ is $\kappa$-pure. Let $N \subseteq L$ be such that $N \in S$ and $L$ is $< \kappa$-generated. Then there exists $N' \in S$ containing $N \cup L$ such that $N'/N$ is free. But then $L/N$ is free because it is a submodule of $N'/N$. □

The next lemma is preparatory to giving a characterization of $\leq \kappa$-generated strongly $\kappa$-free modules in terms of $\kappa$-filtrations.

**1.10 Lemma.** *Let $\kappa$ be a regular uncountable cardinal. Let $M$ be strongly $\kappa$-free and let $S$ be as in Definition 1.8. Then for any elements $N$, $N'$ of $S$ such that $N \subseteq N'$, $N'/N$ is projective.*

PROOF. Suppose that $N$ and $N'$ belong to $S$ and $N \subseteq N'$. Choose $N_1 \in S$ such that $N_1 \supseteq N'$ and $N_1/N$ is free. Then $N$ is a direct summand of $N_1$ and hence of $N'$, so $N'/N$ is projective. □

**1.11 Proposition.** *Let $R$ be any ring and $\kappa$ a regular uncountable cardinal. A sufficient condition for a $\leq \kappa$-generated module $M$ to be strongly $\kappa$-free is that there is a $\kappa$-filtration $\{M_\nu : \nu < \kappa\}$ of $M$ such that for all $\mu < \nu$, $M_{\nu+1}$ and $M_{\nu+1}/M_{\mu+1}$ are free. If $\kappa > |R|$, then this condition is also necessary.*

PROOF. The sufficiency of the condition is clear since we can take $S$ to be the set of all members of the given filtration which are indexed by successor ordinals. Notice that in this case we obtain an $S$ with the additional property that whenever $N \subseteq N'$ are both in $S$ then $N'/N$ is free.

For necessity, assume that $M$ is strongly $\kappa$-free and $\kappa > |R|$. Let $X = \{x_\nu : \nu < \kappa\}$ be a generating set for $M$. Given $S$ as in Definition 1.8, our aim is to construct a $\kappa$-filtration of $M$ so that for all $\mu < \nu$, $M_{\nu+1} \in S$ and $M_{\nu+1}/M_{\mu+1}$ is free. Suppose that we have done this for all $\mu \leq \nu$ for some $\nu < \kappa$. Let $N' \in S$ which contains $\bigcup_{\mu \le \nu} M_\mu$. By Lemma 1.10, for each $\mu < \nu$, $N'/M_{\mu+1}$ is projective, so by Eilenberg's trick (cf. I.2.3), $N'/M_{\mu+1} \oplus F_\mu$ is free for some free module $F_\mu$ on a basis of size $\lambda_\mu$ for some $\lambda_\mu < \kappa$. Now choose $M_{\nu+1}$ to be an element of $\mathcal{S}$ containing $N' \cup \{x_\nu\}$ and such that $M_{\nu+1}/N' \cong R^{(\lambda)}$ for some $\lambda \ge \sup\{\lambda_\mu : \mu < \nu\}$. (Here we use the hypothesis on $\kappa$.) Then

$$M_{\nu+1}/M_{\mu+1} \cong N'/M_{\mu+1} \oplus M_{\nu+1}/N'$$

is free for each $\mu < \nu$. $\square$

Note that we do not say anything about whether the limit stages in the above filtration are free; this is closely related to the fact that we do not know if a strongly $\kappa$-free module is $\kappa$-free. However at limit stages of cofinality $\omega$, the submodules are free.

We conclude this section with a discussion of the existence of $\kappa$-free modules when $R$ is a left perfect ring. It turns out that the analysis is particularly simple in this case. (This includes the case of $R$ a division ring, or, more generally, a semisimple ring.) We shall leave the more complicated — and more interesting — case of non-left-perfect rings — such as $\mathbb{Z}$ — to Chapter VII. We shall not develop the theory of left perfect rings here; we will simply cite some results together with sources for their proofs. The definition of perfect rings is due to H. Bass, who proved the following characterization theorem. A ring satisfying any — hence all — of the equivalent conditions in the following theorem is called *left perfect*. (Recall that "module" means *left R-module*.)

**1.12 Theorem.** *Let $R$ be a ring and let $J$ be its Jacobson radical. The following are equivalent.*

(a) *Every $R$-module has a projective cover;*

(b) *every flat $R$-module is projective;*

(c) *every direct limit of projective $R$-modules is projective;*

(d) *$R$ satisfies the descending chain condition on principal right ideals;*

(e) *$R/J$ is semisimple and for every sequence $\{a_n : n \in \omega\}$ in $J$ there exists $m \in \omega$ such that $a_0 \cdots a_m = 0$;*

(f) $R/J$ is semisimple and every non-zero $R$-module contains a maximal proper submodule;

(g) $R$ has no infinite set of orthogonal idempotents, and every nonzero right $R$-module contains a minimal submodule. $\square$

We will mainly use conditions (c) and (d). The key fact about left perfect rings which is relevant for our purposes is the following.

**1.13 Theorem.** *Let $R$ be left perfect. Then there is a finite set $\{P_1, \ldots, P_n\}$ of projective indecomposable $R$-modules satisfying:*

(i) *the mapping which associates $P_k$ with $P_k/JP_k$ is a one-one correspondence of the projective indecomposable $R$-modules with the simple $R/JR$-modules;*

(ii) *every projective $R$-module is uniquely a direct sum of the modules $P_1, \ldots, P_n$.* $\square$

Note that $n = 1$ if and only if $R/J$ is a simple ring. Note also that each of the projective indecomposable modules $P_k$ is cyclic; this follows, for example, from the fact that $P_k$ is the projective cover of a simple module. Moreover, $R$ is a *finite* direct sum of the $P_k$'s since 1 belongs to a finite direct sum.

**1.14 Example.** $\mathbb{Z}/6\mathbb{Z}$ and $\mathbb{Z}/4\mathbb{Z}$ are left perfect rings. (Since they are finite, they certainly satisfy 1.12 (d).) Every free $\mathbb{Z}/6\mathbb{Z}$ module is of the form $\mathbb{Z}/2\mathbb{Z}^{(\alpha)} \oplus \mathbb{Z}/3\mathbb{Z}^{(\alpha)}$, so $\{\mathbb{Z}/2\mathbb{Z}, \mathbb{Z}/3\mathbb{Z}\}$ is the set of projective indecomposables discussed in 1.13. For any infinite $\lambda$ the $\mathbb{Z}/6\mathbb{Z}$-module

$$\mathbb{Z}/2\mathbb{Z}^{(\lambda)} \oplus \mathbb{Z}/3\mathbb{Z}^{(\lambda^+)}$$

is $\lambda^+$-free and of cardinality $\lambda^+$ but is not free. Every $\mathbb{Z}/4\mathbb{Z}$-module $M$ is of the form $\mathbb{Z}/2\mathbb{Z}^{(\alpha)} \oplus \mathbb{Z}/4\mathbb{Z}^{(\beta)}$, and is free if and only if $\alpha = 0$ if and only if $M$ is projective if and only if $M$ has the property that every finite subset is contained in a free submodule. Here there is only one projective indecomposable, namely $\mathbb{Z}/4\mathbb{Z}$.

We will say that a module $M$ is $\aleph_0$-*free* if every finitely-generated subset of $M$ is contained in a free submodule.

**1.15 Theorem.** *Let $R$ be a left perfect ring.*

(i) *If $R/J$ is simple, then every module which is $\aleph_0$-free is free.*

(ii) *If $R/J$ is not simple, then for every infinite $\lambda$ there is a $\leq \lambda^+$-generated $M$ which is $\lambda^+$-free and not free.*

PROOF. Let $\{P_1, \ldots, P_n\}$ be as in Theorem 1.13.(i) Assume $n = 1$. If $M$ is $\aleph_0$-free then every finite subset of $M$ is contained in a free submodule, so $M$ is the direct limit of projective submodules, and hence is projective by 1.12(c). Thus $M$ is isomorphic to $P_1^{(\alpha)}$ for some cardinal $\alpha$. If $\alpha$ is infinite then $M$ is clearly $\cong R^{(\alpha)}$. If $\alpha$ is finite then $M$ $\aleph_0$-free implies $M$ is free.

(ii) Assume $n \geq 2$. Let $M = P_1^{(\lambda)} \oplus P_2^{(\lambda^+)} \ldots \oplus P_n^{(\lambda^+)}$. Then $M$ is not free, but it is $\lambda^+$-free (in the sense of 1.1); indeed, we can take $C$ to be the set of all direct sums of $\lambda$ copies of each of the $P_k$ in the above decomposition. $\square$

The $M$ we constructed in part (ii) is not strongly $\lambda^+$-free. In fact, there is no strongly $\lambda^+$-free module which is $\leq \lambda^+$-generated, but there is one which is $\leq \lambda^{++}$-generated; also, if $\kappa$ is a limit cardinal, then every $\leq \kappa$-generated $\kappa$-free module (over a left perfect $R$) is free. (See Exercise 2.)

## §2. $\aleph_1$-free abelian groups

As the title of the section indicates, we shall concentrate in this section on abelian groups, which we shall refer to simply as groups; most of what we do generalizes to modules over a p.i.d. We shall give a necessary and sufficient condition, Pontryagin's Criterion, for a group to be $\aleph_1$-free. We shall also introduce a series of strengthenings of the condition of being $\aleph_1$-free. (Recall that a group is $\aleph_1$-free if every countable subgroup is free.) In all of this, the notion of a pure subgroup plays an important role; we shall give the definition for a general ring and then immediately specialize to groups.

**2.1 Definition.** A submodule $N$ of an $R$-module $M$ is called a *pure* submodule if every finite system of equations

$$\sum_{j=1}^{m} r_{ij}x_j = a_i \qquad (i = 1,\ldots,n),$$

where $a_1,\ldots,a_n$ belong to $N$, has a solution in $N$ whenever it has a solution in $M$.

It is well known that if $R$ is a p.i.d. this notion reduces to that of relative divisibility, that is, $N$ is a pure submodule of $M$ if and only if for every $r \in R$ and $a \in N$, $rx = a$ has a solution in $N$ whenever it has one in $M$. (In fact, the notions coincide if $R$ is a Prüfer domain: see Warfield 1969.)

Given a torsion-free group $M$ and a subgroup $N$ of $M$, let $N_* = \{a \in M : ra \in N \text{ for some } r \neq 0 \text{ in } \mathbb{Z}\}$; $N_*$ is called the *pure-closure* of $N$, and is the smallest pure subgroup of $M$ containing $N$. If $S$ is a subset of $M$, $\langle S \rangle_*$ is then the pure-closure of $\langle S \rangle$, the subgroup generated by $S$; it is sometimes referred to as the pure-closure of $S$.

The following facts are proved quite routinely. (See Fuchs 1970, Chapter V. "$N$ is pure in $M$" is a short-hand way of saying that $N$ is a pure subgroup of M.)

**2.2 Lemma.**
   (i) *If $P$ is pure in $N$ and $N$ is pure in $M$, then $P$ is pure in $M$;*
   (ii) *if $\{N_\nu : \nu < \rho\}$ is a continuous chain of pure subgroups of $M$, then $\bigcup_{\nu < \rho} N_\nu$ is pure in $M$;*
   (iii) *if $N$ is a direct summand of $M$, then $N$ is pure in $M$;*
   (iv) *if $N$ is a subgroup of $M$ such that $M/N$ is torsion-free, then $N$ is pure in $M$;*
   (v) *if $N$ is a pure subgroup of a torsion-free group $M$, then $M/N$ is torsion-free;*
   (vi) *if $P \subseteq N \subseteq M$, where $P$ is pure in $M$ and $N/P$ is pure in $M/P$, then $N$ is pure in $M$.* □

**2.3 Theorem. (Pontryagin's Criterion)** *For any group $M$, the following are equivalent.*
   (a) *$M$ is $\aleph_1$-free;*

(b) *M is torsion-free and every finite subset of M is contained in a finitely-generated pure subgroup of M;*

(c) *every finite subset of M is contained in a pure free subgroup of M;*

(d) *every finite rank subgroup of M is free.*

PROOF. (a) $\Rightarrow$ (b): $M$ is torsion-free since otherwise $M$ contains a finite torsion group, and thus cannot be $\aleph_1$-free. Let $S$ be a finite subset of $M$. If $\langle S \rangle_*$ is not finitely-generated, then there is a countably generated subgroup $N$ of $\langle S \rangle_*$ containing $S$ which is not finitely-generated. But then $N$ is not free, since it has finite rank but is not finitely-generated. This contradicts (a).

(b) $\Rightarrow$ (c): This is clear since, by the Fundamental Theorem of finitely-generated abelian groups, every finitely-generated torsion-free group is free.

(c) $\Rightarrow$ (d): If $N$ is a finite rank subgroup of $M$ and $S$ is a maximal independent subset of $N$, then $N$ is contained in any pure subgroup containing $S$, and hence by (c) in a free one; thus $N$ is free.

(d) $\Rightarrow$ (a): Given a countably generated subgroup $A$ of $M$ we must show that $A$ is free. Let $\{a_n : n \in \omega\}$ be a set of generators of $A$, and for each $n$ let $N_n = \langle \{a_m : m < n\} \rangle_*$. Then $A \subseteq \bigcup_{n \in \omega} N_n$, so it is enough to prove that $\bigcup_{n \in \omega} N_n$ is free. But by (d) each $N_n$ is free and hence finitely-generated; so $N_{n+1}/N_n$ is finitely-generated. Now $N_{n+1}/N_n$ is also torsion-free since $N_n$ is pure in $M$; therefore it is free. It follows (as in the proof of 1.7) that $\bigcup_{n \in \omega} N_n$ is free. $\square$

**2.4 Definition.** Let $\kappa$ be an infinite cardinal. A group $M$ is said to be *$\kappa$-separable* if every subset of $M$ of cardinality $< \kappa$ is contained in a free direct summand of $M$. An $\aleph_0$-separable group is sometimes also called *locally free*.

In this section we shall concentrate on $\aleph_0$-separable groups, which we shall call simply *separable*. (Later, in Chapter VIII, we shall have a lot to say about $\aleph_1$-separable groups.) By 2.3, a separable group is $\aleph_1$-free, and a free group is $\kappa$-separable for every $\kappa$. Also, a $\kappa$-separable group is strongly $\kappa$-free.

The reader should be warned that in many places in the literature, the term "separable" is used in a more general sense: for a group which has the property that every finite subset is contained in a *completely decomposable* summand; our separable groups would then be called $\aleph_1$-free separable groups.

**2.5 Proposition.** *A group M is separable if and only if every finite rank pure subgroup is a free summand of M.*

PROOF. The sufficiency of the condition is obvious. For necessity, if $N$ is a finite rank pure subgroup of $M$, it is free and finitely-generated by 2.3 and contained in a free summand, $F$, by hypothesis. We can assume that $F$ is finitely-generated, so $F/N$ is free since $N$ is pure in $F$. Therefore $N$ is a summand of $F$, and hence of $M$. □

**2.6 Corollary.** *A pure subgroup of a separable group is separable.*

PROOF. Let $M$ be separable and $N$ a pure subgroup of $M$. Let $F$ be a finite rank pure subgroup of $N$. Then $F$ is a finite rank pure subgroup of $M$, so it is a free summand of $M$, and hence of $N$. □

**2.7 Lemma.** *M is separable if and only if every element of M is a member of a cyclic summand of M.*

PROOF. Since necessity is obvious from 2.5, we are left to prove, by induction on $n$, that every pure subgroup of $M$ of rank $n$ is free and a summand of $M$. The case $n = 1$ is immediate from the assumption, so assume that result is proved for some $n \geq 1$ and consider a pure subgroup $N$ of $M$ of rank $n + 1$. By assumption, there is an element $0 \neq b \in N$ such that $\langle b \rangle$ is a summand of $M$: $M = \langle b \rangle \oplus L$. Then $N = \langle b \rangle \oplus (L \cap N)$ and $L \cap N$ is of rank $n$; so by inductive hypothesis, $L \cap N$ is free and a summand of $M$ and thus of $L$; it follows that $N$ is free and a summand of $M$. □

**2.8 Theorem.** *For any infinite cardinal $\kappa$, $\mathbf{Z}^\kappa$ is separable, but not strongly $\aleph_1$-free (and hence not free).*

PROOF. To prove that $\mathbf{Z}^\kappa$ is separable, it is enough to prove, by Lemma 2.7, that every element of $\mathbf{Z}^\kappa$ is a member of a cyclic summand of $P$. Let $a \in \mathbf{Z}^\kappa$, let $k = \gcd\{a(n): n \in \omega\}$, and let $b = k^{-1}a$.

Then $1 = \gcd\{b(n): n \in \omega\}$ and there is a finite subset $J$ of $\kappa$ so that $1 = \gcd\{b(n): n \in J\}$. Let $\mathbf{Z}^J$ denote the subgroup of $\mathbf{Z}^\kappa$ consisting of the elements which are identically zero on $\kappa \setminus J$. Then $\mathbf{Z}^J$ is a free group of finite rank, and $b\!\restriction\!J$ generates a pure cyclic subgroup of $\mathbf{Z}^J$, so by 2.5, it generates a summand of $\mathbf{Z}^J$ and hence $\mathbf{Z}^J$ has a basis of the form $\{b\!\restriction\!J, x_1\!\restriction\!J, \ldots, x_n\!\restriction\!J\}$, for some $x_1, \ldots, x_n$ in $\mathbf{Z}^\kappa$. Let $B$ be the subgroup of $\mathbf{Z}^\kappa$ with basis $\{b, x_1, \ldots, x_n\}$, and let

$$A = \{y \in \mathbf{Z}^\kappa : y(j) = 0 \text{ for all } j \in J\}.$$

Then it is easy to see that $\mathbf{Z}^\kappa = A \oplus B$. Moreover, $\langle b \rangle$ generates a summand of $B$, and hence of $\mathbf{Z}^\kappa$, and $a \in \langle b \rangle$.

To show that $\mathbf{Z}^\kappa$ is not strongly $\aleph_1$-free, we will show that $\mathbf{Z}^{(\omega)}$ is not contained in any countable subgroup $H$ of $\mathbf{Z}^\kappa$ such that $\mathbf{Z}^\kappa/H$ is $\aleph_1$-free (cf. 1.9). Here we identify $\mathbf{Z}^\omega$ with the subgroup consisting of all $b \in \mathbf{Z}^\kappa$ such that $b(\nu) = 0$ for all $\nu \geq \omega$. Suppose $\mathbf{Z}^{(\omega)}$ is contained in a countable group $H$ and $\mathbf{Z}^\kappa/H$ is $\aleph_1$-free. Consider any prime $p$. Then the $p$-adic closure of $\mathbf{Z}^{(\omega)}$ in $\mathbf{Z}^\kappa$ is uncountable, but must be contained in $H$. More explicitly, there exists $b \in \mathbf{Z}^\omega \setminus H$ so that for all $n$, $b(n) \in p^n \mathbf{Z}$. Let $L = \langle H, b \rangle_*$, a countable subgroup of $\mathbf{Z}^\kappa$. Then $b + H$ is a non-zero element of $L/H$ which is divisible by all powers of $p$ (since $H$ contains $\mathbf{Z}^{(\omega)}$), so $L/H$ is not free. $\square$

As we said, a separable group is $\aleph_1$-free; in fact even more holds:

**2.9 Corollary.** *A separable group is torsionless. A torsionless group is $\aleph_1$-free.*

PROOF. If $M$ is separable we must show that for any $0 \neq a \in M$, there exists $f \in M^*$ such that $f(a) \neq 0$. But $M = \langle a \rangle_* \oplus D$ for some $D$, and $\langle a \rangle_* \cong \mathbf{Z}$, so projection on the first coordinate is the desired $f$. Any torsionless group is $\aleph_1$-free because it is a subgroup of $\mathbf{Z}^\kappa$ for some $\kappa$ (cf. I.2.1), and by 2.8, $\mathbf{Z}^\kappa$ is $\aleph_1$-free. $\square$

Recall that in I.2.2 we saw that a dual group is torsionless; now we can prove that even more holds.

**2.10 Corollary.** *For any group $M$, $M^*(= \mathrm{Hom}_{\mathbf{Z}}(M, \mathbf{Z}))$ is separable.*

PROOF. By Corollary 2.6, it suffices to show that $M^*$ is a pure subgroup of a separable group. Note that $M^*$ is a subgroup of the separable group $\mathbf{Z}^M$. To see that it is a pure subgroup, suppose $g \in M^*$ and $k$ divides $g$ in $\mathbf{Z}^M$. Then $k$ divides $g(a)$ for all $a \in M$; so $kh = g$ where $h \in M^*$ is defined by $h(a) = g(a)/k$ for all $a \in M$. □

We now prove an important result of Chase.

**2.11 Lemma.** *For any torsionless group $M$, if $\mathrm{Ext}(M, \mathbf{Z})$ is torsion-free, then $M$ is separable.*

PROOF. Let $\sigma \colon M \to M^{**}$ be the canonical map (cf. §I.1); since $M$ is torsionless, $\sigma[M]$ is isomorphic to $M$. Hence, by 2.6 and 2.10, it suffices to prove that $\sigma[M]$ is a pure subgroup of $M^{**}$. Let $M^{\#}$ be the pure-closure of $\sigma[M]$ in $M^{**}$. Then $C \overset{\mathrm{def}}{=} M^{\#}/\sigma[M]$ is a torsion group. If we can show that $C = 0$, then we are done; so it suffices to show that for every prime $p$, $C[p]$ $(\overset{\mathrm{def}}{=} \{c \in C \colon pc = 0\})$ is zero.

Since $\mathrm{Ext}(G, \mathbf{Z}) \neq 0$ for any non-zero torsion group, if we can show that $\mathrm{Ext}(C[p], \mathbf{Z}) = 0$, then we are done. Now the inclusion of $C[p]$ into $C$ induces an epimorphism of $\mathrm{Ext}(C, \mathbf{Z})$ onto $\mathrm{Ext}(C[p], \mathbf{Z})$ (cf. I.1.1). Since $p\,\mathrm{Ext}(C[p], \mathbf{Z}) = 0$, $p\,\mathrm{Ext}(C, \mathbf{Z})$ is contained in the kernel of this epimorphism. Hence there is an epimorphism of $\mathrm{Ext}(C, \mathbf{Z})/p\,\mathrm{Ext}(C, \mathbf{Z})$ onto $\mathrm{Ext}(C[p], \mathbf{Z})$. So it suffices to show that $\mathrm{Ext}(C, \mathbf{Z})/p\,\mathrm{Ext}(C, \mathbf{Z}) = 0$. The short exact sequence

$$0 \to \sigma[M] \overset{\mu}{\longrightarrow} M^{\#} \to C \to 0$$

(where $\mu$ is an inclusion) induces the exact sequence

$$(M^{\#})^* \overset{\mu^*}{\longrightarrow} \sigma[M]^* \to \mathrm{Ext}(C, \mathbf{Z}) \to \mathrm{Ext}(M^{\#}, \mathbf{Z}) \to \mathrm{Ext}(\sigma[M], \mathbf{Z}) \to 0.$$

By Exercise 7, $\mu^*$ is surjective. Thus we can regard $\mathrm{Ext}(C, \mathbf{Z})$ as a subgroup of $\mathrm{Ext}(M^{\#}, \mathbf{Z})$, so

$$\mathrm{Ext}(M^{\#}, \mathbf{Z})/\mathrm{Ext}(C, \mathbf{Z}) \cong \mathrm{Ext}(\sigma[M], \mathbf{Z}) \cong \mathrm{Ext}(M, \mathbf{Z}).$$

Since $M^\#$ is torsion-free, $\text{Ext}(M^\#, \mathbf{Z})$ is divisible (see XII.1.1 for a proof of this standard fact). If there were a non-zero element

$$x + p \, \text{Ext}(C, \mathbf{Z}) \in \text{Ext}(C, \mathbf{Z})/p \, \text{Ext}(C, \mathbf{Z}),$$

we could choose $y \in \text{Ext}(M^\#, \mathbf{Z})$ such that $py = x$ and obtain an element $y + \text{Ext}(C, \mathbf{Z})$ of $\text{Ext}(M^\#, \mathbf{Z})/\text{Ext}(C, \mathbf{Z})$ of order $p$, which is impossible since $\text{Ext}(M, \mathbf{Z})$ is torsion-free. $\square$

In fact, in the next result we shall see that $\text{Ext}(M, \mathbf{Z})$ torsion-free implies that $M$ is torsionless, and give some conditions equivalent to this hypothesis. First we need some more definitions.

**2.12 Definition.** Let $\kappa$ be an infinite cardinal. A group $M$ is called $\kappa$-*coseparable* if it is $(\kappa + \aleph_1)$-free and every subgroup $N$ of $M$ such that $M/N$ is $< \kappa$-generated contains a direct summand $D$ of $M$ such that $M/D$ is $< \kappa$-generated.

A group $M$ is called *finitely-projective* if for every epimorphism $\varphi \colon F \to H$, where $F$ is a finite rank free group, and every homomorphism $\psi \colon M \to H$, there is a homomorphism $\eta \colon M \to F$ such that $\varphi \circ \eta = \psi$. (See also Exercise 9.)

In this section we shall deal only with $\aleph_0$-coseparable groups, which we shall call *coseparable*. (In Chapter XII we shall show that it is not decidable in ZFC whether all $\aleph_1$-coseparable groups are free.) Notice that a dual group need not be coseparable: in fact $\mathbf{Z}^\omega (= \mathbf{Z}^{(\omega)^*})$ is not coseparable (cf. Exercise 16).

**2.13 Theorem.** *For any reduced group $M$, the following are equivalent:*

(1) $\text{Ext}(M, \mathbf{Z})$ *is torsion-free;*

(2) $M$ *is finitely projective;*

(3) $M$ *is coseparable;*

(4) $M$ *is separable and coseparable.*

PROOF. (1) $\Rightarrow$ (2): Note first that $M$ must be torsion-free, because otherwise $\text{Ext}(M, \mathbf{Z})$ has a direct summand isomorphic to

$\mathrm{Ext}(\mathbf{Z}/p\mathbf{Z}, \mathbf{Z})$, which has order $p$. Suppose we are given

$$
\begin{array}{c}
M \\
\downarrow{\psi} \\
0 \to K \to F \xrightarrow{\phi} H \to 0
\end{array}
$$

where $F$ is a finite rank free group and the row is exact. Then $H$ is finitely-generated, so $H = B \oplus P$ where $B$ is finite and $P$ is free. Hence $M = \psi^{-1}[B] \oplus P'$ where $P'$ is free. Since there is no problem defining $\eta$ on $P'$, we can assume, without loss of generality, that $P = 0$ (i.e. $H$ is finite). Now the short exact sequence induces the exact sequence

$$
\mathrm{Hom}(M, F) \xrightarrow{\varphi} \mathrm{Hom}(M, H) \xrightarrow{\delta} \mathrm{Ext}(M, K) \to \mathrm{Ext}(M, F).
$$

The last term is torsion-free since it equals $\mathrm{Ext}(M, \mathbf{Z})^{(n)}$ for some $n$, so $\mathrm{im}(\delta)$ is a pure subgroup of $\mathrm{Ext}(M, K)$. Now $\mathrm{Ext}(M, K)$ is divisible since $M$ is torsion-free, so $\mathrm{im}(\delta)$ is divisible. But $\mathrm{Hom}(M, H)$ is of bounded order, since $H$ is finite, so $\mathrm{im}(\delta)$ must equal $\{0\}$. Therefore $\varphi^*$ is surjective, so there exists $\eta \in \mathrm{Hom}(M, F)$ such that $\varphi^*(\eta) = \psi$.

(2) $\Rightarrow$ (3): First we show that $M$ is $\aleph_1$-free; in fact we prove that $M$ is torsionless. If $a$ is a non-zero element of $M$, we must show that there exists $\eta \in M^*$ such that $\eta(a) \neq 0$. Since $M$ is reduced, there exists $n \in \mathbf{Z} \setminus \{0\}$ such that $a \notin nM$. Let $L$ be a subgroup of $M$ maximal with respect to the property that $nM \subseteq L$ and $a \notin L$. Then $M/L$ is cyclic: otherwise, since it is a direct sum of cyclics — because it has bounded order — we could contradict the maximality of $L$. Therefore there is an epimorphism $\varphi: \mathbf{Z} \to M/L$. Since $M$ is finitely projective, there exists $\eta: M \to \mathbf{Z}$ such that $\varphi \circ \eta$ is the canonical map: $M \to M/L$. Clearly $\eta(a) \neq 0$, since $a + L \neq 0$.

To complete the proof that $M$ is coseparable, consider a subgroup $N$ of $M$ such that $M/N$ is finitely-generated. Let $F$ be a finite rank free group such that there is an epimorphism $\varphi: F \to M/N$. Since $M$ is finitely projective, there is a homomorphism $\eta: M \to F$ such that $\varphi \circ \eta$ equals the canonical map: $M \to M/N$. But then since im($\eta$) is free and finitely-generated, $D \stackrel{\text{def}}{=} \ker(\eta)$ is a direct summand of $M$ contained in $N$, and such that $M/D$ is finitely-generated.

(3) $\Rightarrow$ (1): To see that $\text{Ext}(M, \mathbf{Z})$ is torsion-free, fix $n \neq 0$ in $\mathbf{Z}$ and consider the short exact sequence

$$0 \to \mathbf{Z} \xrightarrow{n} \mathbf{Z} \xrightarrow{\pi} \mathbf{Z}/n\mathbf{Z} \to 0.$$

This sequence induces the exact sequence

$$\text{Hom}(M, \mathbf{Z}) \xrightarrow{\pi^*} \text{Hom}(M, \mathbf{Z}/n\mathbf{Z}) \xrightarrow{\delta} \text{Ext}(M, \mathbf{Z}) \xrightarrow{n^*} \text{Ext}(M, \mathbf{Z})$$

(cf. I.1.1). It is enough to prove that $\pi^*$ is surjective, for then $\delta \equiv 0$, and hence $n^*$ is injective, or multiplication by $n$ in $\text{Ext}(M, \mathbf{Z})$ is one-one. So consider an element $\varphi$ of $\text{Hom}(M, \mathbf{Z}/n\mathbf{Z})$. Then $M/\ker(\varphi)$ is cyclic, so, since $M$ is coseparable, we can write $M = D \oplus F$ where $D \subseteq \ker(\varphi)$ and $F$ is finitely-generated and free. Then it is clear that we can define $\psi \colon M \to \mathbf{Z}$ so that $\pi \circ \psi = \varphi$, i.e. $\varphi = \pi^*(\psi)$.

We have now shown that (1), (2), and (3) are equivalent. If (1) holds, then by the proof of (2) $\Rightarrow$ (3), $M$ is torsionless, so (1) $\Rightarrow$ (4) by Lemma 2.11. Since (4) $\Rightarrow$ (3) is obvious, we are done. $\square$

We now have the following implications (where, from now on, we use the term "coseparable" to mean "reduced coseparable"):

free $\Rightarrow$ coseparable $\Rightarrow$ separable $\Rightarrow$ torsionless $\Rightarrow$ $\aleph_1$-free.

By Exercises 11-12 and 16-18, the last three implications are not reversible. By XIII.1.2, it is consistent that even strongly $\aleph_1$-free of cardinality $\aleph_1$ does not imply torsionless. By XII.2.11, assuming CH there is a coseparable group which is not free. Recently Mekler and Shelah have shown that assuming the consistency of a certain large cardinal, it is consistent that coseparable implies free.

We have one more notion to add to our chain of implications.

**2.14 Definition.** A group $M$ is called *hereditarily separable* if $M$ and all of its subgroups are separable.

There are separable groups which are not hereditarily separable: $\mathbf{Z}^\omega$ is separable, but by Exercise 12, $\mathbf{Z}^\omega$ is not hereditarily separable.

**2.15 Proposition.** *An hereditarily separable group is coseparable.*

PROOF. Let $M$ be hereditarily separable and let $N$ be a subgroup such that $M/N$ is finitely-generated; say $M = N + K$ where $K$ is a finitely-generated group. Since $N$ is separable, we can write $N = D \oplus C$, where $C$ is finitely-generated and contains $K \cap N$. Let $F = K + C$. Clearly $M = D + F$ and $F$ is finitely-generated, so it suffices to prove that $D \cap F = \{0\}$. But if $d = k + c$ (where the elements belong to the obvious subgroups), then $k = d - c$ belongs to $K \cap N$, and hence to $C$; thus $d = 0$ since $D \cap C = \{0\}$. $\square$

By 2.13 and 2.15, hereditarily separable equals "hereditarily coseparable." Now our chain of implications looks like this:

free $\Rightarrow$ hereditarily separable $\Rightarrow$ coseparable $\Rightarrow$
separable $\Rightarrow$ torsionless $\Rightarrow$ $\aleph_1$-free.

Assuming CH, the second implication is not reversible. (See Corollary XII.2.12.) Whether or not the first implication is reversible is not decidable in ZFC. (See section VII.4.)

We conclude this section with a result of a different sort, due to Hill.

**2.16 Theorem.** *Let $M$ be a group such that $M = \bigcup_{n \in \omega} F_n$ where $F_n$ is a pure free subgroup of $F_{n+1}$ for all $n$. Then $M$ is free.*

First a lemma, before we prove 2.16.

**2.17 Lemma.** *Let $\{F_j : j \in \omega\}$ be a family of pure free subgroups of a group $M$ such that $F_0 = \{0\}$, and let $X_j$ be a basis of $F_j$. Suppose we are given a subgroup $C$ of $M$ such that for all $j$, $C + F_j$ is pure in $M$ and $C \cap F_j$ is generated by $C \cap X_j$. Given any countable subgroup $N$ of $M$ there is a subgroup $C'$ containing $C + N$ such that $C'/C$ is countable and for all $j$ $C' + F_j$ is pure in $M$ and $C' \cap F_j$ is generated by $C' \cap X_j$.*

PROOF. Let $\varphi : \omega \to \omega$ such that for all $k$, $\{n : n$ is even and $\varphi(n) = k\}$ and $\{n : n$ is odd and $\varphi(n) = k\}$ are both infinite. We shall define a chain of subgroups $A_n$ such that for all $n$, $A_n/C$ is countable. Let

$A_0 = C + N$. Now suppose that $A_{n-1}$ has been defined for some $n \geq 1$.

*Case 1: n is even.* Say $\varphi(n) = k$. There is a countable group $G_n$ such that $A_{n-1} = C + G_n$. For each $g \in G_n$, if there exists $c_g$ such that $c_g + g \in F_k$, fix one such $c_g$ and choose a finite subset $S_g$ of $X_k$ such that $c_g + g \in \langle S_g \rangle$. Let $Y_n = \bigcup_g S_g \cup (C \cap X_k)$; then $A_{n-1} \cap F_k \subseteq \langle Y_n \rangle$ and $(A_{n-1} + \langle Y_n \rangle)/A_{n-1}$ is countable. Let $A_n = A_{n-1} + \langle Y_n \rangle$.

*Case 2: n is odd.* Say $\varphi(n) = j$. Let $m$ be the largest odd integer $< n$ such that $\varphi(m) = j$, or $m = 0$ if none exists. By induction, $A_m + F_j$ will be pure in $M$. Choose $A_n$ containing $A_{n-1}$ such that $(A_n + F_j)/(A_m + F_j)$ is pure in $M/(A_m + F_j)$.

Let $C' = \bigcup_{n \in \omega} A_n$. Clearly $C'$ contains $C + N$ and $C'/C$ is countable. Moreover, using Case 2 and Lemma 2.2(vi), we can prove by induction that for all $j$ and all odd $n$ such that $\varphi(n) = j$, $A_n + F_j$ is pure in $M$, so $C' + F_j$ is pure in $M$ by 2.2(ii). Finally, notice that for all $k$, $C' \cap F_k$ is generated by $\bigcup\{Y_n : n$ is even, $\varphi(n) = k\} = C' \cap X_k$. $\square$

**Proof of 2.16.** For each $n \in \omega$ let $X_n$ be a basis of $F_n$. Let $\{a_\nu : \nu < \lambda\}$ be an enumeration of $M$. We are going to define by induction on $\nu$ a continuous chain $\{C_\nu : \nu < \lambda\}$ of pure subgroups of $M$ whose union is $M$ and is such that:

(1) $C_0 = \{0\}$;
(2) $C_\nu + F_n$ is pure in $M$;
(3) $C_{\nu+1}/C_\nu$ is countable; and
(4) $C_\nu \cap F_n$ is generated by $C_\nu \cap X_n$.

In fact, if $C_\nu$ has been constructed for some $\nu$ we apply Lemma 2.17 — with $C = C_\nu$ and $N = \langle a_\nu \rangle$ — to obtain $C_{\nu+1}$ $(= C')$. Of course, at limit stages we must take the union of the previous stages in order to have a continuous chain.

We claim that $C_{\nu+1}/C_\nu$ is free for all $\nu$. If so, we are done, for then $M$ is free. We use Pontryagin's Criterion, 2.3. Any finite subset $S$ of $C_{\nu+1}$ is contained in $F_m$ for some $m$. Let $P = C_{\nu+1} \cap (C_\nu + F_m)$; then $P$ is pure in $M$ since it is the intersection of pure subgroups of a torsion-free group. Hence $P/C_\nu$ is a pure subgroup of $C_{\nu+1}/C_\nu$ containing the image of $S$. Moreover, $P/C_\nu$ is contained in $(C_\nu + F_m)/C_\nu$, which is isomorphic to $F_m/(C_\nu \cap F_m)$, and thus free by (4). Hence we have shown that every finite subset of $C_{\nu+1}/C_\nu$ is contained in a pure free subgroup; so by 2.3 $C_{\nu+1}/C_\nu$ is free. $\square$

The following corollary is a precursor of the kind of result — a "compactness" result — that we will deal with in the next section. In that section, it will be strengthened considerably.

**2.18 Corollary.** *Suppose $\lambda$ is an uncountable cardinal of cofinality $\aleph_0$ and $M$ is a $\lambda$-free group of cardinality $\lambda$. Then $M$ is free.*

PROOF. By hypotheses, we can write $M$ as the union of a chain, $\{M_n : n \in \omega\}$, of pure subgroups of cardinality $< \lambda$, which are free, so 2.16 applies. $\square$

## §3. Compactness results

We say that $\kappa$ *has the compactness property*, for $R$, if every $\leq \kappa$-generated $\kappa$-free $R$-module is free. We encountered this property in Theorem II.3.10 where it was proved that if $\kappa$ is an $L_{\lambda\omega}$-compact cardinal, then $M$ is free whenever $M$ is a $\kappa$-free $R$-module (of arbitrary cardinality) and $R$ is a p.i.d. of cardinality $< \lambda$. This result requires some hypothesis on $R$ such as every submodule of a free $R$-module is free; for example, there is a non-free $\lambda^+$-free $\mathbf{Z}/6\mathbf{Z}$-module for every $\lambda$.

In section II.2 we defined weakly compact cardinals. Now we are going to prove that such cardinals have the compactness property, for arbitrary rings. The key fact about weakly compact cardinals which we need is the following. (For the links in the chain which connect the definition to this lemma, we must refer the reader to a text on set theory, e.g. Jech 1978; the lemma is proved by using the $\prod_1^1$-indescribability of weakly compact cardinals, as in Exercise 32.10 of Jech.)

**3.1 Lemma.** *Let $\kappa$ be a weakly compact cardinal and let $\{S_\nu : \nu < \kappa\}$ be a set of stationary subsets of $\kappa$. Then there is a stationary* set $T$ *of regular cardinals* $< \kappa$ *such that for every* $\lambda \in T$ *and every* $\nu < \lambda$, $S_\nu \cap \lambda$ *is stationary in* $\lambda$. $\square$

Under the assumption of the Axiom of Constructibility, the property stated in the Lemma characterizes weakly compact cardinals: see R. B. Jensen 1972.

**3.2 Theorem.** *Let* $\kappa$ *be a weakly compact cardinal. If* $M$ *is a* $\leq \kappa$-*generated module which is* $\kappa$-*free, then* $M$ *is free.*

PROOF. Suppose that the result is false, and $M$ is a counterexample. Let $\{M_\nu : \nu < \kappa\}$ be a $\kappa$-filtration as in Lemma 1.5. Without loss of generality we can assume that $M_0 = \{0\}$. Moreover, we can assume that $M_\nu$ is $\leq (|\nu| + \aleph_0)$-generated for every $\nu$ (e.g., by defining a new filtration made up of the same submodules but with long stretches where the filtration remains fixed at a given submodule until the cardinality of the index catches up to the size of the next member of the original filtration). For this filtration, let $E \subseteq \kappa$ be defined as in (1.6.1). By 1.7, $E$ is stationary in $\kappa$. For each $\nu \in E$, let

$$S_\nu = \{\mu > \nu : \mu < \kappa \text{ and } M_\mu / M_\nu \text{ is not free}\},$$

so by definition of $E$, $S_\nu$ is stationary in $\kappa$. For $\nu \notin E$, let $S_\nu = E$. Let $T$ be as in Lemma 3.1, for this family of stationary sets. Now for $\lambda \in T$, $\{M_\nu : \nu < \lambda\}$ is a $\lambda$-filtration of $M_\lambda$ and for each $\nu \in E \cap \lambda$,

$$S_\nu \cap \lambda = \{\mu > \nu : \mu < \lambda \text{ and } M_\mu / M_\nu \text{ is not free}\},$$

and this set is stationary in $\lambda$. Since $E \cap \lambda$ is also stationary in $\lambda$, Proposition 1.7 implies that $M_\lambda$ is not free, a contradiction. $\square$

See also Exercises 13 and VI.11 for other proofs of 3.2, especially when $R$ is a p.i.d.. (The proof in Exercise 13 uses a property of cardinals which is weaker than being weakly compact.) We shall see in Chapter VII that it is consistent with ZFC that the weakly compact cardinals are the only *regular* cardinals which satisfy the compactness property (for $\mathbf{Z}$). However, Shelah's Singular Compactness Theorem (Theorem 3.5 below) says that every *singular* cardinal has the compactness property. We begin the proof of this theorem by proving a result that says that another hypothesis of "almost-freeness" on a $\leq \lambda$-generated module $M$ is sufficient to imply that $M$ is free when $\lambda$ is singular.

**3.3 Theorem.** *Let $\lambda$ be a singular cardinal and let $M$ be a $\leq \lambda$-generated module such that there is a cardinal $\mu < \lambda$ such that for every $\mu < \kappa < \lambda$, $M$ is strongly $\kappa^+$-free. Then $M$ is free.*

PROOF. By hypothesis $\mathrm{cf}(\lambda) < \lambda$. Choose a continuous increasing sequence $\{\kappa_i : i < \mathrm{cf}(\lambda)\}$ of cardinals whose limit is $\lambda$, and such that $\kappa_0 \geq \max\{\mu, \mathrm{cf}(\lambda)\}$. Choose a generating set $G$ for $M$ of cardinality $\lambda$, which we write as the union of a continuous chain of subsets $G_i$ $(i < \mathrm{cf}(\lambda))$, where $|G_i| = \kappa_i$. We will define subsets $C_i^n$, $X_i^n$ of $M$ for $i < \mathrm{cf}(\lambda)$, $n < \omega$ by induction on $n$ (simultaneously for all $i$). If we let $B_i^n$ be the submodule generated by $X_i^n$, then we require for all $i$ and $n$:

(0) $G_i \subseteq B_i^n \subseteq \langle C_i^n \rangle \subseteq B_i^{n+1}$;

(1) $X_i^n$ is a basis of $B_i^n$, and $X_i^{n+1} \supseteq X_i^n$;

(2) $|C_i^n|, |X_i^n| \leq \kappa_i$ and $C_i^{n+1} \supseteq C_j^n$ for all $j \leq i$;

(3) $\langle C_i^n \rangle \subseteq \langle C_i^{n+3} \cap X_{i+1}^{n+2} \rangle$; and

(4) $\{\langle \bigcup_{n \in \omega} C_i^n \rangle : i < \mathrm{cf}(\lambda)\}$ is a continuous chain.

Suppose for a moment that we can do the construction. For each $i < \mathrm{cf}(\lambda)$, let $C_i = \bigcup_{n \in \omega} C_i^n$. By (4) the $\langle C_i \rangle$'s form a continuous chain. Moreover by (0) and (1), $\langle C_i \rangle = \bigcup_{n \in \omega} B_i^n$ has a basis $X_i = \bigcup_{n \in \omega} X_i^n$; and by (3), $\langle C_i \rangle$ is generated by $C_i \cap X_{i+1}$, so $\langle C_{i+1} \rangle / \langle C_i \rangle$ is free. Therefore $\bigcup_{i < \mathrm{cf}(\lambda)} \langle C_i \rangle$ is free. But by (0) $\bigcup_{i < \mathrm{cf}(\lambda)} \langle C_i \rangle$ contains $\bigcup_{i < \mathrm{cf}(\lambda)} G_i$, so it equals $M$.

Thus it remains to do the inductive construction. Since $M$ is strongly $\kappa_i^+$-free, there is a set $\mathcal{S}_i$ as in Definition 1.8: each $N \in \mathcal{S}_i$ is $\leq \kappa_i$-generated and free and for every subset $X$ of cardinality $\kappa_i$ there is an $N' \in \mathcal{S}_i$ such that $N' \supseteq N \cup X$ and $N'/N$ is free. We will do our construction so that each $B_i^n$ belongs to $\mathcal{S}_i$. To begin, let $B_i^0$ be an element of $\mathcal{S}_i$ which contains $G_i$ and let $X_i^0$ be a basis of $B_i^0$. Let $\{b_{i,\alpha}^0 : \alpha < \kappa_i\}$ be an enumeration of $X_i^0$. Suppose now that for some $n$ and all $i$, $X_i^k$ has been defined for all

$k \leq n$ and $C_i^k$ has been defined for all $k < n$. Suppose also that an enumeration $\{b_{i,\alpha}^k : \alpha < \kappa_i\}$ of $X_i^k$ has been chosen and fixed. Choose $Y_i^n \subseteq X_{i+1}^{n-1}$ so that $\langle Y_i^n \rangle \supseteq B_{i+1}^{n-1} \cap C_i^{n-1}$ and $|Y_i^n| \leq \kappa_i$. Let $C_i^n = X_i^n \cup Y_i^n \cup \bigcup_{j \leq i} C_j^{n-1} \cup \{b_{j,\alpha}^n : j < \mathrm{cf}(\lambda) \text{ and } \alpha < \kappa_i\}$. (Notice that there are only $\kappa_i$ elements $b_{j,\alpha}^n$ since $\kappa_i \geq \mathrm{cf}(\lambda)$; it is here that the singularity of $\lambda$ plays its vital role.) Finally choose $B_i^{n+1} \in \mathcal{S}_i$ so that (0) is satisfied and $B_i^{n+1}/B_i^n$ is free; this is possible by the properties of $\mathcal{S}_i$. Then we can choose a basis $X_i^{n+1}$ of $B_i^{n+1}$ extending $X_i^n$, and enumerate it. This completes the inductive step in the construction. It remains to verify that (3) and (4) hold. As for (3), $C_i^n \subseteq C_{i+1}^{n+1} \subseteq B_{i+1}^{n+2}$, so $C_i^n \subseteq B_{i+1}^{n+2} \cap C_i^{n+2} \subseteq \langle Y_i^{n+3} \rangle$ and $Y_i^{n+3} \subseteq C_i^{n+3} \cap X_{i+1}^{n+2}$. Further, (4) will hold because for every limit $j < \mathrm{cf}(\lambda)$,

$$\langle \bigcup_{n \in \omega} C_j^n \rangle = \bigcup_{n \in \omega} B_j^n = \bigcup_{n \in \omega} \langle \{ b_{j,\alpha}^n : \alpha < \kappa_j \} \rangle =$$
$$\bigcup_{n \in \omega} \bigcup_{i < j} \langle \{ b_{j,\alpha}^n : \alpha < \kappa_i \} \rangle \subseteq \bigcup_{i < j} \langle \bigcup_{n \in \omega} C_i^n \rangle.$$

□

We want to derive from 3.3 the result that if $M$ is $\lambda$-free then $M$ is free. Recall that $M$ is a $\lambda$-free module if and only if $M$ is $\kappa$-free for all regular $\kappa < \lambda$. We will need the fact that a $\kappa^+$-free module is strongly $\kappa$-free. The proof of this fact will require game-theoretic methods in the general case; we follow the approach of Hodges 1981. (For a simpler argument in the case of p.i.d.'s see VII.2.11.)

Fix a regular cardinal $\kappa$ and a module $M$. We define a two-player game called the $\kappa$-*Shelah game* on $M$. Player I and Player II make alternate moves indexed by the elements of $\omega$, with I moving first. Player I's move, $P_n$, is a subset of $M$ of cardinality $< \kappa$; Player II's move, $N_n$, is a $< \kappa$-generated submodule of $M$; moreover we require that for all $n \in \omega$,

$$N_{n-1} \cup P_n \subseteq N_n$$

($N_{-1} = \{0\}$). Player II wins the game if and only if for all $n$, $N_n$ and $N_n/N_{n-1}$ are free; otherwise Player I wins. See section II.5 for a discussion of winning strategies for games.

It is easy to see that $M$ is strongly $\kappa$-free if and only if Player II has a winning strategy in the $\kappa$-Shelah game on $M$. Indeed if $M$ is strongly $\kappa$-free and $\mathcal{S}$ is as in Definition 1.8, then Player II's winning strategy is to choose the $N_n$'s from $\mathcal{S}$ such that $N_n/N_{n-1}$ is free. Conversely, if Player II has a winning strategy we can define $\mathcal{S}$ to consist of all moves by II in some $\kappa$-Shelah game which is played according to the winning strategy.

The $\kappa$-Shelah game is clearly an open game; hence by Theorem II.5.2, this game is determined.

**3.4 Proposition.** *Let $\kappa$ be a regular cardinal and $M$ a module which is $\kappa^+$-free. Then $M$ is strongly $\kappa$-free.*

PROOF. By our remarks above it suffices to prove that Player I does not have a winning strategy in the $\kappa$-Shelah game on $M$. So let us fix a strategy $s$ for I, that is, $s$ is a function which given a finite sequence

$$P_0, N_0, P_1, N_1 \ldots, P_k, N_k$$

of moves, where $P_0 = s(\emptyset)$ and $P_i = s(N_0, \ldots, N_{i-1})$ (for $0 < i \leq k$), chooses Player I's next move, $s(N_0, \ldots, N_k)$. We must show that Player II can defeat $s$. Let $\mathcal{C}$ be as in the definition of $\kappa^+$-free (cf. 1.1). We will construct by induction on $\nu$ a $\kappa$-filtration $\{N_\nu : \nu < \kappa\}$ consisting of submodules of $M$. At each stage we will also pick an element $A_\nu$ of $\mathcal{C}$ which contains $N_\nu$, and a set $\{a_\tau^\nu : \tau < \kappa\}$ of generators of $A_\nu$.

Here is how we do the construction. Fix a bijection $\varphi$ of $\kappa \setminus \{0\}$ with $\kappa \times \kappa$ such that for all $\nu$, if $\varphi(\nu) = (\mu, \tau)$ then $\mu < \nu$. Suppose that $N_\mu$, $A_\mu$, and $\{a_\tau^\mu : \tau < \kappa\}$ have been chosen for each $\mu < \nu$ for some $\nu$. If $\nu$ is a limit ordinal we simply take unions. Otherwise, choose $N_\nu$ so that it contains $a_\tau^\mu$ where $\varphi(\nu - 1) = (\mu, \tau)$, and such that it also contains $s(\emptyset)$ and $s(N_{\mu_0}, \ldots, N_{\mu_k})$ whenever $\mu_0 < \ldots < \mu_k < \nu$ and the latter is defined. (This is possible since there are fewer than $\kappa$ such sequences.) Choose $A_\nu$ in $\mathcal{C}$ to contain $N_\nu$. This completes the inductive step in the construction.

Now let $N = \bigcup_{\nu < \kappa} N_\nu = \bigcup_{\nu < \kappa} A_\nu$. Then $N \in \mathcal{C}$, so $N$ is free; let $B$ be a basis of $N$. There is a cub $C \subseteq \kappa$ such that for $\nu \in C$, $N_\nu$

is generated by $N_\nu \cap B$, so $N_\nu$ is free. Then Player II's strategy to defeat $s$ is to play $N_\nu$ where $\nu \in C$; more exactly, if $\nu_0 < \nu_1 < \ldots$ is a sequence of elements of $C$, then

$$s(\emptyset), N_{\nu_0}, s(N_{\nu_0}), N_{\nu_1}, s(N_{\nu_0}, N_{\nu_1}), \ldots$$

is a winning play for II. Hence $s$ is not a winning strategy for I. $\square$

As an immediate consequence of 3.3 and 3.4 we obtain

**3.5 Theorem.** (**Singular Compactness**) *Let $\lambda$ be a singular cardinal and let $M$ be a $\leq \lambda$-generated module which is $\lambda$-free. Then $M$ is free.* $\square$

Shelah's singular compactness theorem is much more general than Theorem 3.5; that is, it applies for abstract notions of "free" which satisfy only a few rudimentary properties. We shall not attempt to state the theorem — which has applications even outside of algebra — in its full generality (see Shelah 1975a, Ben David 1978), but we shall give a generalization of 3.5 in the context of modules. (See Exercise VII.3 for an application to families of countable sets.) The abstract setting will be described below in paragraph 3.6. In reading it, a useful example to keep in mind is the following. Let $\mathcal{L}$ be a fixed set of $R$-modules and $\mu$ a cardinal so that every module in $\mathcal{L}$ is $\leq \mu$-generated. Say that an $R$-module $M$ is "free" if $M = \bigoplus_{i \in I} L_i$ where each $L_i$ is isomorphic to a member of $\mathcal{L}$. If $M$ is "free", $X$ is a "basis" of $M$ if there are generating sets $S_i$ for the $L_i$ of cardinality $\leq \mu$ such that

$$(*) \qquad X = \{Y : \exists J \subseteq I \text{ s.t. } Y = \bigcup_{i \in J} S_i\}$$

(so $\langle Y \rangle$, the submodule generated by $Y$, $= \bigoplus_{i \in J} L_i$). Notice that the choice of a different representation of $M$ as a direct sum of copies of members of $\mathcal{L}$ or a different choice of $S_i's$ gives rise to a different "basis". Say that $A$ is a "free factor" of $M$ if $A = \langle Y \rangle$ for some member, $Y$, of a "basis", $X$, of $M$; in that case, write $X' = X|A$ if $X'$ is the "basis" of $A$ determined as in $(*)$ with respect to the same $S_i$ used to determine $X$ (but using only $i \in J$).

**3.6.** Suppose we are given a class $\mathcal{F}$ of $R$-modules and for each $M \in \mathcal{F}$, a non-empty family, $\mathcal{B}(M)$, of sets of subsets of $M$. We shall say that $M$ is "free" if $M \in \mathcal{F}$, and that $X$ is a "basis" of $M$ if $X \in \mathcal{B}(M)$. We assume that $\{0\}$ belongs to $\mathcal{F}$, and that $\emptyset$ belongs to every "basis". For some fixed cardinal $\mu \geq \aleph_0$, we require the following properties for every $M \in \mathcal{F}$ and $X \in \mathcal{B}(M)$.

(a) $X$ is closed under unions of chains, i.e., if $C \subseteq X$ such that for all $Y, Y' \in C$, $Y \subseteq Y'$ or $Y' \subseteq Y$, then $\cup C \in X$;

(b) if $Y \in X$ and $a \in M$, there exists $Y' \in X$ such that $Y \subseteq Y'$, $a \in \langle Y' \rangle$ and $|Y'| \leq |Y| + \mu$;

(c) if $Y \in X$, then $\langle Y \rangle \in \mathcal{F}$ and $\{Z \in X : Z \subseteq \langle Y \rangle\} \in \mathcal{B}(\langle Y \rangle)$;
In this case we say that $\langle Y \rangle$ is a "free" factor of $M$ and we suppose we are also given a subset of $\mathcal{B}(\langle Y \rangle) \times \mathcal{B}(M)$; we write $X' = X|\langle Y \rangle$ if $(X', X)$ belongs to that subset.

(d) if $A$ is a "free" factor of $M$ and $X' = X|A$, then $X' = \{Z \in X : Z \subseteq A\}$; moreover, for any $X' \in \mathcal{B}(A)$, there exists $X \in \mathcal{B}(M)$ such that $X' = X|A$;

(e) if $\{M_\nu : \nu < \beta\}$ is a continuous chain of "free" modules and for each $\nu < \beta$, $M_\nu$ is a "free factor" of $M_{\nu+1}$, then $\bigcup_{\nu < \beta} M_\nu$ is "free"; moreover, given $X_\nu \in \mathcal{B}(M_\nu)$ for each $\nu < \beta$ such that $X_\nu = X_\mu|M_\nu$ for all $\nu < \mu < \beta$, then the closure of $\bigcup_{\nu < \beta} X_\nu$ under unions of chains is a "basis" of $\bigcup_{\nu < \beta} M_\nu$.

For a regular cardinal $\kappa$, the notion of $\kappa$-"free" is defined as in Definition 1.1, replacing free by "free"; for singular $\lambda > \mu$, we define $M$ to be $\lambda$-"free" if $M$ is $\kappa$-free for every regular $\kappa < \lambda$ such that $\kappa > \mu$. We define $M$ to be *strongly $\kappa$-"free"* if there is a set $\mathcal{S}$ of $< \kappa$-generated "free" submodules of $M$ containing $\{0\}$ and such that for any subset $S$ of $M$ of cardinality $< \kappa$ and any $N \in \mathcal{S}$, there exists $N' \in \mathcal{S}$ such that $N' \supseteq N \cup S$ and $N$ is a "free" factor of $N'$.

**3.7 Theorem.** *Under the assumptions of 3.6, if $\lambda$ is a singular cardinal $> \mu$ and $M$ is a $\leq \lambda$-generated $\lambda$-"free" module, then $M$ is "free".*

PROOF. The proof of 3.4 applies, *mutatis mutandi*, so we can conclude that $M$ is strongly $\kappa$-"free" for all regular $\kappa$ with $\mu < \kappa < \lambda$.

Then we can mimic the proof of 3.3 with only minor modifications. The sets $X_i^n$ will be sets of subsets of $M$. $B_i^n$ will denote the submodule generated by $\cup X_i^n$. Conditions (0) and (4) remain the same and the others are replaced by

(1′) $X_i^n$ is a "basis" of $B_i^n$, $B_i^n$ is a "free" factor of $B_i^{n+1}$, and $X_i^n = X_i^{n+1} | B_i^n$.

(2′) $|C_i^n| \leq \kappa_i$, $B_i^n$ is $\leq \kappa_i$-generated, and $C_i^{n+1} \supseteq C_j^n$ for all $j < i$;

(3′) for all $n$, there exists $Y_i^n \in X_{i+1}^{n-1}$ of cardinality $\leq \kappa_i$ such that

$$Y_i^{n-1} \subseteq Y_i^n \subseteq C_i^n, \text{ and } C_i^n \subseteq \langle Y_i^{n+3} \rangle.$$

We will indicate the modifications needed in the proof of 3.3. Using (e), (0) and (1′), we see that $\langle C_i \rangle$ has a "basis" $X_i \overset{\text{def}}{=}$ the closure of $\cup X_i^n$ under unions of chains. By (3′), $\langle C_i \rangle$ is generated by $\bigcup_{n \in \omega} Y_i^n \in X_{i+1}$, so $\langle C_i \rangle$ is a "free" factor of $\langle C_{i+1} \rangle$, and therefore, by (e), $\bigcup_{i < \text{cf}(\lambda)} \langle C_i \rangle$ is "free".

In the inductive construction, suppose that $X_i^k (k \leq n)$, $C_i^k (k < n)$ and an enumeration, $\{b_{i,\alpha}^k : \alpha < \kappa_i\}$, of a generating set for $B_i^k$ have been chosen for all $i$. Choose (using (a) and (b)) $Y_i^n \in X_{i+1}^{n-1}$ such that $Y_i^n \supseteq Y_i^{n-1}$, $\langle Y_i^n \rangle \supseteq B_{i+1}^{n-1} \cap C_i^{n-1}$ and $|Y_i^n| \leq \kappa_i$. Let $C_i^n = B_i^n \cup Y_i^n \cup \bigcup_{j \leq i} C_j^{n-1} \cup \{b_{j,\alpha}^n : j < \text{cf}(\lambda) \text{ and } \alpha < \kappa_i\}$. Finally choose $B_i^{n+1} \in \mathcal{S}_i$ containing $C_i^n$ and a "basis", $X_i^{n+1}$, of $B_i^{n+1}$ so that $X_i^n = X_i^{n+1} | B_i^n$ (using (d)). To verify the last part of (3′), notice that $C_i^n \subseteq C_{i+1}^{n+1} \subseteq B_{i+1}^{n+2}$ so $\langle Y_i^{n+3} \rangle \supseteq B_{i+1}^{n+2} \cap C_i^{n+2} \supseteq C_i^n$. $\square$

It is natural to ask whether a weaker notion of $\lambda$-free than the one we have used is sufficient for Theorem 3.5. For example, one may ask if $M$ is free whenever $M$ is $\leq \lambda$-generated ($\lambda$ singular) and every subset of $M$ of size $< \lambda$ is contained in a free submodule. We do not know if this is true in general. (It is for modules over a p.i.d., of course.) However, we can prove a result of this sort for modules over a valuation domain. The reader who is not interested in such rings may safely skip the rest of this section. By the same token, we shall feel free to quote some results about modules over valuation domains, principally from Fuchs-Salce 1985. We begin with a result about modules $N$ of projective dimension $\leq 1$ (i.e. $N$ is isomorphic to a quotient of two free modules.)

**3.8 Lemma.** *Suppose $R$ is a valuation domain and $N$ is an $R$-module of projective dimension $\leq 1$. If $C$ is a pure submodule of $N$ of infinite rank $\kappa$, then $C$ is $\leq \kappa$-generated.*

PROOF. Since $C$ is the union of $\kappa$ pure submodules of countable rank, it suffices to prove that a pure submodule $C$ of rank $\aleph_0$ is $\leq \aleph_0$-generated. But this is true by Corollary 4.5 of Fuchs 1983 □

**3.9 Lemma.** *Let $R$ be a valuation domain and let $B_0 \subseteq \ldots \subseteq B_{n-1}$ be pure free submodules of a torsion-free module $M$, and for each $j$ let $X_j$ be a basis of $B_j$. Then given a $\leq \kappa$-generated submodule $N$ of $M$, there is a $\leq \kappa$-generated submodule $C$ of $M$ containing $N$ such that for all $j = 0, \ldots n - 1$:*

    *(1) $C \cap B_j$ is generated by $C \cap X_j$; and*
    *(2) $(B_j \cap C) + B_{j-1}$ is pure in $M$.*

PROOF. Let $\varphi : \omega \to n$ be chosen so that for all $j = 0, \ldots, n - 1$, $\varphi^{-1}[j]$ is infinite. We shall define by induction on $i$ a chain $\{C_i : i \in \omega\}$ of $\leq \kappa$-generated submodules of $M$ containing $N$. We require first that for all even $k = 2i$ there is a subset $Y_i$ of $X_{\varphi(i)}$ of cardinality $\kappa$ such that for any $m < i$ with $\varphi(m) = \varphi(i)$,

$$(3.9.1) \qquad C_{2m} \cap B_{\varphi(m)} \subseteq \langle Y_n \rangle \subseteq C_k \cap B_{\varphi(i)}.$$

This is possible because $C_{2m} \cap B_{\varphi(m)}$ is pure in $C_{2m}$ and because by Lemma II.5.4 of Fuchs-Salce 1985 a pure submodule of a $\leq \kappa$-generated submodule is $\leq \kappa$-generated.

    Secondly, we require that for odd $k = 2i + 1$,

$$(3.9.2) \quad (B_{\varphi(i)} \cap C_k + B_{\varphi(i)-1})/B_{\varphi(i)-1} \text{ is pure in } B_{\varphi(i)}/B_{\varphi(i)-1}.$$

This is possible by Lemma 3.8 because $B_{\varphi(i)}/B_{\varphi(i)-1}$ has projective dimension $\leq 1$.

Then we let $C = \bigcup_i C_i$. Condition (1) follows from (3.9.1), and condition (2) follows from (3.9.2) because $(B_j \cap C) + B_{j-1}$ is pure in $M$ if and only if $(B_j \cap C) + B_{j-1}$ is pure in $B_j$ if and only if $(B_j \cap C + B_{j-1})/B_{j-1}$ is pure in $B_j/B_{j-1}$. $\square$

Notice that in the following we require that every small subset $S$ is contained in a *pure* free submodule; if $S$ has cardinality $\kappa \geq \aleph_0$, then we can choose the pure free submodule to be $\leq \kappa$-generated.

**3.10 Theorem.** *Let $R$ be a valuation domain. Let $\lambda$ be a singular cardinal and let $M$ be a $\leq \lambda$-generated module such that every subset of cardinality $< \lambda$ is contained in a pure free submodule. Then $M$ is free.*

PROOF. Choose $\{\kappa_i : i < \operatorname{cf}(\lambda)\}$ and $G_i$ as in the proof of 3.3. We will define submodules $C_i^n$ and $B_i^n$ satisfying (0), (2), (4) as in 3.3, where $X_i^n$ is a fixed basis of $B_i^n$. In addition we shall require that:

(5) for all $m < n$, $C_i^n \cap B_{i+1}^m$ is generated by $C_i^n \cap X_{i+1}^m$; and

(6) for all $m < n$, $B_i^m$ and $B_{i+1}^m + (B_{i+1}^{m+1} \cap C_i^n)$ are pure in $M$.

It is not hard to see that the hypothesis and Lemma 3.9 allow us to do the construction.

Now let $C_i = \bigcup_{n \in \omega} C_i^n$. As in 3.3, the $C_i$'s form a continuous chain whose union is $M$, so it remains to show that $C_{i+1}/C_i$ is free for all $i$. But

$$C_{i+1}/C_i = \bigcup_{m \in \omega} [(B_{i+1}^m + C_i)/C_i]$$

and for each $m$, since $C_i = \bigcup_{n > m} C_i^n$, the term in square brackets is isomorphic to

$$B_{i+1}^m / \bigcup_{n > m} (B_{i+1}^m \cap C_i^n)$$

which is free by (5). We claim that (6) implies that for each $m (B_{i+1}^m + C_i)/C_i$ is pure in the next term in the chain. Indeed, if $b \in B_{i+1}^m$, $y \in B_i^{m+1}$, and $r \in R$ such that $ry = b + x$ for some $x \in C_i^n$ for some $n$, then $x \in B_{i+1}^{m+1} \cap C_i^n$, so (6) implies that there exists $\tilde{b} \in B_{i+1}^m$ and $\tilde{x} \in B_{i+1}^{m+1} \cap C_i^n$ such that $r(\tilde{b} + \tilde{x}) = b + x$; hence $r\tilde{b} = b + (x - r\tilde{x})$, which proves the claim.

Now, the union of a countable chain of pure free submodules is free (by Fuchs-Salce 1985 XIV.5.3), so $C_{i+1}/C_i$ is free. $\square$

## EXERCISES

1. Let $M$, $M_\nu$, $M'_\nu$, $E$, $E'$, and $C$ be as in 1.6. Prove that if $\nu \in C \cap E$, then $\nu \in E'$. [Hint: To show $\{\mu > \nu : M'_\mu/M'_\nu$ is not free$\}$ is stationary, consider any cub $C'$, and note that $C \cap C' \cap \{\mu > \nu : M_\mu/M_\nu$ is not free$\} \neq \emptyset$.]

2. Let $R$ be a left perfect ring and $M$ an $R$-module.
(i) If $M$ is $\leq \lambda^+$-generated and strongly $\lambda^+$-free, then $M$ is free. [Hint: as in 1.15, $M$ is projective; show that there must be $\lambda^+$ copies of each projective indecomposable in a decomposition of $M$.]
(ii) If $\kappa$ is a limit cardinal and $M$ is $\kappa$-free, then $M$ is free.
(iii) If $R$ has at least two non-isomorphic projective indecomposables, there is a strongly $\lambda^+$-free module which is $\leq \lambda^{++}$-generated.

3. (Stein's Lemma) Any countable group $A$ can be written as a direct sum: $A = N \oplus F$ where $F$ is free and $N^*(= \text{Hom}(N, \mathbb{Z})) = 0$. $N$ is uniquely determined. [Hint: let $N = \cap\{\ker \varphi : \varphi \in A^*\}$]

4.(i) If $A$ is a separable group and $\sigma : A \to A^{**}$ is the canonical map (cf. §I.1), then $A^{**}/\sigma[A]$ is torsion-free. [Hint: if $f \in A^{**}$ and $nf = \sigma(a)$, $A = \langle b \rangle \oplus D$ for some $b$ such that $a = mb$; show that $n$ divides $m$.]
(ii) Hence, if $A$ is torsionless, $A$ is separable if and only if $\sigma[A]$ is a pure subgroup of $A^{**}$.

5.(i) A product of $\aleph_1$-free groups is $\aleph_1$-free.
(ii) If $0 \to A \to B \to C \to 0$ is a short exact sequence and $A$ and $C$ are $\aleph_1$-free groups, then $B$ is $\aleph_1$-free.

6. Let $F$ be a free module over any ring $R$, and $B$ a submodule of $F$. Prove that $B$ is pure in $F$ if and only if for every finite subset $\{b_1, \ldots, b_n\}$ of $B$, there is a homomorphism $\varphi : F \to B$ such that $\varphi(b_i) = b_i$ for all $i = 1, \ldots, n$.

7. If $N$ is a subgroup of $M^{**}$ containing $\sigma[M]$, then the homomorphism: $N^* \to \sigma[M]^*$ induced by inclusion, is surjective. [Hint: if $z \in \sigma[M]^*$, consider $f \in N^*$ defined by: $\langle f, y \rangle = \langle y, z \circ \sigma \rangle$.]

8. If $\text{Ext}(M, \mathbf{Z})$ is torsion-free, then $N$ is separable whenever $N$ is a subgroup of $M$ such that $M/N$ is of bounded order.

9. $M$ is finitely projective if and only if for every epimorphism $\varphi: L \to H$ and homomorphism $\psi: M \to H$ such that $H$ is finitely-generated, there exists $\eta: M \to L$ such that $\varphi \circ \eta = \psi$.

10. Say that a group $A$ is *hereditarily* $\aleph_1$-*separable* if every subgroup is $\aleph_1$-separable. Show that every hereditarily $\aleph_1$-separable group is $\aleph_1$-coseparable.

11. Let $\mathbf{Z}^{<\omega_1}$ denote the subgroup of $\mathbf{Z}^{\omega_1}$ consisting of all elements which are zero in all but countably many coordinates. Let $G = \mathbf{Z}^{\omega_1}/\mathbf{Z}^{<\omega_1}$ (cf. IX.3.4). Show that $G$ is $\aleph_1$-free but not torsionless; in fact $G^* = 0$. [Hint: to see that $G$ is $\aleph_1$-free, show that every countable subgroup is embeddable in $\mathbf{Z}^{\omega}$; to see that $G^* = 0$, use the slenderness of $\mathbf{Z}$.]

12. Let $A$ be the subgroup of $\mathbf{Z}^{\omega}$ generated by $\mathbf{Z}^{(\omega)}$ and $\{2x : x \in \mathbf{Z}^{\omega}\}$. Prove that $A$ is torsionless but not separable. [Hint: $A^* = (\mathbf{Z}^{\omega})^*$. So for all $f \in A^*$, $f((2, 2, 2, \ldots))$ is divisible by 2.]

13.(i) Let $\kappa$ be a regular cardinal. Suppose $M$ has a $\kappa$-filtration $\{M_\alpha : \alpha < \kappa\}$ so that $M_1$ is projective and for all $\alpha < \beta$, $M_{\beta+1}/M_{\alpha+1}$ is projective. Let $E = \{\alpha : M_{\alpha+1}/M_\alpha \text{ is not projective}\}$. Then $\Gamma_p(M) \stackrel{\text{def}}{=} \tilde{E}$ is an invariant of $M$. Furthermore $M$ is projective if and only if $\Gamma_p(M) = 0$. [Hint: to see $\Gamma_p$ is an invariant consider two filtrations $\{M_\alpha : \alpha < \kappa\}$ and $\{M'_\alpha : \alpha < \kappa\}$. Suppose $M_\alpha = M'_\alpha$ and $M_{\alpha+1}/M_\alpha$ is projective. Choose a $\beta$ so that $M'_{\alpha+1} \subseteq M_\beta$. Since $M_\alpha$ is a direct summand of $M_\beta$, it is also a direct summand of $M'_{\alpha+1}$. Hence $M'_{\alpha+1}/M_\alpha$ is projective. See Exercise 19 for the only if direction.]

(ii) Assume $M$ is as above and for all $\alpha$, $M_{\alpha+1}$ is free and $M_{\alpha+2}/M_{\alpha+1}$ is a free module on at least as many generators as $M_{\alpha+1}$. Then $M$ is free if and only if $\Gamma_p(M) = 0$. [Hint: if $M_{\alpha+1}/M_\alpha$ is projective then $M_{\alpha+2}/M_\alpha$ is free.]

(iii) Let $\kappa$ be a cardinal such that for any stationary subset $S$ of $\kappa$, there is a regular cardinal $\mu < \kappa$ such that $S \cap \mu$ is stationary in

$\mu$. If $M$ is a $\leq \kappa$-generated module which is $\lambda$-free for every $\lambda < \kappa$, then $M$ is projective. (cf. 3.2). In particular if $R$ is a p.i.d., then $\kappa$-free implies free. [Hint: $M$ is strongly $\lambda$-free for all $\lambda < \kappa$. Let $S_\lambda$ be as in the definition of strongly $\lambda$-free. Apply 1.10 to prove $S_\lambda \subseteq S_\mu$ for all $\lambda < \mu < \kappa$.]

(iv) Assume the hypothesis of (iii) and that $|R| < \kappa$. Show that if $M$ is $\lambda$-free for all $\lambda < \kappa$, then $M$ is free.

14. For any group $A$ and any prime $p \in \mathbf{Z}$, $\mathrm{Ext}(A, \mathbf{Z})$ has no $p$-torsion if and only if the canonical map

$$(\dagger_p) \qquad\qquad \mathrm{Hom}(A, \mathbf{Z}) \to \mathrm{Hom}(A, \mathbf{Z}/p\mathbf{Z})$$

is surjective. [Hint: consider the long exact sequence induced by

$$0 \to \mathbf{Z} \xrightarrow{p} \mathbf{Z} \to \mathbf{Z}/p\mathbf{Z} \to 0$$

(cf. section XII.2, prior to 2.7).]

Hence if $A$ is reduced, $A$ is coseparable if and only if the maps $(\dagger_p)$ are surjective for all $p$.

*The next 3 exercises presume a knowledge of* Ext *as a group of equivalence classes of short exact sequences (as, for example, in* Fuchs 1970, *section 50).*

15. For any short exact sequence

$$E: A \xrightarrow{\alpha} B \xrightarrow{\beta} C \to 0$$

of groups and any $n \in \mathbf{Z}$, $n[E] = 0$ in $\mathrm{Ext}(C, A)$ if and only if there exists $\varphi: B \to A$ such that $\varphi \circ \alpha = n \cdot id_A$.

16. (i) $\mathrm{Ext}(\mathbf{Z}^\omega, \mathbf{Z})$ is not torsion-free. [Hint: choose $\varphi: \mathbf{Z}^\omega \to \mathbf{Z}/p\mathbf{Z}$ so that $\varphi(e_n) \neq 0$, for all $n$. Apply 14.]

(ii) $\mathrm{Ext}(\mathbf{Z}^\omega, \mathbf{Z})$ is not a torsion group. [Hint: Write $\mathbf{Z}^\omega$ as $\prod_{n \in \omega} P_n$ where each $P_n \cong \mathbf{Z}^\omega$; consider the map

$$\mathrm{Ext}(\mathbf{Z}^\omega, \mathbf{Z}) \to \mathrm{Ext}(\bigoplus_{n \in \omega} P_n, \mathbf{Z}) \to 0$$

and show that $\text{Ext}(\bigoplus_{n \in \omega} P_n, \mathbf{Z})$ has elements of infinite order.]
(iii) Hence $\mathbf{Z}^\omega$ is separable but not coseparable.

17. Let $E$ be as in 15, where $C$ is torsionless.
(i) If $[E]$ is a non-zero torsion element of $\text{Ext}(C, \mathbf{Z})$, then $B$ is torsionless but not separable.
(ii) If $[E]$ is of infinite order, then $B$ is $\aleph_1$-free but not torsionless.
(iii) Conclude from 16 that torsionless does not imply separable and $\aleph_1$-free does not imply torsionless.

18. In this exercise we construct groups of cardinality $\aleph_1$. Let $B$ be a set of branches in $^{<\omega}2$ of cardinality $\aleph_1$ (cf. II.5.3). Let $G = G_2$ be the smallest pure subgroup of $\mathbf{Z}^{<\omega^2}$ containing $e_\rho$ for every $\rho \in {}^{<\omega} 2$ and $a_\sigma \overset{\text{def}}{=} \sum_{n \in \omega} 2^n e_{\sigma \restriction n}$ for every $\sigma \in B$.
(i) $G$ is separable but not coseparable. [Hint: $G$ is separable because it is a pure subgroup of a separable group; to show $G$ is not coseparable, it is enough to show that there is a homomorphism from $G$ to $\mathbf{Z}/2\mathbf{Z}$ which does not lift to a homomorphism from $G$ to $\mathbf{Z}$ (cf. Exercise 14). Show that there is a function $f: G \to \mathbf{Z}/2\mathbf{Z}$ defined by $f(e_\rho) = \rho(n-1)$, where $n$ is the length of $\rho$. Suppose that $g$ is a lifting of $f$ to a homomorphism from $G$ to $\mathbf{Z}$. Let $\hat{G}$ denote the 2-adic completion of $G$ and denote the unique extension of $g$ to a homomorphism from $\hat{G}$ to $J_2$ (the 2-adic integers) by $g$ as well. Show that $g(a_\sigma) = \sum_{n \in \omega} 2^n g(e_{\sigma \restriction n})$; hence if $\sigma \neq \tau$, $g(a_\sigma) \neq g(a_\tau)$. So the range of $g \restriction G$ is uncountable, which is a contradiction.]
(ii) Conclude, from 17(i), that there is a torsionless group of cardinality $\aleph_1$ which is not separable.
(iii) There is an $\aleph_1$-free group of cardinality $\aleph_1$ which is not torsionless. [Hint: use 17(ii) with $C = G^{(\omega)}$.]

19. If $M$ is projective and $\leq \aleph_1$-generated and $\{M_\nu : \nu < \omega_1\}$ is an $\aleph_1$-filtration of $M$, then there is a cub $C$ so that for all $\nu < \mu$ in $C$, $M_\mu / M_\nu$ is projective. [Hint: use Theorem I.2.4.]

20. Suppose $\kappa$ is a regular limit cardinal. Assume $M$ is a $\leq \kappa$-generated module, which is not $< \kappa$-generated, and $M$ is $\lambda$-free for all $\lambda < \kappa$.

(i) For all $\lambda < \kappa$ there is a set $X$ of cardinality $\lambda^{++}$ which is not contained in any $\leq \lambda$-generated submodule. [Hint: assume not. Choose a strictly increasing chain $N_i$ $(i < \lambda^{++})$ of $\leq \lambda^{++}$-generated submodules so that for all odd $i$, $N_i$ is $\leq \lambda$-generated and $N \overset{\text{def}}{=} \cup N_i$ is free (such a chain exists since $M$ is $\lambda^{+++}$-free). Since $N$ is a free module such that every subset of cardinality $\lambda^+$ is contained in a $\leq \lambda$-generated submodule, $N$ is $\leq \lambda$-generated.]
(ii) The condition that $|R| < \kappa$ can be eliminated from 13(iv).

## NOTES

Fuchs 1960 defined the $\kappa$-free abelian groups, and in Fuchs 1970 (Problem 10) asked for the $\kappa$ such that there exist $\kappa$-free groups which are not $\kappa^+$-free. The $\Gamma$-invariant for abelian groups is formalized in Eklof 1977c and Proposition 1.7 is given. The logical significance of strongly $\kappa$-free for abelian groups is given in Eklof 1974. Proposition 1.11 in the general case is due to Kueker 1981. Theorem 1.12 is due to Bass 1960, and 1.13 to Mueller 1970 and, independently, Sabbagh-Eklof 1971. Example 1.14 and Theorem 1.15 are from Eklof-Mekler 1988.

For a survey of notions of purity, see the notes to section 7 in Kaplansky 1969. Theorem 2.3 is from Pontryagin 1934. The notion of separable occurs in Baer 1937 (in a more general sense) and 2.5 is proved there. The notion of a $\kappa$-separable abelian group is defined in Problem 29 of Fuchs 1960. The fact that $\mathbf{Z}^\kappa$ is separable is due to Specker 1950, and the fact that it is not free to Baer 1937; the proof that it is not strongly $\aleph_1$-free comes from Eklof 1974. Corollary 2.10 and Lemma 2.11 are in Chase 1962a. The notion of $\kappa$-coseparable comes from Griffith 1968 where 2.15 is proved. (See also Griffith 1970.) Hiremath 1978 studied finitely projective groups. Theorem 2.13 is a combination of the work of Chase 1962a, Griffith 1968, Hiremath 1978 and Hausen 1981. Theorem 2.16 is due to Hill 1970. Theorem 2.18 is due essentially to Higman 1951.

Theorem 3.2 (for abelian groups) was observed by many people (e.g. Gregory, Kueker, Mekler) at about the same time; a proof was published in Eklof 1974. The Singular Compactness Theorem is due to Shelah 1975a; our exposition is based on a later proof by Shelah which appears in Hodges 1981. Theorem 3.10 is an improvement (due to Eklof-Fuchs) of Theorem 14 of Eklof-Fuchs 1988; it is based on another proof of singular compactness for abelian groups given by Shelah in an unpublished note.

(See also Eklof 1982.) For more on singular compactness for modules, see Eklof-Fuchs-Shelah 19??.

Exercises. 3:Fuchs 1970; 4, 5(ii), 8:Chase 1962a; 5(i):Chase 1963a; 9:Hausen 1981; 10:Griffith 1968; 11:Reid 1967; 12:Nunke(cited in Chase 1962a); 13:new(with help from Shelah); 16:Nunke 1962b; 17:Reid 1967.

There is another definition of $\kappa$-free which is stronger than any of the notions of $\kappa$-free considered elsewhere in this text. We say a module, $A$, is $\kappa$-*free* if there is an expansion of $A$ by countably many function symbols to an algebra $A_1$ so that if $B_1$ is any subalgebra of $A$ of cardinality $< \kappa$, then $B$, the module underlying $B_1$, is free. It is easy to check that this definition of $\kappa$-free implies the other ones we have considered. This definition is the one which Shelah 1985 uses. If we adopt this definition of $\kappa$-free then it is consistent, assuming the consistency of certain large cardinals, that every $\aleph_2$-free $\mathbf{Z}/6\mathbf{Z}$-module of cardinality $\aleph_2$ is in fact free. Moreover, if $\kappa$ is supercompact, then every $\kappa$-free $\mathbf{Z}/6\mathbf{Z}$-module (of arbitrary cardinality) is free. The relevant result here is Chang's conjecture. Chang's conjecture for $(\lambda, \kappa)$ states (in one formulation):

Suppose $A$ is an algebra, in a countable language, of cardinality $\kappa$ and $\varphi(x)$ is an equation in one variable. If $\varphi(x)$ has exactly $\lambda$ solutions in $A$, for some $\aleph_0 \leq \lambda < \kappa$, then $A$ has a subalgebra $B$ of cardinality $\aleph_1$ such that $\varphi(x)$ has exactly $\aleph_0$ solutions in $B$.

Silver showed that it is consistent, assuming the consistency of a large cardinal, that Chang's conjecture for $(\aleph_1, \aleph_2)$ holds. Assume that Chang's conjecture for $(\aleph_1, \aleph_2)$ holds and that $A$ is a non-free $\aleph_2$-free $\mathbf{Z}/6\mathbf{Z}$-module of cardinality $\aleph_2$. Without loss of generality we can assume $A \cong \mathbf{Z}/2\mathbf{Z}^{(\omega_1)} \oplus \mathbf{Z}/3\mathbf{Z}^{(\omega_2)}$. By Chang's conjecture, if $A_1$ is any expansion of $A$, then $A_1$ has a subalgebra of cardinality $\aleph_1$, which as an abelian group is isomorphic to $\mathbf{Z}/2\mathbf{Z}^{(\omega)} \oplus \mathbf{Z}/3\mathbf{Z}^{(\omega_1)}$; thus $A$ is not $\aleph_2$-free in this new strong sense. Conversely, $\mathbf{Z}/2\mathbf{Z}^{(\omega)} \oplus \mathbf{Z}/3\mathbf{Z}^{(\omega_1)}$ is $\aleph_2$-free if Chang's conjecture for $(\aleph_1, \aleph_2)$ does not hold. Much information is known about the consistency of various forms of Chang's conjecture. It is currently open whether or not it consistent that Chang's conjecture holds for all $(\kappa, \kappa^+)$ or equivalently whether (in this definition of $\kappa$-free) it is consistent that $\aleph_2$-free implies free.

# CHAPTER V
# PURE-INJECTIVE MODULES

A module is called *pure-injective* if it is a direct summand of every module in which it is a pure submodule. A non-zero pure-injective module is not slender; in fact, any non-zero homomorphic image of a pure-injective module is not slender (see Lemma 2.1 below). Thus any module which contains a non-zero homomorphic image of a pure-injective module, or contains a copy of $R^\omega$, is not slender. In Chapter IX we shall prove a theorem of Nunke which is a converse of this observation for modules over a principal ideal domain. In preparation for that we shall, in the first section of this chapter, investigate the structure of pure-injective modules over a p.i.d., and in the second section study the homomorphic images of pure-injectives, which are the cotorsion modules. Pure-injective modules over general rings have been studied in great depth by algebraists and model-theorists (see the Notes at the end of the chapter), but we shall present here only what we need for the later chapters of this work.

## §1. Structure theory

Recall that the definition of a pure submodule of a module was given in IV.2.1. A homomorphism $\varphi\colon A \to B$ is called a *pure embedding* if it is an embedding of modules such that $\varphi[A]$ is a pure submodule of $B$. It is easy to see that a module $N$ is pure-injective if and only if it has the injective property with respect to pure embeddings, i.e., for every pure embedding $\alpha\colon A \to B$ and homomorphism $f\colon A \to N$, there is a homomorphism $g\colon B \to N$ such that $g \circ \alpha = f$ (cf. I.2). Just as is the case for injectives, the class of pure-injectives is closed under direct products and direct summands.

We shall make important use of another characterization of pure-injective modules.

**1.1 Definition.** Let $\lambda$ be an infinite cardinal. A module $N$ is called $\lambda^+$-*algebraically compact* if whenever $S$ is a set of $\lambda$ linear equations over $N$ (in any number, finite or infinite, of variables) such that every finite subset of $S$ has a solution in $N$, then $S$ has a solution in $N$. $N$ is called *algebraically compact* (or, alternately, *equationally compact*) if it is $\lambda^+$-algebraically compact for every infinite $\lambda$.

**1.2 Theorem.** *The following are equivalent, for any R-module $N$:*
  (1) *$N$ is $(|R| + \aleph_0)^+$-algebraically compact;*
  (2) *$N$ is pure-injective;*
  (3) *$N$ is algebraically compact.*

PROOF. (1) $\Rightarrow$ (2): Given a pure submodule $A$ of $B$ and a homomorphism $f: A \to N$, we must show that $f$ can be extended to $g: B \to N$. Let us say that a map $\tilde{g}: \tilde{B} \to N$ is a *partial homomorphism* if $\tilde{B}$ is a submodule of $B$ containing $A$ and for every finite system of linear equations

$$\sum_{j=1}^{m} r_{ij} y_j = b_i \qquad (i = 1, \dots, n)$$

where the $b_i$ belong to $\tilde{B}$, if the system has a solution in $B$ then

$$\sum_{j=1}^{m} r_{ij} y_j = \tilde{g}(b_i) \qquad (i = 1, \dots, n)$$

has a solution in $N$. Notice that $f$ is a partial homomorphism because $A$ is a pure submodule of $B$. By Zorn's Lemma there exists a maximal extension of $f$ to a partial homomorphism $\tilde{g}: \tilde{B} \to N$. We will be done if we show that $\tilde{B} = B$. If not, let $c \in B \setminus \tilde{B}$, and aim to contradict the maximality of $\tilde{g}$.

Let $T$ be the set of all finite systems of equations

$$(\tau) \qquad \sum_{j=1}^{m} r_{ij} y_j = v_i + s_i c \qquad (i = 1, \dots, n)$$

(in unknowns $y_j$, $v_i$) such that for some choice of $v_i = b_i$ in $\tilde{B}$

$$(*) \qquad \sum_{j=1}^{m} r_{ij} y_j = b_i + s_i c \qquad (i = 1, \ldots, n)$$

has a solution in $B$. For each $\tau$ in $T$ fix $\Psi(\tau) = (b_1, \ldots, b_n)$ such that $(*)$ has a solution; also let $\{y_j^{(\tau)} : j = 1, \ldots, m\}$ be a new set of variable symbols. Now form a system of equations, $\mathcal{S}$, with coefficients in $N$ by putting into $\mathcal{S}$ for each $\tau \in T$ the equations

$$(**) \qquad \sum_{j=1}^{m} r_{ij} y_j^{(\tau)} - s_i x = \tilde{g}(b_i) \qquad (i = 1, \ldots, n)$$

where $\Psi(\tau) = (b_1, \ldots, b_n)$. Then $\mathcal{S}$ is finitely solvable in $N$ because $\tilde{g}$ is a partial homomorphism (and because there is no "clash" of the variables $y_j^{(\tau)}$ for different $\tau$). Hence $\mathcal{S}$ has a solution in $N$ by the hypothesis on $N$. (Notice that $|\mathcal{S}| \leq |R| + \aleph_0$.) If $d$ is the "$x$-value" of a solution of $\mathcal{S}$, then we extend $\tilde{g}$ to $h : \tilde{B} + Rc \to N$ by defining $h(c) = d$. (Notice that $h$ is well-defined because if $sc = b \in \tilde{B}$, then $sx = \tilde{g}(b)$ belongs to $\mathcal{S}$.) To see that we have a contradiction, it remains only to check that $h$ is a partial homomorphism. Now if

$$(***) \qquad \sum_{j=1}^{m} r_{ij} y_j = b_i' + s_i c \qquad (i = 1, \ldots, n)$$

has a solution in $B$ (where $b_1', \ldots, b_n' \in \tilde{B}$), then for the corresponding $\tau$ we have put the system $(**)$ into $\mathcal{S}$ (where $\Psi(\tau) = (b_1, \ldots, b_n)$). Moreover, by subtracting $(*)$ from $(***)$, we see that the system

$$\sum_{j=1}^{m} r_{ij} y_j = b_i' - b_i \qquad (i = 1, \ldots, n)$$

has a solution in $B$, and hence

$$\sum_{j=1}^{m} r_{ij} y_j = \tilde{g}(b_i') - \tilde{g}(b_i) \qquad (i = 1, \ldots, n)$$

has a solution in $N$, because $\tilde{g}$ is a partial homomorphism. From these facts, and the choice of $d$, one can readily deduce the existence of a solution of

$$\sum_{j=1}^{m} r_{ij} y_j = \tilde{g}(b_i') + s_i d \qquad (i = 1, \ldots, n)$$

in $N$.

(2) $\Rightarrow$ (3): Suppose that $N$ is pure-injective and that $\mathcal{S}$ is a system of equations over $N$ which is finitely solvable in $N$. Let

$$M = (N \oplus F)/K$$

where $F$ is the free module whose basis is the set of unknowns in the system $\mathcal{S}$, and $K$ is the submodule generated by the equations in $\mathcal{S}$, i.e., if $\sum_{j=1}^{m} r_j y_j = b$ belongs to $\mathcal{S}$, then $(b, \sum_{j=1}^{m} r_j y_j)$ belongs to $K$. Let $\iota$ be the canonical embedding of $N$ into $M$. It is easy to see that $\iota$ is injective, and using the finite solvability of $\mathcal{S}$ in $N$, one can verify that $\iota$ is a pure embedding. Let us identify $N$ with $\iota[N]$. Since $N$ is pure-injective, $M = N \oplus C$ for some $C$. By construction $\mathcal{S}$ has a solution in $M$; the projection of this solution onto $N$ is then a solution in $N$.

(3) $\Rightarrow$ (1) is obvious. $\square$

**1.3 Corollary.** *Every module is embeddable as a pure submodule of a pure-injective module.*

PROOF. Let $\kappa = (|R| + \aleph_0)^+$. Given a module $M$, we construct by induction a chain $\{M_\nu : \nu \leq \kappa\}$ of modules such that $M_0 = M$ and for all $\nu$: (1) $M_\nu$ is a pure submodule of $M_{\nu+1}$; (2) if $\nu$ is a limit, $M_\nu = \bigcup_{\mu < \nu} M_\mu$; (3) every system of linear equations over $M_\nu$ which is finitely solvable in $M_\nu$ is solvable in $M_{\nu+1}$. The construction of $M_{\nu+1}$ given $M_\nu$ can be done as in the proof of (2) $\Rightarrow$ (3) above.

Then $M_\kappa$ is $(|R| + \aleph_0)^+$-algebraically compact, and hence pure-injective. $\square$

It is also possible to prove 1.3 by using reduced products (cf. 1.15). By a more careful construction, one can embed $M$ as a pure submodule of a pure-injective module $\bar{M}$, called a *pure-injective envelope* of $M$, such that whenever $N$ is a pure-injective module containing $M$ as a pure submodule, then there is a pure embedding of $\bar{M}$ into $N$ which is the identity on $M$. (See Warfield 1969.)

Now we turn our attention to principal ideal domains, $R$; to begin with we even assume that $R$ is local — that is, $R$ is a *discrete valuation ring*, or d.v.r. — with maximal ideal $pR$. Recall that a module is called *reduced* if it contains no non-zero divisible submodule. See I.3.1 for the definition of the $p$-adic topology.

**1.4 Proposition.** *Let $R$ be a d.v.r. and $M$ a pure-injective $R$-module. Then*

(i) *$p^\omega M$ is a divisible $R$-module; and*

(ii) *if $M$ is reduced, $M$ is Hausdorff and complete in the $p$-adic topology.*

PROOF. (i) If $b \in p^\omega M$ we need to show that for all $n$ there exists $c_n \in p^\omega M$ such that $p^n c_n = b$. For $n \in \omega$ consider the system $\mathcal{S}_n$ of equations

$$\{p^n x = b\} \cup \{p^m y_m = x : m \in \omega\}.$$

It is enough to show that $\mathcal{S}_n$ is solvable in $M$, for then the "$x$-value" of a solution will serve as $c_n$. Since $M$ is pure-injective, it suffices to show that $\mathcal{S}_n$ is finitely-solvable. But any finite subset of $\mathcal{S}_n$ reduces to

$$\{p^n x = b\} \cup \{p^k y_k = x\}$$

for some $k$, and is solvable because $b \in p^{n+k} M$.

(ii) Thus if $M$ is reduced, $M$ is Hausdorff. Now to see that $M$ is complete in the $p$-adic topology, consider a Cauchy sequence $\{a_n : n \in \omega\}$. Without loss of generality we may assume that for all $n \geq m$, $a_n - a_m \in p^m M$. The system of equations

$$\{x - a_n = p^n y_n : n \in \omega\}$$

is finitely solvable because for any $k$, $x = a_k$ yields a solution of

$$\{x - a_n = p^n y_n : n \leq k\}.$$

Hence it is solvable in $M$, and the "$x$-value" of a solution is the desired limit of the Cauchy sequence. □

We shall say that a module $M$ over a d.v.r. $R$ is *complete* if it is Hausdorff and complete in the $p$-adic topology. We shall show that the reduced pure-injectives over a d.v.r. are precisely the complete modules, and we shall determine the structure of complete modules. Since the structure of divisible modules is known (cf. I.2 for the structure of divisible groups), this will give us a complete structure theory for pure-injectives over a d.v.r. We begin with the structure of complete modules.

One way of obtaining a complete module is to start with a direct sum of cyclic modules, $B$, and take its completion, $\hat{B} = \varprojlim B/p^k B$ with respect to the $p$-adic topology. Then $\hat{B}$ is an $R$-module whose topology is its $p$-adic topology; moreover $B$ is pure in $\hat{B}$, because its $p$-adic toplogy is induced by that on $\hat{B}$, and $\hat{B}/B$ is divisible, because $B$ is dense in $\hat{B}$. (Compare I.3.5.) This suggests the following definition.

**1.5 Definition.** Let $R$ be a d.v.r. and $M$ an $R$-module. A submodule $B$ of $M$ is called a *basic* submodule of $M$ if: (1) $B$ is pure in $M$; (2) $B$ is a direct sum of cyclic modules; and (3) $M/B$ is divisible.

In order to make our exposition as self-contained as possible, we shall give a sketch of the proof of the existence and uniqueness of basic submodules. (For more details, see for example Fuchs 1970 or Kaplansky 1969.)

**1.6 Lemma.** *Let $R$ be a d.v.r. If an $R$-module $M$ is not divisible, then it contains a non-zero pure cyclic submodule.*

PROOF. Let $T = \{x \in M : p^n x = 0 \text{ for some } n\}$, the torsion submodule of $M$. If $T$ is divisible, then $M = T \oplus F$ for some torsion-free, non-divisible submodule $F$. If $c \in F$ such that $c \notin pR$, then $c$ generates a pure cyclic submodule.

If $T$ is not divisible, then there is an element, $b$, of $T$ of order $p$ such that for some $n$, $b \in p^n R \setminus p^{n+1} R$. Let $c \in R$ such that $b = p^n c$. It is not hard to see that $\langle c \rangle$ is pure in $T$, and hence in $M$. □

**1.7 Theorem.** *Let $R$ be a d.v.r. If $M$ is an $R$-module, $M$ possesses a basic submodule, which is unique up to isomorphism.*

PROOF. A subset $Y$ of $M$ is called *pure-independent* if $\langle Y \rangle$, the submodule generated by $Y$, is pure in $M$ and equals $\bigoplus_{y \in Y} \langle y \rangle$, the (internal) direct sum of the cyclic submodules generated by the elements of $Y$; $Y$ is called a *basis* of $M$ if it is pure-independent and in addition $M/\langle Y \rangle$ is divisible. To prove the existence of a basic submodule of $M$ we shall prove the stronger fact that every pure-independent subset of $M$ (e.g., $\emptyset$) is contained in a basis of $M$. In fact, we use Zorn's Lemma to extend the given pure-independent subset of $M$ to a maximal one, $Y$. If we let $B = \langle Y \rangle$, it remains only to check that $M/B$ is divisible. If not, then by 1.6, there exists $c \in M$ such that the cyclic submodule of $M/B$ generated by $c + B$ is pure in $M/B$. But then one may easily verify that $Y \cup \{c\}$ is a pure-independent subset of $M$, contradicting the maximality of $Y$.

This takes care of existence. For uniqueness we must show that if

$$B = \bigoplus_{n \geq 1} R/p^n R^{(\alpha_n)} \oplus R^{(\beta)}$$

is a basic submodule of $M$, then the cardinals $\alpha_n$ and $\beta$ are uniquely determined by $M$. But, in fact,

$$\alpha_n = \dim p^{n-1} M[p]/p^n M[p]$$

the $n$th Ulm invariant of $M$, and

$$\beta = \dim M/(T + pM)$$

where $T$ is the torsion submodule of $M$. □

Now we can derive a structure theorem for complete modules.

**1.8 Corollary.** *Let $R$ be a d.v.r. and $M$ a complete $R$-module. Then $M$ is the completion of a direct sum, $B$, of cyclic modules; and $B$ is uniquely determined, up to isomorphism.*

PROOF. Let $B$ be a basic submodule of $M$. Then property (1) of a basic submodule implies that the $p$-adic topology on $B$ is induced by the $p$-adic topology on $M$. Property (3) implies that $B$ is dense in $M$ in the $p$-adic topology. Thus $M$ is a completion of $B$.

If $M$ is the completion of another submodule $C$ which is also a direct sum of cyclic modules, then $C$ is a basic submodule of $M$ — see the remarks before 1.5 — and hence isomorphic to $B$. $\square$

Finally, we can characterize the reduced pure-injective modules over a d.v.r. as the complete modules.

**1.9 Theorem.** *If $R$ is a d.v.r. and $M$ is a reduced $R$-module, then $M$ is pure-injective if and only if $M$ is complete.*

PROOF. The implication from left to right was proved above in 1.4. For the converse, suppose that $M$ is complete. By 1.3, $M$ is a pure submodule of a pure-injective module $N$. We can suppose that $N$ is reduced, since $p^\omega N$ is the divisible part of $N$ by 1.4, and $p^\omega N \cap M = \{0\}$ since $M$ is pure in $N$ and Hausdorff in the $p$-adic topology. Let $Y$ be a basis of $M$, and $B$ the basic submodule of $M$ it generates. By the proof of 1.7, $Y$ extends to a basis of $N$. Hence there is a basic submodule of $N$ of the form $C = B \oplus B'$ for some $B'$. Since $N$ is the completion of $C$ and $M$ is the completion of $B$, the projection of $C$ onto $B$ extends to a projection of $N$ onto $M$; so $M$ is a direct summand of $N$, and hence pure-injective. $\square$

Now we turn to an arbitrary p.i.d., $R$. We shall prove that every reduced pure-injective $R$-module is a product of pure-injective $R_{(p)}$-modules, where $p$ ranges over the set, $\mathcal{P}$, of primes of $R$. (Here $R_{(p)}$ is the localization of $R$ at the prime ideal $pR$, i.e.,$R_{(p)}$ is the subring of the quotient field, $Q$, of $R$ consisting of all $ab^{-1} \in Q$ such that $b \notin pR$.)

**1.10 Lemma.** *Let $M$ be an $R_{(p)}$-module. Then $M$ is pure-injective as $R_{(p)}$-module if and only if it is pure-injective as $R$-module.*

PROOF. ($\Rightarrow$): Suppose that $M$ is pure-injective as $R_{(p)}$-module. By 1.2 it's enough to prove that every system, $\mathcal{S}$, of $R$-module equations which is finitely solvable in $M$ has a solution in $M$. But every $R$-module equation *is* an $R_{(p)}$-module equation, so this is true since $M$ is algebraically compact as $R_{(p)}$-module by 1.2.

($\Leftarrow$): Suppose that $M$ is pure-injective as $R$-module. Let $M$ be a pure submodule of an $R_{(p)}$-module $N$; then the inclusion of $M$ in $N$ is a pure embedding of $R$-modules as well, so $M$ is a direct summand of $N$ (as $R$-modules and hence as $R_{(p)}$-modules). $\square$

**1.11 Theorem.** *Let $R$ be a p.i.d., and $M$ a reduced $R$-module. Then $M$ is pure-injective if and only if $M$ is isomorphic to a product $\prod_{p \in P} M_p$, where each $M_p$ is a pure-injective $R_{(p)}$-module.*

PROOF. ($\Leftarrow$): By 1.10, a pure-injective $R_{(p)}$-module, $M_p$, is a pure-injective $R$-module. Since a product of pure-injectives is pure-injective, we are done.

($\Rightarrow$): First of all notice that, just as in the proof of 1.4, we can show that $\cap\{rM : r \in R \setminus \{0\}\}$ is a divisible submodule, and hence is zero. Now fix a prime $p$ of $R$, and let

$$M_p = \{u \in M : r|u \text{ if } \gcd(r, p) = 1\}.$$

(Here, $r|u$, read "$r$ divides $u$", means that $u \in rM$.) We claim that

(†)    for every $a \in M$ there exists $a_p \in M_p$ such that $p^n | a - a_p$ for all $n$.

To show this, consider the following system, $\mathcal{S}_{a,p}$, of linear equations:

$$\mathcal{S}_{a,p} = \{p^n y_n = a - x : n \in \omega\} \cup \{r z_r = x : \gcd(r, p) = 1\}.$$

This system is finitely solvable in $M$ because given $n \in \omega$ and $r \in R$ such that $\gcd(r, p) = 1$, there exist $s, t \in R$ such that $sr + tp^n = 1$; if we let $x = sra$, then $r(sa) = x$ and $p^n(ta) = a - x$. Hence $\mathcal{S}_{a,p}$ is solvable in $M$; if we let $a_p$ be the "$x$-value" of a solution, we have the desired element.

Notice that for a given $a$ and $p$, the element $a_p$ is unique, because $\cap\{rM: r \in R \setminus \{0\}\} = \{0\}$. Now we define a function $f: M \to \prod_{p \in \mathcal{P}} M_p$ by: $f(a) = (a_p)_{p \in \mathcal{P}}$, where $a_p$ is as in (†). It is easy to see that $f$ is a homomorphism. Moreover, $f$ is one-one because if $f(a) = (a_p)_{p \in \mathcal{P}} = f(b)$, then for all $p$ and $n$,

$$p^n \text{ divides } a - b = (a - a_p) - (b - a_p),$$

so $a - b$ belongs to $\cap\{rM: r \in R \setminus \{0\}\} = \{0\}$.

To see that $f$ is surjective, let $(a_p)_{p \in \mathcal{P}} \in \prod_{p \in \mathcal{P}} M_p$, and consider the system of equations

$$\{p^n y_{np} = x - a_p : p \in \mathcal{P}, n \in \omega\}.$$

This system is finitely solvable because for any set $\{p_1, \ldots, p_m\}$ of primes, if $x = \sum_{i=1}^{m} a_{p_i}$, then for all $n \in \omega$, $p_i^n | x - a_{p_i}$ for $i = 1, \ldots, m$. Therefore the system has a simultaneous solution, and if $a$ is the "$x$-value" of such a solution, then $f(a) = (a_p)_{p \in \mathcal{P}}$.

So we have proved that $M$ is isomorphic to $\prod_{p \in \mathcal{P}} M_p$. All that remains is to show that each $M_p$ is a pure-injective $R_{(p)}$-module. For $M_p$ to be an $R_{(p)}$-module, we need unique divisibility by elements $q$ of $R \setminus pR$; but divisibility holds because of the definition of $M_p$ and because of the isomorphism $f$; uniqueness holds because $\{x \in M_p: qx = 0\} \subseteq \cap\{rM: r \in R \setminus \{0\}\} = \{0\}$. Finally, $M_p$ is pure-injective as $R$-module since it is a direct summand of a pure-injective, $M$; hence by 1.10 it is pure-injective as $R_{(p)}$-module. $\square$

**1.12 Summary.** Let us sum up what we know about the structure of pure-injective $R$-modules for the case $R = \mathbf{Z}$. In this case, an abelian group $A$ is pure-injective if and only if $A$ is isomorphic to a group of the form $\prod_{p \in \mathcal{P}} A_p \oplus D$ where for each prime $p$, $A_p$ is the completion of a group of the form

$$\bigoplus_{n \in \omega} Z(p^n)^{(\alpha_{p,n})} \oplus \mathbf{Z}_{(p)}^{(\beta_p)}$$

where $Z(p^n)$ is the cyclic group of order $p^n$ ($=$ the cyclic $\mathbf{Z}_{(p)}$-module $\mathbf{Z}_{(p)}/p^n\mathbf{Z}_{(p)}$). We can also replace $\mathbf{Z}_{(p)}$ by its completion, $J_p$ (cf.

I.3.4), and say that $A_p$ is the completion of

$$\bigoplus_{n \in \omega} Z(p^n)^{(\alpha_{p,n})} \oplus J_p^{(\beta_p)}.$$

Moreover, $D$ is of the form

$$\bigoplus_{p \in P} Z(p^\infty)^{(\gamma_p)} \oplus \mathbf{Q}^{(\delta)}$$

(see I.2 for definitions). It follows from the proof of 1.7 and from the structure theory of divisible groups, that the cardinal numbers $\alpha_p$, $n$, $\beta_p$, $\gamma_p$, and $\delta$ form a complete set of invariants for $A$.

We conclude this section with some results about abelian groups which are pure-injective. Here $\mathrm{Hom}(A, B)$ will mean $\mathrm{Hom}_{\mathbf{Z}}(A, B)$.

**1.13 Theorem.** *If $T$ is a torsion abelian group and $C$ is any abelian group, then $\mathrm{Hom}(T, C)$ is a reduced pure-injective group.*

PROOF. Since $T$ is the direct sum, $T = \bigoplus_{p \in P} T_p$ of its $p$-primary parts, and $\mathrm{Hom}(T, C) \cong \prod_{p \in P} \mathrm{Hom}(T_p, C)$, we can assume without loss of generality that $T$ is a $p$-group for some prime $p$. Then $\mathrm{Hom}(T, C)$ is a $\mathbf{Z}_{(p)}$-module because for any $f \in \mathrm{Hom}(T, C)$ and any $m$ relatively prime to $p$, $mf[T] = f[mT] = f[T]$ and the fact that $f[T]$ is a $p$-group imply that there is a unique $g \in \mathrm{Hom}(T, C)$ such that $mg = f$. What we must show is that $\mathrm{Hom}(T, C)$ is a *complete* $\mathbf{Z}_{(p)}$-module.

First of all, $\mathrm{Hom}(T, C)$ is Hausdorff in the $p$-adic topology: indeed, if $f \in p^\omega \mathrm{Hom}(T, C)$ and $a \in T$, then $p^n a = 0$ for some $n$; now $f = p^n g$ for some $g \in \mathrm{Hom}(T, C)$, so $f(a) = p^n g(a) = g(p^n a) = 0$.

Secondly, every Cauchy sequence in $\mathrm{Hom}(T, C)$ has a limit: indeed, let $\{f_n : n \in \omega\}$ be a sequence in $\mathrm{Hom}(T, C)$ such that for all $n$, $f_{n+1} - f_n = p^n g_n$ for some $g_n$; then define $h : T \to C$ by:

$$h(a) = f_1(a) + \sum_{n \geq 1} (f_{n+1} - f_n)(a) = f_1(a) + \sum_{n \geq 1} p^n g_n(a);$$

for all $a \in T$; this is a finite sum because $a$ has finite order; then it is straightforward to check that $h$ is the limit of the sequence. $\square$

See Fuchs 1970 (section 46) for information about the cardinal invariants of $\mathrm{Hom}(T, C)$.

**1.14 Corollary.** *If $D$ is a divisible abelian group and $A$ is any abelian group, then $\mathrm{Hom}(A, D)$ is pure-injective.*

PROOF. Let $T$ be the torsion part of $A$. The short exact sequence

$$0 \to T \to A \to A/T \to 0$$

induces the Cartan-Eilenberg sequence

$$0 \to \mathrm{Hom}(A/T, D) \to \mathrm{Hom}(A, D) \to$$
$$\mathrm{Hom}(T, D) \to \mathrm{Ext}(A/T, D) = 0.$$

The last term is 0 because $D$ is injective. We claim that $\mathrm{Hom}(A/T, D)$ is divisible. If so, then the preceding exact sequence splits, and $\mathrm{Hom}(A, D) \cong \mathrm{Hom}(A/T, D) \oplus \mathrm{Hom}(T, D)$. Now $\mathrm{Hom}(T, D)$ is pure-injective by 1.13, so $\mathrm{Hom}(A, D)$ is pure-injective.

Thus it remains to prove the claim. Let $D'$ be a torsion-free divisible group containing $A/T$. We have the Cartan-Eilenberg sequence

$$\ldots \to \mathrm{Hom}(D', D) \to \mathrm{Hom}(A/T, D) \to \mathrm{Ext}(D'/(A/T), D) = 0.$$

Moreover, $\mathrm{Hom}(D', D)$ is divisible since $D'$ is torsion-free and divisible, so $\mathrm{Hom}(A/T, D)$ must be divisible. $\square$

Note that we have identified the divisible part, $\mathrm{Hom}(A/T, D)$, of $\mathrm{Hom}(A, D)$ and the reduced part, $\mathrm{Hom}(T, D)$.

**1.15 Theorem.** *Let $F$ be a filter on a set $I$ such that there exist sets $X_n \in F$ such that $\bigcap_{n \in \omega} X_n = \emptyset$. Then for any abelian groups $A_i (i \in I)$, the reduced product $\prod_I A_i/F$ is pure-injective.*

PROOF. (The definition of the reduced product $\prod_I A_i/F$ is given in II.3.1.) Let $F$ and $X_n$ be as in the statement of the theorem. Let

$A = \prod_I A_i/F$. By 1.2 it suffices to show that $A$ is $\aleph_1$-algebraically compact. So let $\mathcal{S} = \{s_n: n \in \omega\}$ be a system of equations over $A$ which is finitely solvable in $A$. Thus each $s_n$ is of the form

$$\sum_{j=1}^{m} r_{nj}y_j = (a_n)_F$$

where $m = m(n)$ may depend on $n$, $r_{nj} \in \mathbf{Z}$, and $a_n \in \prod_I A_i$. Let $s_n(i)$ denote the corresponding equation

$$\sum_{j=1}^{m} r_{nj}y_j = a_n(i)$$

in $A_i$. By hypothesis, for each $k \in \omega$ there is a solution in $A$ of $\{s_n: n \leq k\}$, so by definition of the reduced product and because $F$ is closed under finite intersections, there exists a set $Y_k \in F$ such that for all $i \in Y_k$, $\{s_n(i): n < k\}$ has a solution in $A_i$. By replacing $Y_k$ with $Y_0 \cap \ldots \cap Y_k \cap X_k$ we can assume that $\bigcap_{k\in\omega} Y_k = \emptyset$ and $Y_n \supseteq Y_{n+1}$. As well, we can assume $Y_0 = I$.

Now we are ready to define elements $b_j \in \prod_I A_i (j \in \omega)$ such that $y_j = (b_j)_F$ is a solution to $\mathcal{S}$ in $A$. For each $i$, there is a unique $k(i)$ so that $i \in Y_{k(i)} \setminus Y_{k(i)+1}$. Choose the $b_j(i)$ so that they are a solution of the system $\{s_n(i): n < k(i)\}$; this is possible since $i \in Y_{k(i)}$. This will then give us the desired solution of $\mathcal{S}$ because for every $n$

$$\{i \in I: \text{ the } b_j(i) \text{ solve } s_n(i) \text{ in } A_i\}$$

contains $Y_{n+1}$ and hence belongs to $F$. $\square$

**1.16 Corollary.** *For any abelian group* $B$, $B^\omega/B^{(\omega)}$ *is pure-injective.*

PROOF. In the theorem, let $A_i = B$, let $F$ be the cofinite filter on $I = \omega$, and let $X_n = \omega \setminus \{n\}$. $\square$

## §2. Cotorsion groups

Throughout this section we will be dealing with abelian groups, i.e., $\mathbf{Z}$-modules, which we will refer to simply as groups. (Of course, our discussion extends routinely to arbitrary p.i.d.'s.) A group $A$ is called *cotorsion* if $\text{Ext}(J, A) = 0$ for all torsion-free groups $J$. That is to say, $A$ is cotorsion provided that it is a direct summand of every containing group $B$ with the property that $B/A$ is torsion-free. Since $B/A$ torsion-free implies that $A$ is pure in $B$, it follows that every pure-injective group is cotorsion. Also, (see 2.2), a homomorphic image of a cotorsion group is cotorsion. Later we shall show that the cotorsion groups are precisely the homomorphic images of pure-injective groups (see 2.6). This is of interest to us because of the following (which holds over arbitrary rings).

**2.1 Lemma.** *If $M$ contains a non-zero homomorphic image of a pure-injective module, then $M$ is not slender.*

PROOF. Suppose there is a homomorphism $h: N \to M$ such that $N$ is pure-injective and there exists $a \in N$ such that $h(a) \neq 0$. There is a homomorphism $f: R^{(\omega)} \to N$ such that for all $n \in \omega$, $f(e_n) = a$. Since $R^{(\omega)}$ is pure in $R^\omega$, $f$ extends to $g: R^\omega \to N$. Then $h \circ g: R^\omega \to M$ shows that $M$ is not slender. $\square$

**2.2 Lemma.** *For any short exact sequence $0 \to A \to B \to C \to 0$, if $A$ and $C$ are cotorsion then $B$ is cotorsion. Conversely, if $B$ is cotorsion then $C$ is cotorsion.*

PROOF. Both assertions follow immediately from the exactness of the induced sequence $\text{Ext}(J, A) \to \text{Ext}(J, B) \to \text{Ext}(J, C) \to 0$ for any (torsion-free) group $J$. $\square$

Note that it is not true that every subgroup of a cotorsion group is cotorsion; in fact every group is a subgroup of a divisible, hence cotorsion, group. Here is an important way in which cotorsion groups arise:

**2.3 Proposition.** *For any groups $A$ and $B$, $Ext(A, B)$ is cotorsion.*

PROOF. Let $D$ be a divisible group containing $B$. Then the short exact sequence

$$0 \to B \to D \to D/B \to 0$$

induces the exact sequence

$$\text{Hom}(A, D/B) \to \text{Ext}(A, B) \to \text{Ext}(A, D) = 0.$$

The last term is 0 because $D$ is injective. The first term is pure-injective (and hence cotorsion) by 1.14. Hence $\text{Ext}(A, B)$ is the homomorphic image of a cotorsion group, and is thus cotorsion. $\square$

**Example.** Not every group which arises in this way is pure-injective. Let $A$ be $\mathbb{Q}/\mathbb{Z}$ and let $B$ be a reduced $p$-group such that $p^\omega B \neq \{0\}$. The short exact sequence

$$0 \to \mathbb{Z} \to \mathbb{Q} \to A \to 0$$

induces the exact sequence

$$\begin{aligned} \text{Hom}(\mathbb{Q}, B) &\to \text{Hom}(\mathbb{Z}, B) \to \text{Ext}(A, B) \to \\ \text{Ext}(\mathbb{Q}, B) &\to \text{Ext}(\mathbb{Z}, B). \end{aligned}$$

The last term is 0 since $\mathbb{Z}$ is projective, and the first term is 0 since $B$ is reduced. Thus $B \cong \text{Hom}(\mathbb{Z}, B)$ is isomorphic to a subgroup of $\text{Ext}(A, B)$, and, in fact, $B$ is the torsion part of $\text{Ext}(A, B)$ because $\text{Ext}(\mathbb{Q}, B)$ is torsion-free. Therefore $\text{Ext}(A, B)$ cannot be pure-injective since the $p$-torsion part, $T$, of a reduced pure-injective group satisfies $p^\omega T = \{0\}$ (cf. 1.4 and 1.11).

Here is a simple criterion for being cotorsion:

**2.4 Proposition.** *A is cotorsion if and only if* $\text{Ext}(\mathbb{Q}, A) = 0$.

PROOF. Suppose that $\text{Ext}(\mathbb{Q}, A) = 0$ and that $J$ is a torsion-free group. Now $J$ can be embedded in a divisible group, which is a direct sum of copies of $\mathbb{Q}$. Thus by the Cartan-Eilenberg sequence we have an epimorphism

$$\text{Ext}(\mathbb{Q}^{(\kappa)}, A) \to \text{Ext}(J, A \to 0$$

for some cardinal $\kappa$. Since $\text{Ext}(\mathbb{Q}^{(\kappa)}, A) \cong \text{Ext}(\mathbb{Q}, A)^\kappa = 0$, we can conclude that $\text{Ext}(J, A) = 0$. $\square$

**2.5 Lemma.** *If A is reduced and cotorsion, then* $\operatorname{Ext}(\mathbf{Q}/\mathbf{Z}, A) \cong A$.

PROOF. The short exact sequence

$$0 \to \mathbf{Z} \to \mathbf{Q} \to \mathbf{Q}/\mathbf{Z} \to 0$$

induces the exact sequence

$$\operatorname{Hom}(\mathbf{Q}, A) \to \operatorname{Hom}(\mathbf{Z}, A) \to \operatorname{Ext}(\mathbf{Q}/\mathbf{Z}, A) \to \operatorname{Ext}(\mathbf{Q}, A).$$

The first group is 0 because $A$ is reduced; the last is 0 because $A$ is cotorsion. Therefore $A \cong \operatorname{Hom}(\mathbf{Z}, A) \cong \operatorname{Ext}(\mathbf{Q}/\mathbf{Z}, A)$. $\square$

**2.6 Theorem.** *If A is a cotorsion group, then A is the homomorphic image of a pure-injective group. Moreover, if A is torsion-free, then A is pure-injective.*

PROOF. Let $D$ be a divisible group containing $A$; if $A$ is torsion-free choose $D$ to be torsion-free. Then the short exact sequence

$$0 \to A \to D \to D/A \to 0$$

induces the exact sequence

$$\operatorname{Hom}(\mathbf{Q}/\mathbf{Z}, D) \to \operatorname{Hom}(\mathbf{Q}/\mathbf{Z}, D/A) \to$$
$$\operatorname{Ext}(\mathbf{Q}/\mathbf{Z}, A) \to \operatorname{Ext}(\mathbf{Q}/\mathbf{Z}, D).$$

The last term is 0 since $D$ is injective, and the next-to-last is isomorphic to $A$ by 2.5. Hence $A$ is a homomorphic image of $\operatorname{Hom}(\mathbf{Q}/\mathbf{Z}, D/A)$, which is pure-injective by 1.14. If $A$ is torsion-free, then the first term is 0, so $A$ is isomorphic to $\operatorname{Hom}(\mathbf{Q}/\mathbf{Z}, D/A)$. $\square$

**2.7 Corollary.** *If A is a reduced torsion-free non-zero cotorsion group, then A contains a copy of $J_p$ for some prime p.*

PROOF. This follows immediately from Theorem 2.6 and the structure theorem for pure-injectives (cf. 1.12). $\square$

We conclude with a brief discussion of a class of groups which arise, among other places, in the study of endomorphism rings of groups. (See section XIII.3.)

**2.8 Definition.** A group $A$ is called *cotorsion-free* if it does not contain any non-zero subgroups which are cotorsion.

**2.9 Theorem.** *For any group $A$ the following are equivalent:*

(1) $A$ *is cotorsion-free;*

(2) $\mathrm{Hom}(\hat{\mathbf{Z}}, A) = 0$, *where* $\hat{\mathbf{Z}} = \prod_{p \in P} J_p$ *(see I.3);*

(3) $A$ *does not contain a copy of* $\mathbf{Q}$, $Z(p)$, *or* $J_p$ *for any prime* $p$;

(4) $A$ *does not contain any non-zero pure-injective subgroup;*

(5) $A$ *is reduced and torsion-free and does not contain a direct summand isomorphic to* $J_p$ *for any prime $p$;*

(6) $A$ *is reduced and torsion-free and does not contain a subgroup isomorphic to* $J_p$ *for any prime $p$.*

PROOF. (1) $\Rightarrow$ (2) follows immediately from the facts that $\hat{\mathbf{Z}}$ is pure-injective (cf. 1.9 and 1.11) and that a homomorphic image of a pure-injective group is cotorsion. (4) $\Rightarrow$ (3) is easy, and (3) $\Rightarrow$ (4) follows from 2.7 because (3) clearly implies that A is reduced and torsion-free. (3) obviously implies (5), and (6) $\Rightarrow$ (1) follows from 2.7. So it remains to prove (2) $\Rightarrow$ (3) and (5) $\Rightarrow$ (6).

Assume (2). If $A$ contains a copy of $\mathbf{Q}$, then there is a non-zero homomorphism from $\hat{\mathbf{Z}}$ to $A$ by the injectivity of $\mathbf{Q}$. If $A$ contains a copy of $Z(p)$, we can obtain a non-zero homomorphism of $\hat{\mathbf{Z}}$ into $A$ by composing the canonical surjection from $\hat{\mathbf{Z}}$ onto $\hat{\mathbf{Z}}/p\hat{\mathbf{Z}}$ with the projection of $\hat{\mathbf{Z}}/p\hat{\mathbf{Z}}$ — which is a direct sum of copies of $Z(p)$ — onto one of its summands. Finally, $A$ obviously cannot contain a copy of $J_p$. This proves (2) $\Rightarrow$ (3).

Now assume (5), and suppose, to obtain a contradiction, that there is an embedding $f: J_p \rightarrow A$. Since $A$ is reduced, there is a maximal power, $p^k$, of $p$ which divides $f(1)$. Define $g: J_p \rightarrow A$ by $g(x) = p^{-k} f(x)$ for all $x \in J_p$. (Note that $p^k$ divides $f(x)$ because there exists $n \in \omega$ such that $p^k | (x-n)$, so $p^k$ divides $f(x) - n \cdot f(1)$.) Clearly $g$ is an embedding, so we will obtain a contradiction if we show that $\mathrm{im}(g)$ is pure in $A$, for then it will be a direct summand of $A$. But for any $x \in J_p$ there exists $n \in \mathbf{Z}$ such that $p$ divides

$(x - n)$ in $J_p$. Since $p$ divides $ng(1)$ in $A$ only if $p$ divides $n$, we get that $p$ divides $g(x)$ in $A$ only if $p$ divides $g(x)$ in $\mathrm{im}(g)$. $\square$

As a consequence of the theorem and the results of Chapter III, we obtain immediately:

**2.10 Corollary.** (i) *If $A$ is a countable group, then $A$ is cotorsion-free if and only if $A$ is reduced and torsion-free if and only if $A$ is slender.*

(ii) *If $A$ is $\aleph_1$-free, then $A$ is cotorsion-free.* $\square$

By 2.1 and 2.6 we know that a necessary condition for $A$ to be slender is that it be cotorsion-free. That this condition is not sufficient is seen by the example of $\mathbf{Z}^\omega$, which is cotorsion-free. However, Nunke's theorem (IX.2.4) will tell us that $\mathbf{Z}^\omega$ is the *only* additional obstacle to $A$ being slender.

## EXERCISES

In Exercises 1-3, $D$ is an $\omega_1$-incomplete ultrafilter on a cardinal $\kappa$, $M$ is a group, and $M^*$ is the ultrapower $M^\kappa/D$ of $M$ with respect to $D$. (See II.3.1; note that here $M^*$ does *not* mean the dual of $M$.) By Theorem 1.15, $M^*$ is $\aleph_1$-algebraically compact, hence pure-injective. Since $M^*$ is pure-injective, it is determined by cardinal invariants $\alpha_{p,\,n}$, $\beta_p$, $\gamma_p$, $\delta$, as in 1.12.

1. Prove: (a) $\beta_p = \lim_{n \to \infty} \dim p^n M^*/p^{n+1} M^*$;
$(b)\gamma_p = \lim_{n \to \infty} \dim p^n M^*[p]$;
$(c)\delta = \lim_{n \to \infty} \mathrm{rank}\,(n!M^*)$.
[Hint: the inequalities " $\leq$ " hold in any pure-injective group; the opposite inequalities need that $M^*$ is an ultrapower with respect to an $\omega_1$-incomplete ultrafilter: e.g., in (a), use that $p^n M^*/p^{n+1} M^* \cong (p^n M/p^{n+1} M)^\kappa/D$ and that $\beta_p = \dim(M^*/T + pM^*)$, where $T$ is the torsion subgroup of $M^*$.]

2. Prove that for any group $A$, the following are equivalent:
(1) $A \cong A^\omega/D$ for every ultrafilter $D$ on $\omega$;
(2) $A \cong M^\kappa/D$ for some $\omega_1$-incomplete ultrafilter $D$ on some $\kappa$;
(3) $A$ is pure-injective, and if $\alpha_{p,\,n}$, $\beta_p$, $\gamma_p$ and $\delta$ are the invariants of $A$, as in 1.12, then 1(a), (b), and (c) hold as well as:

(d) if $\sigma = \alpha_{p,\,n}$, $\beta_p$, or $\gamma_p$, then either $\sigma$ is finite, or $\sigma^\omega = \sigma$;

(e) either $\delta = 0$ or $\delta^\omega = \delta$.

[Hint: For (2) $\Rightarrow$ (3), use the fact that $|I^\omega/D| = |I^\omega/D|^\omega$ if $I$ is infinite (see Exercise II.4).]

3. If $A \cong \prod_{n\in\omega} B_n/D$ for any groups $B_n$ and some non-principal ultrafilter $D$ on $\omega$, then $A \cong A^\omega/U$ for every ultrafilter $U$ on $\omega$. [Hint: Use Exercise II.14 and Exercise 2 above.]

4. The cardinal invariants of $\mathbf{Z}^\omega/\mathbf{Z}^{(\omega)}$ are: $\alpha_{p,\,n} = 0$; $\beta_p = 2^{\aleph_0}$; $\gamma_p = 0$; $\delta = 2^{\aleph_0}$ (cf. 1.12 and 1.16).

In Exercises 5-7, let $F$ be a filter on $I$. Let us say that $F$ is *weakly $\kappa$-complete* if the intersection of $< \kappa$ elements of $F$ is non-empty. (The definition of $\kappa$-complete is given in II.2.4.)

5. Generalize Theorem 1.15 as follows: Let $F$ be a $\kappa$-complete filter which is not weakly $\kappa^+$-complete. Then for any modules $M_i$ ($i \in I$), the reduced product $N = \prod_{i\in I} M_i/F$ satisfies: any system of $\kappa$ linear equations over $N$ is solvable in $N$ whenever every subsystem of size $< \kappa$ is solvable in $N$.

6. (i) If $F$ is $\kappa$-complete and $M$ is pure-injective, then $N = M^I/F$ satisfies: any system of $\kappa$ linear equations over $N$ is solvable in $N$ whenever every finite subsystem is solvable.

(ii) Suppose $F$ is weakly $\kappa^+$-complete and $M$ is such that there is a system of $\kappa$ linear equations over $M$ which is not solvable in $M$ but is such that every subsystem of size $< \kappa$ is solvable. Then $M^I/F$ has the same property.

7. Let $A$ be an abelian group.

(i) If $A$ is pure-injective, then for all filters $F$, $A^I/F$ is pure-injective.

(ii) If $A$ is not pure-injective, then $A^I/F$ is pure-injective if and only if $F$ is not weakly $\omega_1$-complete. [Hint: use 5 & 6.]

8. Suppose $M$ is an $R$-module and $\kappa = (|R| + \aleph_0)$. Let $I$ be the set of finite subsets of $\kappa$ and $F$ the filter generated by $\{\{X \in I : \alpha \in X\} : \alpha < \kappa\}$. Show that the diagonal embedding $\delta_M : M \to M^I/F$ is

a pure embedding and every finitely solvable system of $\kappa$ equations whose parameters are from $\delta_M[M]$ is solvable in $M^I/F$. (This can be used to give an alternate proof of 1.2 (2) $\Rightarrow$ (3).)

9. If $A$ is cotorsion-free and $\varphi: \mathbb{Z}^\omega \to A$, then for all $x \in A \setminus \{0\}$, $\{n \in \omega: \varphi(e_n) = x\}$ is finite. [Hint: assume $\varphi$ and $x$ form a counterexample. Let $B = \langle x \rangle_*$ and choose $p$ so that $B$ is Hausdorff in the $p$-adic topology. Let $\hat{B}$ denote the $p$-adic closure of $B$ in $A$. Choose $\zeta \in \hat{\mathbb{Z}}_p$ so that $\zeta x \notin \hat{B}$ (and hence not in $A$). Show $\zeta x \in \mathrm{rge}(\varphi)$.]

## NOTES

The theory of pure-injective modules was developed by Kaplansky, Łoś, Maranda, and Fuchs, among others; see Fuchs 1970, Chapter VI, for more on this subject; see also Warfield 1969. The notion of a basic subgroup is due to Kulikov; see Fuchs 1970, Chapter VI. See Prest 1988 for an exposition of the important role of pure-injectivity in the model theory of modules. Our proofs of 1.2, 1.4 and 1.11 are inspired by model-theoretic approaches to the subject (particularly Eklof-Fisher 1972). Theorem 1.13 is due to Fuchs and Harrison. Theorem 1.15 is in Mycielski 1964 in more general form. (For modules over a general ring $R$, there are ultrafilters with respect to which arbitrary ultraproducts will be pure-injective; this is due to Keisler 1964b and Kunen 1972; see also Eklof 1977b (section 10).) Corollary 1.16 is due to Balcerzyk 1959.

Cotorsion groups were discovered independently by Fuchs, Harrison and Nunke; for more on this subject see Fuchs 1970, Chapter IX. The notion of cotorsion-free comes from Göbel 1975, where it is called *stout*. The various parts of Theorem 2.9 are proved in Göbel 1975, Göbel-Wald 1979, and Dugas-Göbel 1982a.

Exercises. 1-3:Eklof 1973; 4:Balcerzyk 1959; 5:Mycielski 1964; 7:Dugas-Göbel 1979a.

# CHAPTER VI
# MORE SET THEORY

In Chapter II we dealt with methods and results in Zermelo-Frankel set theory. In this chapter we turn to results which are (relatively) consistent with ZFC but not provable in ZFC. Our approach is axiomatic: we discuss certain principles which have been proved to be consistent with ZFC, and derive consequences of them, but we do not give the consistency proofs here; for that, we refer the reader to standard texts on set theory.

Some of the principles we discuss are consequences of the Axiom of Constructibility. Kurt Gödel introduced the Axiom of Constructibility in connection with his proof that the Axiom of Choice and the Continuum Hypothesis are relatively consistent with ZF. Very roughly, it says that the universe of sets, denoted V, is equal to the smallest transitive submodel of V satisfying ZF and containing all the ordinals; this smallest model is denoted L, and consequently the Axiom of Constructibility is abbreviated as "V = L". We will give a precise formulation of the Axiom of Constructibility in section 3, but we will not use the axiom directly. Instead, we will use certain combinatorial consequences of the Axiom of Constructibility that were isolated by Ronald Jensen. Among these are the diamond principles, with which we begin in section 1; there are also weak forms of the diamond principles which are consequences of (a weak form of) the Continuum Hypothesis. In section 3 we will consider other consequences of V = L. Section 4 deals with some other principles, namely Martin's Axiom and the Proper Forcing Axiom, which are inconsistent with the Axiom of Constructibility, or even with the Continuum Hypothesis. For the material in sections 3 and 4 we need some notions from model theory, which are introduced in section 2.

## §1. Prediction principles

In many inductive constructions the possible obstacles to the success of the construction can be predicted; then, knowing the obstacles, one can avoid them.

The first use of a prediction principle that we are aware of is the construction by Dushnik-Miller 1940 of a dense subset of $\mathbb{R}$ which has no (non-trivial) order-automorphisms. Their proof uses the fact that if $X$ is a dense subset of $\mathbb{R}$ and $\varphi: X \to X$ is an order-automorphism, then $\varphi$ has a unique extension to an order-automorphism $\hat{\varphi}$ of $\mathbb{R}$. Furthermore, any order-automorphism of $\mathbb{R}$ is determined by its value on $\mathbb{Q}$. Since there are $(2^{\aleph_0})^{\aleph_0} = 2^{\aleph_0}$ functions from $\mathbb{Q}$ to $\mathbb{R}$, there is a list $\{\varphi_i : i < 2^{\aleph_0}\}$ of all the order-automorphisms of $\mathbb{R}$. Hence if $X$ is a dense subset of $\mathbb{R}$, the set of order-automorphisms of $X$ is contained in $\{\varphi_i {\restriction} X : i < 2^{\aleph_0}\}$. The construction of the desired set $X$ is an induction of order type $2^{\aleph_0}$ where at step $i + 1$, $\varphi_i$ is prevented from being an order-automorphism of $X$. More exactly two disjoint sets $A_i$, $B_i$ are constructed by induction on $i$ so that the desired set, $X$, will be $\cup A_i$ and so that $\cup B_i$ will form a set of forbidden elements. Let $A_{-1} = \mathbb{Q}$ and $B_{-1} = \emptyset$. Take unions at limit ordinals. At stage $i+1$, choose $a_i \notin B_i$ so that $\varphi_i(a_i) \notin A_i \cup \{a_i\}$. Let $A_{i+1} = A_i \cup \{a_i\}$ and $B_{i+1} = B_i \cup \{\varphi_i(a_i)\}$.

In this construction we were able to enumerate all the functions that we wanted to kill and carry on an induction long enough to kill them all. In many cases, this will not be possible; it is here that the diamond principles may be useful.

**1.1 Definition.** Let $E$ be a stationary subset of a regular uncountable cardinal $\kappa$. By $\Diamond_\kappa(E)$ we mean the following principle:

> there is a family $\{W_\alpha : \alpha \in E\}$ of sets such that for each $\alpha \in E$, $W_\alpha \subseteq \alpha$, and for all $X \subseteq \kappa$, $\{\alpha \in E : W_\alpha = X \cap \alpha\}$ is stationary in $\kappa$.

A family $\{W_\alpha : \alpha \in E\}$ satisfying these hypotheses will be called a $\Diamond_\kappa(E)$-*sequence*. The principle $\Diamond_\kappa(\kappa)$ is denoted $\Diamond_\kappa$, $\Diamond_{\omega_1}(E)$ is denoted $\Diamond(E)$, and $\Diamond_{\omega_1}$ is denoted simply $\Diamond$.

Thus a $\Diamond_\kappa(E)$-sequence "predicts" what the intersection of $X$ with $\alpha$ will be, and no matter what $X$ is chosen, it is correct for

a relatively large set of $\alpha$'s. R. B. Jensen 1972 proved that ZF +
$V = L$ implies that $\Diamond_\kappa(E)$ holds for all regular uncountable $\kappa$ and
all stationary $E \subseteq \kappa$; hence $\Diamond_\kappa(E)$ is consistent with ZFC. (But
it is not provable from ZFC.) It is obvious that if $E \subseteq E'$ and
$\Diamond_\kappa(E)$ holds, then $\Diamond_\kappa(E')$ holds as well. Also, if $\tilde{E}_1 = \tilde{E}_2$, then
$\Diamond_\kappa(E_1)$ holds if and only if $\Diamond_\kappa(E_2)$ holds. The next lemma gives
some convenient alternate formulations of the diamond principles
in terms of $\kappa$-filtrations (see II.4.12).

**1.2 Lemma.** *Assume* $\Diamond_\kappa(E)$. *Then*

(i) *for any set $A$ of cardinality $\kappa$ and any $\kappa$-filtration $\{A_\nu : \nu < \kappa\}$
of $A$, there is a family $\{Y_\alpha : \alpha \in E\}$ such that for each $\alpha \in E$,
$Y_\alpha \subseteq A_\alpha$, and for all $X \subseteq A$, $\{\alpha \in E : Y_\alpha = X \cap A_\alpha\}$ is stationary
in $\kappa$; and*

(ii) *for any sets $A$ of cardinality $\kappa$ and $B$ of cardinality $\leq \kappa$,
and any $\kappa$-filtrations $\{A_\nu : \nu \in \kappa\}$ and $\{B_\nu : \nu \in \kappa\}$ of $A$ and $B$
respectively, there is a family $\{g_\alpha : \alpha \in E\}$ such that for each $\alpha \in
E$, $g_\alpha$ is a function : $A_\alpha \to B_\alpha$ and for all functions $f : A \to B$,
$\{\alpha \in E : f \restriction A_\alpha = g_a\}$ is stationary in $\kappa$.*

PROOF. (i) There is a cub $C$ such that for all $\nu \in C$, if $\nu^+$ denotes
the successor of $\nu$ in $C$, then $|A_{\nu^+} \setminus A_\nu| = |\nu^+ \setminus \nu|$ (cf. Exercise II.18).
Then define by transfinite induction a bijection $\theta : \kappa \to A$ such that
for all $\nu \in C$ $\theta[\nu] = A_\nu$. If $\{W_\alpha : \alpha \in E\}$ is a $\Diamond_\kappa(E)$-sequence,
and we let $Y_\alpha = \theta[W_\alpha]$ for $\alpha \in E \cap C$ and let $Y_\alpha$ be arbitrary for
$\alpha \in E \setminus C$, then $\{Y_\alpha : \alpha \in E\}$ has the desired properties.

(ii) The key is to regard functions from $A$ to $B$ as subsets of
$A \times B$. Apply (i) to the $\kappa$-filtration $\{A_\nu \times B_\nu : \nu \in \kappa\}$ of $A \times B$ to get
the family $\{Y_\alpha : \alpha \in E\}$. If $Y_\alpha(\subseteq A_\alpha \times B_\alpha)$ is a function : $A_\alpha \to B_\alpha$,
let $g_\alpha$ be $Y_\alpha$; otherwise let $g_\alpha$ be arbitrary. Now given $f : A \to B$,
there is a cub $C$ such that for $\nu \in C$ $f[A_\nu] \subseteq B_\nu$ (cf. II.4.12), and
hence $f \cap (A_\alpha \times B_\alpha) = f \restriction A_\alpha$. It is then clear from the choice of
$\{Y_\alpha : \alpha \in E\}$ that $\{g_\alpha : \alpha \in E\}$ has the desired properties. $\square$

In the appropriate contexts, we refer to $\{X_\alpha : \alpha \in E\}$ and $\{g_\alpha : \alpha \in
E\}$, with the properties given in the lemma, as $\Diamond_\kappa(E)$-sequences.

**1.3 Proposition.** *If $\kappa = \lambda^+$, $\Diamond_\kappa$ implies that $2^\lambda = \lambda^+$.*

PROOF. Let $\{W_\alpha : \alpha \in \kappa\}$ be a $\diamondsuit_\kappa$-sequence. For every $X \subseteq \lambda$ there exists a stationary set of $\alpha \in E$ such that $X \cap \alpha = W_\alpha$. Hence there exists an $\alpha > \lambda$ such that $X \cap \alpha = W_\alpha$; but in this case, $X = X \cap \alpha$, so every subset of $\lambda$ occurs in the sequence $\{W_\alpha : \alpha \in \kappa\}$. Thus the cardinality, $2^\lambda$, of $\mathcal{P}(\lambda)$ is $\leq \kappa = \lambda^+$. $\square$

Gregory 1976 and Shelah 1981c have shown that the Generalized Continuum Hypothesis implies $\diamondsuit_\kappa$ for every $\kappa = \lambda^+$ where $\lambda \geq \aleph_1$. (See Exercises 13 – 14.) However it is not the case that GCH implies $\diamondsuit$.

For examples of the uses of diamond, we turn to the notion of a ladder system, which was defined in II.4.13.

**1.4 Proposition.** *Suppose that $\diamondsuit(E)$ holds for some subset $E$ of* $\lim(\omega_1)$. *Then for any ladder system $\eta$ on $E$, there is a 2-coloring of $\eta$ which cannot be uniformized.*

PROOF. In Lemma 1.2(ii) let $A = \omega_1$, $A_\nu = \nu$, and $B = B_\nu = 2$. Let $\{g_\alpha : \alpha \in E\}$ be the corresponding $\diamondsuit(E)$-sequence. We think of the functions $g_\alpha$ as predicting functions $f : \omega_1 \to 2$, and we define the coloring $c_\alpha$ so as to kill the possibility that $f$ will become the first coordinate of a uniformizing pair $(f, f^*)$. For each $\alpha \in E$, define

$$c_\alpha(k) = \begin{cases} 1 & \text{if } \{n : g_\alpha(\eta_\alpha(n)) = 0\} \text{ is infinite} \\ 0 & \text{otherwise} \end{cases}$$

for all $k \in \omega$ (so that $c_\alpha$ is a constant function). We claim that $c = \{c_\alpha : \alpha \in E\}$ cannot be uniformized. Indeed, suppose there is a uniformization $(f, f^*)$. Then by Lemma 1.2(ii) there exists $\alpha \in E$ (even a stationary set of such $\alpha$) so that $f \upharpoonright \alpha = g_\alpha$. We consider two cases: either there exists $f(\eta_\alpha(n)) = 0$, for infinitely many $n$, or not. In the first case, we have defined $c_\alpha(k) = 1$ for all $k$, so for infinitely many $n$, $f(\eta_\alpha(n)) \neq c_\alpha(n)$; in the second case, we have defined $c_\alpha(k) = 0$ for all $k$, so that there are infinitely many $n$ such that $f(\eta_\alpha(n)) \neq c_\alpha(n)$. Thus in either case we have a contradiction of the definition of a uniformization. $\square$

In our second application, we use a diamond principle to predict cubs; we shall make use of it in Chapter VIII.

**1.5 Proposition.** *Let $E_0$ be a subset of* $\lim(\omega_1)$ *and assume* $\Diamond(E_0)$. *Given stationary subsets $E_0$ and $E_1$ of $\omega_1$, there is a ladder system $\eta$ on $E_0$ so that for every cub $C$, there exists $\alpha \in E_0$ such that for all $n$, $\eta_\alpha(n) \in C \cap E_1$.*

PROOF. Let $\{W_\alpha : \alpha \in E_0\}$ be a $\Diamond(E_0)$-sequence. We use these predictions of the cub $C$ to define our ladders. If $W_\alpha \cap E_1$ is cofinal in $\alpha$, let $\eta_\alpha$ be a ladder whose range is contained in $W_\alpha \cap E_1$; otherwise, let $\eta_\alpha$ be arbitrary. Now, given a cub $C$, $C \cap E_1$ is unbounded in $\omega_1$ since $E_1$ is stationary in $\omega_1$. Let

$$D = \{\delta \in \omega_1 : \delta = \sup Y \text{ for some subset } Y \text{ of } C \cap E_1 \text{ s. t. } \delta \notin Y\}.$$

(That is, $D$ is the set of limit points of the closure of $C \cap E_1$.) Then it is easy to see that $D$ is a cub, so by definition of a $\Diamond(E_0)$-sequence there exists $\delta \in D \cap E_0$ such that $C \cap \delta = W_\delta$. By the choice of $\delta$ in $D$, $W_\delta \cap E_1 (= C \cap \delta \cap E_1)$ is cofinal in $\delta$, so we have defined $\eta_\delta$ so that for all $n$, $\eta_\delta(n) \in C \cap E_1$. $\square$

Devlin and Shelah have isolated a weaker prediction principle which can in many constructions be used in place of diamond. (In Chapter XIII we will introduce a prediction principle that is provable in ZFC and consider when and how it can be used in place of diamond.)

**1.6 Definition.** Let $E$ be a stationary subset of a regular uncountable cardinal $\kappa$. By $\Phi_\kappa(E)$ we mean the following principle:

given for each $\alpha \in E$ a "partition" function $P_\alpha : \mathcal{P}(\alpha) \to 2 = \{0, 1\}$, there is a function $\rho : E \to 2$ such that for all $X \subseteq \kappa$ $\{\alpha \in E : P_\alpha(X \cap \alpha) = \rho(\alpha)\}$ is stationary in $\kappa$.

A function $\rho$ with these properties is called a *weak diamond function* (relative to $\{P_\alpha : \alpha \in E\}$). If $E$ satisfies $\Phi_\kappa(E)$ we say that $E$ has the weak diamond property or that $E$ is *non-small*; otherwise we say that $E$ is *small*.

Thus $P_\alpha$ partitions the subsets of $\alpha$ into two classes, and $\Phi_\kappa(E)$ allows us to predict, not what $X \cap \alpha$ will be, but only whether it will belong to the first or second member of the partition determined by $P_\alpha$. In many cases, this will suffice, since we get to chose the partition functions to fit our needs. By the same coding techniques as in Lemma 1.2 we can prove the following lemma. (We use $^XY$ to denote the set of all functions from $X$ to $Y$.)

**1.7 Lemma.** *Assume* $\Phi_\kappa(E)$. *Then*

(i) *For any set $A$ of cardinality $\kappa$, any $\kappa$-filtration $\{A_\nu : \nu < \kappa\}$ of $A$ and any family of functions $P_\alpha : \mathcal{P}(A_\alpha) \to 2$ ($\alpha \in E$), there is a function $\rho : E \to 2$ such that for all $X \subseteq A$, $\{\alpha \in E : P_\alpha(X \cap A_\alpha) = \rho(\alpha)\}$ is stationary in $\kappa$;*

(ii) *For any sets $A$ of cardinality $\kappa$, $B$ of cardinality $\leq \kappa$, any $\kappa$-filtrations $\{A_\nu : \nu \in \kappa\}$ and $\{B_\nu : \nu \in \kappa\}$ of $A$ and $B$ respectively and any family of functions $P_\alpha : {}^{A_\alpha}B_\alpha \to 2$ ($\alpha \in E$), there is a function $\rho : E \to 2$ such that for all functions $f : A \to B\{\alpha \in E : P_\alpha(f \upharpoonright A_\alpha) = \rho(\alpha)\}$ is stationary in $\kappa$.* $\square$

Now we will reprove the conclusion of Proposition 1.4 but with a weaker hypothesis.

**1.8 Proposition.** *Assume that $\Phi_{\omega_1}(E)$ holds for some subset $E$ of $\omega_1$. Then for any ladder system $\eta$ on $E$ there is a 2-coloring of $\eta$ which cannot be uniformized.*

PROOF. In Lemma 1.7(ii) let $A = \omega_1$, $A_\nu = \nu$, and $B = B_\nu = 2$. For each $\alpha \in E$ let $P_\alpha : {}^\alpha 2 \to 2$ be defined by:

$$P_\alpha(h) = \begin{cases} 1 & \{n : h(\eta_\alpha(n)) = 0\} \text{ is infinite} \\ 0 & \text{otherwise} \end{cases}$$

Let $\rho$ be the corresponding weak diamond function. Then let $c_\alpha(k) = \rho(\alpha)$ for all $k \in \omega$. We claim that $c = \{c_\alpha : \alpha \in E\}$ cannot be uniformized. Indeed, suppose there is a uniformization $(f, f^*)$. Then by Lemma 1.7(ii) there exists $\alpha \in E$ so that $\rho(\alpha) = P_\alpha(f \upharpoonright \alpha)$. We consider two cases: either $f(\eta_\alpha(n)) = 0$, for infinitely many $n$, or not. In the first case $P_\alpha(f \upharpoonright \alpha) = 1$, so $c_\alpha(k) = 1$ for all $k$, and thus for infinitely many $n$, $f(\eta_\alpha(n)) \neq c_\alpha(n)$; in the second case, $P_\alpha(f{\restriction}\alpha) = 0$ so $c_\alpha(k) = 0$ for all $k$, and thus there are infinitely many $n$ such that $f(\eta_\alpha(n)) \neq c_\alpha(n)$. In either case we have a contradiction of the definition of a uniformization. $\square$

Devlin and Shelah 1978 proved that $\Phi_{\omega_1}(\omega_1)$ is a consequence of a weak form of the Continuum Hypothesis, namely $2^{\aleph_0} < 2^{\aleph_1}$. In fact, the converse holds as well:

**1.9 Proposition.** $\Phi_{\omega_1}(\omega_1)$ implies $2^{\aleph_0} < 2^{\aleph_1}$.

PROOF. Suppose that $\Phi_{\omega_1}(\omega_1)$ holds, but $2^{\aleph_0} = 2^{\aleph_1}$. Let $C = \omega_1 \setminus \omega$. Then there is a one-one function $\Psi: {}^C 2 \to {}^\omega 2$. For each $\alpha > \omega$ define $P_\alpha$ as follows: if $\sigma: \alpha \to 2$, $P_\alpha(\sigma) = 0$ if and only if there exists $\tau: \omega_1 \to 2$ such that $\sigma \subseteq \tau$, $\tau(\alpha) = 0$ and $\Psi(\tau{\restriction}C) = \sigma{\restriction}\omega$. Now let $\rho$ be a weak diamond function for this family of partition functions. Define $\eta: \omega_1 \to 2$ by: $\eta{\restriction}\omega = \Psi((1 - \rho){\restriction}C)$ and $\eta{\restriction}C = (1 - \rho){\restriction}C$. By hypothesis, there exists $\alpha > \omega$ so that $P_\alpha(\eta{\restriction}\alpha) = \rho(\alpha)$; we will see that this leads to a contradiction. First suppose that $\rho(\alpha) = 1$. Then $\eta(\alpha) = (1 - \rho)(\alpha) = 0$; but then $\tau = \eta$ satisfies the properties which imply that $P_\alpha(\eta{\restriction}\alpha) = 0$, a contradiction. Second, suppose that $\rho(\alpha) = 0$. Then by definition of $P_\alpha$, there is a function $\tau$ extending $\eta{\restriction}\alpha$ such that $\tau(\alpha) = 0$ and $\Psi(\tau{\restriction}C) = \eta{\restriction}\omega = \Psi((1-\rho){\restriction}C)$; the latter implies that $\tau{\restriction}C = \eta{\restriction}C$ since $\Psi$ is one-one; thus $\tau = \eta$ which is a contradiction since $0 = \tau(\alpha) = \eta(\alpha) = (1 - \rho)(\alpha) = 1$. $\square$

Thus $\omega_1$ is non-small if and only $2^{\aleph_0} < 2^{\aleph_1}$. In some constructions we may need more than this, namely a large number of non-small subsets of $\omega_1$. In that case we can appeal to the following theorem.

**1.10 Theorem.** Suppose $\kappa = \lambda^+$ and $\Phi_\kappa(E)$ holds. Then there is a decomposition of $E$, $E = \amalg_{\beta \in \kappa} E_\beta$, into $\kappa$ disjoint non-small subsets $E_\beta$.

PROOF. Let $F = \{X \subseteq \kappa: \kappa \setminus X \text{ is small}\}$. By Proposition II.4.8 it is enough to prove that $F$ is a $\kappa$-complete filter on $\kappa$. Now if $X \subseteq Y$ and $X \in F$ then $Y \in F$ since if $\kappa \setminus X$ is small, then $\kappa \setminus Y \subseteq \kappa \setminus X$ is also small. So it remains to show that $F$ is closed under intersection of families of size $\lambda$. For this it suffices to show that if $\{S_\nu : \nu < \lambda\}$ is a family of small subsets of $\kappa$, then $S \stackrel{\text{def}}{=} \bigcup_{\nu < \lambda} S_\nu$ is still small. For each $\nu < \lambda$ let $P^\nu = \{P_\alpha^\nu : \alpha \in S_\nu\}$ be a family of partition functions which witnesses to the failure of $\Phi_\kappa(S_\nu)$, i.e., for any $\rho : S_\nu \to 2$ there exists $Z \subseteq \kappa$ such that $\{\alpha \in S_\nu : P_\alpha^\nu(Z \cap \alpha) = \rho(\alpha)\}$ is thin.

Choose a bijection $\theta : \kappa \times \lambda \to \kappa$. Then there is a cub $C$ so that for $\alpha \in C$, $\theta[\alpha \times \lambda] = \alpha$ (cf. II.4.12). We will now define partition functions $P_\alpha(\alpha \in S)$ which we will use to show that $S$ is small. Given $\alpha \in S \cap C$ let $\nu$ be minimal so that $\alpha \in S_\nu$; then for any $Y \subseteq \alpha$ let

$$P_\alpha(Y) = P_\alpha^\nu(\{\mu \in \kappa : \theta(\mu, \nu) \in Y\}).$$

(If $\alpha \in S \setminus C$, let $P_\alpha$ be arbitrary.) Now, given $\rho : S \to 2$ we know that $\rho {\restriction} S_\nu$ is not a weak diamond function for $\{P_\alpha^\nu : \alpha \in S_\nu\}$, so there exists $Z_\nu \subseteq \kappa$ such that there is a cub $C_\nu$ such that $C_\nu \cap \{\alpha \in S_\nu : P_\alpha^\nu(Z_\nu \cap \alpha) = \rho(\alpha)\} = \emptyset$. Let

$$Z = \{\theta(\mu, \nu) : \mu \in Z_\nu, \nu < \lambda\}.$$

We shall show that $\rho$ is not a weak diamond function for $\{P_\alpha : \alpha \in S\}$ by showing that

$$C \cap \bigcap_{\nu < \lambda} C_\nu \cap \{\alpha \in S : P_\alpha(Z \cap \alpha) = \rho(\alpha)\}$$

is empty. Suppose that $\alpha$ belongs to $C \cap S$. Let $\nu$ be minimal so that $\alpha \in S_\nu$. By definition of $P_\alpha$, $P_\alpha(Z \cap \alpha) = P_\alpha^\nu(Z_\nu \cap \alpha)$. But then if $\alpha \in C_\nu$, $P_\alpha^\nu(Z_\nu \cap \alpha) \neq \rho(\alpha)$ so $\alpha$ does not belong to the last set in the intersection. $\square$

In fact, one can prove that the filter, $F$, in the proof above is a normal filter; that is, it is closed under diagonal intersections.

**1.11 Corollary.** *If $2^{\aleph_0} < 2^{\aleph_1}$, then there are $\aleph_1$ pairwise disjoint non-small subsets of $\omega_1$.*

PROOF. Since $2^{\aleph_0} < 2^{\aleph_1}$ implies $\omega_1$ is non-small, we can apply 1.10 with $E = \omega_1$. $\square$

There is a result analogous to 1.10 for diamond, which we leave as an exercise. (See Exercise 3.)

We conclude this section by defining another prediction principle:

**1.12 Definition.** Let $\kappa$ be a regular cardinal and $E$ be a stationary subset of $\kappa$. By $\Diamond_\kappa^*(E)$ we mean the following principle:

> there is a family $\{S_\alpha : \alpha \in E\}$ such that each $S_\alpha$ is a subset of $\mathcal{P}(\alpha)$ of cardinality $< \kappa$ and for all $X \subseteq \kappa$ there is a cub $C \subseteq \kappa$ such that $X \cap \alpha \in S_\alpha$ for all $\alpha \in C \cap E$.

Then $\{S_\alpha : \alpha \in E\}$ is called a $\Diamond_\kappa^*(E)$-*sequence*. $V = L$ implies $\Diamond_\kappa^*(E)$ for every stationary subset $E$ of $\omega_1$. We leave it as an exercise to show that $\Diamond_\kappa^*(E)$ implies $\Diamond_\kappa(E')$ for every $E'$ such that $E \cap E'$ is stationary in $\kappa$. (See Exercise 2.)

## §2. Models of set theory

At several points in the rest of the book, we will need to work with countable models of set theory. This will occur when we make use of the axioms MA and PFA (introduced in section 4) and when we use the Black Box (introduced in Chapter XIII). In order to prepare for that, we will spend a little time here investigating the properties of such models. Our discussion of the basic notions of model theory will be rather informal and intuitive, and confined to the context of models of set theory; for a more general and more careful introduction to the basic concepts, the reader may want to consult a text in model theory such as Chang-Keisler 1973 or Bell-Slomson 1969.

The axioms of ZFC can be expressed in a language, $\mathcal{L}_\in$, which has one primitive symbol, namely " $\in$ ", a binary relation symbol. (See the beginning of section II.1.) The *atomic formulas* of $\mathcal{L}_\in$ are those of the form

$$x \in y \quad \text{or} \quad x = y$$

where $x$ and $y$ are variable symbols. The *formulas* of $\mathcal{L}_\in$ are described by the following rules: every atomic formula is a formula; if $\varphi$ and $\psi$ are formulas, then so are

$$\neg\varphi, \ (\varphi \wedge \psi), \ (\varphi \vee \psi), \ \exists x\varphi \ \text{and} \ \forall x\varphi.$$

A formula is a *sentence* if there are no free occurrences of variable symbols, i.e., every occurrence of a variable symbol, say $x$, lies within the scope of a quantifier, $\exists x$ or $\forall x$.

We shall consider only *standard* models for $\mathcal{L}_\in$; that is, a *model* (*for* $\mathcal{L}_\in$) is a set $N$, where the epsilon-symbol is interpreted by the binary relation

$$\{(u, v) \in N \times N : u \in v\}.$$

When we consider whether or not a formula is true in the model $N$, we understand that the quantifiers range over elements of $N$; e.g., to say that $\forall x \exists y \ y \in x$ is true in $N$ means that every member of $N$ has a member which belongs to $N$. The connectives "$\neg$", "$\wedge$" and "$\vee$" mean, respectively, "not", "and", and "or" (the latter in the non-exclusive sense, that at least one of the components is true). With these understandings, it should be intuitively clear what it means to say that a sentence $\varphi$ is true in a model $N$, denoted

$$N \models \varphi.$$

In this case we also say that $N$ is a *model of* $\varphi$.

For example, suppose that $a, \{a\}, \{a, b\}$ are members of $N_1$ but $b \notin N_1$ (so, in particular, $N_1$ is not transitive). Then

$$N_1 \models \exists x_0 \exists x_1 (\neg x_0 = x_1 \wedge \forall z ((z \in x_0 \wedge z \in x_1) \vee (\neg z \in x_0 \wedge \neg z \in x_1))).$$

Informally the sentence says that there are two different sets, $x_0$ and $x_1$, which have exactly the same members. (This means that $N_1$ is not a model of Extensionality.) Indeed, if we take $x_0$ to be $\{a\}$ and $x_1$ to be $\{a, b\}$, then the conditions on $x_0$ and $x_1$ are satisfied in $N_1$: $z$ ranges over all the elements of $N_1$, which do not include $b$. Note that if $N$ is transitive, then $N$ *is* a model of Extensionality.

It is not too hard to see that all the axioms of ZFC except Infinity are true in the model $V_\omega$ (see section II.1). (More precisely, we mean that, assuming the axioms of ZFC, we can prove that $V_\omega$ has the property asserted.) This shows that the axiom of Infinity is independent of the other axioms. If $\kappa$ is a regular uncountable cardinal, then $H(\kappa)$ is a model of all the axioms of ZFC except Power Set. (See page 21 for the definition of $H(\kappa)$.) We denote the axioms of ZFC minus Power Set by ZFC$^-$. (If $\kappa$ is singular and uncountable, then $H(\kappa)$ is a model of the axioms of ZFC except Power Set, Union, and Replacement, and, in fact, weak forms of the latter two hold.)

If $\varphi$ is a formula of $\mathcal{L}_\in$ which is not a sentence, then we cannot ask whether $\varphi$ is true in $N$ until we specify a value for each of the free variables. We use the following notation. If $\varphi$ is a formula whose free variables are among $\{x_0, \ldots, x_n\}$, and $a_0, \ldots, a_n$ are elements of $N$, we write

$$N \models \varphi[a_0, \ldots, a_n]$$

if $\varphi$ is true in $N$ when every free occurrence of $x_i$ is given the value $a_i$. In this case, we say that $a_0, \ldots, a_n$ *satisfy* $\varphi$ in $N$. For example, if $\varphi$ is the formula

$$(\neg x_0 = x_1 \wedge \forall z((z \in x_0 \wedge z \in x_1) \vee (\neg z \in x_0 \wedge \neg z \in x_1)))$$

then the free variables of $\varphi$ are $x_0$ and $x_1$, and if $N_1$ is as above, we have

$$N_1 \models \varphi[\{a\}, \{a, b\}].$$

Often we will follow a more informal usage, and write, for example,

$$N_1 \models \text{“}\{a\} \text{ and } \{a, b\} \text{ have exactly the same members''}.$$

We will need to consider expansions, $\mathcal{L}$, of $\mathcal{L}_\in$ where we add other primitive symbols: constant symbols, relation symbols, and/or function symbols. To simplify notation we will only consider a representative example: consider a language $\mathcal{L}$ with primitive symbols

$\in$ , $c$ (a constant symbol), $p$ (a 3-ary relation symbol), and $f$ (a binary function symbol). Before we can define the atomic formulas of $\mathcal{L}$ we must define the *terms*. Any variable or constant symbol is a term; if $t_1$ and $t_2$ are terms, then so is $ft_1t_2$. An *atomic formula* of $\mathcal{L}$ is one of the form

$$pt_1t_2t_3, \quad t_1 \in t_2 \text{ or } \quad t_1 = t_2$$

where $t_1$, $t_2$ and $t_3$ are terms. Formulas of $\mathcal{L}$ are built from atomic formulas by the same rules as before. To define a (standard) model for $\mathcal{L}$ we must give more than just a set $N$: we must also specify an element $C$ of $N$, a subset $P \subseteq N \times N \times N$, and a function $F: N \times N \to N$. We will sometimes denote such a model by $(N, C, P, F)$. Usually we will abuse notation, and refer to the model as $N$, assuming that we have in mind a fixed interpretation, $C$, $P$, and $F$, of the symbols $c$, $p$, and $f$, respectively. Then for example we write

$$N \models \exists x(pcxc \wedge \neg pxcc)$$

if there is an element $a \in N$ such that $(C, a, C) \in P$ but $(a, C, C) \notin P$.

We say that $N$ is a *submodel* of $N'$ if $N$ is a subset of $N'$ and (in the case of $\mathcal{L}$), the interpretations of the symbols $c$, $p$, and $f$ in $N$ are the restriction to $N$ of their interpretations in $N'$, i.e., (with the obvious notation) $C = C'$, $P = P' \cap (N \times N \times N)$, and $F = F' {\restriction} N \times N$. Moreover, we require that $N$ be closed under the function(s), i.e., $F[N \times N] \subseteq N$. We say that $N$ is an *elementary submodel* of $N'$ (denoted $N \preceq N'$ or sometimes $N \prec N'$ if $N \neq N'$) if it is a submodel, and for every formula $\varphi$ with free variables among $\{x_0, \ldots, x_n\}$ and every sequence of elements $a_0, \ldots, a_n$ of $N$,

$$N \models \varphi[a_0, \ldots, a_n] \text{ iff } \quad N' \models \varphi[a_0, \ldots, a_n],$$

that is, $a_0, \ldots, a_n$ satisfy the same first-order properties in $N$ as they do in $N'$.

A useful criterion is given by the following lemma of Tarski-Vaught (cf. Chang-Keisler 1973, p.108 ).

**2.1 Lemma.** *Suppose $N$ is a submodel of $N'$. Then $N$ is an elementary submodel of $N'$ if for all $n \in \omega$, all formulas $\varphi$ with free variables among $\{x_0, \ldots, x_n\}$, and all $a_0, \ldots, a_{n-1}$ in $N$, if $N' \models \exists x_n \varphi[a_0, \ldots, a_{n-1}]$ then there exists $a_n \in N$ such that $N' \models \varphi[a_0, \ldots, a_n]$.* $\square$

If $A$ is a common subset of models $N'$ and $N$, we write $N' \equiv_A N$ if for every formula $\varphi$ with free variables among $\{x_0, \ldots, x_n\}$ and every sequence of elements $a_0, \ldots, a_n$ of $A$,

$$N \models \varphi[a_0, \ldots, a_n] \text{ if and only if } N' \models \varphi[a_0, \ldots, a_n].$$

Note that $N' \preceq N$ if and only if $N'$ is a submodel of $N$ and $N' \equiv_{N'} N$.

The following corollary implies that the collection of countable elementary submodels of a model $N'$ is a closed unbounded set of subsets of $N'$ (cf. VI.4.8). For the first part, it is important that $\mathcal{L}$ is a countable language, i.e., it has only countably many primitive symbols.

**2.2 Corollary.** (i) *For any model $N'$ and any countable subset $X$ of $N'$, there is a countable elementary submodel $N$ of $N'$ which contains $X$.*

(ii) *If $N_n \preceq N'$ and $N_n$ is a submodel of $N_{n+1}$ for all $n \in \omega$, then $\bigcup_n N_n \preceq N'$.*

PROOF. (i) We can assume that $X$ contains all the interpretations, $C$, of constant symbols in $\mathcal{L}$. Then we let $N$ be a subset of $N'$ which contains $X$ and is closed under the functions, $F$, on $N'$ which interpret function symbols of $\mathcal{L}$ and which satisfies the property that for all formulas $\varphi$ with free variables among $\{x_0, \ldots, x_n\}$, and all $a_0, \ldots, a_{n-1}$ in $N$, if $N' \models \exists x_n \varphi[a_0, \ldots, a_{n-1}]$ then there exists $a_n \in N$ such that $N' \models \varphi[a_0, \ldots, a_n]$. It is clear that there is such an $N$ which is countable; by 2.1, $N \preceq N'$.

(ii) We will only sketch the proof. Let $N = \bigcup_n N_n$. We prove by induction on the length of the formula $\varphi$ that if $N' \models \varphi[a_0, \ldots, a_k]$ and $a_0, \ldots, a_k \in N$, then $N \models \varphi[a_0, \ldots, a_k]$. The crucial case is when $\varphi$ is of the form $\exists x_{k+1}\theta$; then there is some $n$ so that $a_0, \ldots, a_k \in N_n$. Since $N_n \preceq N'$, there is $a_{k+1} \in N_n$ such that $N_n \models \theta[a_0, \ldots, a_{k+1}]$. Hence $N' \models \theta[a_0, \ldots, a_{k+1}]$. So by induction $N \models \theta[a_0, \ldots, a_{k+1}]$. $\square$

We are mainly interested in countable elementary submodels of $H(\kappa)$ for some regular uncountable $\kappa$ (or in countable elementary submodels of some expansion of $H(\kappa)$ to a model for an expansion, $\mathcal{L}$, of $\mathcal{L}_\in$). These will be models of ZFC$^-$ since $H(\kappa)$ is a model of ZFC$^-$. We are going to prove some properties of these submodels which will be useful in the sequel; they are also illustrative of similar properties that will be employed in the sequel without explicit comment.

**2.3 Lemma.** *Let $N$ be a countable elementary submodel of $H(\kappa)$ for some regular $\kappa > \aleph_1$.*

(i) *Suppose that $\varphi$ is a formula with free variable $x_0$ (possibly mentioning elements of $N$) such that in any model of ZFC$^-$, there is one and only one element that satisfies $\varphi$. If $a$ belongs to $H(\kappa)$ and satisfies $\varphi$, then $a$ belongs to $N$.*

(ii) *$\omega$ and $\omega_1$ belong to $N$ and $\omega \subseteq N$.*

(iii) *If $a$, $A$, $B$, $f$ are members of $N$, $a \in A$, and $f: A \to B$, then $f(a) \in N$.*

(iv) *If $S \in N$ and $S$ is countable, then $S \subseteq N$.*

(v) *For every $\alpha \in \omega_1 \cup \{\omega_1\}$, $\alpha \cap N$ is an ordinal.*

(vi) *If $\{A_\alpha : \alpha \in \omega_1\}$ is an indexed family of sets which belongs to $N$, then $A_\alpha$ belongs to $N$ for every $a \in \omega_1 \cap N$.*

PROOF. (i) Since $N$ is a model of ZFC$^-$, there is an element $b \in N$ such that $N \models \varphi[b]$. Since $N$ is an elementary submodel of $H(\kappa)$, we have $H(\kappa) \models \varphi[b]$. But then we must have $a = b$ because only one element satisfies $\varphi$ in $H(\kappa)$.

(ii) This follows immediately from (i) since $\omega$, $\omega_1$ belong to $H(\kappa)$ and $\omega$ (respectively, $\omega_1$) is the unique element which satisfies the formula $\varphi$ which says that $x_0$ is the first infinite (respectively, uncountable) ordinal. A similar argument shows that each $n \in \omega$ belongs to $N$.

(iii) When we say that $f: A \to B$, we mean that this holds in V, the class of all sets. But then

$$H(\kappa) \models \text{``} f \text{ is a function from A to } B\text{''},$$

since $H(\kappa)$ is transitive and $f$, $A$ and $B$ belong to $H(\kappa)$. Hence

$$N \models \text{``} f \text{ is a function from A to } B\text{''}$$

since $N$ is an elementary submodel of $H(\kappa)$. Let $\varphi$ be the formula which says that $(x_1, x_2) \in x_0$. Then there is an element $b \in B \cap N$ such that

$$N \models \varphi[f, a, b].$$

Consequently $H(\kappa) \models \varphi[f, a, b]$, so $b = f(a)$.

(iv) Since $S$ is countable there is a function $g: \omega \to S$ which maps onto $S$. Now $g$ belongs to $H(\kappa)$: it is hereditarily of cardinality $< \kappa$, since $S$ is. There is a formula $\varphi$ of $\mathcal{L}_\in$ which expresses the fact that $x_2$ is a function which maps $x_0$ onto $x_1$. Since $H(\kappa) \models \exists x_2 \varphi[\omega, S]$, we also have $N \models \exists x_2 \varphi[\omega, S]$. Therefore there is an element $f$ of $N$ such that $N \models \varphi[\omega, S, f]$. But then $H(\kappa) \models \varphi[\omega, S, f]$ and hence $f$ is *really* a function which maps $\omega$ onto $S$ (i.e., this is true in V). For any element $s$ of $S$ there exists $n \in \omega$ such that $f(n) = s$. But then, by (iii) and (ii), $f(n) \in N$.

(v) To show that $\alpha \cap N$ is an ordinal it suffices to show that it is transitive, i.e., if $\gamma \in \beta \in \alpha \cap N$, then $\gamma \in \alpha \cap N$. But $\gamma$ belongs to $\alpha$ since every ordinal is transitive, and $\gamma$ belongs to $N$ by (iv) since $\beta$ is countable.

(vi) The indexed family $\{A_\alpha: \alpha \in \omega_1\}$ is really a function $f$ on $\omega_1$ such that $f(\alpha)$ is $A_\alpha$. Now the result follows immediately from (ii) and (iii). $\square$

Notice that although $\omega_1$ belongs to $N$, not every element of $\omega_1$ belongs to $N$; in fact $\omega_1 \cap N$ is a countable set. However, from the point of view of $N$, $\omega_1$ *is* an uncountable set; that is,

$$N \models \text{``there is no function from } \omega \text{ onto } \omega_1\text{''}.$$

When we say that a group, $A$, belongs to $N$, we mean that both the underlying set, $A$, and the addition function $+: A \times A \to A$ belong to $N$ — or, equivalently, the ordered pair $(A, +) \in N$. In that case, $A \cap N$ is closed under addition by 2.3(iii). Moreover, by 2.3(i), $0 \in N$ since 0 is the unique element of $A$ which satisfies $x + x = x$; similarly $A \cap N$ is closed under additive inverses. Thus $A \cap N$ is a subgroup of $A$.

## §3. L, the constructible universe

The Axiom of Constructibility, $V = L$, has already been mentioned in the introduction to this chapter. After the preceding section, it is fairly easy to give a precise definition of L. For any model $N$, let $DEF(N)$ consist of those $Y \subseteq N$ such that there is a formula $\varphi(x, y_0, \ldots, y_n)$ and elements $a_1, \ldots, a_n \in N$ so that $Y = \{a: N \models \varphi[a, a_1, \ldots, a_n]\}$. The class L is built up by induction on the ordinals in the same way as V was except that DEF replaces the power set operation; by transfinite induction on ordinals define: $L_0 = \emptyset$; $L_{\alpha+1} = DEF(L_\alpha)$; and for limit ordinals $\alpha$, $L_\alpha = \bigcup_{\beta < \alpha} L_\beta$. The intuition is that DEF and its iterates form a first order version of the power set operation. It is not hard to see that L is a transitive class which contains all the ordinals.

Although L is a proper class, and not a set, the notions of truth and satisfaction given in the previous section still make sense. There is a formula $L(x)$ such that for any set $a$, $a \in L$ if and only if $L[a]$ holds (in V). Gödel showed that every axiom of ZFC holds in L, assuming that every axiom of ZF holds in V; that is, L is an *inner model* of ZFC — a transitive class containing *Ord* such that every axiom of ZFC is valid. Furthermore, if $M$ is any inner model of ZFC, then $L \subseteq M$ and for any $a \in M$, $L[a]$ holds in $M$ if and only if $a \in L$. In particular, $\forall x\, L(x)$ is valid in L; this sentence, $\forall x\, L(x)$, is called the Axiom of Constructibility, and denoted $V = L$. So the axiom $V = L$ is consistent with ZFC: it is true in the model L. (But it is not provable from ZFC, as will become clear in section 4.)

In considering L, we are interested both in principles true in L, i.e consequences of $V = L$, and the relation between L and V even when they are not equal. Certain facts are clear. Since L is transitive and contains *Ord*, a set is an ordinal in L if and only if it is an ordinal (in V). As well, any cardinal, $\kappa$, in V is also a cardinal in L, since if

$$L \models \text{``}g \text{ is a bijection between } \alpha \text{ and } \kappa\text{''}$$

for some ordinal $\alpha < \kappa$, then $g$ shows, in V, that $\kappa$ is not a cardinal. (The reverse implication need not be true: an ordinal in V which is not a cardinal may be a cardinal in L.) Similarly a regular cardinal (in V) is a regular cardinal in L. Whether a singular cardinal can be a regular cardinal in L is a little delicate: if there is a singular cardinal which is regular in L, then certain large cardinal properties must hold (see the discussion after 3.16).

In order to distinguish between elements defined in L and those defined in V, it is usual to use a superscript L. For example, $\omega_1^L$ denotes the ordinal which L "thinks" is the first uncountable cardinal, i.e.

$$L \models \text{``}\omega_1^L \text{ is the first uncountable ordinal''}.$$

So $\omega_1^L$ is either $\omega_1$, itself, or a countable ordinal. Similarly if $\kappa$ is a cardinal, then $\kappa^{+L}$ is the successor cardinal to $\kappa$ in L.

Gödel showed that GCH is true in L, or, otherwise said, that GCH is a consequence of $ZF + V = L$. In section 1, we discussed the diamond principles, $\Diamond(E)$, which follow from $V = L$ (for stationary $E$); here we will concentrate on other aspects of the constructible universe, particularly the useful principle, $E(\kappa)$.

**3.1 Definition.** For a regular cardinal $\kappa$, let $E(\kappa)$ denote the following statement:

> There is a non-reflecting stationary subset of $\kappa$ consisting of ordinals of cofinality $\omega$.

(See definition II.4.6.)

We will state without proof a theorem of R. B. Jensen 1972. The parenthetical "$V = L$" indicates that this is a theorem proved under the hypotheses $ZFC + V = L$.

**3.2 Theorem.** (V = L) *For any regular cardinal $\kappa$, E($\kappa$) holds if and only if $\kappa$ is not weakly compact.* □

The principle E($\kappa^+$) can be established indirectly by use of another principle, $\square_\kappa$ (read "square kappa").

**3.3 Definition.** Let $\kappa$ be a cardinal. By $\square_\kappa$, we mean the following principle:

There is a sequence $(C_\alpha: \alpha < \kappa^+$ and $\alpha$ is a limit ordinal) such that:

(i) $C_\alpha$ is closed and unbounded in $\alpha$;
(ii) if cf($\alpha$) < $\kappa$ then $|C_\alpha| < \kappa$;
(iii) if $\gamma$ is a limit point of $C_\alpha$, then $C_\gamma = C_\alpha \cap \gamma$.

Note in the definition above that if cf($\alpha$) = $\kappa$, then (ii) and (iii) imply that the order type of $C_\alpha$ is $\kappa$. Again we have a result of R. B. Jensen 1972.

**3.4 Theorem.** (V = L) *For every infinite cardinal $\kappa$, $\square_\kappa$ is true.* □

From $\square_\kappa$, we can derive E($\kappa^+$) in a very strong way:

**3.5 Theorem.** *Suppose that $\kappa$ is an infinite cardinal and $\square_\kappa$ holds. If E is a stationary subset of $\kappa^+$, then E is the disjoint union*

$$E = \amalg_{\beta \leq \kappa} E_\beta$$

*where each $E_\beta$ is a non-reflecting subset of $\kappa^+$.*

PROOF. It is clearly enough to show that lim($\kappa^+$), the set of limit ordinals in $\kappa^+$, is a disjoint union of $\kappa$ non-reflecting subsets. Let $(C_\alpha: \alpha < \kappa^+)$ be a sequence guaranteed by $\square_\kappa$. For each $\beta \leq \kappa$, let $E_\beta = \{\alpha:$ the order type of $C_\alpha$ is $\beta\}$. By (ii) and the remark after the definition of $\square_\kappa$, lim($\kappa^+$) = $\bigcup_{\beta \leq \kappa} E_\beta$. We claim that each $E_\beta$ is non-reflecting. Fix $\beta$. Consider a limit ordinal $\alpha$. Since there is at most one element $\gamma$ of $C_\alpha$ so that the order type of $C_\alpha \cap \gamma$ is $\beta$, the intersection of $E_\beta$ and the limit points of $C_\alpha$ has cardinality at most 1. □

In particular, if $E$ = the set of limit ordinals of cofinality $\omega$, then 3.5 immediately implies $E(\kappa^+)$ because at least one of the $E_\beta$'s must be stationary by II.4.5. One advantage that $\Box_\kappa$ has over $E(\kappa^+)$ is that it remains true in V as long as the cardinals don't change: a stationary set in L may cease to be stationary, so a particular witness to $E(\kappa^+)$ may not be stationary, but a $\Box_\kappa$-sequence in L will retain all its properties in V (except it may no longer go as far as $\kappa^+$, if $\kappa^{+L} \neq \kappa^+$).

**3.6 Corollary.** *Suppose $\kappa$ is a cardinal and for some ordinal $\mu$, $\mu$ is a cardinal in L and $\mu^{+L} = \kappa$. Then $E(\kappa)$ holds.*

PROOF. The sequence witnessing $\Box_\mu$ in L also witnesses $\Box_\mu$ in V, so $\Box_\kappa$ holds in V, and we can apply 3.5. Alternately, we can use the fact that any non-reflecting set in L is non-reflecting in V, and that in L — by 3.5 — and hence in V, $\kappa$ can be partitioned into $|\mu|$ non-reflecting sets. $\Box$

This corollary can be used to illustrate, in a very simple form, two of the themes of set theory in the last 15 years; namely that principles true in L may affect V and that the consistency of set-theoretic statements even about small cardinals may imply (and be equivalent to) the consistency of the existence of large cardinals.

**3.7 Proposition.** *Suppose there is no inaccessible cardinal in L. Then for every uncountable regular cardinal $\kappa$, $E(\kappa)$ holds.*

PROOF. Since every regular cardinal $\kappa$ is a regular cardinal in L and by hypothesis there is no regular limit cardinal in L, there is some ordinal $\mu$ so that $\kappa = \mu^{+L}$. $\Box$

If we assumed only that there was no inaccessible cardinal (in V) we could not hope to obtain such a result: it is consistent — assuming the consistency of a Mahlo cardinal (to be defined later) — that $E(\aleph_2)$ does not hold. If we have a model in which $E(\aleph_2)$ does not hold, then either there is no inaccessible cardinal or there is a least inaccessible cardinal $\lambda$. In either case, by considering V or $V_\lambda$, we have a model in which $E(\aleph_2)$ holds and there is no inaccessible. Later (after 3.13), we will see that the existence of a Mahlo cardinal is equiconsistent with "there is a regular uncountable cardinal $\kappa$ such that $E(\kappa)$ fails". Now we want to consider a less well known set-theoretic principle.

**3.8 Definition.** Suppose $\kappa$ and $\lambda$ are infinite cardinals. By $*(\kappa, \lambda)$ we mean the following principle:

> There exists a family $\{S_\alpha : \alpha < \kappa^+\}$ of subsets of $\kappa$ each of cardinality $\lambda$ such that for each $I \subseteq \kappa^+$ of cardinality $\kappa$, there exists $\{S_i^* : i \in I\}$ where $S_i^* \subseteq S_i$, $|S_i \setminus S_i^*| < \lambda$ and $S_i^* \cap S_j^* = \emptyset$ if $i \neq j$.

This principle will be used in VII.2 to construct groups which are $\kappa^+$-free but not strongly $\kappa^+$-free. The following is a theorem of ZFC.

**3.9 Theorem.** *If $\kappa$ is a regular cardinal, then $*(\kappa, \kappa)$ holds.*

PROOF. By induction on $\alpha < \kappa^+$, choose functions $S_\alpha : \kappa \to \kappa$ such that for all $\alpha < \beta$, there is $\gamma$ with the property that $S_\alpha(\delta) < S_\beta(\delta)$ whenever $\delta > \gamma$. We claim that we can let $\{S_\alpha : \alpha < \kappa^+\}$ be the desired family of subsets of $\kappa \times \kappa$, a set of cardinality $\kappa$. Note that for $\alpha \neq \beta$, $|S_\alpha \cap S_\beta| < \kappa$. Given $I \subseteq \kappa^+$ of cardinality $\kappa$, choose a well-ordering, $\prec$, of $I$ of order-type $\kappa$; we can let $S_i^* = S_i \setminus (\cup\{S_j : j \in I, j \prec i\})$ for all $i \in I$. $\square$

To get a result for singular cardinals, we assume $V = L$.

**3.10 Theorem.** $(V = L)$ *If $\kappa$ is a singular cardinal then $*(\kappa, \operatorname{cf}(\kappa))$ holds.*

The proof of this theorem requires some notions from model theory which go beyond those discussed in the previous section. We include a proof, since it is not readily available elsewhere, but the reader who is willing to accept it on faith may skip the proof of 3.10 as well as the following result of R. B. Jensen 1972 which we cite without proof.

**3.11 Theorem.** (V = L) *Suppose that N is a model of cardinality* $\aleph_1$ *and U is a relation symbol whose interpretation in N has cardinality* $\aleph_0$. *Then for all infinite cardinals* $\kappa$ *there is a model M of cardinality* $\kappa^+$ *which is elementarily equivalent to N such that the interpretation of U in M has cardinality* $\kappa$. $\square$

**Proof of 3.10.** Let the model $N$ have as its universe $\omega \cup W$, where $W$ is a set of $\aleph_1$ infinite almost disjoint subsets of $\omega$ (cf. II.5.5). Let $<$ be the usual ordering on $\omega$ and let $<_1$ be an ordering of $W$ of order type $\omega_1$. Notice that each element of $W$ is cofinal in $\omega$. Let $U$ be a unary relation symbol whose interpretation is $\omega$. Finally let $g$ be a binary function from $W$ to $\omega$ so that for all $v \in W$, the family of sets $\{\{n \in u : g(v, u) < n\} : u <_1 v\}$ is disjoint. Let $N = (\omega \cup W; \omega, <, <_1, g, \in)$. Suppose $M = (\omega^* \cup W^*; \omega^*, <^*, <_1^*, g^*, \in^*)$ is as guaranteed by Theorem 3.11. Since for $v, v' \in W^*$, $v = v'$ if and only if $\{x : x \in^* v\} = \{x : x \in^* v'\}$, there is no harm in assuming $\in^* \restriction \omega^* \times W^*$ is $\in$ (i.e., replace $M$ with an isomorphic copy). Note that each $v \in W^*$ is cofinal in $\omega^*$, since each $v \in W$ is cofinal in $\omega$.

Suppose $I$ is a bounded subset of $W^*$, i.e., there is some $v_0$ such that $v <_1^* v_0$ for all $v \in I$. Since $(\{x \in v : g^*(v_0, v) <^* x\} : v \in I)$ is a disjoint family of non-empty subsets of a set of cardinality $\kappa$, $|I| \le \kappa$. So the cofinality of $<_1^*$ is $\kappa^+$, and every subset of cardinality $\kappa$ is bounded. We want to show that $\omega^*$ has cofinality $\mathrm{cf}(\kappa)$. Once we do this we enumerate $W^*$ as $\{v_\alpha : \alpha < \kappa^+\}$. Since each $v_\alpha$ is cofinal in $\omega^*$, we can choose $S_\alpha$ a cofinal subset of $v_\alpha$ so that the order type of $S_\alpha$ is $\mathrm{cf}(\kappa)$. Using $g^*$, we can show that if $I \subseteq \kappa^+$ of cardinality $\kappa$, then there exists $\{S_i^* : i \in I\}$ where $S_i^* \subseteq S_i$, $|S_i \setminus S_i^*| < \mathrm{cf}(\kappa)$ and $S_i^* \cap S_j^* = \emptyset$ if $i \ne j$.

Suppose that $\mathrm{cf}(<^*) \ne \mathrm{cf}(\kappa)$. Choose an increasing chain $A_j (j < \mathrm{cf}(\kappa))$ so that $\omega^* = \cup A_j$ and each $A_j$ has cardinality $< \kappa$. For each $v \in W^*$ there is $j_v$ so that $A_{j_v} \cap v$ is cofinal in $<^*$. So there is $j_0$ and some subset $I$ of $W^*$ of cardinality $\kappa$ so that $j_v = j_0$ for all $v \in I$. Suppose $I$ is bounded by $v_0$. Then $\{\{x \in v \cap A_{j_0} : g^*(v_0, v) <^* x\} : v \in I\}$ is a disjoint family of cardinality $\kappa$ of non-empty subsets of a set of cardinality $< \kappa$, which is a contradiction. $\square$

It is perhaps more natural to view $\Box_\kappa$ as a statement about $\kappa^+$, rather than as a statement about $\kappa$. This suggests trying to generalize it to regular limit cardinals; we encounter a difficulty if there is a stationary set of regular cardinals. A cardinal $\kappa$ is said to be a *Mahlo* cardinal if $\kappa$ is strongly inaccessible and $\{\mu: \mu < \kappa$ and $\mu$ is a regular cardinal$\}$ is stationary in $\kappa$. We call $\kappa$ *weakly Mahlo* if the demand that it be strongly inaccessible is omitted. Notice that a (weakly) Mahlo cardinal must be a limit cardinal. Since any singular cardinal contains a closed unbounded set consisting of ordinals which are not cardinals, any Mahlo cardinal is regular. Also any (strong) limit cardinal, $\kappa$, contains a closed unbounded set of (strong) limit cardinals, so a weakly Mahlo cardinal $\kappa$ is the limit of $\kappa$ weakly inaccessible cardinals and a Mahlo cardinal $\kappa$ is the limit of $\kappa$ inaccessible cardinals. It is not hard to show (if an appropriate definition of weakly compact is used) that a weakly compact cardinal $\kappa$ is the limit of $\kappa$ Mahlo cardinals.

An ordinal is a *singular limit ordinal* if it is a limit ordinal and not a regular cardinal. We state without proof the following result of Prikry-Solovay.

**3.12 Theorem.** $(V = L)$ *For every regular cardinal $\lambda$ there is a sequence $(C_\alpha: \alpha$ is a singular limit ordinal $< \lambda)$ satisfying the following properties:*

*For every $\alpha$, $C_\alpha$ is a cub in $\alpha$ of order type $< \alpha$ and whenever $\beta$ is a limit point of $C_\alpha$, then $\beta$ is a singular limit ordinal and $C_\beta = C_\alpha \cap \beta$.* $\Box$

If $\lambda = \kappa^+$, then it is possible to derive (in ZFC) $\Box_\kappa$ from the principle above. So 3.12 is a generalization of Theorem 3.4. We want to derive the analog of 3.5 from 3.12. There are some obvious restrictions. If $\lambda$ is a Mahlo cardinal, then 3.12 implies nothing about the stationary set of regular cardinals. As well, if $\lambda$ is weakly compact then there is *no* stationary set which is non-reflecting (cf. IV.3.1).

**3.13 Theorem.** $(V = L)$ *Suppose $\kappa$ is a regular cardinal and $E$ is a stationary subset of $\kappa$ consisting of singular limit ordinals. Then*

*E is a disjoint union*

$$E = \amalg_{\alpha < \kappa} E_\alpha$$

*such that for all $\alpha < \kappa$ and all $\beta \in E_\alpha$, $\alpha < \beta$ and such that if $E_\alpha \cap \mu$ is stationary in $\mu$, then $\mu$ is a regular cardinal. Moreover, if $\kappa$ is not a Mahlo cardinal, then each $E_\alpha$ is non-reflecting.*

PROOF. We can assume that $E$ is the set of singular limit ordinals. Let $(C_\alpha : \alpha$ is a singular limit ordinal $< \kappa)$ be as in 3.12. Define $E_\alpha = \{\beta :$ order type of $C_\beta$ is $\alpha\}$. Then as in Theorem 3.5, we can show that $E_\alpha \cap \delta$ is not stationary in any singular limit ordinal $\delta$. □

Arguing as we did in the proofs of 3.6 and 3.7, we can show that if $\kappa$ is a regular cardinal and E($\kappa$) does not hold, then $\kappa$ is a Mahlo cardinal in L. Combining this observation with the previously mentioned result that it is consistent that E($\aleph_2$) does not hold if it is consistent that there is a Mahlo cardinal, we can conclude that it is consistent that there is a Mahlo cardinal if and only if it is consistent that there is a regular cardinal $\kappa$ so that E($\kappa$) does not hold.

In VII we will see that E($\kappa$) is sufficient (although not necessary) to construct a $\kappa$-free non-free group of cardinality $\kappa$. On the other hand, if $\kappa$ is an inaccessible cardinal such that every stationary subset is stationary in some regular cardinal $\mu < \kappa$, then every $\kappa$-free group of cardinality $\kappa$ is free (cf. the proof of IV.3.2). To find a property of a successor cardinal $\kappa$ such that $\kappa$-free implies $\kappa^+$-free, is more delicate. (See Exercises VII.21 – 23) It is consistent that there exists a regular cardinal $\kappa$ such that $\kappa$-free implies $\kappa^+$-free if and only if it is consistent that there is a *reflection cardinal*, that is, a regular cardinal $\lambda$ on which there is a normal $\lambda$-complete ideal $I$ with the property that if $X \notin I$, then $\{\alpha : X \cap \alpha$ is stationary in $\alpha\} \notin I$. The consistency of the existence of a weakly compact cardinal implies the consistency of the existence of a reflection cardinal which in turn implies the consistency of the existence of a Mahlo cardinal; neither of these implications is reversible. Also, the conclusion of IV.3.1 implies the consistency of the existence of a weakly compact cardinal.

**3.14.** In II and above, we considered various sorts of large cardinals. Roughly speaking, the large cardinal axioms fall into three classes: those which are consistent with V = L; those which are consistent with "the universe is L-like"; and those which are inconsistent with any L-like principle.

It is straightforward to see that if $\kappa$ is weakly inaccessible or weakly Mahlo, then it is weakly inaccessible or weakly Mahlo in L. Since GCH is true in L, in fact such a $\kappa$ must be strongly inaccessible or Mahlo. This observation allows one to show that "∃ a weakly inaccessible cardinal" implies ZFC is consistent (cf. the end of section II.1): one begins with $\kappa$ which is weakly inaccessible; in L, $\kappa$ is strongly inaccessible, so $(V_\kappa)^L \models$ ZFC. Less obvious than the foregoing, but still true, is the fact that if $\kappa$ is weakly compact then $\kappa$ is weakly compact in L. So each of the statements "∃ an inaccessible cardinal", "∃ a Mahlo cardinal" and "∃ a weakly compact cardinal" is consistent with V = L, providing that it is consistent with ZFC.

Although we will not consider any large cardinal notions stronger than weakly compact which are consistent with V = L, there are such notions. On the other hand, there are large cardinal notions which are known to be inconsistent with V = L. For example, the existence of a measurable cardinal implies that V ≠ L. In the years since Scott 1961 first proved this result, the reason for the inconsistency of "∃ a measurable cardinal" and V = L, has been clarified by the following theorem of Gaifman (cf. Jech 1978, p. 337):

**3.15 Theorem.** *If there is a measurable cardinal, then the uncountable cardinals form a set of order-indiscernibles for L. That is:*

> *if $\varphi(x_1, \ldots, x_n)$ is a formula, and $\kappa_1 < \ldots < \kappa_n$ and $\mu_1 < \ldots < \mu_n$ are increasing sequences of uncountable cardinals (in V), then $\varphi[\kappa_1, \ldots, \kappa_n]$ is true in L if and only if $\varphi[\mu_1, \ldots, \mu_n]$ is true in L.* □

If the conclusion of the theorem holds, let

$$0^{\#} \overset{\text{def}}{=} \bigcup_{n \in \omega} \{\varphi \colon \varphi \text{ is a formula with free variables among}$$
$$\{x_1, \ldots, x_n\} \text{ and } L \models \varphi[\aleph_1, \ldots, \aleph_n]\}.$$

$0^{\#}$ has certain properties such that the conclusion of the theorem is equivalent to the existence of a set of formulas with these properties; there is at most one set of formulas with these properties. The conclusion of the theorem is therefore abbreviated "$0^{\#}$ exists". (It is not true that "$0^{\#}$ exists" implies the consistency of the existence of a measurable cardinal.)

Notice that if $0^{\#}$ exists, then whenever $\kappa$ is a cardinal, $\kappa$ is a regular cardinal in L (because there is a regular cardinal, $\mu$, in L). So if $0^{\#}$ exists, then $V \neq L$. (In fact, a little more argument would show that if $0^{\#}$ exists and $\kappa$ is a cardinal, then $\kappa$ is weakly compact in L.) It is not true that $V \neq L$ is equivalent to the existence of $0^{\#}$, but there is a weak sense in which this is true. R. B. Jensen proved the following remarkable theorem which says that either $0^{\#}$ exists or V is not very far from L.

**3.16 Theorem.** (Covering Lemma) *The following are equivalent:*

   (a) $0^{\#}$ *does not exist;*
   (b) *for every uncountable set of ordinals $X$, there is $Y \in L$ such that $X \subseteq Y$ and $|X| = |Y|$;*
   (c) *for every singular cardinal $\kappa$, $\kappa^{+L} = \kappa^{+}$.* $\square$

An immediate consequence of the Covering Lemma is that if $0^{\#}$ does not exist, then $\square_{\kappa}$ holds for every singular cardinal $\kappa$ and hence $E(\kappa^{+})$ holds. So the failure of $E(\kappa^{+})$ for some singular cardinal $\kappa$ implies that it is consistent that there is a proper class of weakly compact cardinals (since if $0^{\#}$ exists then there is a proper class of weakly compact cardinals in L). It can be shown that $0^{\#}$ exists if and only if there is a singular cardinal $\kappa$ so that $\kappa$ is regular in L.

One of the principal concerns of set theory over the past few years has been to find inner models which will allow the existence of large cardinals but still be L-like in the sense that many of the properties of L, in particular $\square_\kappa$, hold. If $0^\#$ does not exist, then the inner models are L, but there are inner models which contain many measurable cardinals. There is an analog of the Covering Lemma which is true for these inner models. As was the case for L, the existence of a singular cardinal $\kappa$ so that $E(\kappa^+)$ does not hold implies the failure of the analog of the Covering Lemma. The following theorem (cf. Mitchell 1984) gives some idea of the strength of the existence of a singular cardinal $\kappa$ so that $E(\kappa^+)$ does not hold.

**3.17 Theorem.** *If it is consistent that there is a singular cardinal $\kappa$ so that $E(\kappa^+)$ does not hold, then it is consistent that there is a proper class of measurable cardinals.* $\square$

This theorem is close to the optimal result. If we assume the existence of an $L_{\omega_1\omega}$-compact cardinal, then there is a limit to the cardinals for which $E(\kappa)$ holds:

**3.18 Theorem.** *Suppose $\kappa$ is an $L_{\omega_1\omega}$-compact cardinal. Then for every $\lambda \geq \kappa$, $E(\lambda)$ does not hold.*

PROOF. We will show (VII.1.4) that if $E(\lambda)$ holds for a regular cardinal $\lambda$, then there is a $\lambda$-free group which is not free. But by II.3.10, any $\kappa$-free group is free. $\square$

Theorem 3.18 can be proved directly, but the first proof was the one we have given above.

## §4. MA and PFA

The axioms that we introduce in this section are consistent with ZFC but not with ZFC + V = L. In fact, they arise out of the method of forcing discovered by P. Cohen, who used it to prove that $\neg$CH is consistent with ZFC. The method of forcing begins with a partially ordered set $\mathbf{P}$ and constructs a model $V^{\mathbf{P}}$ of ZFC whose properties depend on the choice of $\mathbf{P}$. We will not use the method of forcing (directly) in this book, but the consistency of the axioms discussed here is proved using forcing, and the axioms themselves refer to partially ordered sets. So we need to begin with some terminology involving partially ordered sets.

**4.1 Definition.** Let $\mathbf{P} = (\mathbf{P}, \leq)$ be a partially ordered set (*poset*, for short). We will assume that all posets have a least element, 0; i.e., $0 \leq p$ for all $p \in \mathbf{P}$. Two elements, $p$ and $q$, of $\mathbf{P}$ are called *compatible* if there exists $r \in \mathbf{P}$ such that $r \geq p$ and $r \geq q$; otherwise they are *incompatible*. A subset $\mathcal{G}$ of $\mathbf{P}$ is *directed* if for all $p, q \in \mathcal{G}$, there exists $r \in \mathcal{G}$ such that $r \geq p$ and $r \geq q$. A subset $D$ of $\mathbf{P}$ is *dense* (in $\mathbf{P}$) if for every $p \in \mathbf{P}$ there exists $q \in D$ such that $p \leq q$.

An *antichain* in $\mathbf{P}$ is a subset $A$ of $\mathbf{P}$ such that any two distinct elements of $A$ are incompatible. We say that $\mathbf{P}$ satisfies the *countable chain condition* (or, $\mathbf{P}$ is c.c.c., for short) if every antichain in $\mathbf{P}$ is countable. (Yes, it should really be called the countable antichain condition!)

The reader should be warned that in many sources (e.g., Jech 1978), the linear ordering is reversed in the above definitions; for example $p$ and $q$ are called compatible if there exists $r$ such that $r \leq p$ and $r \leq q$.

The model $\mathbf{V^P}$ constructed by the forcing method from $\mathbf{P}$ is an extension of the starting model, V, and contains a set $\mathcal{G}$ — a so-called "generic" set — which is a directed subset of $\mathbf{P}$ and has non-zero intersection with every $D \in$ V which is a dense subset of $\mathbf{P}$. This property is reflected in the following axioms, which are sometimes referred to as "internal forcing axioms".

**4.2 Definition.** Let $\kappa$ be an infinite cardinal. By $\mathrm{MA}(\kappa)$ we mean the following principle:

> for every c.c.c. poset $\mathbf{P}$ and every family $\mathcal{D} = \{D_\alpha : \alpha \in \kappa\}$ of dense subsets of $\mathbf{P}$, there is a directed subset $\mathcal{G}$ of $\mathbf{P}$ such that for all $\alpha \in \kappa$, $\mathcal{G} \cap D_\alpha \neq \emptyset$.

(We say that $\mathcal{G}$ *meets* each $D_\alpha$ and call $\mathcal{G}$ a $\mathcal{D}$-*generic* subset of $\mathbf{P}$.) By MA (Martin's Axiom) we mean: $\mathrm{MA}(\kappa)$ for all $\kappa < 2^{\aleph_0}$.

We leave it as an exercise to check that $\mathrm{MA}(\aleph_0)$ is a theorem of ZFC (cf. Exercise 5). The next result implies that $\mathrm{MA}(2^{\aleph_0})$

is not consistent with ZFC. It is a theorem (of Solovay-Tenenbaum 1971/Martin-Solovay 1970) that MA$+\neg$CH is consistent with ZFC; in particular, MA$(\aleph_1) + 2^{\aleph_0} = \aleph_2$ is consistent with ZFC.

**4.3 Proposition.** MA$(\kappa)$ *implies that* $2^{\aleph_0} > \kappa$.

PROOF. It suffices to prove that for any family $\{f_\alpha : \alpha < \kappa\}$ of functions from $\omega$ into $2 = \{0, 1\}$, there is a function $g : \omega \to 2$ which is not in the family. Our poset will consist of finite approximations to the desired function: $\mathbf{P}$ is the set of all $p : S \to 2$, where $S$ is a finite subset of $\omega$. Make $\mathbf{P}$ into a poset by defining $q \leq p$ if and only if $p$ is an extension of $q$. Thus $p$ and $q$ are compatible if and only if they agree on $\mathrm{dom}(p) \cap \mathrm{dom}(q)$, for then $p \cup q \geq p, q$. Obviously $\mathbf{P}$ is c.c.c. since $\mathbf{P}$ is countable. Given a family $\{f_\alpha : \alpha < \kappa\}$, for each $\alpha < \kappa$, define

$$D_\alpha = \{p \in \mathbf{P} : \exists n \in \mathrm{dom}(p) \text{ s.t. } p(n) \neq f_\alpha(n)\}.$$

$D_\alpha$ is dense in $\mathbf{P}$, because given any $q \in \mathbf{P}$, choose $m \notin \mathrm{dom}(q)$ and let $p = q \cup \{(m, 1 - f_\alpha(m))\}$; then $q \leq p$ and $p \in D_\alpha$. For each $n \in \omega$, define

$$E_n = \{p \in \mathbf{P} : n \in \mathrm{dom}(p)\}.$$

Clearly $E_n$ is dense in $\mathbf{P}$. Now, by MA$(\kappa)$, there exists a directed subset $\mathcal{G}$ of $\mathbf{P}$ which meets each $D_\alpha$ and each $E_n$. Since the elements of $\mathcal{G}$ are mutually compatible, $\cup \mathcal{G}$ is a function, $g$, and since $\mathcal{G}$ meets each $E_n$, the domain of $g$ is $\omega$. However, since $\mathcal{G}$ meets each $D_\alpha$, $g$ is a function: $\omega \to 2$ which does not equal any $f_\alpha$. □

When $\kappa = \aleph_0$, 4.3 is Cantor's result that $2^{\aleph_0} > \aleph_0$. It can be proved that MA $+ \neg$CH implies that $2^\kappa = 2^{\aleph_0}$ for all $\kappa < 2^{\aleph_0}$. (See Exercise 12.)

Although it was not the case in the previous result, usually the hardest part of any proof from Martin's Axiom is the verification that $\mathbf{P}$ is c.c.c. Rather than make that verification directly, we shall make use of the criterion in the following lemma, which often simplifies the combinatorics involved. First we need some definitions,

which will also be used when we come to the Proper Forcing Axiom
(PFA).

**4.4 Definition.** A subset $W$ of a poset $\mathbf{P}$ is called *predense above*
$p$ if for every $r \geq p$ there is an element of $W$ which is compatible
with $r$. $W$ is called *predense* (in $\mathbf{P}$) if it is predense above 0, i.e.,
$W$ is predense above every element of $\mathbf{P}$.

   Given a set $N$ (which will, in practice, be a countable model
of set theory), an element $p$ of $\mathbf{P}$ is called $N$-*generic* if for every
$D \in N$ which is a dense subset of $\mathbf{P}$, $D \cap N$ is predense above $p$.

   Obviously, every dense subset of $\mathbf{P}$ is predense, and if $A$ is predense in $\mathbf{P}$, then

$$\bar{A} \overset{\text{def}}{=} \{p \in \mathbf{P} : p \geq q \text{ for some } q \in A\}$$

is dense in $\mathbf{P}$. If $A$ is a maximal antichain in $\mathbf{P}$, i.e., an antichain
which is not included in any larger antichain, then $A$ is predense
in $\mathbf{P}$. Notice also that 0 is $N$-generic if and only if for every dense
subset $D$ of $\mathbf{P}$ which belongs to $N$, $D \cap N$ is predense in $\mathbf{P}$.

   Given a poset, $\mathbf{P}$, we say that a cardinal $\kappa$ is *large enough* (for $\mathbf{P}$)
if $\kappa$ is regular and of cardinality $> \aleph_1$ and the set of dense subsets
of $\mathbf{P}$ is an element of $\mathrm{H}(\kappa)$. In that case, $\mathbf{P}$, every element of $\mathbf{P}$, and
every dense subset of $\mathbf{P}$ belongs to $\mathrm{H}(\kappa)$.

**4.5 Lemma.** *Let $\mathbf{P}$ be a poset and $\kappa$ a cardinal large enough for*
$\mathbf{P}$. *The following are equivalent*:
   (1) $\mathbf{P}$ *is c.c.c.*;
   (2) *for all countable $N \preceq \mathrm{H}(\kappa)$, 0 is $N$-generic*;
   (3) *every countable subset of $\mathrm{H}(\kappa)$ is contained in a countable*
   $N \preceq \mathrm{H}(\kappa)$ *such that 0 is $N$-generic*.

PROOF. $(1) \Rightarrow (2)$ : Let $D \in N$ such that $D$ is dense in $\mathbf{P}$. Since
$\mathrm{H}(\kappa)$ is a model of ZFC$^-$ and $N$ is an elementary submodel of
$\mathrm{H}(\kappa)$, there exists $A \in N$ such that $A$ is maximal with respect to
the property that it is an antichain in $\mathbf{P}$ contained in $D$. Since
$D$ is dense, $A$ is actually a maximal antichain of $\mathbf{P}$. By (1), $A$ is

countable; hence $A \subseteq N$ (cf. 2.3(iv)). Therefore every element of **P** is compatible with some element of $A \subseteq D \cap N$.

(2) $\Rightarrow$ (3) is immediate from Corollary 2.2(i).

(3) $\Rightarrow$ (1) : It suffices to show that every maximal antichain $A$ in **P** is countable. Given $A$, choose a countable $N \preceq H(\kappa)$ such that $\{A, \mathbf{P}\} \subseteq N$ and 0 is $N$-generic. We claim that $A \subseteq N$, which will imply that $A$ is countable. Now $A$ is predense in **P**, so $\bar{A}$ is a dense subset of **P** which belongs to $N$ because it is definable from $A$. Let $r \in A$. Since 0 is $N$-generic, there is a member, $p$, of $\bar{A} \cap N$ which is compatible with $r$. But $p \geq q$ for some $q \in A \cap N$, by definition of $\bar{A}$ and since $p \in N$. Thus $r$ is compatible with $q$, so $r = q$ since $A$ is an antichain, and hence $r$ belongs to $N$. $\square$

We will now put this lemma to work in the following applications of Martin's Axiom. The first one should be compared with Proposition 1.4. (See also Exercise II.20.)

**4.6 Proposition.** $(MA(\aleph_1))$ *Let $E \subseteq \lim(\omega_1)$ be a stationary subset of $\omega_1$ and $\eta$ a ladder system on $E$. If $c$ is an $\omega_1$-coloring of $\eta$ such that for all $\delta \in E$ and all $n \in \omega$, $c_\delta(n) \leq \eta_\delta(n)$, then $c$ can be uniformized.*

PROOF. Our partial order **P** will consist of finite approximations to a uniformization of $c$. Let **P** be the set of all pairs $(p, p^*)$ such that $p^*: S \to \omega$ where $S$ is a finite subset of $E$, and $p$ is a restriction to a finite domain of a function $f_p: \omega_1 \to \omega_1$ such that for every $\alpha \in \omega_1$, $f_p(\alpha) \leq \alpha$ and for every $\delta \in S$, $f_p(\eta_\delta(n)) = c_\delta(n)$ if $n \geq p^*(\delta)$. Partially order **P** by: $(p, p^*) \leq (q, q^*)$ if and only if $q$ is an extension of $p$ and $q^*$ is an extension of $p^*$.

For $\delta \in E$, let $D_\delta = \{(p, p^*) \in \mathbf{P}: \delta \in \mathrm{dom}(p^*)\}$. We claim that $D_\delta$ is dense in **P**. Indeed, given any $(p, p^*) \in \mathbf{P}$ and $\delta \notin \mathrm{dom}(p^*)$ we can find $k$ such that if $n \geq k$, then

$$\eta_\delta(n) \notin \{\eta_\gamma(m): \gamma \in \mathrm{dom}(p^*), m \in \omega\} \cup \mathrm{dom}(p)$$

(because if $\gamma \neq \delta$ $\mathrm{rge}(\eta_\gamma) \cap \mathrm{rge}(\eta_\delta)$ is finite, since the two ladders have different limits). Let $\tilde{p} = p^* \cup \{(\delta, k)\}$; we can alter $f_p$ to show that $(p, \tilde{p})$ belongs to **P** and hence to $D_\delta$.

For $\alpha \in \omega_1$, let $D'_\alpha = \{(p,p^*) \in \mathbf{P} : \alpha \in \mathrm{dom}(p)\}$. It is easy to see that $D'_\alpha$ is dense in $\mathbf{P}$. If $\mathbf{P}$ is c.c.c., there is a directed subset $\mathcal{G}$ of $\mathbf{P}$ which meets each $D_\delta$ and each $D'_\alpha$. Let $f^* = \cup\{p^* : (p,p^*) \in \mathcal{G}$ for some $p\}$. Then $f^*$ is a function — because $\mathcal{G}$ is directed — with domain $E$ — because $\mathcal{G}$ meets every $D_\delta$. Define $f = \cup\{p : (p,p^*) \in \mathcal{G}$ for some $p^*\}$. Then $f$ is a function with domain $\omega_1$ — because $\mathcal{G}$ meets each $D'_\alpha$ — and $(f,f^*)$ is a uniformization of $c$.

So it remains to prove that $\mathbf{P}$ is c.c.c. Let $\kappa$ be large enough for $\mathbf{P}$. We shall show that $0$ is $N$-generic whenever $N = \bigcup_{i \in \omega} N_i$ where $N_i \prec N_{i+1} \prec H(\kappa)$ and $N_i \cap \omega_1 < N_{i+1} \cap \omega_1$. Given such an $N$, let $\alpha_i = N_i \cap \omega_1$ and $\alpha = N \cap \omega_1$. Suppose that $(p,p^*) \in \mathbf{P}$ and $D \in N$ is a dense subset of $\mathbf{P}$. Let $f_p$ be as in the definition of $\mathbf{P}$, so that $p$ is a restriction of $f_p$. Choose $i$ so that $D \in N_i$, $\mathrm{dom}(p) \cap \alpha \subseteq \alpha_i$, and $\mathrm{dom}(p^*) \cap \alpha \subseteq \alpha_i$. Notice that $p[\alpha_i] \subseteq \alpha_i$. Let $Y = \{\beta : \beta \in \alpha_i$ and $\beta \in \mathrm{rge}(\eta_\delta)$ for some $\delta \in \mathrm{dom}(p^*) \setminus \alpha_i\}$. Then $Y$ is finite since $\delta > \alpha_i$ if $\delta \in \mathrm{dom}(p^*) \setminus \alpha_i = \mathrm{dom}(p^*) \setminus \alpha$. Since $N_i$ is a model of ZFC$^-$, there exists $q \in N_i$ such that $q = p{\restriction}\alpha_i \cup f_p{\restriction}Y$ (cf. Exercise 4). Let $q^* = p^*{\restriction}\alpha_i$. Then $(q,q^*) \in N_i$ so since $N_i \models$ "$D$ is dense in $\mathbf{P}$", there exists $(r,r^*) \in D \cap N_i$ such that $(q,q^*) \le (r,r^*)$. Note that $\mathrm{dom}(r), \mathrm{dom}(r^*) \subseteq \alpha_i$; it follows that by choice of $q$ and $q^*$, $(p \cup r, p^* \cup r^*) \in \mathbf{P}$. Therefore $(r,r^*) \in D \cap N$ is compatible with $(p,p^*)$. $\square$

The significance of the next application will become clear shortly when we discuss applications of PFA and NPA.

**4.7 Proposition.** $(\mathrm{MA}(\aleph_1))$ *Suppose $E \subseteq \lim(\omega_1)$ and $\eta$ is a ladder system on $E$ such that for every cub $C$ there exists $\delta \in E \cap C$ such that for all but finitely many $n$, $\eta_\delta(n) \in E \cap C$. Then $E$ is the disjoint union of stationary sets $E_0$ and $E_1$ such that for every cub $C$, there exists $\delta \in E_0 \cap C$ such that for all but finitely many $n$, $\eta_\delta(n) \in E_1 \cap C$.*

PROOF. We claim that it suffices to prove that there is a function $g : E \to 2$ such that for every $\delta \in E$ there exists $m_\delta$ so that for all $n \ge m_\delta$, if $\eta_\delta(n) \in E$ then $g(\eta_\delta(n)) = 1 - g(\delta)$. Indeed, if such a $g$ exists, define $S_i = \{\delta \in E : g(\delta) = i\}$ for $i = 0, 1$. Notice that by hypothesis, if $\delta \in S_i$, then for all sufficiently large $n$, $\eta_\delta(n) \in S_{1-i}$. We will show that there is $i \in \{0, 1\}$ so that $E_0 = S_i$ and $E_1 = S_{1-i}$ are as desired. Suppose not; then there are cubs $C_0$, $C_1$ so that for $i = 0, 1$, if $\delta \in C_i \cap S_i$ then there are infinitely many $n$ so that $\eta_\delta(n) \notin S_{1-i}$. Consider $\delta \in C_0 \cap C_1 \cap E$. If $\delta \in S_i$, then for all but finitely many $n$, $\eta_\delta(n) \in S_{1-i}$. This contradicts the choice of $C_0$ and $C_1$. (Note that this argument also shows that $E_0$ and $E_1$ are stationary.)

Our poset will consist of finite approximations to the desired $g$. Let **P** be the set of all pairs $(p, p^*)$ such that $p : S \rightarrow 2$ and $p^* : S \rightarrow \omega$ where $S$ is a finite subset of $E$ and there is a function $\psi_p : E \rightarrow 2$ extending $p$ such that for all $\delta \in S$ and all $n \geq p^*(\delta)$, if $\eta_\delta(n) \in E$ then $\psi_p(\eta_\delta(n)) = 1 - \psi_p(\delta)$. Partially order **P** by $(p, p^*) \leq (q, q^*)$ if and only if $p \subseteq q$ and $p^* \subseteq q^*$.

If $\delta \in E$, let $D_\delta = \{(p, p^*) \in \mathbf{P} : \delta \in \text{dom}(p)\}$. To see that $D_\delta$ is dense in **P**, suppose $(p, p^*) \in \mathbf{P}$ and $\delta \notin \text{dom}(p)$. Let $\psi_p$ be as in the definition. As in the proof of 4.6, we can find $k$ such that if $n \geq k$ then

$$\eta_\delta(n) \notin \cup \{\text{rge}(\eta_\gamma) : \gamma \in \text{dom}(p)\} \cup \text{dom}(p).$$

Hence $(p \cup \{(\delta, \psi_p(\delta))\}, p^* \cup \{(\delta, k)\})$ is a member of **P** which extends $(p, p^*)$.

The proof that **P** is c.c.c. is almost identical to that in 4.6. Then by $\text{MA}(\aleph_1)$ there is a $\{D_\delta : \delta \in E\}$-generic set $\mathcal{G}$. If

$$g \stackrel{\text{def}}{=} \cup \{p : (p, p^*) \in \mathcal{G} \text{ for some } p^*\}$$

then $g$ is the desired function. $\square$

We now turn to a principle which strengthens $\text{MA}(\aleph_1)$.

**4.8 Definition.** A collection $\mathcal{C}$ of countable subsets of $H(\kappa)$ is called a *cub* in $H(\kappa)$ if every countable subset of $H(\kappa)$ is contained in some member of $\mathcal{C}$ and if $\mathcal{C}$ is closed under unions of countable chains. (Compare Exercise II.15.)

For example, by 2.2, the set of all countable elementary sub-models of $H(\kappa)$ is a cub.

A poset **P** is called *proper* if for some $\kappa$ large enough for **P**, there is a cub $\mathcal{C}$ of countable elementary submodels of $H(\kappa)$ such that for all $N \in \mathcal{C}$ and all $q \in N \cap \mathbf{P}$ there exists $p \geq q$ such that $p$ is $N$-generic. By PFA($\lambda$) we mean the principle:

> for every proper poset **P** of cardinality $\lambda$ and every family $\mathcal{D} = \{D_\alpha : \alpha \in \omega_1\}$ of dense subsets of **P**, there is a directed subset $\mathcal{G}$ of **P** such that for all $\alpha \in \omega_1$, $\mathcal{G} \cap D_\alpha \neq \emptyset$.

By PFA we mean: PFA($\lambda$) for all $\lambda$.

It can be shown that if **P** is proper for some $\kappa$ large enough for **P**, then **P** is proper for all $\kappa$ large enough for **P**. PFA stands for "Proper Forcing Axiom." Note that the parenthetical $\lambda$ has a different meaning in PFA($\lambda$) than the $\kappa$ in MA($\kappa$) : in the former it refers to the size of the poset; in the latter, to the number of dense subsets. By Lemma 4.5, every c.c.c. poset is proper; hence PFA implies MA($\aleph_1$). Veličković 19?? has proved that PFA implies that $2^{\aleph_0} = \aleph_2$; hence PFA implies MA $+ \neg$CH. Shelah 1982 has proved that PFA($\aleph_1$) is consistent with ZFC and that — assuming the consistency of the existence of a large (supercompact) cardinal — PFA is consistent with ZFC. The consistency of PFA implies the consistency of some large cardinals. However, many consequences of PFA can be shown to be consistent with ZFC without assuming the consistency of any large cardinals: see Mekler 1983.

The following result shows that Proposition 1.5 is not a theorem of ZFC. (See also 4.11, below.) If $\beta$ belongs to a set $C$ and $\beta < \sup C$, $\beta^+$ will denote the next element of $C$. (In context, there will be no ambiguity.)

**4.9 Proposition.** (PFA) *For any stationary set $E \subseteq \lim(\omega_1)$ and any ladder system $\eta$ on $E$, there is a cub $C$ such that:*

    (1) *for all $\delta \in E$ there exists $m_\delta$ such that for all $n \geq m_\delta$, $\eta_\delta(n) \notin C$;*

    (2) *for all $\beta \in C$, $\beta^+ \in \mathrm{succ}(\omega_1)$; and*

    (3) *for all $\beta \in C \cap E$, $\beta^+ = \beta + 1$.*

PROOF. The poset **P** will consist of countable approximations to the cub we seek. Recall (from II.4.1) that a subset $p$ of $\omega_1$ is called *closed* if sup $X \in p$ for all countable subsets $X$ of $p$. Let **P** be the set of all countable closed subsets $p$ of $\omega_1$ such that: for every $\delta \in E$ there exists $m_\delta$ such that for all $n \geq m_\delta$, $\eta_\delta(n) \notin p$; sup $p \notin E$; for all $\beta \in p \setminus \{\text{sup } p\}$, $\beta^+ \in \text{succ}(\omega_1)$; and for all $\beta \in (p \cap E) \setminus \{\text{sup } p\}$, $\beta^+ = \beta + 1$. (Here, $\beta^+$ denotes the next element of $p$.) Partially order **P** by the relation of end-extension, i.e., $p \leq q$ if and only if $q \cap (\text{sup } p) = p$.

If $\mu \in \omega_1$, let $D_\mu = \{p \in \mathbf{P} : \mu < \text{sup } p\}$; $D_\mu$ is dense in **P** because for any $p \in \mathbf{P}$, if $\mu \geq$ sup $p$, then $p \cup \{\mu + 1\} \in D_\mu$ and $p \leq p \cup \{\mu + 1\}$. If **P** is proper, then there exists a directed subset $\mathcal{G}$ of **P** which meet each $D_\mu$. Let $C = \cup \mathcal{G}$; because $\mathcal{G}$ meets every $D_\mu$, $C$ is clearly unbounded. To see that it is closed, consider a countable subset $X$ of $C$; if $\mu = $ sup $X$, choose $p \in \mathcal{G}$ such that $\mu <$ sup $p$. It suffices to shows that $X \subseteq p$ because $p$ is closed. But for any $\alpha \in X$, there exists $q, r \in \mathcal{G}$ such that $\alpha \in q$ and $p$, $q \leq r$; then $\alpha \in p$ since $q \subseteq r$ and since $r$ is an end-extension of $p$. Furthermore, $C$ has the property that its intersection with every ladder is finite; indeed, if $\delta \in E$ and $p \in \mathcal{G}$ such that sup $p > \delta$, then $\eta_\delta(n) \in C$ if and only if $\eta_\delta(n) \in p$.

So it remains to prove that **P** is proper. Let $\mathcal{C}$ be the cub of all countable elementary submodels $N$ of $H(\kappa)$ such that $N = \bigcup_{i \in \omega} N_i$ where $N_i \prec N_{i+1} \prec H(\kappa)$ and $N_i \cap \omega_1 < N_{i+1} \cap \omega_1$. Fix an $N \in \mathcal{C}$ and let $\alpha_i = N_i \cap \omega_1$ and $\alpha = N \cap \omega_1$. Let $q \in N \cap \mathbf{P}$. Let $\{D_n : n \in \omega\}$ be an enumeration of the dense subsets of **P** which belong to $N$. Without loss of generality, we can assume $q \in N_0$ and $D_n \in N_{n+1}$. We are going to define a chain

$$q_0 \leq q_1 \leq \cdots \leq q_n \leq \cdots$$

of elements of **P** such that $q_0 = q$ and $q_{n+1} \in D_n \cap N_{n+1}$. Note that $\sup(\bigcup_n q_n) = \alpha$ because for all $\mu < \alpha$ there exists $n$ such that $q_{n+1} \in D_\mu$. If $\alpha \notin E$, then $p \overset{\text{def}}{=} \bigcup_n q_n \cup \{\alpha\}$ will be the $N$-generic element we seek — because for any $r \geq p$ and any $D_n \in N$, $r$ is compatible with $q_{n+1} \in D_n \cap N_{n+1}$. If $\alpha \in E$ we need to impose an additional condition on the inductive construction: namely, that $\eta_\alpha(m) \notin q_n$ whenever $\eta_\alpha(m) \geq \alpha_0$. This will insure that $\bigcup_n q_n$ has finite intersection with $\eta_\alpha$; then the $N$-generic $p$ we seek will be $\bigcup_n q_n \cup \{\alpha, \alpha+1\}$.

So it remains to define the chain. Suppose that $q_n \in N_n$ has been chosen; note that $\sup q_n < \alpha_n$. Let

$$\nu_n = \max(\{\eta_\alpha(m): \eta_\alpha(m) < \alpha_{n+1}\} \cup \{\alpha_n\})$$

and let $\tilde{q}_n = q_n \cup \{\nu_n + 1\}$. Then $\tilde{q}_n \in \mathbf{P} \cap N_{n+1}$, and since

$$N_{n+1} \models \text{``} D_n \text{ is dense in } \mathbf{P} \text{''},$$

there exists $q_{n+1} \geq \tilde{q}_n$ such that $q_{n+1} \in D_n \cap N_{n+1}$. Notice that by the definition of end extension and by the inductive choice of $q_n$, $q_{n+1} \cap \mathrm{rge}(\eta_\alpha) \subseteq \alpha_0$. $\square$

In fact, Proposition 4.9 can be proved from $\mathrm{PFA}(\aleph_1)$ alone: see Exercise 8. The following is an *ad hoc* definition. NPA stands for "Non-Properness Axiom."

**4.10 Definition.** By NPA we mean the principle:

> $\mathrm{MA}(\aleph_1) +$ there is a stationary set $E \subseteq \lim(\omega_1)$ and a ladder system $\eta$ on $E$ such that for every cub $C$ there exists $\delta \in C \cap E$ such that $\eta_\delta(n) \in C \cap E$ for all $n \in \omega$.

By 4.9, NPA and PFA are inconsistent. However, $\mathrm{NPA} + 2^{\aleph_0} = \aleph_2$ is consistent with ZFC; in fact it holds in the standard Solovay-Tenenbaum c.c.c. extension of L, since every cub in the extension contains a cub in L (cf. Eklof 1983, Theorem 0.8). As a consequence of 4.7, a weak form of Proposition 1.5 holds in a model of NPA:

**4.11 Proposition.** (NPA) *There is a stationary subset $E$ of* $\lim(\omega_1)$ *and a ladder system $\eta$ on $E$ such that $E$ is the disjoint union of stationary sets $E_0$ and $E_1$ such that for every cub $C$, there exists $\delta \in E_0 \cap C$ such that for all but finitely many $n$, $\eta_\delta(n) \in E_1 \cap C$.* $\square$

In fact, in the Solovay-Tenenbaum model a stronger form of 4.11 is true; the words "for all but finitely many" can be replaced by "all".

Now we have the following diagram of axioms, where the axioms in a branch of the tree are mutually consistent, and, in fact, each axiom implies the one below (in the presence of ZFC), but different branches are mutually inconsistent.

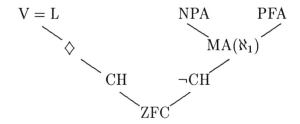

We conclude with one additional axiom which implies $\Diamond$.

**4.12 Definition.** Let $E$ be a subset of $\omega_1$. A poset $\mathbf{P}$ is called *E-complete* if there exists a cub $\mathcal{C}$ of countable elementary submodels of $H(\kappa)$ — for some $\kappa$ large enough for $\mathbf{P}$ — such that for any $N \in \mathcal{C}$ with $N \cap \omega_1 \in E$, whenever

$$p_0 \leq p_1 \leq \ldots \leq p_n \leq \ldots$$

is a chain of elements of $\mathbf{P} \cap N$ such that for all dense subsets $D$ of $\mathbf{P}$ which belong to $N$ there exists $n$ with $p_n \in D$, then there exists $q \in \mathbf{P}$ such that $q \geq p_n$ for all $n \in \omega$. By $\mathrm{Ax}(S) + \Diamond^*(\omega_1 \setminus S)$ we mean the principle:

> there is a stationary and co-stationary subset $S$ of $\omega_1$ such that: (1) for all proper posets $\mathbf{P}$ of cardinality $\aleph_1$ which are $(\omega_1 \setminus S)$-complete, and all families $\mathcal{D}$ of $\aleph_1$ dense subsets of $\mathbf{P}$, there is a $\mathcal{D}$-generic subset of $\mathbf{P}$; and (2) $\Diamond^*(\omega_1 \setminus S)$ holds (hence for all $E \subseteq \omega_1$, if $E \setminus S$ is stationary, $\Diamond(E)$ holds; cf. Exercise 2)

Shelah 1982 has shown that $\mathrm{Ax}(S) + \Diamond^*(\omega_1 \setminus S)$ is consistent with ZFC + GCH. In a model of $\mathrm{Ax}(S) + \Diamond^*(\omega_1 \setminus S)$ we obtain a mix of properties implied by PFA and by $V = L$, depending on which stationary set is involved. For example we have the following. (Compare 1.4 and 4.6.)

**4.13 Proposition.** $(\mathrm{Ax}(S) + \Diamond^*(\omega_1 \setminus S))$ *There is a stationary and co-stationary subset $S$ of $\omega_1$ such that:*

(1) *for any stationary $E \subseteq S$ and any ladder system $\eta$ on $E$, if $c$ is an $\omega_1$-coloring of $\eta$ such that for all $\delta \in E$ and all $n \in \omega$, $c_\delta(n) \leq \eta_\delta(n)$, then $c$ can be uniformized; and*

(2) *for any $E$ such that $E \setminus S$ is stationary, and any ladder system $\eta$ on $E$, there is a 2-coloring of $\eta$ which cannot be uniformized.*

PROOF. The proof of (2) is exactly like the proof of 1.4 since $\Diamond(E)$ holds if $E \setminus S$ is stationary.

To prove (1), let $\mathbf{P}$ consist of all pairs $(p, p^*)$ for which there is $\alpha \in \omega_1$ such that $(p, p^*)$ is a uniformization of $c$ "up to $\alpha$", i.e., $p: \alpha \to \omega_1$ and $p^*: E \cap \alpha + 1 \to \omega$ such that for all $\delta \in E \cap \alpha + 1$ and all $n \geq p^*(\delta)$, $p(\eta_\delta(n)) = c_\delta(n)$.

If $D'_\alpha$ is defined as in the proof of 4.6, then it is not hard to see that it is a dense subset of $\mathbf{P}$. If $\mathbf{P}$ is proper and $(\omega_1 \setminus S)$-complete, then there is a directed subset $\mathcal{G}$ of $\mathbf{P}$ which meets these dense subsets, and, as in 4.6, we obtain the desired uniformization.

So it remains to prove that $\mathbf{P}$ is proper and $(\omega_1 \setminus S)$-complete. Let $\kappa$ be large enough for $\mathbf{P}$. Let $\mathcal{C}$ be the cub of all countable elementary submodels $N$ of $\mathrm{H}(\kappa)$ such that $N = \bigcup_{i \in \omega} N_i$ where $N_i \prec N_{i+1} \prec \mathrm{H}(\kappa)$ and $N_i \cap \omega_1 < N_{i+1} \cap \omega_1$. Fix an $N \in \mathcal{C}$ and let $\alpha_i = N_i \cap \omega_1$ and $\alpha = N \cap \omega_1$. Let $(q, q^*) \in N \cap \mathbf{P}$. To show that $\mathbf{P}$ is proper, we must show that there exists $(p, p^*) \geq (q, q^*)$ such that $(p, p^*)$ is $N$-generic. Let $\{D_n : n \in \omega\}$ be an enumeration of the dense subsets of $\mathbf{P}$ which belong to $N$. Without loss of generality we can assume that $(q, q^*) \in N_0$ and that $D_n \in N_{n+1}$. We will define by induction a chain

$$(q_0, q_0^*) \leq (q_1, q_1^*) \leq \cdots \leq (q_n, q_n^*) \leq \cdots$$

of elements of $N$ such that $(q_0, q_0^*) = (q, q^*)$ and $(q_{n+1}, q_{n+1}^*) \in D_n \cap N_{n+1}$. If $(p, p^*) \stackrel{\text{def}}{=} (\bigcup_n q_n, \bigcup_n q_n^*)$ is an element of $\mathbf{P}$, then it is the $N$-generic extension of $(q, q^*)$ we seek. If $\alpha \notin E$, then $(p, p^*)$ will be an element of $\mathbf{P}$, without any further care being taken. (In fact, this observation means that $\mathbf{P}$ is $(\omega_1 \setminus S)$-complete.) But if $\alpha \in E$, we may already have committed ourselves to the value of $p(\eta_\alpha(n))$ for all $n \in \omega$. In order to insure that our commitment allows $(p, p^*)$ to be in $\mathbf{P}$ we add an additional requirement to our inductive construction. We will require that $q_n(\eta_\alpha(m))$ is defined and equals $c_\alpha(m)$ whenever $\alpha_0 \leq \eta_\alpha(m) < \alpha_n$.

So it remains to do the inductive construction. Suppose that $(q_n, q_n^*) \in N_n$ has been chosen. Since $S_n \stackrel{\text{def}}{=} \{\eta_\alpha(m) : \alpha_n \leq \eta_\alpha(m) < \alpha_{n+1}\}$ is finite, there is an element $(\tilde{q}_n, \tilde{q}_n^*) \in \mathbf{P} \cap N_{n+1}$ which extends $(q_n, q_n^*)$ and satisfies: $\tilde{q}_n(\beta) = c_\alpha(\beta)$ for all $\beta \in S_n$. (We use the facts that $N_{n+1}$ is a model of ZFC$^-$ and that every element of $\mathrm{dom}(q_n)$ is $< \alpha_n$.) Now, since

$$N_{n+1} \models \text{``}D_n \text{ is dense in } \mathbf{P}\text{''},$$

there exists $(q_{n+1}, q_{n+1}^*) \in D_n \cap N_{n+1}$ such that $(q_{n+1}, q_{n+1}^*) \geq (\tilde{q}_n, \tilde{q}_n^*)$. This completes the inductive construction. $\square$

## EXERCISES

1. Let $\Diamond'(E)$ denote the following principle:

> there is a family $\{S_\alpha : \alpha \in E\}$ such that for each $\alpha \in E$ $S_\alpha$ is a countable set of subsets of $\alpha$, and such that for all $X \subseteq \omega_1$, $\{\alpha \in E : X \cap \alpha \in S_\alpha\}$ is stationary in $\omega_1$.

Prove that $\Diamond'(E)$ is equivalent to $\Diamond(E)$ (in ZFC). [Hint: Assuming $\Diamond'$, show, as in 1.2, that there are subsets $Y_{\alpha,n}$ of $\alpha \times \omega$ such that for all $X \subseteq \omega_1 \times \omega$, $\{\alpha \in E : X \cap (\alpha \times \omega) \in \{Y_{\alpha,n} : n \in \omega\}\}$ is stationary. Then show that for some $n$, $\{\{\nu \in \alpha : (\nu, n) \in Y_{\alpha,n}\} : \alpha \in E\}$ is a $\Diamond$-sequence.]

2. $\Diamond_\kappa^*(E)$ implies $\Diamond_\kappa(E')$ for every $E' \subseteq \kappa$ such that $E \cap E'$ is stationary.

3. If $\diamondsuit_\kappa(E)$ holds and $\kappa = \lambda^+$, then there is a decomposition of $E$, $E = \amalg_{\beta\in\kappa}E_\beta$, into $\kappa$ disjoint subsets such that for all $\beta \in \kappa$, $\diamondsuit_\kappa(E_\beta)$ holds. [Compare 1.10.]

4. Suppose $N$ is a countable elementary submodel of $H(\kappa)$ for some uncountable $\kappa$, and $f \in H(\kappa)$ such that dom $f = \{a_1, \ldots, a_n\}$ and $a_i \in N$ and $f(a_i) \in N$ for all $i = 1, \ldots, n$. Prove that $f \in N$. [Hint: the axioms of ZFC$^-$ imply that $f$ exists.]

5. Show that MA($\aleph_0$) is a theorem of ZFC. [Hint: build a chain of elements of **P** so that the $n$th one belongs to the $n$th dense subset.]

6. Show that Proposition 4.6 can be proved — under the assumption MA($\aleph_1$) — when the hypothesis on $c$ is weakened to: $c_\delta(n) \le \eta_\delta(n) + \omega$. Generalize this further.

7. What goes wrong with the proof of 4.13(1) if we try to weaken the hypothesis on $c$ to: $c_\delta(n) < \delta$? [Compare Exercise II.20].

8. Prove Proposition 4.9 using only PFA($\aleph_1$). [Hint: let **P** consist of all pairs $(f, f')$ such that $f = g{\restriction}X$, $f' = g'{\restriction}g[X]$ where $g$ enumerates in increasing order a countable closed subset of $\omega_1$, $X$ is a finite subset of dom $g$, dom $g' = $ rge $g$, and for all $\delta \in$ dom $g'$, if $n \ge g'(\delta)$, then $\eta_\delta(n) \notin$ rge $g$.]

9. Use $\diamondsuit(E)$ to prove that there is a ladder system $\eta$ on $E$ such that for every cub $C$ there exists $\delta \in C \cap E$ such that $\eta_\delta(n) \in C \cap E$ for infinitely many $n \in \omega$.

10. Suppose $N'$ is a model for an expansion $\mathcal{L}$ of $\mathcal{L}_\in$ which contains a binary relation symbol $<$, and suppose that the interpretation of $<$ in $N'$ well-orders $N'$. Prove that if $N_1$ and $N_2$ are elementary submodels of $N'$, then $N_1 \cap N_2$ is an elementary submodel of $N'$. [Hint: use 2.1; observe that if $N = N_1$ or $N_2$, and $N' \models \exists x\varphi[a_0, \ldots, a_{n-1}]$, then the $<$-minimal element $a \in N'$ such that $N' \models \exists x\varphi[a_0, \ldots, a_{n-1}, a]$ belongs to $N$.]

11. The following property of $\kappa$ is equivalent to $\kappa$ being weakly compact (cf. Drake 1974, p. 292):

$\kappa$ is strongly inaccessible and any structure $N = (V_\kappa, \in, \ldots)$ has a proper elementary extension $N' = (A, \in, \ldots)$ where $\kappa \in A$ and $A$ is transitive.

Using this characterization of weakly compact, prove that if $\kappa$ is weakly compact, $|R| < \kappa$ and $M$ is a $\kappa$-free module of cardinality $\kappa$, then $M$ is free. [Hint: Choose a $\kappa$-filtration $\{M_\alpha : \alpha < \kappa\}$ of $M$ and consider the model $(V_\kappa, \in, M, \{M_\alpha : \alpha < \kappa\})$; then $M_\kappa \in A$ and is free; but $M = M_\kappa$.]

12. Prove that MA + $\neg$CH implies $2^\kappa = 2^{\aleph_1}$ for all $\kappa < 2^{\aleph_0}$ in the following sequence of steps:

If $\kappa < 2^{\aleph_0}$, choose an almost disjoint family $\{A_\alpha : \alpha < \kappa\}$ of subsets of $\omega$ (cf. II.5.5.) For any $X \subseteq \kappa$, define functions $c_{\alpha,X} : A_\alpha \to 2$ such that $c_{\alpha X} \equiv 0$ if $\alpha \notin X$ and $c_{\alpha,X} \equiv 1$ if $\alpha \in X$.

(i) Assuming MA($\kappa$), show that there exists $g_X : \omega \to 2$ such that for all $\alpha < \kappa$, if $\alpha \notin X$, then $g_X(n) = c_{\alpha,X}(n)$ for all but finitely many $n \in A_\alpha$, and if $\alpha \in X$, then $g_X(n) = c_{\alpha,X}(n)$ for infinitely many $n \in A_\alpha$.

(ii) If $\theta : \mathcal{P}(\kappa) \to {}^\omega 2$ is such that $\theta(X) = g_X$ for all $X$, show that $\theta$ is one-one.

The next two exercises give a proof of $\diamondsuit_{\mu^+}$ from GCH for $\mu \geq \aleph_1$.

13. Suppose that $\lambda = 2^\mu = \mu^+$ and $\kappa$ is a regular cardinal $< \mu$. Let $E = \{\alpha : \alpha < \lambda$ and $\mathrm{cf}(\alpha) = \kappa\}$.

(a) There is a list $\{X_\alpha : \alpha < \lambda\}$ of all the bounded subsets of $\lambda$, i.e., subset contained in $\tau$ for some $\tau < \lambda$.

(b) Assume $\mu = \mu^\kappa$. For $\alpha \in E$, let $S_\alpha = \{\bigcup Y : Y \subseteq \{X_\beta \cap \alpha : \beta < \alpha\}$ and $|Y| \leq \kappa\}$. Then $\{S_\alpha : \alpha \in E\}$ is a $\diamondsuit_\lambda^*(E)$-sequence. [Hint: for all $W \subseteq \lambda$, define a cub $C_1$ as follows. Let $0 \in C_1$. If $\alpha \in C_1$, let $\alpha^+$, the successor of $\alpha$ in $C_1$, be the least ordinal $\beta > \alpha$ so that for some $\gamma < \beta$, $X_\gamma = W \cap \alpha$. Then the limit points of $C_1$ form the desired cub.]

(c) Assume $\mu$ is singular, $\mathrm{cf}(\mu) = \rho \neq \kappa$, and for every $\delta < \mu$, $\delta^\kappa < \mu$. For each $\alpha \in E$, fix $\{\alpha_i : i < \kappa\}$ a ladder on $\alpha$. Also fix an increasing sequence of sets $\{U_j^\alpha : j < \rho\}$ such that $\alpha = \bigcup U_j^\alpha$ and for all $j$, $|U_j^\alpha| < \mu$. Let $S_\alpha = \{\cup Y: Y \subseteq \{X_\beta \cap \alpha: \beta \in U_j^\alpha\}$, for some $j$ and $|Y| \leq \kappa\}$. Then $\{S_\alpha: \alpha \in E\}$ is a $\diamondsuit^*(E)$ sequence. [Hint: consider $W \subseteq \lambda$ and $C_1$ defined as above. Suppose $\delta \in E$ is a limit point of $C_1$. Then there is a set $\{\gamma_i: i < \kappa\}$ of ordinals $< \delta$, so that if $I \subseteq \kappa$ and $|I| = \kappa$ then $\bigcup_{i \in I} X_{\gamma_i} = W \cap \delta$. Since $\rho \neq \kappa$, there is $j < \rho$ so that $|U_j^\delta \cap \{\gamma_i: i < \kappa\}| = \kappa$.]

14. Assume GCH. Suppose $\mu$ is a cardinal and $\mathrm{cf}(\mu) = \rho$. Then $\diamondsuit^*(\{\alpha < \mu^+: \mathrm{cf}(\alpha) \neq \rho\})$ holds. [Hint: consider two cases: (b) $\rho = \mu$; (c) $\rho \neq \mu$.]

## NOTES

The diamond principles are due to R. B. Jensen (See R. B. Jensen 1972 and Devlin 1977.) The weak diamond principles are from Devlin-Shelah 1978, as is 1.8 (and implicitly 1.4). Proposition 1.5 is from Eklof 1983. Theorem 1.10 is essentially due to Devlin-Shelah 1978.

For historical notes and references for model theory see Chang-Keisler 1973. For more on the Axiom of Constructibility, see Devlin 1977 and Devlin 1984.

3.1 through 3.4, 3.5, 3.11 and 3.13 are due to R. B. Jensen (See R. B. Jensen 1972 and Devlin 1977.) Theorem 3.5 is due to Magidor and, independently, Solovay. 3.8 and 3.9 are due to Shelah (cf. Mekler-Shelah 1987). Theorem 3.10 is due to Litman and Shelah. Theorem 3.12 is due to Prikry-Solovay 1975. Theorem 3.15 is due to Gaifman; for more on this result and on $0^\#$ see Jech 1978, section 30. Theorem 3.16 is from Devlin-Jensen 1975.

The basic facts about MA are from Martin-Solovay 1970 and Solovay-Tenenbaum 1971. PFA and $\mathrm{Ax}(S)$ are due to Shelah 1982. Lemma 4.5 follows Mekler 1984. Proposition 4.6 is from Devlin-Shelah 1978. Proposition 4.7 is essentially from Eklof 1983 as is 4.9. The idea of Definition 4.10 comes from Eklof 1983, but the notation NPA does not appear elsewhere. Mekler 1983 contains an axiom PFA$^-$ which can be shown consistent without assuming the consistency of any large cardinals and which has many of the consequences of PFA.

Exercises.    1:Kunen; 2:Jensen; 8:Mekler 1983; 9:Eklof 1983; 12: Martin-Solovay 1970; 13(b):Gregory 1976; 13(c):Shelah 1981c; 14:Gregory 1976 (regular $\mu$) and Shelah 1981c (singular $\mu$).

That the result in 14 is sharp is due to R. B. Jensen in the case of $\mu = \aleph_0$ (cf. Devlin-Johnsbråten 1974), and to Shelah for $\mu \geq \aleph_1$ (cf. Steinhorn-King 1980 for $\mu$ regular, and Shelah 1984c for $\mu$ singular).

# CHAPTER VII
# ALMOST FREE MODULES REVISITED

In this chapter we continue with the study, begun in Chapter IV, of $\kappa$-free modules. In the first section, we construct $\leq \kappa$-generated $\kappa$-free $R$-modules which are not free; for most $\kappa$ we require the hypothesis $V = L$. Combined with the results of IV.1, this gives a complete answer to the question of the existence of such modules, assuming $V = L$.

For the rest of the chapter we consider mostly $\mathbb{Z}$-modules. In section 2, we compile information about $\kappa$-free $\mathbb{Z}$-modules, often under the hypothesis $V = L$. Without the hypothesis $V = L$, these questions are more difficult, so in sections 3 and 3A we take a different approach, by giving a purely set-theoretic characterization of those $\kappa$ for which there is a $\kappa$-free abelian group of cardinality $\kappa$ which is not free. (This is a theorem of ZFC.) As a by-product of the machinery developed there, we are able to show that whenever there is a $\kappa$-free abelian group of cardinality $\kappa$ which is not free, then there is one which is strongly $\kappa$-free and not free. We also discuss generalizations of the results of this section to other varieties, including varieties of modules over non-left-perfect rings. Finally in section 4, we make use of this machinery to prove that it is independent of ZFC whether every hereditarily separable abelian group of cardinality $\aleph_1$ is free.

There is some overlap between the results in sections 1 and 2 and those in section 3; many of the results in the first two sections can also be proved, and improved, by using the set-theoretic analysis. In general, it is easier to work with the set-theoretic objects than with their group-theoretic analogues. Thus the work necessary to prove the set-theoretic characterization yields more than the characterization itself.

## §1. $\kappa$-free modules revisited

Let $\kappa$ be an uncountable cardinal. Our goal in this section is to construct $\kappa$-free $R$-modules which are $\leq \kappa$-generated and not free.

Of course, by IV.3.2 and IV.3.5, for this to be possible $\kappa$ must be regular and not weakly compact. Since we have already disposed of the case of left perfect rings at the end of section IV.1, we shall confine ourselves here to rings which are not left perfect. In that case we shall show that the existence of such modules, which are even strongly $\kappa$-free, is consistent with ZFC.

Our method of construction will be based on IV.1.5 and IV.1.11. We will define by transfinite induction a $\kappa$-filtration $\{M_\nu : \nu < \kappa\}$ such that for all $\nu$, $M_\nu$ is free and $< \kappa$-generated and for all $\nu < \mu$, $M_\mu / M_{\nu+1}$ is free. In order to insure that the union of the filtration is not free we shall make $M_\mu / M_\nu$ non-free whenever $\nu$ belongs to a certain stationary set, $E$ (cf. IV.1.7). The crucial step in the construction will be the definition of $M_{\nu+1}$ when $\nu$ belongs to $E$. The "template" on which we build $M_{\nu+1}$ is provided by the following result; (2) is sometimes referred to as the "Construction Principle", or "(CP)" (cf. 3A.16).

**1.1 Proposition.** *The following are equivalent:*

(1) *$R$ is not left perfect;*

(2) *there are free modules $K \subseteq F$ of rank $\aleph_0$ such that $F/K$ is not projective, but $K$ is the union of an increasing chain of free submodules, $K_n (n \in \omega)$ such that for all $n$, $K_{n+1}/K_n$ is free of rank 1 and $F/K_n$ is free.*

PROOF. We will leave the proof of (2) $\Rightarrow$ (1) as an exercise, since we do not need it. For the other implication, assume $R$ is not left perfect; then there is an infinite descending chain $a_0 R \supset a_0 a_1 R \supset \dots \supset a_0 a_1 \cdots a_n R \supset \dots$ of principal right ideals (cf. IV.1.12). Let $F$ be the free $R$-module with basis $\{x_k : k \in \omega\}$; let $K_n$ be the submodule generated by $\{x_k - a_k x_{k+1} : k \leq n\}$ and let $K$ be the union of the $K_n$'s. It is easy to see that each $K_n$ is free (with the given generating set as basis) and that $K$ is free (with $\{x_k - a_k x_{k+1} : k \in \omega\}$ as basis). Moreover, $K_n$ is a summand of $F$, with complementary summand generated by the linearly independent set $\{x_k : k > n\}$. To finish the proof we must show that $K$ is not a direct summand of $F$. Suppose, to obtain a contradiction, that it is; then because $\{x_k - a_k x_{k+1} : k \in \omega\}$ is a basis of $K$, a summand of $F$, there is a homomorphism $\varphi : F \to F$ such that for all $k$, $\varphi(x_k - a_k x_{k+1}) = x_k$. For each $m$, let $\varphi(x_m) = \sum_k r_{m,k} x_k$. So

$$x_m = \varphi(x_m - a_m x_{m+1}) = \sum_k (r_{m,k} - a_m r_{m+1,k}) x_k.$$

Comparing coefficients we see that

(1.1.1)          $a_m r_{m+1,k} = r_{m,k}$ if $m \neq k$; and

(1.1.2)          $a_m r_{m+1,m} = r_{m,m} - 1$

Then

$$a_0 a_1 \cdots a_{m-1} a_m r_{m+1,m} = a_0 a_1 \cdots a_{m-1} (r_{m,m} - 1)$$

by (1.1.2) and this equals $r_{0,m} - a_0 a_1 \cdots a_{m-1}$ by (1.1.1), which finally equals $-a_0 a_1 \cdots a_{m-1}$ because for sufficiently large $m$, $r_{0,m} = 0$.

Thus for sufficiently large $m$

$$a_0 a_1 \cdots a_{m-1} R \subseteq a_0 a_1 \cdots a_{m-1} a_m R,$$

which contradicts the choice of an infinite descending chain of right ideals. $\square$

**1.2 Corollary.** *Suppose $R$ is not left-perfect. Given any cardinals $\lambda_n \geq \aleph_0$ ($n \in \omega$), there exist free modules $K' \subseteq F'$ such that $K'$ is not a summand of $F'$ but $K'$ is the union of a chain of free submodules $K'_n$ such that for all $n$, $F'/K'_n$ is free, $K'_0$ is free of rank $\lambda_0$, and $K'_{n+1}/K'_n$ is free of rank $\lambda_{n+1}$.*

PROOF. Let $F$, $K$ and $K_n$ be as in 1.1. Note that $K_0$ and $K_{n+1}/K_n$ are free of rank 1. Let $H_n$ be a free module of rank $\lambda_n$. Let $H = \bigoplus_{n \in \omega} H_n$, $K'_n = K_n \oplus \bigoplus_{k \leq n} H_n$, $K' = K \oplus H$, $F' = F \oplus H$. $\square$

We will now undertake the construction of a $\kappa$-free module in the case when $\kappa = \aleph_1$; this case, which can be carried out in ZFC, will serve as a paradigm for the general construction.

**1.3 Theorem.** *If $R$ is not left perfect, then there exists a $\leq \aleph_1$-generated $\aleph_1$-free, strongly $\aleph_1$-free module which is not projective.*

PROOF. Let us fix a subset $E$ of $\lim(\omega_1)$ which is stationary in $\aleph_1$. We shall define by transfinite induction an increasing chain of modules $M_\nu (\nu < \omega_1)$ satisfying for all $\nu < \mu < \omega_1$ :
  (a) $M_\nu$ is a free module of rank $\aleph_0$;
  (b) if $\nu$ is a limit ordinal, $M_\nu = \bigcup_{\tau < \nu} M_\tau$;
  (c) if $\nu \notin E$, $M_\mu/M_\nu$ is free of rank $\aleph_0$;
  (d) if $\nu \in E$, $M_{\nu+1}/M_\nu$ is not projective.
Assuming for the moment that we can do this, let us see what we get. Let $M = \bigcup_{\nu < \omega_1} M_\nu$. Clearly $M$ is $\leq \aleph_1$-generated, since each $M_\nu$ is $\leq \aleph_0$-generated. $M$ is $\aleph_1$-free since $\mathcal{C} = \{M_\nu : \nu < \omega_1\}$ is a cub of $< \aleph_1$-generated free submodules of $M$ (cf. IV.1.1). Moreover, (c) and Proposition IV.1.11 imply that $M$ is strongly $\aleph_1$-free since $E$ consists of limit ordinals.

Finally, to show that $M$ is not free we shall show that if $\nu \in E$, then for every $\mu > \nu$, $M_\mu/M_\nu$ is not projective. This is sufficient because if $M$ were projective, then, by Kaplansky's result (I.2.4), $M$ is the direct sum of countably generated projectives, so there is a cub $C$ such that for $\nu < \mu$, where $\nu, \mu \in C$, $M_\mu/M_\nu$ is projective (cf. Exercise IV.19) Now for every $\mu > \nu$

$$0 \to M_{\nu+1}/M_\nu \to M_\mu/M_\nu \to M_\mu/M_{\nu+1} \to 0$$

splits because $M_\mu/M_{\nu+1}$ is free by (c); thus

$$M_\mu/M_\nu \cong M_{\nu+1}/M_\nu \oplus M_\mu/M_{\nu+1}$$

so $M_\mu/M_\nu$ cannot be projective since $M_{\nu+1}/M_\nu$ is not projective by (d).

It remains to do the construction so that (a) through (d) are satisfied. We begin by letting $M_0$ be a free module of rank $\aleph_0$. Now suppose that $M_\mu$ has been defined for all $\mu < \delta$ for some $\delta < \omega_1$ such that (a) – (d) hold for all $\nu < \mu < \delta$. If $\delta$ is a limit ordinal, we let $M_\delta$ be the union of the $M_\mu$'s, as required by (b). Properties (b) and (d) are easily verified for all $\nu < \mu \leq \delta$; (a) will follow from (c). To check (c), let $\nu < \delta$ be such that $\nu \notin E$; we must show that $M_\delta/M_\nu$ is free. For this we use the fact that

(1.3.1)    $E$ is non-reflecting.

(See II.4.6.) In fact, there exists $C \subseteq \delta$ of order type $\omega$ such that sup $C = \delta$ and $C$ consists of successor ordinals; then $C \cap E = \emptyset$. Now $M_\delta = \cup\{M_\tau : \tau \in C, \tau \geq \nu\}$, and for all $\sigma < \tau$ in $C$, $M_\tau/M_\sigma$ is free and $M_\sigma/M_\nu$ is free if $\sigma \geq \nu$; therefore $M_\delta/M_\nu$ is free.

If $\delta = \nu + 1$ and $\nu \notin E$, we let $M_\delta = M_\nu \oplus F$, where $F$ is a free module of rank $\aleph_0$. In this case it is routine to verify that properties (a) through (d) continue to hold.

The final, and critical, case is when $\delta = \nu + 1$ and $\nu \in E$. In this case we use the fact that

(1.3.2)    if $\nu \in E$, then $\mathrm{cf}(\nu) = \omega$.

Hence there is a strictly increasing sequence $\{\tau_n : n \in \omega\}$ of successor ordinals such that $\sup\{\tau_n : n \in \omega\} = \nu$. For convenience, let $\tilde{M}_n$ denote $M_{\tau_n}$. Since $\tau_n \notin E$, $\tilde{M}_{n+1}/\tilde{M}_n$ is free of rank $\aleph_0$ for all $n$. Now apply Corollary 1.2 with $\lambda_n = \aleph_0$ for all $n$ to get modules $F'$, $K'$ and $K'_n$ as given there. Then since $\tilde{M}_0 \cong K'_0$ and $\tilde{M}_{n+1}/\tilde{M}_n$ is free and isomorphic to $K'_{n+1}/K'_n$ for all $n$, we can define an isomorphism $f: K' \to M_\nu$ such that for all $n$, $f(K'_n) = \tilde{M}_n$. Now we define $M_{\nu+1}$ so that $f$ extends to an isomorphism of $F'$ onto $M_{\nu+1}$, that is $M_{\nu+1}$ is the pushout of

$$
\begin{array}{c}
M_\nu \\
\uparrow {\scriptstyle f} \\
K' \subseteq F'
\end{array}
$$

(Alternately, choose $M_{\nu+1}$ and $g$ so that $g: F' \to M_{\nu+1}$ extends $f$ and is a one-one onto set map; then let the module structure of $M_{\nu+1}$ be induced by $g$). Properties (a), (b) and (d) are clear. As for (c), by construction $M_{\nu+1}/\tilde{M}_n \cong F'/K'_n$ is free for all $n$; if $\mu < \nu$ and $\mu \notin E$, then $\mu < \tau_n$ for some $n$ and then $M_{\nu+1}/M_\mu$ is free because $M_{\nu+1}/\tilde{M}_n$ and $\tilde{M}_n/M_\mu$ are free. $\square$

What is special about $\aleph_1$ in the above proof? Only the fact that the properties (1.3.1) and (1.3.2) hold for *every* subset $E$ of $\lim(\aleph_1)$. In some models of ZFC there is no stationary subset of $\aleph_2$ with these properties. But we have seen in VI.3.2 that $V = L$ implies that such a stationary subset does exist in every cardinal where it is consistent that it exists. Thus we can prove the following more general theorem. Note that all we need for the proof is that $E(\kappa)$ holds for the given $\kappa$.

**1.4 Theorem.** $(V = L)$ *If $R$ is not left perfect, then for every regular cardinal $\kappa$ which is not weakly compact there exists a $\leq \kappa$-generated $\kappa$-free, strongly $\kappa$-free module which is not projective.*

PROOF. We shall concentrate on the modifications that need to be made to the proof of Theorem 1.3. We begin by fixing a subset $E$ of $\lim(\kappa)$ which is stationary in $\kappa$ and satisfies (1.3.1) and (1.3.2); such a set exists by VI.3.2. We shall define an increasing chain of modules $M_\nu$ $(\nu < \kappa)$ satisfying, for all $\nu < \mu < \kappa$, (b) and (d) as in the proof of 1.3 and also:

(a') $M_\nu$ is a free module of rank $|\nu| + \aleph_0$; and

(c') if $\nu \notin E$, $M_\mu/M_\nu$ is free of rank $|\mu| + \aleph_0$.

As before, if we can do this construction then $M = \bigcup_{\nu<\kappa} M_\nu$ will be the module sought for. Notice that $\Gamma(M) = \tilde{E} \neq 0$.

The construction begins by letting $M_0$ be a free module of rank $\aleph_0$. Suppose that $M_\mu$ has been defined for all $\mu < \delta$ for some $\delta < \kappa$ such that (a'), (b), (c') and (d) hold. If $\delta$ is a limit ordinal we let $M_\delta$ be the union of the $M_\mu$'s. Then (a') follows from (c'); now by (1.3.1) there is a cub $C$ in $\delta$ such that $C \cap E = \emptyset$. (If $\mathrm{cf}(\delta) = \omega$, then $C$ is a sequence of order type $\omega$ as in the proof of 1.3.) Then if $\nu \in \delta \setminus E$, $M_\delta = \bigcup\{M_\tau : \tau \in C, \tau \geq \nu\}$ and, as before, $M_\delta/M_\nu$ is free.

If $\delta = \nu + 1$ and $\nu \notin E$, let $M_\delta = M_\nu \oplus F_\nu$, where $F_\nu$ is a free module of rank $|\nu| + \aleph_0$.

In the critical case when $\delta = \nu + 1$ and $\nu \in E$, we use (1.3.2). Let $\{\tau_n : n \in \omega\}$ and $\tilde{M}_n$ be as in the proof of 1.3. Then we apply Corollary 1.2 with $\lambda_n = |\tau_n| + \aleph_0$ and finish the construction as before. $\square$

By a simple trick we can produce large numbers of $\kappa$-free modules under the hypotheses of Theorem 1.4.

**1.5 Corollary.** (V = L) *If $R$ is not left perfect, then for any regular $\kappa$ which is not weakly compact there exist $2^\kappa$ pairwise non-isomorphic $\leq \kappa$-generated $\kappa$-free, strongly $\kappa$-free modules.*

PROOF. Let $E$ be a stationary subset of $\kappa$ as in the proof of 1.4. We can write $E$ as the disjoint union

$$E = \amalg_{\nu < \kappa} E_\nu$$

of $\kappa$ stationary subsets $E_\nu$ (cf. II.4.9). Now let $\{X_i : i \in I\}$ be a family of $2^\kappa$ non-empty subsets of $\kappa$ (so $|I| = 2^\kappa$). For each $i \in I$, let

$$S_i = \cup\{E_\nu : \nu \in X_i\}.$$

Then for $i \neq j$, $\tilde{S}_i \neq \tilde{S}_j$ because either there exists $\nu \in X_i \setminus X_j$ or there exists $\nu \in X_j \setminus X_i$; in either case since $E_\nu$ is stationary there is no cub $C$ such that $S_i \cap C = S_j \cap C$.

Now for each $i$, $S_i$ is stationary; moreover it satisfies (1.3.1) and (1.3.2) because it is a subset of $E$. Thus we can construct, as in 1.4, a $\leq \kappa$-generated, $\kappa$-free, strongly $\kappa$-free module $M_i$ such that $\Gamma(M_i) = \tilde{S}_i$. But then $M_i$ cannot be isomorphic to $M_j$ if $i \neq j$ because $\Gamma(M)$ is an invariant of $M$. $\square$

The modules constructed in 1.5 are non-isomorphic because they have different $\Gamma$-invariants; later, in Chapter VIII, we shall consider the question of constructing non-isomorphic groups with the same $\Gamma$-invariant.

As mentioned in section VI.3, Theorem 1.4 cannot be proved in ZFC. (See also the results of Magidor and Shelah mentioned after

Theorem 2.6.) However, the case of Theorem 1.4 where $\kappa = \aleph_{n+1}$ for $n \in \omega$ *is* provable in ZFC; we will give the proof for $R = \mathbf{Z}$ in the next section; for the general case see Theorem 3.13.

We can also use Theorem 1.4 to obtain a partial result about the consistency strength of the assumption that there is a cardinal $\kappa$ such that $\kappa$-free implies free for $\mathbf{Z}$-modules. We have already seen in II.3.11 that $\kappa$ has this property if $\kappa$ is $L_{\omega_1\omega}$-compact. The proof of 1.4 shows that if $E(\kappa^+)$ holds for every singular cardinal $\kappa$, then there is no $\kappa$ such that $\kappa$-free implies free. Hence by VI.3.17, if it is consistent that there exists $\kappa$ such that $\kappa$-free implies free, then it is consistent that there is a proper class of measurable cardinals.

## §2. κ-free abelian groups

Now we shall specialize to the case when $R = \mathbf{Z}$ and gather more information about the existence of $\kappa$-free groups (i.e., $\mathbf{Z}$-modules). As usual, most, if not all, of the results of this section will generalize to arbitrary p.i.d.'s with appropriate modifications.

First of all, recall that our definitions simplify in this case, because every submodule of a free module is free and also because in the uncountable case we can speak of cardinality rather than number of generators. If $\kappa$ is an uncountable cardinal, a group $A$ of arbitrary cardinality is $\kappa$-free if and only if every subgroup of cardinality $< \kappa$ is free; $A$ is strongly $\kappa$-free if and only if every subset of size $< \kappa$ is contained in a free subgroup $B$ such that $A/B$ is $\kappa$-free. Thus a strongly $\kappa$-free group is $\kappa$-free. A group is $\aleph_0$-free if it is torsion-free, which is equivalent to saying that every finitely-generated subgroup is free.

The definition of $\Gamma(A)$, for a $\kappa$-free group A of cardinality $\kappa$ (an uncountable cardinal) also simplifies (cf. the discussion after IV.1.6). We say that a subgroup $B$ of A is $\kappa$-*pure* in A if $A/B$ is $\kappa$-free. If $A = \bigcup_{\nu < \kappa} A_\nu$ is a $\kappa$-filtration of $A$, then $\Gamma(A) = \tilde{E}$, where

$$E = \{\nu < \kappa : A_\nu \text{ is not } \kappa\text{-pure in } A\}.$$

Notice also that (by refining the filtration) we can always obtain a

$\kappa$-filtration such that if $A_\nu$ is not $\kappa$-pure in $A$, then $A_{\nu+1}/A_\nu$ is not free.

A group $A$ is strongly $\kappa$-free if and only if every subset of cardinality $< \kappa$ is contained in a free $\kappa$-pure subgroup. Hence, a group $A$ of cardinality $\kappa$ is strongly $\kappa$-free if and only if $A$ has a $\kappa$-filtration $\{A_\nu : \nu < \kappa\}$ such that for all $\nu$, $A_{\nu+1}$ is free and $\kappa$-pure in $A$ (cf. IV.1.11). If a group $A$ is (strongly) $\kappa$-free, then every subgroup of $A$ is (strongly) $\kappa$-free. Hence, every (strongly) $\kappa$-free group of cardinality $\kappa$ is free if and only if every (strongly) $\kappa$-free group is $\kappa^+$-free; if this holds, we say that (strongly) $\kappa$-free implies $\kappa^+$-free, and write "(strongly) $\kappa$-free $\Rightarrow \kappa^+$-free". In these terms the results of IV.3 say that if $\kappa$ is weakly compact or singular, then $\kappa$-free $\Rightarrow \kappa^+$-free, and Theorem 1.4 says that otherwise (assuming V = L) strongly $\kappa$-free does not imply $\kappa^+$-free.

We are going to investigate more fully when these implications hold and when they fail. For the rest of this section $\kappa$ will denote a regular uncountable cardinal, unless otherwise specified. In order to extend the method of section 1 we make the following definition.

**2.1 Definition.** For any infinite cardinal $\lambda$ we say that F($\lambda$) *holds* if there is a free group $F$ of rank $\lambda$ and a subgroup $K$ such that $F/K$ is not free but if $K'$ is any subgroup of $K$ of rank $< \lambda$ such that $K/K'$ is free, then $F/K'$ is free.

Proposition 1.1 implies that F($\aleph_0$) holds. Indeed, if $F$, $K$ and $K_n$ are as in 1.1, then any finitely-generated $K'$ is contained in $K_n$ for some $n$; if $K/K'$ is free, then so is $K_n/K'$ and hence so is $F/K'$ since the latter is isomorphic to $F/K_n \oplus K_n/K'$.

**2.2 Lemma.** F($\kappa$) *holds if and only if $\kappa$-free does not imply $\kappa^+$-free.*

PROOF. ($\Leftarrow$) Let $A$ be a $\kappa$-free group of cardinality $\kappa$ which is not free. Let $\pi: F \to A$ be an epimorphism of a free group of rank $\kappa$ onto $A$, and let $K = \ker(\pi)$. We claim that the pair $K$, $F$ attests that F($\kappa$) holds. Now $F/K$ is not free since it is isomorphic to $A$. If $K'$ is a subgroup of $K$ of rank $< \kappa$, there is a direct summand $F'$ of $F$ of rank $< \kappa$ which contains $K'$. We have a short exact sequence

$$0 \to (K \cap F')/K' \to F'/K' \to F'/(K \cap F') \to 0$$

which splits because $F'/(K \cap F')$ is isomorphic to a subgroup of $A$ of cardinality $< \kappa$ and hence is free. If $K/K'$ is free, then so is $(K \cap F')/K'$, and hence so is $F'/K'$. But then $F/K' \cong F'/K' \oplus F/F'$ is also free.

($\Rightarrow$) Suppose that $\kappa$-free implies $\kappa^+$-free, and suppose that $K \subseteq F$ is a pair of free groups of rank $\kappa$ such that for every summand $K'$ of $K$ of rank $< \kappa$, $F/K'$ is free. To prove that $F/K$ is free, it suffices to prove that $F/K$ is $\kappa$-free. But if $F'$ is any subgroup of $F$ of cardinality $< \kappa$, there is a summand $K'$ of $K$ of rank $< \kappa$ containing $F' \cap K$; then there is a well-defined embedding of $(F' + K)/K$ into $F/K'$, showing that $(F' + K)/K$ is free. $\square$

Notice that if the pair $K$, $F$ attests to the fact that $F(\kappa)$ holds, then there is a $\kappa$-filtration, $K = \bigcup_{\nu < \kappa} K_\nu$, of $K$ such that for all $\nu$, $F/K_\nu$ is free; indeed just take any $\kappa$-filtration such that each $K_\nu$ is a summand of (the free group) $K$.

Recall (see II.4.6) that if $E$ is a subset of $\kappa$, $E'$ is defined to be $\{\delta \in \kappa : E \cap \delta \text{ is stationary in } \delta\}$, and $E$ is said to be non-reflecting in $\kappa$ if $E' = \emptyset$.

Now we are ready for our generalization of the basic construction of section 1.

**2.3 Theorem.** *Let $\kappa$ be a regular uncountable cardinal such that there is a non-reflecting stationary subset $E$ of $\kappa$ such that for every $\sigma \in E$, $\sigma$ is a limit ordinal and $F(\mathrm{cf}(\sigma))$ holds. Then there is a strongly $\kappa$-free group $M$ of cardinality $\kappa$ such that $\Gamma(M) = \tilde{E}$ (and hence $M$ is not free).*

PROOF. We proceed as in the proof of 1.4 to define by induction an increasing chain of groups $M_\nu$ satisfying (a'), (b), (c'), and (d). The only difference is in the critical case when $\delta = \nu + 1$ and $\nu \in E$. Let $\lambda = \mathrm{cf}(\nu)$, and let $C = \{\tau_\mu : \mu < \lambda\}$ be a cub in $\nu$ such that for all $\mu$, $\tau_\mu \notin E$. (Remember, $E$ is non-reflecting.) Let $\tilde{M}_\mu = M_{\tau_\mu}$. Since $F(\lambda)$ holds, there is a free group $F$ of rank $\lambda$ and a subgroup $K$

such that $F/K$ is not free but there is a $\lambda$-filtration, $K = \bigcup_{\mu<\lambda} K_\mu$, such that for all $\mu$ $F/K_\mu$ is free. We can arrange, by adding a free module, that $K_0$ has rank equal to the rank of $\tilde{M}_0$, and for all $\mu$, $K_{\mu+1}/K_\mu$ has rank equal to the rank of $\tilde{M}_{\mu+1}/\tilde{M}_\mu$. (See the proof of 1.2.) Thus there is an isomorphism, $f$, of $K$ with $M_\nu$ which takes each $K_\mu$ to $\tilde{M}_\mu$. As in 1.3 and 1.4 we define $M_{\nu+1}$ so that $f$ extends to an isomorphism of $F$ with $M_{\nu+1}$. Then the verification of the desired properties (a') – (d) is just as in 1.3. $\square$

**2.4 Corollary.** *If $\kappa$-free does not imply $\kappa^+$-free, then strongly $\kappa^+$-free does not imply $\kappa^{++}$-free.*

PROOF. By the Singular Compactness Theorem (IV.3.5), $\kappa$ is regular. Let $E = \{\nu < \kappa^+ : \mathrm{cf}(\nu) = \kappa\}$. Then by II.4.7, $E$ is a non-reflecting stationary subset of $\kappa^+$. Moreover, for all $\nu$ in $E$, $\kappa = \mathrm{cf}(\nu)$, so the hypothesis and Lemma 2.2 imply that $\mathrm{F}(\mathrm{cf}(\nu))$ holds. Therefore Theorem 2.3 applies, and the conclusion of 2.4 is true. $\square$

**2.5 Corollary.** *For all $n \in \omega$, strongly $\aleph_{n+1}$-free does not imply $\aleph_{n+2}$-free.*

PROOF. The case $n = 0$ is Theorem 1.3. The result follows by induction on $n$, using 2.4. $\square$

Does 2.5 hold for larger cardinals $\kappa$? Of course, we have a complete answer if we assume $V = L$. We also have the following results of Magidor and Shelah, which we state without proof.

**2.6 Theorem.**
(i) *For all $n \in \omega$, $\aleph_{\omega n+1}$-free does not imply $\aleph_{\omega n+2}$-free.*
(ii) *If $\lambda$ is a regular cardinal such that $\mathrm{F}(\lambda)$ holds, then $\aleph_{\lambda+1}$-free does not imply $\aleph_{\lambda+2}$-free.* $\square$

By 2.6(i) and 2.4, $\mathrm{F}(\aleph_{\omega n+m+1})$ holds for all $n, m \in \omega$. Thus $\aleph_{\omega^2+1}$ is the first regular cardinal for which the first part of 2.6 does not give any information. If we define inductively $\kappa_0 = \aleph_1$, $\kappa_{n+1} = \aleph_{\kappa_n+1}$, then 2.6(ii) implies that $\mathrm{F}(\kappa_n)$ holds for all $n \in \omega$. The supremum of the $\kappa_n$'s is the first cardinal fixed point, i.e., the first cardinal $\kappa$ such that $\kappa = \aleph_\kappa$. The second part of 2.6 does not give any information about cardinals above this fixed point.

Magidor and Shelah 19?? have proved that 2.6 is best possible in the following sense: assuming the consistency of the existence of certain large cardinals, they have proved that (i) there is a model of ZFC + GCH in which $\aleph_{\omega^2+1}$-free implies $\aleph_{\omega^2+2}$-free; and (ii) there is a model of ZFC + GCH in which the first cardinal fixed point, $\kappa$, has the property that $\kappa$-free implies free (for groups of arbitrary cardinality).

Another consequence of 2.3 gives information about which elements $\tilde{E}$ of $D(\kappa)$ can arise as $\Gamma_\kappa(A)$ for some strongly $\kappa$-free $A$ of cardinality $\kappa$. (Notice that if $A$ is $\kappa$-free but not strongly $\kappa$-free, then $\Gamma_\kappa(A) = 1 = \tilde{\kappa}$.) Let $W$ be the class of all limit ordinals $\sigma$ whose cofinality is a weakly-compact cardinal. ($W$ is a proper class — not a set — but when we intersect it with a set, $E$, we obtain a set.)

**2.7. Corollary.** (V = L) *Let $E$ be a non-reflecting stationary subset of $\kappa$ such that $E \cap W$ is not stationary in $\kappa$. Then there is a strongly $\kappa$-free $A$ of cardinality $\kappa$ such that $\Gamma(A) = \tilde{E}$.*

PROOF. By intersecting $E$ with a cub we can assume that every element of $E$ is a limit ordinal whose cofinality is not weakly compact. Then the result follows from 2.3 and 1.4. □

This naturally raises the question of whether the converse of 2.7 is true. In fact, it turns out that the condition $(E \cap W) = 0$ is necessary (and also sufficient if $\kappa$ is a successor cardinal), but the condition that $E$ is non-reflecting is not necessary: we will break $E$ up into a disjoint union of non-reflecting stationary sets (using VI.3.5), apply 2.3 to each one of them, and then put the resulting groups together as a direct sum. This leads us to consider the $\Gamma$-invariant of a direct sum of groups. Some caution is necessary, since we may not have $(\bigcup_{\nu<\kappa} E_\nu) = \bigcup_{\nu<\kappa} \tilde{E}_\nu$ in $D(\kappa)$. (Consider for example $E_\nu = \{\nu\}$ for all $\nu < \kappa$.) The problem is that the intersection of $\kappa$ cubs need not be a cub. However, the diagonal intersection of $\kappa$ cubs is a cub, and this explains the following result. Here, $(\nu, \kappa)$ denotes $\{\mu < \kappa : \mu > \nu\}$.

**2.8 Lemma.** *If $\{A_\nu : \nu < \kappa\}$ is a family of $\kappa$-free groups of cardinality $\kappa$, and $\Gamma(A_\nu) = \tilde{E}_\nu$ for all $\nu$, then $\Gamma(\bigoplus_{\nu < \kappa} A_\nu) = \tilde{S}$, where*

$$S = \bigcup_{\nu < \kappa} E_\nu \cap (\nu, \kappa).$$

PROOF. We may suppose that $A_\nu$ has a $\kappa$-filtration $A_\nu = \bigcup_{\mu < \kappa} A_{\nu,\mu}$ such that $E_\nu = \{\mu < \kappa : A_{\nu,\mu}$ is not $\kappa$-pure in $A_\nu\}$. Let

$$A'_{\nu,\mu} = \begin{cases} 0 & \text{if } \mu \leq \nu \\ A_{\nu,\mu} & \text{if } \mu > \nu \end{cases}.$$

Then $A_\nu = \bigcup_{\mu < \kappa} A'_{\nu,\mu}$ is a $\kappa$-filtration such that

$$E_\nu \cap (\nu, \kappa) = \{\mu < \kappa : A'_{\nu,\mu} \text{ is not } \kappa\text{-pure in } A_\nu\}.$$

Moreover, if we let $A = \bigoplus_{\nu < \kappa} A_\nu$, and $B_\mu = \bigoplus_{\nu < \kappa} A'_{\nu,\mu}$, then $A = \bigcup_{\mu < \kappa} B_\mu$ is a $\kappa$-filtration of $A$. (Notice that each $B_\mu$ has cardinality $< \kappa$ because it equals $\bigoplus_{\nu < \mu} A'_{\nu,\mu}$.) For this filtration it is easy to compute that

$$S = \{\mu < \kappa : B_\mu \text{ is not } \kappa\text{-pure in } A\}. \quad \square$$

Notice that if $\lambda < \kappa$ and $\Gamma(A_\nu) = \tilde{E}_\nu$ for all $\nu < \lambda$, then $\Gamma(\bigoplus_{\nu < \lambda} A_\nu) = (\widetilde{\bigcup_{\nu < \lambda} E_\nu})$, since without loss of generality we can assume that $E_\nu \cap \lambda = \emptyset$ for all $\nu < \lambda$.

**2.9 Lemma.** *Suppose $\lambda$ is a regular cardinal such that $\mathrm{F}(\lambda)$ fails. If $K \subseteq F$ are free groups (of arbitrary cardinality) such that $K = \bigcup_{\mu < \lambda} K_\mu$ (not necessarily continuous) and $F/K_\mu$ is free for all $\mu$, then $F/K$ is free.*

PROOF. The proof is by induction on the cardinality, $\rho$, of $F$. If $\rho \leq \lambda$, then $F/K$ is free since $\mathrm{F}(\lambda)$ does not hold. Suppose that $\rho > \lambda$ and the lemma is true for smaller cardinalities than $\rho$. Then

$F/K$ is $\rho$-free, so it is free if $\rho$ is singular (by IV.3.5); thus we may suppose that $\rho$ is regular. Choose a $\rho$-filtration $F = \bigcup_{\nu<\rho} F_\nu$ such that each $F_\nu$ is a direct summand of $F$ and if $(F_\nu + K)/K$ is not $\rho$-pure in $F/K$ then $(F_{\nu+1} + K)/(F_\nu + K)$ is not free. For each $\mu < \lambda$ let

$$C_\mu = \{\nu < \rho \colon (F_\nu + K_\mu)/K_\mu \text{ is } \rho\text{-pure in } F/K_\mu\}.$$

Since $F/K_\mu$ is free, $C_\mu$ is a cub in $\rho$; hence $C = \bigcap_{\mu<\lambda} C_\mu$ is a cub in $\rho$ since $\lambda < \rho$. Now $F/K = \bigcup_{\nu<\rho}(F_\nu + K)/K$ (a continuous union) and for each $\nu < \rho$

$$((F_{\nu+1} + K)/K)/((F_\nu + K)/K) \cong F_{\nu+1}/(F_\nu + (K \cap F_{\nu+1})).$$

Moreover $F_\nu + (K \cap F_{\nu+1})) = \bigcup_{\mu<\lambda}(F_\nu + (K_\mu \cap F_{\nu+1})$, and if $\nu \in C$, then for all $\mu < \lambda$,

$$F_{\nu+1}/(F_\nu + (K_\mu \cap F_{\nu+1})) \cong (F_{\nu+1} + K_\mu)/(F_\nu + K_\mu)$$

is free, by definition of $C_\mu$. Thus by induction (since $|F_{\nu+1}| < \rho$), $F_{\nu+1}/(F_\nu + (K \cap F_{\nu+1}))$ is free for every $\nu \in C$. It follows that $F/K$ is free. $\square$

Let $\mathcal{R}$ be the class of all regular cardinals. We are now ready to state and prove the theorem characterizing the possible $\Gamma$-invariants of a strongly $\kappa$-free group.

**2.10 Theorem.** *Let $\kappa$ be a regular uncountable cardinal and $\tilde{E}$ an element of $D(\kappa)$. If*

*(1) there is a strongly $\kappa$-free $A$ of cardinality $\kappa$ with $\Gamma(A) = \tilde{E}$,*
*then*
*(2) $E \cap W$ and $E' \cap \mathcal{R}$ are not stationary in $\kappa$.*
*Moreover, if we assume $V = L$, then (2) implies (1).*

PROOF. (1) $\Rightarrow$ (2): Let $A = \bigcup_{\nu<\kappa} A_\nu$ be a $\kappa$-filtration of $A$ such that $E = \{\nu < \kappa \colon A_\nu \text{ is not } \kappa\text{-pure in } A\} \subseteq \lim(\kappa)$. Without loss of generality we may suppose that if $\nu \in E$, then $A_{\nu+1}/A_\nu$ is not free.

For any $\nu \in E$, let $\lambda = \mathrm{cf}(\nu)$ and $f: \lambda \to \nu$ an increasing function such that $\sup(\mathrm{im}(f)) = \nu$ and every element of $\mathrm{im}(f)$ is a successor ordinal; then $A_\nu = \bigcup_{\mu < \lambda} A_{f(\mu)}$ and for all $\mu < \lambda$, $A_{\nu+1}/A_{f(\mu)}$ is free. Hence Lemma 2.9 implies that $\mathrm{F}(\lambda)$ must hold, so $\lambda$ cannot be weakly-compact. Thus $E \cap W = \emptyset$.

Now we must show that $E' \cap \mathcal{R}$ is non-stationary. We might as well suppose that $\kappa$ is a limit cardinal, since otherwise $\mathcal{R}$ is not unbounded in $\kappa$. Then there is a cub $C$ such that for all $\nu \in C$, $|A_\nu| = \nu$ and if $\mu < \nu$ then $|A_\mu| < \nu$. It suffices to show that $E' \cap \mathcal{R} \cap C = \emptyset$; if not, let $\lambda \in E' \cap \mathcal{R} \cap C$; then $A_\lambda$ is a group of cardinality $\lambda$ with a $\lambda$-filtration $\{A_\nu : \nu < \lambda\}$ such that $\{\nu < \lambda : \nu \in E\}$ is stationary in $\lambda$; but then by IV.1.7, $A_\lambda$ is not free, which is a contradiction.

(2) $\Rightarrow$ (1): Suppose first that $\kappa$ is a successor cardinal, say $\kappa = \lambda^+$, and that $E \cap W$ is non-stationary. By VI.3.5 we can write $E$ as the disjoint union

$$E = \amalg_{\nu < \lambda} E_\nu$$

of non-reflecting subsets $E_\nu$. Since $E_\nu \cap W$ is not stationary in $\kappa$, Corollary 2.7 implies that there exists $A_\nu$ such that $\Gamma(A_\nu) = \tilde{E}_\nu$; then Lemma 2.8, and the remark following it, completes the proof in this case.

So we may suppose that $\kappa$ is a limit cardinal. Since the infinite cardinals form a cub in $\kappa$ we may assume that every member of $E$ is an infinite cardinal. Also we can suppose that $E' \cap \mathcal{R} = \emptyset$. Suppose first that $E$ consists only of regular cardinals. Since every singular cardinal is the sup of a cub of ordinals which are not regular cardinals, we have that $E' \subseteq \mathcal{R}$; but then $E' = \emptyset$, i.e., $E$ is non-reflecting. Then Corollary 2.7 gives us the desired $A$.

Now suppose that $E$ consists only of singular cardinals. By VI.3.13 we can write $E$ as a disjoint union

$$E = \amalg_{\nu < \kappa} E_\nu$$

where $E_\nu \cap (\nu + 1) = \emptyset$ and $E'_\nu \subseteq \mathcal{R}$ for all $\nu$; but then since

$E' \cap \mathcal{R} = \emptyset$, each $E_\nu$ is non-reflecting, so we can apply Corollary 2.7 and Lemma 2.8 to finish in this case.

Finally, for an arbitrary $E$ consisting of cardinals we can write $E = E_r \cup E_s$, where $E_r$ (resp. $E_s$) consists of the regular (resp. singular) cardinals in $E$, and apply Lemma 2.8 and the two cases above. $\square$

Note that the failure of (2) implies the existence of relatively large cardinals: if $E \cap W$ is stationary in $\kappa$ then there must be many weakly compact cardinals below $\kappa$; moreover, if $E' \cap \mathcal{R}$ is stationary in $\kappa$ then $\kappa$ must be a Mahlo cardinal. (See section VI.3.) In ZFC, it is possible to prove, for $0 < n \in \omega$, that every element of $D(\aleph_n)$ is realizable as the $\Gamma$-invariant of an $\aleph_n$-free group. See Exercises 15 – 17 for more on this and other results provable in ZFC.

So far we have been dealing with the question of when strongly $\kappa$-free $\Rightarrow$ $\kappa^+$-free, and refinements thereof. In the next section we shall prove (Theorem 3.3) that if $\kappa$-free does not imply $\kappa^+$-free, then strongly $\kappa$-free does not imply $\kappa^+$-free. Now we shall turn briefly to the question of when $\kappa$-free $\Rightarrow$ strongly $\kappa$-free.

The following is a special case of IV.3.4, but since the proof is so much simpler in this case, we give a self-contained argument here.

**2.11 Lemma.** *Let $\kappa$ be a regular cardinal and $A$ a group (of arbitrary cardinality) which is $\kappa^+$-free. Then $A$ is strongly $\kappa$-free.*

PROOF. Suppose false; that is, there is a subset $X$ of $A$ of cardinality $< \kappa$ which is not contained in a $\kappa$-pure subgroup of cardinality $< \kappa$. Define by induction on $\nu < \kappa$ a continuous chain $\{B_\nu : \nu < \kappa\}$ of subgroups of $A$ of cardinality $< \kappa$ such that $X \subseteq B_0$ and for all $\nu$, $B_{\nu+1}/B_\nu$ is not free. This is possible since, for all $\nu$, $B_\nu$ is not $\kappa$-pure by choice of $X$. Let $B = \bigcup_{\nu < \kappa} B_\nu$. Then $|B| \leq \kappa$ and $B$ is not free by IV.1.7. But this contradicts the fact that $A$ is $\kappa^+$-free. $\square$

**2.12 Theorem.** *If $\kappa$ is a limit cardinal, or $\kappa = \lambda^+$ where $F(cf(\lambda))$ fails, then $\kappa$-free implies strongly $\kappa$-free.*

PROOF. Suppose first that $\kappa$ is a limit cardinal, and that $A$ is a group (of arbitrary cardinality) which is $\kappa$-free. Let $X$ be a subset of $A$ of infinite cardinality $\lambda < \kappa$. Since $A$ is $\lambda^{++}$-free, $A$ is strongly $\lambda^+$-free by Lemma 2.11, and hence there is a $\lambda^+$-pure subgroup $B$ of $A$ of cardinality $\lambda$ containing $X$. We will be done if we can show that $B$ is in fact $\kappa$-pure in $A$. Now if $H$ is any subgroup of $A$ containing $B$ and of cardinality $< \kappa$, then $H$ is free, so $H = H' \oplus H''$, where $H'$ has cardinality $\lambda$ and contains $B$; then $B$ is a summand of $H'$, and hence of $H$; thus we have proved that $B$ is $\kappa$-pure in $A$.

Now suppose that $\kappa = \lambda^+$, where $\mathrm{F}(\mathrm{cf}(\lambda))$ fails. If $A$ is $\kappa$-free, consider a subset $X$ of $A$ of cardinality $\lambda$. Let $\gamma = \mathrm{cf}(\lambda)$, and write $X = \bigcup_{\nu<\gamma} X_\nu$ where $|X_\nu| = \tau_\nu < \lambda$. Now $A$ is strongly $\tau_\nu^+$-free by 2.11, so by induction on $\nu$ we can define a chain of subgroups $B_\nu$ of $A$ such that $X_\nu \subseteq B_\nu$, $|B_\nu| = \tau_\nu$ and $B_\nu$ is $\tau_\nu^+$-pure in $A$. As in the first part of this argument we can then conclude that each $B_\nu$ is $\kappa$-pure in $A$. Let $B = \bigcup_{\nu<\gamma} B_\nu$; it suffices to prove that $B$ is $\kappa$-pure in $A$. But if $B \subseteq F \subseteq A$ and $|F| = \lambda$, then $F$ is free, and Lemma 2.9 implies that $F/B$ is free since $\mathrm{F}(\gamma)$ fails. $\square$

By IV.3.2, $\mathrm{F}(\gamma)$ fails if $\gamma$ is weakly compact. Hence, if there is a $\kappa$-free group which is not strongly $\kappa$-free then $\kappa$ is the successor of a cardinal whose cofinality is not weakly compact. We conclude this section with the proof of the converse of this result assuming $V = L$.

Recall from VI.3.8 the definition of $*(\kappa, \lambda)$. Recall also that we proved there that if $\kappa$ is regular then $*(\kappa, \kappa)$ holds and that assuming $V = L$, $*(\kappa, \mathrm{cf}(\kappa))$ holds if $\kappa$ is singular.

**2.13 Theorem.** *If $*(\kappa, \lambda)$ and $\mathrm{F}(\lambda)$ hold, then there is a $\kappa^+$-free group which is not strongly $\kappa^+$-free.*

PROOF. Let $\{S_i : i \in \kappa^+\}$ be as in the definition of $*(\kappa, \lambda)$ and let $K \subseteq F$ attest that $\mathrm{F}(\lambda)$ holds. Let $K_i \subseteq F_i (i \in \kappa^+)$ be disjoint copies of $K \subseteq F$. Let $B_i$ be a basis of $K_i$, which we can assume is indexed by $S_i$ since $K_i$ has rank $\lambda$:

$$B_i = \{b_x^i : x \in S_i\}.$$

Let $F = \bigoplus_{i<\kappa^+} F_i$, and let $N$ be the subgroup of $F$ generated by all elements of the form $b_x^i - b_y^j$ where $x \ (\in S_i)$ equals $y \ (\in S_j)$. Let $A$ be $F/N$. Then $A$ is not strongly $\kappa^+$-free because $C = \bigoplus_{i<\kappa^+} K_i/N$ is a subgroup of $A$ of rank $\kappa$ which we claim cannot be contained in a $\kappa^+$-pure subgroup; indeed $A/C \cong \bigoplus_{i<\kappa^+} F_i/K_i$ and each $F_i/K_i$ is non-free and of rank $\lambda \leq \kappa$; so if $C \subseteq C' \subseteq A$ such that $C'$ has rank $\kappa$, then $A/C'$ is not $\kappa^+$-free.

So it remains to prove that $A$ is $\kappa^+$-free. Now if $L$ is a subgroup of $A$ of rank $\kappa$, then for some subset $I$ of $\kappa^+$ of cardinality $\kappa$, we can regard $L$ as a subgroup of $\bigoplus_{i \in I} F_i/(N \cap \bigoplus_{i \in I} K_i)$. For this $I$, let $\{S_i^*: i \in I\}$ be as in the definition of $*(\kappa, \lambda)$. Let $K_i'$ be the subgroup of $K_i$ with basis $\{b_x^i: x \in S_i \setminus S_i^*\}$. Then $N \cap \bigoplus_{i \in I} K_i = N \cap \bigoplus_{i \in I} K_i'$. Also $F_i = K_i' \oplus C_i$ for some $C_i$ since $K_i'$ has rank $< \lambda$ and $K_i \subseteq F_i$ attests to F($\lambda$). One may verify easily that $\bigoplus_{i \in I} F_i/(N \cap \bigoplus_{i \in I} K_i')$ is isomorphic to $\bigoplus_{i \in I} C_i \oplus \tilde{K}$, where $\tilde{K}$ is a free group on a basis equinumerous with $\bigcup_{i \in I}(S_i \setminus S_i^*)$. Hence $L$ is isomorphic to a subgroup of a free group and is thus free. $\square$

**2.14 Corollary.** (V = L) *For any uncountable cardinal $\mu$, $\mu$-free implies strongly $\mu$-free if and only if $\mu$ is a limit cardinal or $\mu = \kappa^+$ where the cofinality of $\kappa$ is weakly compact.*

PROOF. ($\Leftarrow$) follows from Theorem 2.12 since by IV.3.2, F(cf($\kappa$)) fails when cf($\kappa$) is weakly compact.

($\Rightarrow$) It suffices to verify that $*(\kappa, \lambda)$ is true for some $\lambda$ such that F($\lambda$) holds whenever $\mu = \kappa^+$ and the cofinality of $\kappa$ is not weakly compact. If $\kappa$ is regular we can let $\lambda = \kappa$ by VI.3.9. If $\kappa$ is not regular, we can let $\lambda = $ cf($\kappa$) by VI.3.10.

**2.15 Example.** Assume V = L and suppose there is a weakly compact cardinal $\lambda$. Let $\kappa = \lambda^{++}$. Then $\lambda^+ = $ cf($\lambda^+$) is not weakly compact — since it is not a limit cardinal — so by 2.14 there is a $\kappa$-free group $A$ of cardinality $\kappa$ which is not strongly $\kappa$-free (and hence $\Gamma(A) = 1$). However, $W$ is stationary in $\kappa$ by II.4.7; this shows that the implication (1) $\Rightarrow$ (2) of 2.10 is not correct if the word "strongly" is removed from (1).

Theorem 2.13 says in particular that if $*(\lambda, \aleph_0)$ holds, then $\lambda^+$-free does not imply free. If there is a cardinal $\kappa$ such that $\kappa$-free implies free, then for all $\lambda \geq \kappa$, $*(\lambda, \aleph_0)$ fails (and so the conclusion of VI.3.11 fails for all $\lambda \geq \kappa$, with $\mathrm{cf}(\lambda) = \omega$).

## §3. Transversals and $\lambda$-systems

The preceding sections have provided a great deal of information about the existence of $\kappa$-free and strongly $\kappa$-free groups which are not free. But many questions remain open and seem hard. In this section we take a different tack, which is to connect these questions about groups to questions about pure combinatorial set theory. In fact, Shelah has shown (see 3.2 below) that there is an exact translation of these group-theoretic questions into set-theoretic ones. In order to explain that, we begin with some definitions.

Let $\mathcal{S} = \{s_i : i \in I\}$ be an indexed family of sets. A *transversal* for $\mathcal{S}$ is a one-one function $T : I \rightarrow \cup \mathcal{S}$ such that for all $i \in I$, $T(i) \in s_i$. The "marriage theorem" of P. Hall 1935 says that if $I$ is finite, a necessary and sufficient condition for $\mathcal{S}$ to have a transversal is that $|\cup_{i \in K} s_i| \geq |K|$ for all subsets $K$ of $I$. M. Hall 1948 showed that if $I$ is arbitrary but each $s_i$ is finite, then $\mathcal{S}$ has a transversal if and only if every finite subfamily of $\mathcal{S}$ has a transversal.

From now on, $\mathcal{S}$ will be a family of *countably infinite* sets.

**3.1 Definition.** We say that $\mathcal{S}$ is *free* if $\mathcal{S}$ has a transversal, and that $\mathcal{S}$ is $\lambda$-*free* if every subset of $\mathcal{S}$ of cardinality $< \lambda$ has a transversal.

Shelah adopted this terminology when he gave an axiomatic treatment of "free" (satisfied by the notion in 3.1) under which he could prove the Singular Compactness Theorem (cf. IV.3.5 and IV.3.7). The first of the following two theorems shows that the connection between "free" families and free groups goes deeper than that. The second is a purely group-theoretic consequence which shows the power of the method described here.

**3.2 Theorem.** *For any uncountable cardinal $\lambda$, there is a $\lambda$-free abelian group which is not $\lambda^+$-free if and only if there is a family*

*S of countable sets which is λ-free but not λ⁺-free (in the sense of Definition 3.1).*

**3.3 Theorem.** *For any uncountable cardinal* $\lambda$, *if there is a λ-free abelian group which is not λ⁺-free, then there is a strongly λ-free abelian group which is not λ⁺-free.*

Throughout this section, $\lambda$ will denote a regular uncountable cardinal.

Note that (for some $\lambda$, e.g. $\aleph_{\omega^2+1}$) there are models of ZFC in which the hypothesis of 3.3 holds and other models of ZFC in which the hypothesis of 3.3 fails (cf. the comments following Theorem 2.6); what this result is telling us is that in *any* model of ZFC, if λ-free does not imply λ⁺-free, then strongly λ-free does not imply λ⁺-free. Similar comments apply to 3.2. These results can be extended to modules over arbitrary non-left-perfect rings: see Corollary 3.13.

In this section we will outline the proofs of these theorems (leaving some of the details to section 3A and the Exercises). For better or worse, these proofs are not direct, but make use of another, fairly complicated but natural, notion — that of a λ-system. In order to motivate that notion, we begin with a dissection of a non-free λ-free group; our aim is to reduce the reason for non-freeness to its ultimate, countable, cause. (The reader might want to start by thinking of $\lambda = \aleph_2$ or $\aleph_3$ for a non-trivial but not too complicated example.)

**3.4.** Let $A$ be a λ-free group of cardinality $\lambda$ which is not free. Fix a λ-filtration

$$A = \bigcup_{\alpha < \lambda} B_\alpha$$

such that if $B_\alpha$ is not λ-pure in $A$ (i.e., $A/B_\alpha$ is not λ-free), then $B_{\alpha+1}/B_\alpha$ is not free. Let

$$E_\emptyset = \{\alpha < \lambda : B_\alpha \text{ is not } \lambda\text{-pure in } A\}$$

and for each $\alpha \in E_\emptyset$, let $\lambda_\alpha(< \lambda)$ be minimal such that $B_{\alpha+1}/B_\alpha$ has a subgroup of cardinality $\lambda_\alpha$ which is not free. Notice that $\lambda_\alpha$

is regular by IV.3.5. If $\lambda_\alpha$ is uncountable, choose $A_\alpha \subseteq B_{\alpha+1}$ of cardinality $\lambda_\alpha$ such that

$$C_\alpha \overset{\text{def}}{=} (A_\alpha + B_\alpha)/B_\alpha$$

is not free. Then $C_\alpha$ is $\lambda_\alpha$-free, and we can choose a $\lambda_\alpha$-filtration of $A_\alpha$,

$$A_\alpha = \bigcup_{\beta < \lambda_\alpha} B_{\alpha,\beta},$$

such that if $(B_{\alpha,\beta} + B_\alpha)/B_\alpha$ is not $\lambda_\alpha$-pure in $C_\alpha$, then

$$(B_{\alpha,\beta+1} + B_\alpha/B_\alpha)/(B_{\alpha,\beta} + B_\alpha/B_\alpha) \cong (B_{\alpha,\beta+1} + B_\alpha)/(B_{\alpha,\beta} + B_\alpha)$$

is not free. Since $C_\alpha$ is not free,

$$E_\alpha \overset{\text{def}}{=} \{\beta < \lambda_\alpha : B_{\alpha,\beta} + B_\alpha/B_\alpha \text{ is not } \lambda_\alpha\text{-pure in } C_\alpha\}$$

is stationary in $\lambda_\alpha$. For each $\beta \in E_\alpha$, choose $\lambda_{\alpha,\beta}$ $(< \lambda_\alpha)$ minimal such that there is a subgroup $A_{\alpha,\beta}$ of $B_{\alpha,\beta+1}$ of cardinality $\lambda_{\alpha,\beta}$ so that

$$C_{\alpha,\beta} \overset{\text{def}}{=} (A_{\alpha,\beta} + B_{\alpha,\beta} + B_\alpha)/(B_{\alpha,\beta} + B_\alpha)$$

is not free. If $\lambda_{\alpha,\beta}$ is uncountable, continue by choosing a $\lambda_{\alpha,\beta}$-filtration of $A_{\alpha,\beta} \dots$ and so forth.

Notice that every sequence

$$\lambda = \lambda_\emptyset > \lambda_\alpha > \lambda_{\alpha,\beta} > \lambda_{\alpha,\beta,\gamma} > \dots$$

is finite — since the cardinal numbers are well-ordered — and terminates with an $\aleph_0$. Thus, starting with the $\lambda$-free group $A$ of cardinality $\lambda$ we obtain what is called a $\lambda$-system in the following definition.

**3.5 Definition.** (1) The set of all functions

$$\eta : n = \{0, \dots, n-1\} \to \lambda$$

$(n \in \omega)$ is denoted $^{<\omega}\lambda$; the domain, $n$, of $\eta$ is denoted $\ell(\eta)$ and called the *length* of $\eta$; we identify $\eta$ with the sequence

$$\langle \eta(0), \eta(1), \dots, \eta(n-1)\rangle.$$

Define a partial ordering on $^{<\omega}\lambda$ by: $\eta_1 \leq \eta_2$ if and only if $\eta_1$ is a restriction of $\eta_2$. This makes $^{<\omega}\lambda$ into a *tree* (cf. II.5.3). For any $\eta = \langle \alpha_0, \ldots, \alpha_{n-1} \rangle \in {}^{<\omega}\lambda$, $\eta^\frown \langle \beta \rangle$ denotes the sequence $\langle \alpha_0, \ldots, \alpha_{n-1}, \beta \rangle$. If $S$ is a subtree of $^{<\omega}\lambda$, an element $\eta$ of $S$ is called a *final node* of $S$ if no $\eta^\frown \langle \beta \rangle$ belongs to $S$. Denote the set of final nodes of $S$ by $S_f$.

(2) A $\lambda$-*set* is a subtree $S$ of $^{<\omega}\lambda$ together with a cardinal $\lambda_\eta$ for every $\eta \in S$ such that $\lambda_\emptyset = \lambda$, and:

(a) for all $\eta \in S$, $\eta$ is a final node of $S$ if and only if $\lambda_\eta = \aleph_0$;

(b) if $\eta \in S \setminus S_f$, then $\eta^\frown \langle \beta \rangle \in S$ implies $\beta \in \lambda_\eta$, $\lambda_{\eta^\frown \langle \beta \rangle} < \lambda_\eta$ and $E_\eta \overset{\text{def}}{=} \{\beta < \lambda_\eta : \eta^\frown \langle \beta \rangle \in S\}$ is stationary in $\lambda_\eta$.

(3) A $\lambda$-*system* is a $\lambda$-set together with a set $B_\eta$ for each $\eta \in S$ such that $B_\emptyset = \emptyset$, and for all $\eta \in S \setminus S_f$:

(a) for all $\beta \in E_\eta$, $\lambda_{\eta^\frown \langle \beta \rangle} \leq |B_{\eta^\frown \langle \beta \rangle}| < \lambda_\eta$;

(b) $\{B_{\eta^\frown \langle \beta \rangle} : \beta \in E_\eta\}$ is a continuous chain of sets, i.e. if $\beta < \beta'$ are in $E_\eta$, then $B_{\eta^\frown \langle \beta \rangle} \subseteq B_{\eta^\frown \langle \beta' \rangle}$; and if $\sigma$ is a limit point of $E_\eta$, then $B_{\eta^\frown \langle \sigma \rangle} = \cup \{B_{\eta^\frown \langle \beta \rangle} : \beta < \sigma, \beta \in E_\eta\}$.

(4) For any $\lambda$-system $\Lambda = (S, \lambda_\eta, B_\eta : \eta \in S)$, and any $\eta \in S$, let $\bar{B}_\eta = \cup \{B_{\eta \restriction m} : m \leq \ell(\eta)\}$. Say that a family $\mathcal{S}$ of countable sets is *based on* $\Lambda$ if $\mathcal{S}$ is indexed by $S_f$, and for every $\eta \in S_f$, $s_\eta \subseteq \bar{B}_\eta$.

(5) Given a $\lambda$-free group $A$ of cardinality $\lambda$ which is not free, let $\Lambda(A)$ denote a $\lambda$-system which is constructed as in 3.4. (We will identify $\lambda_{\alpha, \beta}$ with $\lambda_{\langle \alpha, \beta \rangle}$.) In particular, $(A_\eta + \langle \bar{B}_\eta \rangle)/\langle \bar{B}_\eta \rangle$ is, for every $\eta$, a group of cardinality $\lambda_\eta$ which is $\lambda_\eta$-free but not free. Note also that if $\eta = \nu^\frown \langle \gamma \rangle$ and if $\delta \in E_\nu$ is such that $\gamma < \delta$, then $A_\eta \subseteq B_{\nu^\frown \langle \delta \rangle}$. If $\Lambda(A) = (S, \lambda_\eta, B_\eta : \eta \in S)$, let $\Lambda'(A) = (S, \lambda_\eta, B'_\eta : \eta \in S)$, where $B'_\eta = B_\eta \times \omega$; it is also a $\lambda$-system. Notice that there are some non-canonical choices involved in defining $\Lambda(A)$, namely the choices of the $A_\eta$, so we shall not attempt to regard $\Lambda(A)$ as an invariant of $A$.

A subtree $S$ of $^{<\omega}\lambda$ is said to have *height* $n$ if all the final nodes of $S$ have length $n$. A $\lambda$-set or $\lambda$-system is said to have height $n$ if its associated subtree $S$ has height $n$. A $\lambda$-set of height 1 is essentially just a stationary subset of $\lambda$, so the notion of a $\lambda$-set may be regarded as a generalization of that of a stationary set. Exercise 2 is a generalization of the fact that if a stationary set is a union of countably many sets, then one of them must be stationary. Not every $\lambda$-set has a height, of course, but Exercise 2(b) says that every one contains a "sub-$\lambda$-set" which has a height.

**3.6 Lemma.** *If $S = \{s_\eta : \eta \in S_f\}$ is based on the $\lambda$-system $\Lambda = (S, \lambda_\eta, B_\eta : \eta \in S)$, then $S$ has cardinality $\leq \lambda$ and is not free.*

PROOF. It is obvious that $|S_f| \leq \lambda$. Now, to obtain a contradiction, suppose that $T$ is a transversal for $S$. Let

$$Y = \{\alpha < \lambda : \exists \eta \in S_f (\eta(0) = \alpha \text{ and } T(s_\eta) \in B_\alpha)\}.$$

For each $\alpha \in Y$, choose $\varphi(\alpha) = T(s_\eta)$ for some $\eta$ such that $\eta(0) = \alpha$ and $T(s_\eta) \in B_\alpha$. If $Y$ is a stationary subset of $\lambda$, then Fodor's Lemma (cf. Exercise II.19) implies that there exists a stationary subset of $\lambda$ on which $\varphi$ is constant. This contradicts the fact that $T$ is one-one, so since $E_\emptyset = \{\alpha < \lambda : \eta(0) = \alpha \text{ for some } \eta \in S_f\}$ is stationary in $\lambda$, there must exist $\beta_0 \in E_\emptyset$ such that for all $\eta \in S_f$, $\beta_0 = \eta(0)$ implies $T(s_\eta) \notin B_{\beta_0}$. Fix such a $\beta_0$ and let

$$Y' = \{\gamma < \lambda_{\beta_0} : \exists \eta \in S_f (\eta(0) = \beta_0, \eta(1) = \gamma, \text{ and } \quad T(s_\eta) \in B_{\beta_0, \gamma})\}.$$

As before, Fodor's Lemma implies that $Y'$ is not stationary in $\lambda_{\beta_0}$, so there must exist $\beta_1$ such that $\langle \beta_0, \beta_1 \rangle \in S$ and for all $\eta \in S_f$, $\eta \upharpoonright 2 = \langle \beta_0, \beta_1 \rangle$ implies that $T(s_\eta) \notin B_{\beta_0} \cup B_{\beta_0, \beta_1}$. Continuing in this way we obtain an infinite descending sequence of cardinals

$$\lambda > \lambda_{\beta_0} > \lambda_{\beta_0, \beta_1} > \lambda_{\beta_0, \beta_1, \beta_2} > \ldots$$

which is a contradiction. $\square$

Now we are in a position to prove that if there is a $\lambda$-free group which is not $\lambda^+$-free, then there is a $\lambda$-free family which is not $\lambda^+$-free. (See 3.5(5) for the definition of $\Lambda'(A)$.)

**3.7 Proposition.** *If $A$ is a $\lambda$-free group of cardinality $\lambda$ which is not free, then there is a family $S$ based on $\Lambda'(A)$ which is $\lambda$-free but not free.*

PROOF. By the previous lemma it suffices to define a $\lambda$-free family based on $\Lambda'(A)$. Extending the notation and the construction described in 3.4 in the obvious fashion, we have for each $\eta \in S_f$, a countable subgroup $A_\eta$ of $A$ such that

$$A_\eta + \langle \bar{B}_\eta \rangle / \langle \bar{B}_\eta \rangle$$

is not free. Now there is a countable subset $t_\eta$ of $\bar{B}_\eta$ such that $A_\eta \cap \langle \bar{B}_\eta \rangle$ is contained in the subgroup generated by $t_\eta$. Let $s_\eta = t_\eta \times \omega$. It suffices to show that $S \overset{\text{def}}{=} \{s_\eta : \eta \in S_f\}$ is $\lambda$-free.

Let $I$ be a subset of $S_f$ of cardinality $< \lambda$ and let $F$ be the subgroup of $A$ generated by $\cup \{\bar{B}_\eta \cup A_\eta : \eta \in I\}$. Since $A$ is $\lambda$-free, $F$ is free; let $X$ be a basis of $F$.

If $Y$ is a subset of $X$ and $\eta \in S_f$, say that $\eta$ *depends* on $Y$ if there exists $a \in A_\eta \setminus \langle \bar{B}_\eta \rangle$ such that $a \in \langle \bar{B}_\eta \cup Y \rangle$. We claim that if $Y' \subseteq X$ is countable, then

(3.7.1) $\qquad \{\eta \in S_f : \eta \text{ depends on } Y \cup Y' \text{ but not on } Y\}$

is countable. Assuming this for the moment, we can write $X$ as the union of a continuous chain: $X = \cup_{\nu < \sigma} X_\nu$ such that $X_0 = \emptyset$ and for all $\nu$, $|X_{\nu+1} \setminus X_\nu| \le \aleph_0$ and

(3.7.2) $\qquad$ if $\eta \in I$ depends on $X_\nu$, then $t_\eta \subseteq \langle X_\nu \rangle$.

Note that, by the finitary character of dependence, for all $\eta$ there is a largest $\nu$ such that $\eta$ does not depend on $X_\nu$. We claim that

(3.7.3) $\qquad$ if $\eta \in I$ doesn't depend on $X_\nu$, then $t_\eta \not\subseteq \langle X_\nu \rangle$.

Assuming this for the moment, let us construct a transversal for $\{s_\eta : \eta \in I\}$. For each $\nu < \sigma$, choose a transversal $T_\nu$ for

$$\{s_\eta : \eta \text{ depends on } X_{\nu+1} \text{ but not on } X_\nu\}$$

so that for all such $\eta$, $T_\nu(s_\eta) \in (\langle X_{\nu+1} \rangle \times \omega) \setminus (\langle X_\nu \rangle \times \omega)$; this is possible by (3.7.2), (3.7.3) and Exercise 1. Then $T = \bigcup_{\nu < \sigma} T_\nu$ is the desired transversal.

Thus it remains to prove (3.7.1) and (3.7.3). To prove (3.7.1) it suffices to prove that for any $y \in \langle Y' \rangle$ there is at most one $\eta \in S_f$ which does not depend on $Y$ but is such that there exists $b \in \bar{B}_\eta$, $u \in \langle Y \rangle$ and $a \in A_\eta \setminus \langle \bar{B}_\eta \rangle$ such that $a = b + u + y$. Indeed, if there is another $\nu \in S_f$ (not depending on $Y$) with $c \in \bar{B}_\nu$, $v \in \langle Y \rangle$, and $\tilde{a} \in A_\nu \setminus \langle \bar{B}_\nu \rangle$ such that $\tilde{a} = c + v + y$, then

$$(3.7.4) \qquad a - \tilde{a} = (b - c) + (u - v).$$

Now, if $\nu <_\ell \eta$ (where $<_\ell$ denotes the lexicographic ordering — for the definition, see the proof of 4.9), then $\bar{B}_\nu \subseteq \bar{B}_\eta$, and $A_\nu \subseteq \langle \bar{B}_\eta \rangle$, so (3.7.4) implies that $\eta$ depends on $Y$, a contradiction.

We shall prove (3.7.3) by showing that

$$A_\eta + \langle X_\nu \rangle / \langle X_\nu \rangle \text{ is not isomorphic to } A_\eta + \langle t_\eta \cup X_\nu \rangle / \langle t_\eta \cup X_\nu \rangle$$

In fact, $(A_\eta + \langle X_\nu \rangle) / \langle X_\nu \rangle$ is free because it is a subgroup of $F / \langle X_\nu \rangle$. On the other hand,

$$(A_\eta + \langle t_\eta \cup X_\nu \rangle) / \langle t_\eta \cup X_\nu \rangle \cong A_\eta / A_\eta \cap \langle t_\eta \rangle \cong (A_\eta + \langle \bar{B}_\eta \rangle) / \langle \bar{B}_\eta \rangle$$

which is not free by choice of $A_\eta$. (The last isomorphism is by choice of $t_\eta$; the next-to-last is by choice of $t_\eta$ and because $\eta$ does not depend on $X_\nu$.) $\square$

Having gone from a $\lambda$-free group to a $\lambda$-free family, we now want to go the other way. We begin by describing a general construction of a group from a family.

**3.8.** Given an indexed family $\mathcal{S} = \{s_i : i \in J\}$ of countably infinite sets, let us choose and fix an enumeration of each $s_i$:

$$s_i = \{x_{i,n} : n \in \omega\}.$$

Thus $\bigcup \mathcal{S} = \{x_{i,n} : i \in J, n \in \omega\}$, but, of course, $x_{i,n}$ may equal $x_{j,m}$ for $(i, n) \neq (j, m)$. Let $Y = \{y_i^n : i \in J, n \in \omega\}$ be a set of symbols disjoint from $\bigcup \mathcal{S}$, and let $F$ be the free group on the set $Y \cup (\bigcup \mathcal{S})$. Let $K$ be the subgroup of $F$ generated by all elements of the form

$$(3.8.1) \qquad\qquad 2y_i^{n+1} - y_i^n + x_{i,n}.$$

The construction is non-canonical, since it depends on the choices of the enumerations of the $s_i$, but for our purposes it will do no harm to denote by $A(\mathcal{S})$ the group $F/K$. Our aim is to show that, under suitable hypotheses on $S$, $A(\mathcal{S})$ is strongly $\lambda$-free and not $\lambda^+$-free.

**3.9 Lemma.** *If $\mathcal{S}$ is based on a $\lambda$-system, then $A(\mathcal{S})$ is not free.*

PROOF. Suppose, to the contrary, that $\mathcal{S} = \{s_\eta \colon \eta \in J = S_f\}$ is based on the $\lambda$-system $\Lambda = (S, \lambda_\eta, B_\eta \colon \eta \in S)$ and that $A = A(\mathcal{S})$ is free. For each $\alpha < \lambda$, let $A_\alpha$ be the subgroup generated by (the images of)

$$\{y_\eta^n \colon \eta \in S_f, \eta(0) < \alpha, n \in \omega\} \cup \bigcup \{s_\eta \colon \eta \in S_f, \eta(0) < \alpha\}.$$

Then this defines a $\lambda$-filtration of $A$. Since $A$ is free, $A = \oplus\{Z_i \colon i \in I\}$ for some index set $I$ and $Z_i \cong \mathbf{Z}$. Then

$$C \overset{\mathrm{def}}{=} \{\alpha < \lambda \colon \exists I_\alpha \subseteq I \text{ s.t. } A_\alpha = \oplus\{Z_i \colon i \in I_\alpha\}\}$$

is a cub. Moreover

$$C' \overset{\mathrm{def}}{=} \{\alpha < \lambda \colon \text{ for all } x_{\tau,n} \in \cup\{B_\eta \colon \eta(0) < \alpha\} \; \exists \sigma, m \text{ s.t.}$$
$$x_{\tau,n} = x_{\sigma,m} \text{ and } \sigma(0) < \alpha\}$$

is a cub. Hence (by 3.5(2)(b)) there exists a limit point $\beta_0$ of $C \cap C'$ such that $\beta_0$ is a limit point of $E_\emptyset$. We claim that $\tau = \langle \beta_0 \rangle \notin S_f$. Indeed, if $\tau \in S_f$, then we would have $y_\tau^0 \in A \setminus A_{\beta_0}$ divisible by all powers of 2 mod $A_{\beta_0}$, because

$$s_\tau \subseteq \bar{B}_\tau = B_{\beta_0} = \bigcup \{B_\alpha \colon \alpha < \beta_0, \alpha \in E_\emptyset\}$$

and so $s_\tau \subseteq A_{\beta_0}$ because $\beta_0 \in C'$; this contradicts the choice of $\beta_0 \in C$.

Now $A_{\beta_0+1}/A_{\beta_0}$ is free and of cardinality $\lambda_{\beta_0}$. We continue by choosing a $\lambda_{\beta_0}$-filtration $\{A_{\beta_0,\nu}/A_{\beta_0}: \nu < \lambda_{\beta_0}\}$ of $A_{\beta_0+1}/A_{\beta_0}$ by defining $A_{\beta_0,\nu}$ to be the subgroup of $A_{\beta_0+1}$ generated by (the images of)

$$\{y_\eta^n: \eta \in S_f, \eta(0) \le \beta_0, \eta(1) < \nu, n \in \omega\} \cup$$
$$\bigcup\{s_\eta: \eta \in S_f, \eta(0) \le \beta_0, \eta(1) < \nu\}.$$

Then, as above, we obtain $\beta_1$ such that $\langle \beta_0, \beta_1 \rangle \in S \setminus S_f$ and $A_{\beta_0,\beta_1+1}/A_{\beta_0,\beta_1}$ is free. Continuing in this way we obtain a contradiction, as in 3.6. $\square$

In order to get $A(\mathcal{S})$ to be (strongly) $\lambda$-free, we shall need some additional hypotheses on $\mathcal{S}$.

**3.10 Definition.** An indexed family $\mathcal{S}$ of countable sets based on a $\lambda$-system $\Lambda$ is said to have the *reshuffling property* if for $\alpha < \lambda$ and every subset $I$ of $S_f$ such that $|I| < \lambda$, there is a well-ordering $<_I$ of $I$ such that for every $\tau$, $\eta \in I$, $s_\eta \setminus \bigcup_{\nu <_I \eta} s_\nu$ is infinite, and $\tau(0) \le \alpha < \eta(0)$ implies that $\tau <_I \eta$.

**3.11 Proposition.** *If $\mathcal{S}$ has the reshuffling property, then $A(\mathcal{S})$ is strongly $\lambda$-free of cardinality $\lambda$.*

PROOF. It is not hard to see that $A \overset{\text{def}}{=} A(\mathcal{S})$ has cardinality $\lambda$. Let $A_\alpha$ be defined as in 3.9. It will suffice to prove that for all $\alpha \in \lambda \cup \{-1\}$, $A/A_{\alpha+1}$ is $\lambda$-free. (Notice that $A_0 = \{0\}$.)

Recall the construction in 3.8: $A$ is generated by $\{y_\eta^n: n \in \omega, \eta \in S_f\}$ subject to the relations $2y_\eta^{n+1} = y_\eta^n - x_{\eta,n}$, where $s_\eta = \{x_{\eta,n}: n \in \omega\}$. To prove that $A/A_{\alpha+1}$ is $\lambda$-free, it suffices to prove that if $\alpha < \beta$ then $A_\beta/A_{\alpha+1}$ is free. Let $I = \{\eta \in S_f: \eta(0) < \beta\}$, and let $<_I$ be the well-ordering given by the reshuffling property for $I$ and $\alpha$. Notice that $A_\beta$ is generated by $\{y_\eta^n: \eta \in I, n \in \omega\} \cup \bigcup_{\eta \in I} s_\eta$. We claim that the cosets of

$$Z \overset{\text{def}}{=} \{y_\eta^n: \alpha < \eta(0) < \beta, x_{\eta,n} \in s_\eta \setminus \bigcup_{\nu <_I \eta} s_\nu\}$$

are a basis of $A_\beta / A_{\alpha+1}$. Indeed, by comparing coefficients in $F$ and using the fact that $\nu(0) \leq \alpha < \eta(0)$ implies $\nu <_I \eta$, it is not hard to see that the cosets are independent. Now let $G$ be the subgroup of $A_\beta / A_{\alpha+1}$ that they generate; we will prove, by transfinite induction with respect to the well-ordering $<_I$, that the coset of every $x_{\eta,n}$ and every $y_\eta^n$ ($\eta \in I$, $n \in \omega$) is in $G$. Indeed, suppose $x_{\eta,n} \in \bigcup_{\nu <_I \eta} s_\nu$; then there is a first $m > n$ such that $x_{\eta,m} \in s_\eta \setminus \bigcup_{\nu <_I \eta} s_\nu$; so $y_\eta^m \in Z$; we might as well suppose that $m = n + 1$; but then $y_\eta^n = 2y_\eta^m + x_{\eta,n}$ by (3.8.1). Now by induction $x_{\eta,n} + A_{\alpha+1}$ belongs to $G$, so $y_\eta^n + A_{\alpha+1} \in G$. Once we have shown that all the $y_\eta^n + A_{\alpha+1}$ belong to $G$, (3.8.1) implies that all the $x_{\eta,n} + A_{\alpha+1}$ are in $G$. $\square$

We will give the proof of the following important result in a separate following section, since it is quite long.

**3.12 Theorem.** *If there is a family of countable sets of cardinality $\lambda$ which is $\lambda$-free but not free, then there is a $\lambda$-system $\Lambda$ and a family $S$ of countable sets based on $\Lambda$ which has the reshuffling property.* $\square$

This is the last piece of the proof of 3.2 and 3.3.

**Proofs of 3.2 and 3.3.** Proposition 3.7 implies one direction of 3.2. The other direction follows immediately from 3.12, 3.11 and 3.9. As for 3.3, if there is a $\lambda$-free group which is not $\lambda^+$-free, then by 3.2 (or 3.7) there is a $\lambda$-free family of cardinality $\lambda$ which is not free. So by 3.12, 3.11 and 3.9, there is a strongly $\lambda$-free group of cardinality $\lambda$ which is not free. $\square$

Let us sum up the situation for $\mathbb{Z}$-modules and consider what we can say about modules over other rings:

**3.13 Corollary.** *For any regular uncountable $\lambda$, the following are equivalent:*

*(1) There is a $\lambda$-free abelian group of cardinality $\lambda$ which is not free.*

*(2) There is a strongly $\lambda$-free abelian group of cardinality $\lambda$ which is not free.*

*(3) There is a $\lambda$-free family of cardinality $\lambda$ which is not free.*

(4) *There is a $\lambda$-free family based on a $\lambda$-system.*

(5) *For every non-left-perfect ring $R$, there is a $\leq \lambda$-generated strongly $\lambda$-free $R$-module which is not projective.* $\square$

PROOF. We have already shown that (1) through (4) are equivalent. The proof of (5) implies (1) is a triviality since in (5) we quantify over *all* non-left-perfect rings — including $\mathbf{Z}$. The proof of (4) $\Rightarrow$ (5) requires some modification of the above arguments. Given a ring $R$ which is not left perfect, let $\{a_n: n \in \omega\}$ be as in the proof of Proposition 1.1. Given an indexed family $\mathcal{S} = \{s_i: i \in I\}$, enumerate the $s_i$ as in 3.8 and define $Y$ as in 3.8. Let $F$ be the free $R$-module on $Y \cup (\bigcup \mathcal{S})$. Let $K$ be the submodule of $F$ generated by all elements of the form

$$y_i^n - a_n y_i^{n+1} + x_{i,n}$$

and let $A(\mathcal{S}) = F/K$. The proof of 3.11 proceeds very much as before but that of 3.9 requires more care because a submodule of a projective module is not necessarily projective. We will sketch the changes needed. Suppose that $A(\mathcal{S})$ is projective; by Kaplansky's result (I.2.4), it is a direct sum of countably-generated projectives; say

$$A(\mathcal{S}) = \bigoplus \{P_i: i \in I\}$$

for some index set $I$ and countably-generated projectives $P_i$. Define a $\lambda$-filtration as in 3.9; then, as in 3.9,

$$C \overset{\text{def}}{=} \{\alpha < \lambda : \exists I_\alpha \subseteq I \text{ s.t. } A_\alpha = \bigoplus \{P_i: i \in I_\alpha\}\}$$

is a cub. Let $C'$ and $\beta_0$ be as in 3.9; there exists $\zeta > \beta_0$ such that $A_\zeta/A_{\beta_0}$ is projective; moreover, by 3.11, $A_\zeta/A_{\beta_0+1}$ is free. But then $A_{\beta_0+1}/A_{\beta_0}$ is isomorphic to a summand of $A_\zeta/A_{\beta_0}$ and hence projective. If $\beta_0 \in S_f$, we have a contradiction of the construction and (the proof of) Proposition 1.1; so $\beta_0$ does not belong to $S_f$. Since $A_{\beta_0+1}/A_{\beta_0}$ is projective, we can repeat the preceding argument and obtain a contradiction as in 3.9. $\square$

We have said that a $\lambda$-set is a generalization of a stationary set. Now we will give the appropriate generalization of the notion of the diamond principle for a stationary set.

**3.14 Definition.** Let $\Lambda = (S, \lambda_\eta, B_\eta : \eta \in S)$ be a $\lambda$-system. By $\Diamond(\Lambda)$ we mean the following principle:

> There is a family $\{C_\eta : \eta \in S\}$ such that for any $X \subseteq \bigcup_{\eta \in S} B_\eta$ there is a $\lambda$-set $S' \subseteq S$ such that for all $\eta \in S'$, $X \cap \bar{B} = \bar{C}_\eta$ (where $\bar{C}_\eta = \bigcup \{C_{\eta \restriction m} : m \leq \ell(\eta)\}$ as in 3.5(4)).

We leave the proof of the following to section 3A.

**3.15 Theorem.** *Let $\lambda$ be a regular uncountable cardinal. Then $\Diamond(\Lambda)$ holds for every $\lambda$-system $\Lambda$ if and only if for every regular cardinal $\kappa \leq \lambda$ and every stationary subset $E$ of $\kappa$, $\Diamond(E)$ holds.* $\square$

**3.16 Corollary.** $(\mathrm{V} = \mathrm{L})$ *For every regular uncountable $\lambda$ and every $\lambda$-system $\Lambda$, $\Diamond(\Lambda)$ holds.* $\square$

## §3A. Reshuffling $\lambda$-systems

In this appendix to section 3 we will prove Theorem 3.12: if there is a family of countable sets of cardinality $\lambda$ which is $\lambda$-free but not free, then there is a $\lambda$-system $\Lambda$ and a family $\mathcal{S}$ of countable sets based on $\Lambda$ so that $\mathcal{S}$ has the reshuffling property. After we prove this theorem, we shall consider the question of constructing $\lambda$-free algebras for varieties of algebras in general. We shall define a property called the strong construction principle, denoted $(\mathrm{CP}+)$ which some varieties satisfy. The construction of $\lambda$-free algebras from a $\lambda$-system, $\Lambda$, and a family, $\mathcal{S}$, of countable sets based on $\Lambda$ for varieties which satisfy $(\mathrm{CP}+)$ is similar to the construction of $\lambda$-free groups, but there are some differences. This section is not required for anything that follows.

Our proof of Theorem 3.12 will proceed by induction on $\lambda$. In order to carry out the arguments by induction it is helpful to make some definitions. First of all, given a subtree $S$ of ${}^{<\omega}\lambda$ and an element $\eta$ of $S$, let $S^\eta = \{\nu \in S : \eta \leq \nu\}$. We will liberalize the definition of subtree so that $S^\eta$ is a subtree, i.e., we allow an initial node ($\eta$ in this case) other than $\emptyset$. We also liberalize the definitions of $\lambda$-set and $\lambda$-system so that given a $\lambda$-system $\Lambda = (S, \lambda_\eta, B_\eta : \eta \in S)$ and a node $\eta$ of $S$, we can refer to the $\lambda_\eta$-system $\Lambda^\eta = (S^\eta, \lambda_\nu, B'_\nu : \nu \in S^\eta)$. (Define $B'_\eta = \emptyset$ and $B'_\nu = B_\nu$ if $\nu \neq \eta$.) If $I$ is a subset of $S_f$, let $I^\eta$ denote $I \cap S^\eta$.

Let $S$ be a $\lambda$-set. We call a subset $I$ of $S_f$ *non-small* if $\{\nu \in S : \nu \leq \eta$ for some $\eta \in I\}$ contains a subtree which, with the associated cardinals, is a $\lambda$-set; otherwise it is *small*. (Since this terminology is used only in this section, it should not cause any confusion with the unrelated notion in VI.1.6.) Note that $I$ is small if and only if $\{\alpha < \lambda : I^{\langle\alpha\rangle}$ is a non-small subset of the $\lambda_{\langle\alpha\rangle}$-system $\Lambda^{\langle\alpha\rangle}\}$ is non-stationary in $\lambda$. In particular, $I$ is small if $|I| < \lambda$.

Let $\mathcal{S}$ be a family of countable sets based on a $\lambda$-system $\Lambda$. For $\nu \in S_f$ and $k \leq \ell(\eta)$, let $s_\nu^k = s_\nu \cap B_{\nu\restriction k}$. So $s_\nu$ is the union of $s_\nu^1, \ldots, s_\nu^n$, where $n$ is the length of $\nu$. (We will shortly make an assumption which implies that $s_\eta^1, \ldots, s_\eta^n$ are disjoint.) Given $\eta \in S$, let $s_\nu^\eta = \cup\{s_\nu^k : k > \ell(\eta)\}$. Let $\mathcal{S}^\eta = \{s_\nu^\eta : \nu \in S^\eta\}$; it is a family of countable sets based on $\Lambda^\eta$.

Finally, let us define an $\aleph_0$-system to consist of the subtree $\{\emptyset\}$.

**3A.1 Proposition.** *Suppose $\Lambda$ is a $\lambda$-system and $\mathcal{S}$ is a family of countable sets based on $\Lambda$. For $I \subseteq S_f$, if $\{s_\eta : \eta \in I\}$ has a transversal, then $I$ is small.*

PROOF. This is a strengthening of 3.6. Suppose $I$ is not small. If $\Lambda = (S, \lambda_\eta, B_\eta : \eta \in S)$, choose a subtree $T$ of $S$ so that $T_f \subseteq I$ and $\Lambda_1 = (T, \lambda_\eta, B_\eta : \eta \in T)$ is a $\lambda$-system. Then by 3.6, $\{s_\eta : \eta \in T_f\}$ does not have a transversal. $\square$

We first introduce some assumptions that any $\lambda$-system $\Lambda = (S, \lambda_\eta, B_\eta : \eta \in S)$ and family of countable sets $\mathcal{S}$ based on $\Lambda$ will be assumed to satisfy. We leave to the reader to verify that any $\Lambda$ and $\mathcal{S}$ can be modified in an essentially harmless way so as to satisfy the assumptions. (The exact meaning of this statement will become clear in the proof of 3A.5, but, for example, we want to retain the property of $\lambda$-freeness if we already have it.)

1. For $\eta$, $\nu \in S$, if $B_\eta \cap B_\nu \neq \emptyset$, then there are $\tau \in S$ and $\alpha$, $\beta$ so that $\eta = \tau^\frown \langle \alpha \rangle$ and $\nu = \tau^\frown \langle \beta \rangle$.

2. For $\eta$, $\nu \in S_f$ and $k$, $m \in \omega$, if $s_\eta^k \cap s_\nu^m \neq \emptyset$ then $k = m$, $\ell(\eta) = n = \ell(\nu)$ for some $n$ and for all $j \neq k$ $\eta(j) = \nu(j)$.

Assumption 2 guarantees that $s_\eta^1, \ldots, s_\eta^n$ are disjoint.

3. For each $k$, $\{s_\eta^k : \eta \in S_f\}$ has a tree structure. Namely, for each $\eta$ there is an enumeration $t_0^{k\eta}, t_1^{k\eta}, \ldots$ of $s_\eta^k$ so that for all $\nu$, $\eta \in S_f$ and $n \in \omega$, if $t_{n+1}^{k\eta} \in s_\nu^k$, then $t_n^{k\eta} \in s_\nu^k$. (Hint: replace $B_\nu$ by $^{<\omega}B_\nu$ and $s_\nu^k$ by a set of finite sequences.)

Whenever we list the elements of $s_\eta^k$, we will assume the list satisfies property 3.

We will now define what it means for a $\lambda$-system and a family of countable sets based on the $\lambda$-system to be beautiful. The proof of 3.12 then divides into two parts. First we will show that if there is a family of countable sets of cardinality $\lambda$ which is $\lambda$-free but not free, then there is a $\lambda$-system $\Lambda$ and $\mathcal{S}$ a family of countable sets based on $\Lambda$ so that $\Lambda$ and $\mathcal{S}$ are beautiful. Secondly we will show that if $\Lambda$ and $\mathcal{S}$ are beautiful then $\mathcal{S}$ has the reshuffling property.

**3A.2 Definition.** Suppose $\Lambda = (S, \lambda_\eta, B_\eta : \eta \in S)$ is a $\lambda$-system and $\mathcal{S}$ is a family of countable sets based on $\Lambda$. We define by induction on $\lambda$ what it means for $\Lambda$ and $\mathcal{S}$ to be *beautiful*. Any $\aleph_0$-system and family of sets based on it is beautiful. Suppose now that $\lambda$ is uncountable. We say that $\Lambda$ and $\mathcal{S}$ are beautiful if they satisfy properties 1, 2 and 3 and the following.

4. $\mathcal{S}$ is $\lambda$-free.

5. For all $\alpha \in E_\emptyset$, $\Lambda^{\langle \alpha \rangle}$ and $\mathcal{S}^{\langle \alpha \rangle}$ are beautiful.

6. One of the following three possibilities holds:

(a) Every $\gamma \in E_\emptyset$ has cofinality $\omega$ and there is an increasing sequence of ordinals $\{\gamma_n : n \in \omega\}$ approaching $\gamma$ such that for all $\eta \in S_f$ if $\eta(0) = \gamma$ then $s_\eta^1 = \{\langle \gamma_n, t_n \rangle : n \in \omega\}$ for some $t_n$'s.

Moreover, these enumerations of the $s_\eta^1$ satisfy the tree property of assumption 3.

(b) There is an uncountable cardinal $\kappa$ and an integer $m > 0$ so that for all $\gamma \in E_\emptyset$ the cofinality of $\gamma$ is $\kappa$ and for all $\eta \in S_f$, $\lambda_{\eta \restriction m} = \kappa$. Further, for each $\gamma \in E_\emptyset$ there is a strictly increasing continuous sequence $\{\gamma_\rho : \rho < \kappa\}$ cofinal in $\gamma$ such that for all $\eta \in S_f$ if $\eta(0) = \gamma$ then $s_\eta^1 = \{\gamma_{\eta(m)}\} \times X$ for some $X$.

(c) Each $\gamma \in E_\emptyset$ is a regular cardinal and $\lambda_{\langle \gamma \rangle} = \gamma$. Further, for every $\eta \in S_f$ if $\eta(0) = \gamma$ then $s_\eta^1 = \{\eta(1)\} \times X$ for some $X$.

For example, if $E(\kappa)$ holds for a regular cardinal $\kappa$, then we can construct a beautiful $\kappa$-system as follows. Let $E$ be a stationary non-reflecting set consisting of ordinals of cofinality $\omega$; view $E$ as a $\kappa$-system of height 1. Let $\eta$ be a tree-like ladder system (see XII.3); if $s_{\langle \delta \rangle} \overset{\text{def}}{=} \eta_\delta[\omega] \times \{0\}$, then $E$ and $\mathcal{S} \overset{\text{def}}{=} \{s_{\langle \delta \rangle} : \delta \in E\}$ are beautiful. For other examples see Exercises 5 – 10.

**3A.3 Lemma.** *If there is a family of countable sets of cardinality $\lambda$ which is $\lambda$-free but not free then there is a $\lambda$-system $\Lambda = (S, \lambda_\eta, B_\eta : \eta \in S)$ and $\mathcal{S}$ a family of countable sets based on $\Lambda$ satisfying 1, 2 and 3 above so that for all $\eta \in S$, $\mathcal{S}^\eta$ is $\lambda_\eta$-free. Further for all $\gamma \in E_\emptyset$, $\lambda_{\langle \gamma \rangle} \leq \gamma$.*

PROOF. The argument is analogous to the construction of a $\lambda$-system from a $\lambda$-free non-free abelian group (see Exercise 3). $\square$

**3A.4 Lemma.** *Suppose $S$ is a $\lambda$-set and $\kappa$ is a cardinal so that for all $\eta \in S$, $\kappa \neq \lambda_\eta$. If $\varphi : S_f \to \kappa$, then there is a sub-$\lambda$-set $S'$ of $S$ and $\gamma < \kappa$ such that for all $\eta \in S_f'$, $\varphi(\eta) < \gamma$.*

PROOF. The proof is by induction on $\lambda$. If $\kappa > \lambda$ the conclusion is clear since $|S_f| \leq \lambda$. So assume that $\kappa < \lambda$. By induction, for each $\alpha \in E_\emptyset$ there exists $\gamma_\alpha < \kappa$ and a sub-$\lambda_{\langle \alpha \rangle}$-set $T_\alpha$ of $S^{\langle \alpha \rangle}$ such that for all $\eta \in (T_\alpha)_f$, $\varphi(\eta) < \gamma_\alpha$. Define $\psi : E_\emptyset \setminus \kappa \to \kappa$ by: $\psi(\alpha) = \gamma_\alpha$. Then $\psi$ is regressive, so by Fodor's Lemma (II.4.11), there exists $\gamma$ and a stationary subset $E_\emptyset'$ of $E$ such that for $\alpha \in E'$, $\psi(\alpha) = \gamma$. Then let $S' = \cup\{T_\alpha : \alpha \in E'\} \cup \{\emptyset\}$

**3A.5 Theorem.** *Suppose there exists a family of countable sets of cardinality $\lambda$ which is $\lambda$-free but not free. Then there is a $\lambda$-system $\Lambda$ and a family of countable sets $\mathcal{S}$ based on $\Lambda$ so that $\Lambda$ and $\mathcal{S}$ are beautiful.*

PROOF. The theorem is proved by induction on regular cardinals $\lambda$. In fact we prove something stronger: that given a $\lambda$-system $\Lambda' = (S', \lambda_\eta, B'_\eta : \eta \in S')$ and family $\mathcal{S}' = \{s'_\eta : \eta \in S'_f\}$ based on $\Lambda'$, we can transform it into a $\lambda$-system $\Lambda = (S, \lambda_\eta, B_\eta : \eta \in S)$ and family $\mathcal{S} = \{s_\eta : \eta \in S_f\}$ so that there is a one-one map $\psi$ of $S$ into $S'$ and for each $\eta \in S_f$ a level-preserving bijection $\theta_\eta$ from $s_\eta$ to $s'_{\psi(\eta)}$ so that for any $\eta$, $\nu$ in $S$, if $x \in s'_{\psi(\eta)}$, $y \in s'_{\psi(\nu)}$ and $x \neq y$, then $\theta_\eta^{-1}(x) \neq \theta_\nu^{-1}(y)$. (Actually these maps will not be explicitly defined, but their definition will be evident; it is the existence of these maps which allows us to apply the inductive hypothesis.)

By Lemma 3A.3 and the inductive hypothesis, we can assume that we are given a $\lambda$-system $\Lambda'$ and a family of countable sets $\mathcal{S}' = \{s'_\eta : \eta \in S'_f\}$ based on $\Lambda'$ such that: $\mathcal{S}'$ is $\lambda$-free; $\Lambda'$, $\mathcal{S}'$ satisfy 1, 2 and 3; and for all $\alpha \in E'_\emptyset$, $\Lambda'^{\langle\alpha\rangle}$, $\mathcal{S}'^{\langle\alpha\rangle}$ are beautiful. There are several cases to consider.

*Case 0: there is $\beta \in E'_\emptyset$ so that $\{\eta \in S'_f : s'^1_\eta \subseteq B'_\beta\}$ is not small.* By restricting to a sub-$\lambda$-set of $S'$, we can assume that for all $\eta \in S'_f$, $s'^1_\eta \subseteq B'_\beta$. Choose $\varphi$ a one-one map from $E'_\emptyset$ to the limit ordinals of cofinality $\omega$ which are greater than $\beta$. For each $\eta \in S'$, let $\psi(\eta) = \langle \varphi(\eta(0)) \rangle ^\frown \eta \restriction (0, \ell(\eta))$. Let $S = \psi[S']$. For each $\gamma \in \mathrm{rge}(\varphi)$, choose an increasing sequence $\{\gamma_n : n \in \omega\}$ cofinal in $\gamma$. For $\eta \in S'_f$, let $t'_0, t'_1, \ldots$ be the enumeration of $s'^1_\eta$ guaranteed by property 3. Let $s^1_{\psi(\eta)} = \{t_0, t_1, \ldots\}$ where $\varphi(\eta(0)) = \gamma$ and $t_n = \langle \gamma_n, \ldots, \gamma_0, t'_n \rangle$. Let $s_{\psi(\eta)} = s^1_{\psi(\eta)} \cup s'^2_\eta \ldots \cup s'^{\ell(\eta)}_\eta$. Let $\mathcal{S} = \{s_\eta : \eta \in S_f\}$. To define the $\lambda$-system $\Lambda = (S, \lambda_\eta, B_\eta : \eta \in S)$, let $\lambda_{\psi(\eta)} = \lambda_\eta$ for $\eta \in S'$ and let $B_{\langle\gamma\rangle} = {}^{<\omega}(\gamma \cup B'_\beta)$ and $B_{\psi(\eta)} = B_\eta$ for $\eta \in S'$ of length greater than 1. (Here and later, if necessary, correct the $B_{\langle\gamma\rangle}$'s and the $s^1_\eta$'s so that (1) holds.)

*Case 1: not case 0, but $\{\gamma \in E'_\emptyset : \mathrm{cf}(\gamma) = \omega\}$ is stationary in $\lambda$.* This case is like case 0, but simpler.

*Case 2: not case 0 or case 1 and for some uncountable cardinal $\kappa < \lambda$, $\{\gamma \in E'_\emptyset : \mathrm{cf}(\gamma) = \kappa\}$ is stationary.* Without loss of generality we can assume that every $\gamma \in E'_\emptyset$ has cofinality $\kappa$. For each $\gamma \in E'_\emptyset$, choose $\{\gamma_\rho : \rho < \kappa\}$ a strictly increasing continuous sequence cofinal in $\gamma$. Assume, in order to get a contradiction, that for all $m < \omega$, $\{\eta \in S'_f : \lambda_{\eta \restriction m} = \kappa\}$ is small. In this case we can assume that for all $\eta \in S'$, $\lambda_\eta \neq \kappa$. Define $\varphi : S'_f \to \kappa + 1$ as follows: if $\eta(0) = \gamma$, let $\varphi(\eta)$ be the least $\rho$ so that $s'^1_\eta \subseteq B'_{\langle \gamma_\rho \rangle}$ if there is such a $\rho$, and $\varphi(\eta) = \kappa$ otherwise. It is easy to see that $\{\eta : \varphi(\eta) < \kappa\}$ is not small. (In fact it must contain $\{\eta : \eta(0) = \gamma$ and $\gamma$ is a limit point of $E_\emptyset\}$.) By Lemma 3A.4, for some $\rho < \kappa$, $X = \{\eta : \varphi(\eta) < \rho\}$ is not small. Let $E = \{\gamma : \{\eta \in X : \eta(0) = \gamma\}$ is not small in $S'^{\langle \gamma \rangle}\}$. By hypothesis, $E$ is stationary in $\lambda$. By Fodor's Lemma there is $E_1$ a stationary subset of $E$ and $\beta_1 < \lambda$ so that $\beta_1 = \gamma_\rho$ for all $\gamma \in E_1$. Choose $\beta_1 < \beta \in E_\emptyset$. By the choice of $\beta$, $\{\eta \in S'_f : s'^1_\eta \subseteq B'_\beta\}$ is not small. This is a contradiction, since case 0 was assumed not to hold. Hence there is some $m$, so that $\{\eta \in S'_f : \lambda_{\eta \restriction m} = \kappa\}$ is not small. So we can assume for all $\eta \in S'_f$, that $\lambda_{\eta \restriction m} = \kappa$.

After making these assumptions we can choose a sub-$\lambda$-set $S$ of $S'$, and $m$ so that for all $\eta \in S$, $\lambda_{\eta \restriction m} = \kappa$. For $\eta \in S$ let $\lambda_\eta = \lambda'_\eta$. For $\eta \in S_f$ and $\gamma = \eta(0)$, let $s_\eta = (\{\gamma_{\eta(m)}\} \times s'^1_\eta) \cup \ldots \cup s'^{\ell(\eta)}_\eta$. Finally we can let $B_{\langle \gamma \rangle} = \gamma \times B'_\alpha$ and $B_\eta = B'_\eta$ if $\ell(\eta) > 1$.

*Case 3: not case 0, 1, or 2.* For $\gamma \in E_\emptyset$, let $\varphi(\gamma) = \mathrm{cf}(\gamma)$ if $\mathrm{cf}(\gamma) < \gamma$ and 0 otherwise. By Fodor's lemma, $\varphi$ must be constant on a stationary set. Since we are not in any of the previous cases, $\varphi$ must have constant value 0. That is $\{\gamma \in E_\emptyset : \gamma$ is a regular cardinal$\}$ must be stationary. So we can assume that every $\gamma \in E_\emptyset$ is a regular cardinal. By 3A.3, we can assume that $\lambda_{\langle \gamma \rangle} \leq \gamma$, for all $\gamma \in E_\emptyset$. If there is a stationary subset of $E_\emptyset$ such that $\lambda_{\langle \gamma \rangle} < \gamma$, then as in case 2, we can get a contradiction (i.e. show that case 0 occurs). So we can assume that $\lambda_{\langle \gamma \rangle} = \gamma$, for all $\gamma \in E_\emptyset$. The rest of the proof is similar to that of case 2. $\square$

The theorem above can be strengthened to take into account the height of the $\lambda$-system. If we begin with a family of countable sets based on a $\lambda$-system of height $n$, then we end up with a $\Lambda$ and $S$ which are beautiful so that $\Lambda$ has height $n$.

**3A.6 Theorem.** *Suppose* $\Lambda = (S, \lambda_\eta, B_\eta : \eta \in S)$ *is a $\lambda$-system and $S$ is a family of countable sets based on $\Lambda$. If $\Lambda$ and $S$ are beautiful, then $S$ has the reshuffling property.*

PROOF. The theorem is proved by induction on regular cardinals. For the purposes of the induction, we strengthen the reshuffling property. We show that if $I$ is any small subset of $S_f$, then for every $\alpha < \lambda$ there is a well-ordering $<_I$ of $I$ such that for every $\tau, \eta \in I$, $s_\eta \setminus \bigcup_{\nu <_I \eta} s_\nu$ is infinite and $\tau(0) \le \alpha < \eta(0)$ implies that $\tau <_I \eta$. In the degenerate case, $\lambda = \aleph_0$, there is nothing to prove. Assume the theorem is true for all regular $\mu < \lambda$ and $\lambda$ is uncountable. We claim it is enough to prove the following statement.

$(*)$    Suppose $X \subseteq E_\emptyset$ is a non-stationary set and $\alpha < \lambda$. Then there is a well-ordering $<_X$ of $S_X \overset{\text{def}}{=} \{\eta \in S_f : \eta(0) \in X\}$ so that for all $\eta \in S_X$, $s_\eta \setminus \bigcup_{\nu <_X \eta} s_\nu$ is infinite and if $\tau(0) \le \alpha < \eta(0)$ then $\tau <_X \eta$.

Suppose for the moment we have proved $(*)$ and that $I$ is a small subset of $S_f$ and $\alpha < \lambda$. Let $X = \{\beta \in E_\emptyset : \beta \le \alpha$ or $I^{\langle\beta\rangle}$ is non-small in $S^{\langle\beta\rangle}\}$. Then $X$ must be non-stationary. Let $<_X$ be the ordering guaranteed by $(*)$ and for each $\beta \notin X$ let $<_\beta$ be the well-ordering of $I^{\langle\beta\rangle}$ guaranteed by the inductive hypothesis. Then we can define a well-ordering $<_I$ on $I$ by $\eta <_I \nu$ if and only if either: $\eta, \nu \in S_X$ and $\eta <_X \nu$; or $\eta \in S_X$ and $\nu \notin S_X$; or $\beta, \gamma \notin X$, $\eta \in I^{\langle\beta\rangle}$, $\nu \in I^{\langle\gamma\rangle}$ and $\beta < \gamma$; or $\beta \notin X$, $\eta, \nu \in I^{\langle\beta\rangle}$ and $\eta <_\beta \nu$. Using property 2 we can check that this definition works, so we are reduced to proving $(*)$.

    Let $X$ be a non-stationary subset of $E_\emptyset$ and choose a closed unbounded subset $C$ of $\lambda$ so that $C \cap X = \emptyset$. We can further assume that the first element of $C$ is 0 and the second element is $\alpha + 1$. For $\beta \in C$, let $\beta^*$ denote the successor of $\beta$ in $C$. For each

$\beta \in C$ let $T_\beta$ be a transversal of $\{s_\eta : \eta \in S_f, \beta < \eta(0) < \beta^*\}$ which exists since $\mathcal{S}$ is $\lambda$-free. For $\beta < \gamma < \beta^*$, let $I_\gamma = \{\eta \in S_f : \eta(0) = \gamma$ and $T_\beta(s_\eta) \notin s_\eta^1\}$. It is a small subset of $S^{\langle \gamma \rangle}$ by 3A.1, and hence, by induction there is a corresponding well-ordering $<_\gamma$ of $I_\gamma$ given by the reshuffling property. We can assume that $T_\beta \restriction \{s_\eta : \eta \in I_\gamma\}$ is the transversal *associated to* the ordering $<_\gamma$, i.e., $T_\beta(\nu)$ is the first (with respect to the tree ordering given by property 3) element of $s_\nu^k \setminus \bigcup_{\eta <_\gamma \nu} s_\eta$, where $k$ is the least integer such that the latter set is infinite.

There are three cases to consider, corresponding to the three possibilities in clause 6 of the definition of beautiful.

*Case a: every $\gamma \in E_\emptyset$ has cofinality $\omega$.* Define a relation $R$ on $S_X$ as follows: let $\nu R \eta$ if and only if for some $\beta \in C$, $\beta < \nu(0)$, $\eta(0) < \beta^*$ and $T_\beta(s_\eta) \in s_\nu$. (Roughly the relation $R$ can be thought of as saying that $\nu$ may cause trouble if it precedes $\eta$ in the well-ordering.) Since $s_\nu$ is countable and $T_\beta$ is one-one, for each $\nu$ there are at most countably many $\eta$ so that $\nu R \eta$. For each $\beta \in C$, choose a sequence of disjoint countable sets, $(U_\delta^\beta : \delta < \mu)$, such that $\{\eta \in S_X : \beta < \eta(0) < \beta^*\} = \bigcup_{\delta < \mu} U_\delta^\beta$ and for all $\nu \in U_\delta^\beta$, $\{\eta : \nu R \eta\} \subseteq \bigcup_{\rho \leq \delta} U_\rho^\beta$. For each $\beta \in C$ and each $\delta < \mu$, let $<_\delta^\beta$ be an ordering of order type $\omega$ of $\{\eta(0) : \eta \in U_\delta^\beta\}$.

Given $\nu \in U_\rho^\beta$ and $\eta \in U_\delta^\gamma$ define $\nu <_X \eta$ if and only if either: $\beta < \gamma$; or $\beta = \gamma$ and $\rho < \delta$; or $\beta = \gamma$, $\rho = \delta$ and $\nu(0) <_\delta^\beta \eta(0)$; or $\beta = \gamma$, $\rho = \delta$, $\nu(0) = \eta(0)$, $\nu \in I_{\nu(0)}$ and $\eta \notin I_{\nu(0)}$; or $\beta = \gamma$, $\rho = \delta$, $\eta(0) = \nu(0)$, $\eta, \nu \in I_{\nu(0)}$ and $\nu <_{\nu(0)} \eta$; or $\beta = \gamma$, $\rho = \delta$, $\eta(0) = \nu(0)$, $\eta, \nu \in I \setminus I_{\nu(0)}$ and $\nu$ is less than $\eta$ in some fixed well-ordering of $S_X$.

We will verify in detail that this well-ordering works. The proof in case (b) is similar: there we will only define the well-ordering and leave the verifications to the reader. (It would be possible to simplify the proof in this case slightly, but the simplification is not available in case (b).) By the choice of $C$, if $\eta(0) \leq \alpha < \nu(0)$ then $\eta <_X \nu$.

For $\beta \in C$ and $\eta \in S_X$ so that $\beta < \eta(0) < \beta^*$, let the *level* of $\eta$ be the (unique) $k$ so that $T_\beta(s_\eta) \in s_\eta^k$. Suppose first that the level of $\eta$ is $k > 1$ (so that $\eta \in I_{\eta(0)}$). Suppose that $\nu <_X \eta$ and $s_\nu \cap s_\eta^k \neq \emptyset$. Then by property 2, $\nu(0) = \eta(0)$. Suppose $\eta \in U_\delta^\beta$, $s_\eta^k = \{t_n : n \in \omega\}$ and $T_\beta(\eta) = t_n$. If $\nu \notin U_\delta^\beta$, we claim that $s_\nu \cap s_\eta^k \subseteq \{t_0, \ldots, t_{n-1}\}$. Indeed, if $\nu \in U_\rho^\beta (\rho < \delta)$ and $t_m \in s_\nu$ for some $m \geq n$, then by property 3 of the definition of beautiful, $t_n \in s_\nu$; so $\eta \in U_\tau^\beta$ for some $\tau \leq \rho$, a contradiction. Now assume $\nu \in U_\delta^\beta$, and so $\nu <_{\eta(0)} \eta$. By the choice of $T_\beta$ and property 3, $\nu <_{\eta(0)} \eta$ implies that $s_\nu \cap s_\eta^k \subseteq \{t_0, \ldots, t_{n-1}\}$. So we have shown that if $\nu <_X \eta$, then $s_\nu \cap s_\eta^k \subseteq \{t_0, \ldots, t_{n-1}\}$.

Suppose now that the level of $\eta$ is 1, $\eta(0) = \gamma$, $s_\eta^1 = \{\langle \gamma_n, t_n \rangle : n < \omega\}$ and $T_\beta(\eta) = \langle \gamma_n, t_n \rangle$, where $\beta \in C$ and $\beta < \gamma < \beta^*$. Choose $m$ so that $\beta < \gamma_m$. If $\nu(0) < \beta$, then $s_\nu \cap s_\eta^1 \subseteq \{\langle \gamma_j, t_j \rangle : j < m\}$. Suppose $\eta \in U_\delta^\beta$. As above, we can show that if $\beta < \nu(0) < \beta^*$, $\nu \notin U_\delta^\beta$ and $\nu <_X \eta$ then $s_\nu \cap s_\eta^1 \subseteq \{\langle \gamma_0, t_0 \rangle, \ldots, \langle \gamma_{n-1}, t_{n-1} \rangle\}$. By property 2, if $\nu(0) = \eta(0)$ and $\nu \neq \eta$ then $s_\nu \cap s_\eta^1 = \emptyset$. Let $\rho^j$ $(j < k)$ enumerate the finitely many ordinals in $U_\delta^\beta$ which are $<_\delta^\beta \gamma$. Define $\rho^{j*}$ to be $\sup\{\rho_i^j : i \in \omega, \rho_i^j < \gamma\}$. Choose $r$ so that $\sup\{\rho^{j*} : j < k\} < \gamma_r$. Hence if $n_0 = \max\{n, m, r\}$ and $\nu <_X \eta$, then $s_\nu \cap s_\eta^1 \subseteq \{\langle \gamma_0, t_0 \rangle, \ldots, \langle \gamma_{n_0}, t_{n_0} \rangle\}$.

*Case b: for some uncountable cardinal $\kappa$ and $m \in \omega$, every* $\gamma \in E_\emptyset$ *has cofinality $\kappa$ and if $\eta \in S_f$, then $\lambda_{\eta \restriction m} = \kappa$.* For every $\gamma \in X$ and $\rho < \kappa$, let $I_{\rho\gamma} = \{\eta \in S_f : \eta(0) = \gamma$ and $\eta(m) < \rho\} \cup I_\gamma$. Since $I_{\rho\gamma}$ is a small subset of $S_f^{\langle \gamma \rangle}$, there is a well-ordering $<_{\rho\gamma}$ of $I_{\rho\gamma}$ satisfying the inductive hypothesis. Let $T_{\rho\gamma}$ be the transversal associated to $I_{\rho\gamma}$. Let $T_\beta^\rho$ be the function defined on the $\{s_\eta : \eta \in S_f, \beta < \eta(0) < \beta^*\}$ which is $T_{\rho\gamma}$ on $\{s_\eta : \eta \in I_{\rho\gamma}\}$ for $\beta < \gamma < \beta^*$ and $T_\beta$ otherwise. Note that $T_\beta^\rho$ is a transversal.

As before, we define a relation $R$ : let $\nu R \eta$ if there is $\beta \in C$ and $\rho < \kappa$ so that $\beta < \eta(0)$, $\nu(0) < \beta^*$, and $T_\beta^\rho(s_\eta) \in s_\nu$. For any $\nu$ there are at most $\kappa$ many $\eta$ so that $\nu R \eta$. For each $\beta \in C$, choose a sequence, $(U_\delta : \delta < \mu)$, of disjoint sets of cardinality at most $\kappa$ such that $\{\eta \in S_X : \beta < \eta(0) < \beta^*\} = \bigcup_{\delta < \mu} U_\delta^\beta$ and for all $\nu \in U_\delta^\beta$, $\{\eta : \nu R \eta\} \subseteq \bigcup_{\xi \leq \delta} U_\xi^\beta$. For each $\beta \in C$ and each $\delta < \mu$, let $<_\delta^\beta$ be an ordering of order type $\kappa$ of $\{\eta(0) : \eta \in U_\delta^\beta\} = Z_{\beta\delta}$. For each $\gamma \in Z_{\beta\delta}$, choose $\rho_\gamma < \kappa$ such that $\sup(\{\tau^* \in U_\delta^\beta : \tau <_\delta^\beta \gamma\} \cup \{\beta\}) < \gamma_{\rho_\gamma}$. Here $\tau^*$ is $\sup\{\tau_i : i < \kappa, \tau_i < \gamma\}$.

Suppose $\nu \in U_\xi^\beta$, $\eta \in U_\delta^\sigma$ and $\gamma = \nu(0)$. Define $\nu <_X \eta$ if and only if $\beta < \sigma$; or $\beta = \sigma$ and $\xi < \delta$; or $\beta = \sigma$, $\xi = \delta$ and $\nu(0) <_\delta^\beta \eta(0)$; or $\beta = \sigma$, $\xi = \delta$, $\eta(0) = \nu(0)$, $\nu \in I_{\rho_\gamma\gamma}$ and $\eta \notin I_{\rho_\gamma\gamma}$; or $\beta = \sigma$, $\xi = \delta$, $\eta(0) = \nu(0)$, $\eta$, $\nu \in I_{\rho_\gamma\gamma}$ and $\nu <_{\rho_\gamma\gamma} \eta$; or $\beta = \sigma$, $\xi = \delta$, $\eta(0) = \nu(0)$, $\eta$, $\nu \in I \setminus I_{\rho_\gamma\gamma}$ and $\nu$ is less than $\eta$ in some fixed well-ordering of $S_X$.

In the verification in this case we assign $\eta$ level 1 if $\eta \notin I_{\rho_\gamma\gamma}$ (where $\gamma = \eta(0)$), and otherwise $\eta$ has level $k$ ($> 1$) if $T_{\rho_\gamma\gamma}(\eta) \in s_\eta^k$.

*Case c: every $\gamma \in E_\emptyset$ is a regular cardinal and $\lambda_{\langle\gamma\rangle} = \gamma$.* We first note that for any $\beta < \lambda$, $E_\emptyset \cap \beta$ is not stationary in $\beta$. If $\beta$ is not a regular cardinal, then the set of regular cardinals less than $\beta$ is not stationary in $\beta$; if $\beta$ is a regular cardinal and $E_\emptyset \cap \beta$ were stationary in $\beta$, then $\{s_\eta : \eta(0) < \beta\}$ would not have a transversal.

Since $X$ is non-stationary and $X \subseteq E_\emptyset$, we can choose for each $\gamma \in X$ a closed unbounded subset $C_\gamma$ of $\gamma$ so that for $\beta \neq \gamma \in X$, $C_\beta \cap C_\gamma = \emptyset$ (see Exercise II.27). We can assume that if $\alpha < \gamma$, then the first element of $C_\gamma$ is greater than $\alpha$. For $\rho \in C_\gamma$, let

$$J_\rho = \{\eta \in S_f : \eta(0) = \gamma, \eta(1) < \rho \text{ and for all } \sigma \in C_\gamma \cap \rho, \sigma < \eta(1)\}.$$

Notice that $J_\rho$ is empty unless $\rho$ is either the first element of $C_\gamma$ or a successor in $C_\gamma$. (If $\rho$ is the successor of $\tau$ in $C_\gamma$, then $J_\rho = \{\eta : \eta(0) = \gamma$ and $\tau < \eta(1) < \rho\}$.) Since $J_\rho$ is a small subset of $S^{\langle\gamma\rangle}$, there is a well-ordering $<_\rho$ of $J_\rho$ satisfying the inductive hypothesis so that:

> if $\rho$ is the successor of $\sigma$ in $C_\gamma$, then $<_\rho$ is the restriction of a well-ordering of $\{\eta : \eta(0) = \gamma$ and $\eta(1) < \rho\}$ satisfying the induction hypothesis, in $S$, with respect to $\sigma$ (i.e., $\sigma$ plays the role of $\alpha$).

For $\rho \in C_\gamma$, let $U_\rho = J_\rho \cup \{\eta \in S_f : \eta(0) = \gamma$ and $\eta(1) = \rho\}$. Let $<_{U_\rho}$ be any well-ordering of $U_\rho$ so that $J_\rho$ is an initial segment of $U_\rho$

and the restriction of the ordering to $J_\rho$ is $<_\rho$. If $\rho$ is not in any $C_\gamma$ let $U_\rho = \emptyset$. Since the cubs are disjoint, $U_\rho$ is well defined. Notice that by our choice of the cubs, $U_\alpha$ is empty. Furthermore, for any $\rho < \alpha$, if $\eta \in U_\rho$ then $\eta(0) \leq \alpha$.

We now can define the required well-ordering. Let $\prec$ be the well-ordered sum of the $U_\rho$; i.e., $S_X = \sum_{\rho < \lambda} U_\rho$. By the remark above, this well-ordering satisfies the part of the inductive hypothesis concerning $\alpha$. Let us check that this ordering satisfies the first part of the inductive hypothesis. There are two cases to consider. First suppose that $\eta \in J_\rho$ for some $\rho \in C_\gamma$. Then

$$(s_\eta \setminus s_\eta^1) \setminus \bigcup_{\nu \prec \eta} s_\nu = (s_\eta \setminus s_\eta^1) \setminus \bigcup \{s_\nu \setminus s_\nu^1 : \nu \prec \eta, \nu(0) = \gamma\}.$$

By the choice of $<_\rho$, this last set is infinite. The second case occurs when $\eta(0) = \gamma$ and $\eta(1) = \rho \in C_\gamma$. If $\nu \prec \eta$, and $\nu(1) = \rho$, then $\nu(0) = \gamma$. (We chose the disjoint cubs in order to be able to make this assertion.) So $s_\nu^1$ and $s_\eta^1$ are, by clause 2 in the definition of beautiful, disjoint. $\square$

This completes the proof of Theorem 3.12.

In order to extend the results of the last section to varieties of algebras other than abelian groups, it is necessary to use a slightly stronger version of the reshuffling property. Fortunately, the proof we have given requires only minor modification in order to prove the stronger result. Rather than giving the modification, we will just state the new result.

**3A.7 Theorem.** *Suppose there is a $\lambda$-free non-free family of countable sets of cardinality $\lambda$. Then there is a $\lambda$-system $\Lambda$, a family of countable sets $S$ based on $\Lambda$, and a natural number $n$ so that $\Lambda$ has height $n$ and for all $\eta \in S_f$, $s_\eta$ is the disjoint union of infinite sets $s_\eta^1, \ldots, s_\eta^n$, and $S$ satisfies the following stronger form of the reshuffling property:*

(1) *If $I$ is a subset of $S_f$ of cardinality less than $\lambda$ and $\eta_0 \in I$, then there is a well-ordering, $<_I$, of $I$ so that $\eta_0$ is the first element of $<_I$ and for all $\eta \in I$ there is $k$ so that $s_\eta^k \cap \bigcup_{\nu <_I \eta} s_\nu$ is finite.*

(2) *If $I$ is a subset of $S_f$ of cardinality less than $\lambda$ and $\alpha < \lambda$, then there is a well-ordering $<_I$ of $I$ so that for all $\eta \in I$ there is $k$ so that $s_\eta^k \cap \bigcup_{\nu <_I \eta} s_\nu$ is finite and if $\nu(0) \leq \alpha < \eta(0)$ then $\nu <_I \eta$.*
□

The construction of $\lambda$-free groups can be generalized to many varieties, particularly varieties of modules over non-left-perfect rings and many varieties of groups (non-abelian, of course). The property which is needed to generalize the construction is (CP+), the strong construction principle:

**3A.8 Definition.** A variety $\mathcal{V}$ of algebras satisfies the strong construction principle, (CP+), if for every $0 < n < \omega$ there are countably generated free algebras $H \subseteq K \subseteq L$ and a partition of $\omega$ into $n$ infinite blocks $s^1, \ldots, s^n$ so that

> (1) $H$ is freely generated by $\{h_m : m < \omega\}$, and for every subset $J \subseteq \omega$ if for some $k$, $J \cap s^k$ is finite then the algebra generated by $\{h_m : m \in J\}$ is a free factor of $L$; and
>
> (2) $L = K * F(\omega)$ and $H$ is not a free factor of $L$.

Here $*$ denotes free product, $F(\omega)$ is the free algebra on $\aleph_0$ generators, and $H$ is a free factor of $L$, denoted $H|L$, means that there is a free algebra $G$ so that $H * G = L$. In a variety of modules, free product is the same as direct sum. The examples of greatest concern here are varieties of modules. No great harm will be done, if "free factor" is always read as "direct summand with free complementary summand", "free product" as "direct sum" and "variety" as "variety of modules over some ring".

**3A.9 Proposition.** *If $R$ is a ring which is not left perfect, then the variety of $R$-modules satisfies* (CP+).

PROOF. Given a ring $R$ which is not left perfect, let $\{a_m : m \in \omega\}$ be as in the proof of Proposition 1.1. Let $K$ be the module freely generated by the elements $\{y_m : m < \omega\}$. Let

$$h_m = a_m y_{m+1} - y_m.$$

Let $H$ be the free module with basis $\{h_m : m \in \omega\}$, and let $L = K \oplus R^{(\omega)}$. Since $K/H$ is not projective, condition (2) is satisfied. Given $0 < n < \omega$, let $s^1, \ldots, s^n$ be any partition of $\omega$ into infinite blocks. If $J \cap s^k$ is finite, then $J$ is a co-infinite subset of $\omega$, and $\{h_m : m \in J\} \cup \{y_m : m \notin J\}$ is easily seen to be a set of free generators of $K$. $\square$

One difficulty encountered in trying to show the existence of (strongly) λ-free algebras in a general variety is the decision of what definition of λ-free to adopt. Since subalgebras of free algebras may not be free, it will not do to require that every subalgebra generated by fewer than λ elements is free. (Compare the discussion in IV.1.) There are several possible definitions and the exact relation among them is not fully understood. However the algebras we shall build will satisfy the strongest possible requirements. For a variety which satisfies the strong construction principle, we describe a general construction of an algebra from a family of countable sets. In the special case of abelian groups, the construction will be the one given in 3.8.

**3A.10.** Suppose $\mathcal{V}$ is a variety which satisfies (CP+). Let $\mathcal{S} = \{s_i : i \in I\}$ be an indexed family of countable sets. Assume each $s_i$ is the disjoint union of infinite sets $s_i^1, \ldots, s_i^n$. Let $H$, $K$, $L$, $\{h_n : n \in \omega\}$, $s^1, \ldots, s^n$ be as in the definition of (CP+). For each $i$, fix an enumeration of $s_i$,

$$s_i = \{x_{i,m} : m \in \omega\}$$

so that for each $k$, $s_i^k = \{x_{i,m} : m \in s^k\}$. For each $i \in I$, choose $L_i$ a copy of $L$. Assume that the various copies of $L$ are disjoint, and disjoint from $\bigcup \mathcal{S} = \{x_{i,m} : m \in s_i, \, i \in I\}$. Let $K_i$, $H_i$ and $\{h_{i,m} : m \in \omega\}$ denote the corresponding copies of $K$, $H$, $\{h_m : m \in \omega\}$ in $L_i$. Let $G$ be the algebra freely generated by $\bigcup \mathcal{S}$. Let $F$ be $G * *_{i \in I} L_i$. Let $\theta$ be the smallest congruence which identifies $h_{i,m}$ and $x_{i,m}$ for all $i$ and $m$. Let $A(\mathcal{S})$ be $F/\theta$.

Just as in 3.11, we can prove the following lemma.

**3A.11 Lemma.** *Suppose $\mathcal{V}$ is a variety which satisfies* (CP+). *If a $\lambda$-system $\Lambda$ and a family of countable sets $\mathcal{S}$ satisfy the conclusion of* 3A.7, *then $A(\mathcal{S})$ is strongly $\lambda$-free. In particular if $I$ is a subset of $S_f$ of cardinality less than $\lambda$ and $\mathcal{S}_1 = \{s_\eta : \eta \in I\}$, then $A(\mathcal{S}_1)$ is free.* $\square$

What is perhaps surprising is the fact that the proof that $A(\mathcal{S})$ is not free has at least two subtleties.

**3A.12 Lemma.** *Suppose $\mathcal{V}$ is a variety which satisfies* (CP+). *If a $\lambda$-system $\Lambda$ and a family of countable sets $\mathcal{S}$ satisfy the conclusion of* 3A.7, *then $A(\mathcal{S})$ is not free.*

PROOF. Suppose, to the contrary, that $\mathcal{S} = \{s_\eta : \eta \in S_f\}$ is based on the $\lambda$-system $\Lambda = (S, \lambda_\eta, B_\eta : \eta \in S)$ and that $A = A(\mathcal{S})$ is free. For each $\alpha < \lambda$, let $A_\alpha$ be the subalgebra generated by (the images of)

$$\bigcup\{L_\eta : \eta \in S_f, \eta(0) < \alpha\} \cup \bigcup\{s_\eta : \eta \in S_f, \eta(0) < \alpha\}.$$

Then this defines a $\lambda$-filtration of $A$. As in the proof of 3.11 we can show that for all $\alpha < \beta$, $A_{\alpha+1}|A_\beta$. Note that there is a closed unbounded set $D$, so that for all $\alpha \in D$, $A_\alpha$ is the subalgebra generated by (the images of)

$$\bigcup\{L_\eta : \eta \in S_f, \eta(0) < \alpha\} \cup ((\bigcup \mathcal{S}) \cap \bigcup\{B_\eta : \eta \in S, \eta(0) < \alpha\}).$$

Since $A$ is free, there is a closed unbounded set $C \subseteq \lambda$, so that for all $\alpha < \beta$, $\alpha, \beta \in C$, $A_\alpha|A_\beta$.

We first claim that for $\alpha \in C$, $A_\alpha|A_{\alpha+1}$. (For modules this is automatic.) By the construction, $A_{\alpha+1} = B * G$, where $B \supseteq A_{\alpha+1}$ and is generated by $\mu$ elements over $A_{\alpha+1}$ and $G$ is a free algebra of rank $\mu$. Here $\mu$ is either 0, or an infinite cardinal less than $\lambda$. Choose $\beta \in C$ so that $\alpha < \beta$. Since $A_\alpha|A_\beta$ and $A_{\alpha+1}|A_\beta$, there is a free algebra $H$ of rank $\mu$ so that $A_\alpha|(B * G * H)$. But there is an isomorphism of $B * G * H$ with $B * G$ which fixes $B$ and hence $A_\alpha$. So $A_\alpha|A_{\alpha+1}$.

By 3.5(2)(b), there exists a limit point $\beta_0$ of $C \cap D$ such that $\beta_0$ is a limit point of $E_\emptyset$. We claim that $\tau = \langle \beta_0 \rangle \notin S_f$. Assume, for the sake of arriving at a contradiction, that $\tau \in S_f$. Choose a well-ordering, $<$, of $\{\eta \in S_f : \eta(0) \leq \beta_0\}$ as guaranteed by (i) of 3A.7 with $\tau$ as the first element. Using this well-ordering we can see that $A_{\beta_0+1} = L_\tau * G$ for some free algebra $G$. More exactly, choose $G_\eta$ a complementary factor in $L_\eta$ of the algebra generated by $s_\eta \setminus \bigcup_{\nu<\eta} s_\nu$. Then $A_{\beta_0+1} = *_{\eta(0)\leq\beta_0} G_\eta$, and $G_\tau = L_\tau$. Similarly, $H_\tau | A_{\beta_0}$. Since $A_{\beta_0} | A_{\beta_0+1}$, we have a contradiction.

Now $A_{\beta_0+1} = A_{\beta_0} * F$, where $F$ is free and of cardinality $\lambda_{\beta_0}$. We can construct a $\lambda_{\beta_0}$-filtration of $A_{\beta_0+1}$ by letting $A_{\beta_0\alpha}$ be generated by $A_{\beta_0}$ together with

$$\bigcup\{L_\eta : \eta \in S_f, \eta(0) = \beta_0 \text{ and } \eta(1) < \alpha\} \ \cup$$
$$\bigcup\{s_\eta : \eta \in S_f, \eta(0) = \beta_0 \text{ and } \eta(1) < \alpha\}.$$

We define $D_1$ a closed unbounded subset of $\lambda_{\beta_0}$ in the same manner as we defined $D$. By comparing the filtration with a $\lambda_{\beta_0}$-filtration of $F$, we can show that there is a closed unbounded set $C_1$ so that if $\alpha < \beta$, $\alpha$, $\beta \in C_1$ then $A_{\beta_0\alpha} | A_{\beta_0\beta}$. As above we obtain $\beta_1$ such that $\langle\beta_0, \beta_1\rangle \in S \setminus S_f$ and $\beta_1 \in C_1 \cap D_1$. Continuing in this way we obtain a contradiction, as in 3.6. $\square$

Combining these results we have the following. (Compare with 3.13.)

**3A.13 Theorem.** *Suppose $\mathcal{V}$ is a variety of algebras which satisfies* (CP+). *Then for every cardinal $\lambda$ for which there is a $\lambda$-free non-free abelian group of cardinality $\lambda$, there is a non-free (strongly) $\lambda$-free algebra which is $\leq \lambda$-generated.* $\square$

For any variety of algebras, we can define the *incompactness spectrum* of the variety as the class of uncountable cardinals $\lambda$ such that there is a $\leq \lambda$-generated algebra which is $\lambda$-free but not free. By the Singular Compactness Theorem — which holds in this more general setting — we know that any member of an incompactness spectrum is a regular cardinal. Theorem 3A.13 can be restated in terms of the incompactness spectrum.

**3A.14 Theorem.** *Suppose $\mathcal{V}$ is a variety that satisfies* (CP+).
*The incompactness spectrum of $\mathcal{V}$ contains that of abelian groups.*
□

For any ring $R$, let $\text{Inc}(R)$ denote the incompactness spectrum
of the variety of $R$-modules. Also, define $\text{Inc}'(R)$ to be the set
of uncountable $\lambda$ such that there is a $\leq \lambda$-generated $R$-module
which is $\lambda$-free but not *projective*. Obviously $\text{Inc}'(R) \subseteq \text{Inc}(R)$ and
$\text{Inc}(\mathbf{Z}) = \text{Inc}'(\mathbf{Z})$. If $R$ is left-perfect, then by IV.1.15 and Exercise
IV.2, $\text{Inc}'(R) = \emptyset$ and $\text{Inc}(R)$ is either empty or equals *Succ*, the
class of successor cardinals.

**3A.15 Proposition.** *Let $R$ be a ring which is not left perfect.*
  (i) $\text{Inc}(\mathbf{Z}) \subseteq \text{Inc}'(R)$.
  (ii) (V = L) $\text{Inc}(\mathbf{Z}) = \text{Inc}(R) = \text{Inc}'(R)$.

PROOF. Part (i) is a restatement of part of 3.13. As for (ii), by
IV.3.2 and IV.3.5, $\text{Inc}(R)$ is contained in the class of all regular
cardinals which are not weakly compact; but by 1.4, this latter
class is contained in $\text{Inc}(\mathbf{Z})$. □

For example, if $R = \mathbf{Z} \oplus \mathbf{Z}$, it is provable in ZFC that $\text{Inc}(R) =$
$\text{Inc}(\mathbf{Z}) \cup Succ$ and $\text{Inc}'(R) = \text{Inc}(\mathbf{Z})$ (cf. Exercise 20). It is tempting
to conjecture that it is provable *as a theorem of ZFC* that for any
ring $R$ which is not left-perfect, $\text{Inc}(R)$ is either $\text{Inc}(\mathbf{Z})$ or $\text{Inc}(\mathbf{Z}) \cup$
$Succ$, and $\text{Inc}'(R) = \text{Inc}(\mathbf{Z})$. (Note that in some models of ZFC,
e.g., a model of V = L + there is no inaccessible cardinal, $\text{Inc}(\mathbf{Z}) =$
$Succ =$ the class of regular uncountable cardinals.)

As we know from modules, not every variety satisfies (CP+).
One might wonder if there is anything that one can say about varieties which fail to satisfy (CP+). We do not know the answer
to this question, but more information is available about varieties
which satisfy the construction principle, (CP).

**3A.16 Definition.** A variety $\mathcal{V}$ of algebras satisfies the construction principle, (CP), if there are countably generated free algebras
$H \subseteq K \subseteq L$ such that:

(1) $H$ is freely generated by $\{h_m : m < \omega\}$, and for every finite subset $J \subseteq \omega$ the algebra generated by $\{h_m : m \in J\}$ is a factor of $L$; and

(2) $L = K * F(\omega)$ and $H$ is not a free factor of $L$.

Every variety *of groups* which satisfies (CP) also satisfies (CP+), but this is not true for varieties in general. Assuming V = L, any variety which satisfies (CP) has (strongly) $\lambda$-free $\leq$ $\lambda$-generated algebras for every regular cardinal $\lambda$ which is not weakly compact. (Use 3A.12 and the $\mathcal{S}$ built from a set witnessing E($\lambda$) — cf. the comment before 3A.3.) Without any set-theoretic assumption, it can be shown that a variety in a countable language satisfies (CP) if and only if for some cardinal $\lambda$ there is a strongly $\lambda$-free algebra of cardinality $\lambda$ which is not free. On the other hand, if $\mathcal{V}$ is a variety in a countable language which does not satisfy (CP), then the incompactness spectrum of $\mathcal{V}$ is either empty or consists of the class of successor cardinals.

**3A.17.** In the study of the incompactness spectrum of a variety, there are two sorts of non-free $\lambda$-free algebras, $A$, which may occur; those which are *essentially free* — i.e., $A * F(\kappa)$ is free for some $\kappa$ — and those which are *essentially non-free*. The *essentially non-free spectrum* of a variety $\mathcal{V}$ is the class of cardinals $\lambda$ for which there is an essentially non-free $\leq$ $\lambda$-generated algebra which is $\lambda$-free. For modules over a ring $R$, we have denoted the essentially non-free spectrum by Inc$'(R)$. For every $n$, we can define a principle (CP$_n$) by clauses (i) and (ii) of 3A.8. In this notation, (CP) is (CP$_1$) and (CP+) is equivalent to "(CP$_n$), for all $n$". It is conjectured that for any variety $\mathcal{V}$, $\lambda$ is in the essentially non-free incompactness spectrum if and only if for some $n$, (CP$_n$) holds and there is a $\lambda$-free family of countable sets based on a $\lambda$-system of height $n$.

We conclude this section with the

**Proof of 3.15.** If $\Diamond(\Lambda)$ holds for all $\Lambda$, then certainly $\Diamond(E)$ holds for all $E$, since a stationary set $E \subseteq \lambda$ together with a $\lambda$-filtration can be regarded as a $\lambda$-system of height 1. The proof of the other direction is by induction on $\lambda$ and makes use of the notions defined at the beginning of this section. For $\lambda = \aleph_1$, $\Diamond(\Lambda)$ is the same as $\Diamond(E_\emptyset)$. Now suppose that $\lambda > \aleph_1$. Given a $\lambda$-system $\Lambda$, for each $\alpha \in E_\emptyset$ we have by induction a family $\{C_\nu : \nu \in S^{\langle \alpha \rangle}\}$ satisfying the theorem for the $\lambda_\alpha$-system $\Lambda^{\langle \alpha \rangle}$. Let $\{C_{\langle \alpha \rangle} : \alpha \in E_\emptyset\}$ be a $\Diamond_\lambda(E_\emptyset)$-sequence (cf. VI.1.1). Let $C_\emptyset = \emptyset$. The union of these different families gives a family $\{C_\eta : \eta \in S\}$. Now given any $X \subseteq \bigcup\{B_\eta : \eta \in S\}$, for each $\alpha \in E_\emptyset$, by considering

$$X_\alpha \stackrel{\text{def}}{=} X \cap \bigcup\{B_\nu : \nu \in S^{\langle \alpha \rangle} \setminus \{\langle \alpha \rangle\}\},$$

we obtain a $\lambda_{\langle \alpha \rangle}$-set $S'_\alpha \subseteq S^{\langle \alpha \rangle}$ such that for all $\nu \in S'_\alpha \setminus \{\langle \alpha \rangle\}$, $X \cap (\bigcup\{B_{\nu \restriction n} : 2 \leq n \leq \ell(\nu)\}) = \bigcup\{C_{\nu \restriction n} : 2 \leq n \leq \ell(\nu)\}$. Moreover, $\{\alpha \in E_\emptyset : X \cap B_{\langle \alpha \rangle} = C_{\langle \alpha \rangle}\}$ is a stationary subset, $Y$, of $\lambda$. Then $S' \stackrel{\text{def}}{=} \bigcup\{S'_\alpha : \alpha \in Y\}$ is the desired $\lambda$-set. $\square$

## §4. Hereditarily separable groups

We return now to the theme of section 2 of Chapter IV: $\aleph_1$-free groups. In that section we introduced a series of successive strengthenings of the notion of $\aleph_1$-free. Now we want to add one more notion to the list. As in the previous sections, we shall, for convenience, deal with groups, i.e., $\mathbf{Z}$-modules, rather than modules over an arbitrary p.i.d.

**4.1 Definition.** A group $A$ is called a *Whitehead group* (or *W-group* for short) if $\text{Ext}(A, \mathbf{Z}) = 0$.

Now a free group is clearly a W-group. Also, if $B$ is a subgroup of a group $A$, then by I.1.1,

$$\text{Ext}(A, \mathbf{Z}) \to \text{Ext}(B, \mathbf{Z}) \to 0$$

is exact, so if $A$ is a W-group then $B$ is also a W-group. By Theorem IV.2.13, every W-group is separable. (See also XII.1.3.) Hence every W-group is hereditarily separable, and we have the following chain of implications. (Compare with the chain given in section IV.2.)

free $\Rightarrow$ W-group $\Rightarrow$ hereditarily separable

Whitehead's Problem asks if the first arrow is reversible. We shall see in Chapter XII that the answer is "yes" for countable groups, and "it depends on the set theory" for uncountable groups. Here we shall concentrate on the second arrow and show that its reversibility also depends on the set theory. In fact, these results are later, and perhaps more difficult, results of Shelah than those regarding the first arrow. It may be advisable for the reader who is not already comfortable with proofs using Martin's Axiom and diamond principles, to first read Chapter XII, section 1, before tackling the proofs in this section.

We begin with a useful characterization of hereditarily separable. A subgroup $T$ of $A$ is said to be *large* (or *essential*) if every non-zero subgroup of $A$ has non-zero intersection with $T$. Then it is not hard to see that $A$ has a large cyclic torsion subgroup $T$ if and only if $A$ is isomorphic to a subgroup of $\mathbf{Q}/\mathbf{Z}$ and there is a *finite* set $P$ of primes such that the order of every element of $A$ is divisible only by primes in $P$; this is also equivalent to saying that $A$ is isomorphic to a subgroup of $\bigoplus_{p \in P} Z(p^\infty)$, where $P$ is a *finite* set of *distinct* primes.

**4.2 Lemma.** *For any $\aleph_1$-free group $A$, $A$ is hereditarily separable if and only if $A'$ is separable for every subgroup $A'$ of $A$ such that $A/A'$ has a large cyclic torsion subgroup.*

PROOF. ($\Leftarrow$) Let $B$ be an arbitrary subgroup of $A$. It suffices to show that if $Z$ is a pure cyclic subgroup of $B$, then $Z$ is a summand of $B$ (see IV.2.7). Let $Z_*$ denote the pure closure of $Z$ in $A$, i.e.,

$$Z_* = \{a \in A : na \in Z \text{ for some } n \neq 0\};$$

since $A$ is $\aleph_1$-free, $Z_*$ is also a cyclic group (cf. IV.2.3). We can assume that $B$ is maximal with respect to the property that $Z_* \cap B = Z$, for if $Z$ is a summand of an extension of $B$, then it is a summand of $B$. By the maximality of $B$, $A/B$ has no non-zero subgroup which has zero intersection with $(Z_* + B)/B$. Now $(Z_* + B)/B$ is a cyclic torsion group, so by the hypothesis, $Z$ is a summand of $B$. $\square$

We are now going to describe a group $G$, and then show that it is not a W-group and that, assuming $MA + \neg CH$, it is hereditarily separable. (See Exercise IV.18 for a similar construction.)

**4.3.** Let $\Sigma$ be a subset of $^\omega 2$ of cardinality $\aleph_1$; so $\Sigma$ is a set of functions $\sigma: \omega \to 2 = \{0,1\}$. Let $\Phi = {}^{<\omega}2$; i.e., $\Phi$ consists of all functions $\rho: n \to 2$, for some $n \in \omega$. Let $\Phi_m$ be the set of elements of $\Phi$ whose domains are $\leq m$. Let $V$ be the $\mathbb{Q}$-module with basis $\Sigma \cup \Phi$. Let $\{p_n: n \in \omega\}$ be an enumeration without repetition of the set of primes. If $n \geq 1$, let $s_n = p_0 \cdot p_1 \cdots p_{n-1}$, and let $s_0 = 1$. For each $\sigma \in \Sigma$ and $n \in \omega$, let

$$z_{\sigma,n} = s_{n+1}^{-1}\left(\sigma - \sum_{i=0}^{n} s_i \sigma{\restriction}i\right) \in V.$$

For convenience, let $z_{\sigma,-1} = \sigma$. Let $G$ be the $\mathbb{Z}$-submodule of $V$ generated by

$$\Sigma \cup \Phi \cup \{z_{\sigma,n}: \sigma \in \Sigma, n \in \omega\}.$$

Let $H$ be the countable subgroup of $G$ generated by $\Phi$, and $H_m$ the subgroup of $H$ generated by $\Phi_m$. Notice that

$$p_n z_{\sigma,n} - z_{\sigma,n-1} = -\sigma{\restriction}n \in H_n$$

for all $n \geq 0$. So for all $\sigma \in \Sigma$, $\sigma + H$ is divisible in $G/H$ by all primes. Moreover, $H$ is a pure subgroup of $G$, and

$$G/H = \bigoplus_{\sigma \in \Sigma} \langle \sigma + H \rangle_*$$

where $\langle \sigma + H \rangle_*$, the pure closure of $\sigma + H$ in $G/H$, is generated by $\{z_{\sigma,n} + H: n \in \omega\}$ and is isomorphic to $R$, the group of rationals with square-free denominators.

There are at least two other useful ways to define $G$. If we consider the group $\mathbb{Z}^\Phi$, then $G$ is isomorphic to a pure subgroup of this group. The isomorphism is defined by taking each $\rho \in \Phi$ to $e_\rho$, taking $\sigma \in \Sigma$ to $\sum_{i \in \omega} s_i e_{\sigma \restriction i}$ and taking $z_{\sigma,n}$ to $\sum_{i=n+1}^{\infty} s_{n+1}^{-1} s_i \sigma \restriction i$.

Another (and perhaps more useful) representation of $G$ is via generators and relations. Let $F$ be the group freely generated by $\Phi \cup \{z_{\sigma,n} : n \in \omega \cup \{-1\}\}$. Let $K$ be the subgroup of $F$ generated by the elements $w_{\sigma,n}$, where

$$w_{\sigma,n} \overset{\text{def}}{=} z_{\sigma,n-1} - \sigma \restriction n - p_n z_{\sigma,n}.$$

It is easy to verify that $K$ is freely generated by the elements $w_{\sigma,n}(n \in \omega, \sigma \in \Sigma)$. As well, $F/K \cong G$, via the obvious map from $F$ to $G$.

**4.4 Lemma.** *Let $G$ be as in 4.3. Let $\sigma_1, \ldots, \sigma_r \in \Sigma$ and $m \in \omega$ such that for all $i \neq j$, $\sigma_i \restriction m + 1 \neq \sigma_j \restriction m + 1$. Let the pure closure of $H_m + \langle \sigma_1, \ldots, \sigma_r \rangle$ be denoted by $A$. Then*
  (i) *$A$ is free; in fact, $\Phi_m \cup \{z_{\sigma_1,m}, \ldots, z_{\sigma_r,m}\}$ is a basis; and*
  (ii) *$A$ is a direct summand of $G$.*
*Moreover, every finite subset of $G$ is contained in a subgroup of $G$ of the form of $A$, so $G$ is separable.*

PROOF. Part (i) and the last statement are easily verified by direct computation (by considering coefficients in $V$). As for part (ii), by expanding $\{\sigma_1, \ldots, \sigma_r\}$ we can assume that $\sigma_1, \ldots, \sigma_r$ are such that $\{\sigma_1 \restriction m + 1, \ldots, \sigma_r \restriction m + 1\} = \{\tau \restriction m + 1 : \tau \in \Sigma\}(= \Phi_{m+1} \setminus \Phi_m)$. But then the subgroup of $G$ generated by

$$\{z_{\sigma,n} : \sigma \in \Sigma, n \geq m + 1\}$$

is a complementary summand of $A$. (Notice in particular that for any $\tau \in \Sigma$, if $\tau \restriction m + 1 = \sigma_i \restriction m + 1$, then

$$z_{\tau,m} = p_{m+1} z_{\tau,m+1} - (p_{m+1} z_{\sigma_i,m+1} - z_{\sigma_i,m}).)$$

(The lemma can also be established by considering the representation of $G$ as a subgroup of $\mathbf{Z}^\Phi$.) □

**4.5 Theorem.** *The group $G$ defined in 4.3 is not a W-group.*

PROOF. Let $F$ and $K$ be as in 4.3. Let $h$ be the homomorphism from $K$ to $\mathbf{Z}$ defined by $h(w_{\sigma,n}) = \sigma(n)$. We will show that $h$ does not extend to $F$, and so $\mathrm{Ext}(G, \mathbf{Z}) \neq 0$. Suppose that $h$ has an extension to $F$ which we also denote by $h$. Since $\Sigma$ is uncountable and $\mathbf{Z}$ is countable, there are $\sigma, \tau \in \Sigma$ so that $h(\sigma) = h(\tau)$. An easy induction shows that if $\sigma \restriction n = \tau \restriction n$ then $h(z_{\sigma,n-1}) = h(z_{\tau,n-1})$. Choose $n$ maximal so that $\sigma \restriction n = \tau \restriction n$. For definiteness assume $\sigma(n) = 0$ and $\tau(n) = 1$.

$$0 = h(w_{\sigma,n}) = h(z_{\sigma,n-1}) - h(\sigma \restriction n) - p_n h(z_{\sigma,n}), \text{ and}$$

$$1 = h(w_{\tau,n}) = h(z_{\tau,n-1}) - h(\tau \restriction n) - p_n h(z_{\tau,n}).$$

So $1 = h(w_{\tau,n} - w_{\sigma,n}) = -p_n h(z_{\tau,n} - z_{\sigma,n})$, which is a contradiction.
$\square$

**4.6 Theorem.** (MA $+ \neg$CH) *The group $G$ defined in 4.3 is hereditarily separable.*

PROOF. By 4.2 and IV.2.7, it suffices to prove that if $B$ is a subgroup of $G$ such that $G/B$ has a large cyclic torsion subgroup and $Z$ is a pure cyclic subgroup of $B$, then $Z$ is a direct summand of $B$.

To establish the setting for an application of Martin's Axiom, let $\mathbf{P}$ consist of all functions $f: S \to Z$ where $S$ is a finite rank pure subgroup of $B$ containing $Z$, and $f \restriction Z$ is the identity on $Z$. Partially order $\mathbf{P}$ by: $f \leq f'$ if and only if $f'$ is an extension of $f$. For each $b \in B$, let $D_b = \{f \in \mathbf{P}: b \in \mathrm{dom}(f)\}$. Since $B$ is $\aleph_1$-free, each $D_b$ is dense in $\mathbf{P}$. If $\mathcal{G}$ is a directed subset of $\mathbf{P}$ which has non-empty intersection with each $D_b$, then $\cup \mathcal{G}$ will be a function: $B \to Z$ which is the identity on $Z$. So all we need to do is to show that $\mathbf{P}$ is c.c.c.

Let $\mathcal{S} = \{f_i: S_i \to Z: i \in I\}$ be an uncountable subfamily of $\mathbf{P}$, and for each $i$, let $S_{i,*}$ be the pure closure of $S_i$ in $G$; denote by the same symbol $f_i$ the unique extension of $f_i$ to a homomorphism: $S_{i,*} \to Q$, where $Q$ is the divisible hull of $Z$ (hence an injective group). Since each $D_b$ is dense, we can assume that $\mathcal{S}$ has the following property: for each $i \in I$ there exists $m = m_i \in \omega$ such that if $\Phi_{m+1} \setminus \Phi_m = \{\rho_1, \ldots, \rho_r\}$, then each $S_{i,*}$ is the pure closure of $H_m + \langle \sigma_{i,1}, \ldots, \sigma_{i,r} \rangle$ for some $\sigma_{i,\ell}$ such that for each $\ell$, $\sigma_{i,\ell} \restriction m+1 = \rho_\ell$. Replacing $I$ by an uncountable subset, we can assume that $m$ does not depend on $i$. Moreover, we can choose $m$ large enough so that if $p_n$ divides the order of an element of $G/B$ then $n \leq m$.

Furthermore, again by replacing $I$ by an uncountable subset, we can assume that for all $i, j \in I$, $f_i(\rho) = f_j(\rho)$ for all $\rho \in \Phi_m$, and that $f_i(\sigma_{i,\ell}) = f_j(\sigma_{j,\ell})$ for all $\ell = 1, \ldots, r$. (This is because $Q$ is countable and $\Phi_m$ is finite.) Thinning some more we can also assume that

$$(*) \qquad \text{for all } i, j \in I, \ z_{\sigma_{i,\ell},m} + B = z_{\sigma_{j,\ell},m} + B$$

for all $\ell = 1, \ldots, r$. (This is because $G/B$ and hence $(G/B)^r$ is countable.) We claim that any two elements of (this new thinned out) $\mathcal{S}$ are compatible.

For simplicity of notation, let us write $z_{i,\ell}$ instead of $z_{\sigma_{i,\ell},m}$. (Recall that $m$ is fixed.) By Lemma 4.4(i) we know that

$$Y_i \overset{\text{def}}{=} \Phi_m \cup \{z_{i,\ell} : \ell = 1, \ldots, r\}$$

is a basis of $S_{i,*}$. Moreover, for any $i, j$ the set theoretic union of $Y_i$ and $Y_j$ is linearly independent (cf. the description of $G/H$ in 4.3; note that we may have $z_{i,\ell} = z_{j,\ell}$ for some $\ell$, in which case we count this element only once in $Y_i \cup Y_j$). Therefore since $f_i$ and $f_j$ agree on $\Phi_m$ and since $f_i(z_{i,\ell}) = f_j(z_{j,\ell})$ for all $\ell$, $f_i \cup f_j$ extends (uniquely) to a well-defined homomorphism on $S_{i,*} + S_{j,*} = \langle Y_i \cup Y_j \rangle$, and thus also uniquely to a homomorphism $g: K \to Q$, where $K$ is the pure closure of $S_{i,*} + S_{j,*}$ in $G$. It suffices to prove that $g \restriction (K \cap B)$ takes values in $\mathbf{Z}$ (and not just in $\mathbf{Q}$), for then $g \restriction (K \cap B) \in \mathbf{P}$ and is a common extension of $f_i$ and $f_j$.

Now, a routine computation, comparing coefficients in $V$, shows that $K = S_{i,*} + S_{j,*} + L$, where $L = \langle z_{i,\ell} - z_{j,\ell} : \ell = 1, \ldots, r \rangle_*$, the pure closure in $G$ of the subgroup generated by the $z_{i,\ell} - z_{j,\ell}$. Because of $(*)$, $z_{i,\ell} - z_{j,\ell} \in B$ for all $\ell$. Therefore, by the choice of $m$, $L$ must be a subgroup of $B$, since the order of each element of $(L + B)/B$ is not divisible by any prime dividing the order of an element of $G/B$. Finally, it is easy to see that $K = S_{i,*} + L = S_{j,*} + L$. Thus $K \cap B = S_i + L$. Now, by the thinning out process, we have $g(z_{i,\ell} - z_{j,\ell}) = 0$ for all $\ell$, so $g[L] = \{0\}$. Hence it is clear that $g{\restriction}(K \cap B)$ takes values in $Z$. $\square$

Thus far we have shown that it is consistent with ZFC that there are hereditarily separable groups which are not free and, in fact, are not even W-groups. In XII we will show that $MA + \neg CH$ implies that there are non-free W-groups. So it is consistent that neither of the first two arrows reverses. (See also Exercise XII.C.16.) Now we turn to the other side of the picture, and show that assuming $V = L$, every hereditarily separable group *is* free. We will use the diamond prediction principles for $\lambda$-systems (cf. 3.14) to construct, given a non-free group $A$, a subgroup $A'$ which does not admit a projection onto a pure (cyclic) subgroup, and hence is not separable. As usual, this requires a killing lemma which we use, together with the prediction principle, in the construction. We begin with the statement of the killing principle, but defer its proof till the end. Throughout the rest of the section $p$ will denote a fixed prime.

**4.7 Lemma.** *Let $M$ be a free group and $N' \subseteq N$ subgroups of $M$ such that $M/N$ is countable and not free and $N/N' \cong Z(p^\infty)$. Furthermore assume that $N$ is a pure subgroup of $M$. Let $b \in N \setminus N'$ such that $\langle b \rangle$ is pure in $N$ and $pb \in N'$. Given a homomorphism $h : N' \to \langle pb \rangle$ such that $h(pb) = pb$, there exists $M' \subseteq M$ such that $M' \cap N = N'$, $M' + N = M$, and $h$ does not extend to a homomorphism $: M' \to \langle pb \rangle$.*

Before proving the main theorem, we isolate another useful fact, which we use in the non-crucial steps in the construction when we do not need to kill a function.

**4.8 Lemma.** *Let $M$ be a group and $N' \subseteq N$ subgroups of $M$ such that $N/N'$ is divisible. If $M'$ is a maximal subgroup of $M$ such that $M' \cap N = N'$, then $M' + N = M$, and hence $M/M' \cong N/N'$.*

PROOF. $(M' + N)/M'$ is a subgroup of $M/M'$ which is isomorphic to the divisible group $N/N'$, and hence is a direct summand of $M/M'$ :

$$M/M' = (M' + N)/M' \oplus C/M'.$$

It suffices to prove that $C = M'$; if not, let $c \in C \setminus M'$. Then $M' + \langle c \rangle$ contradicts the maximality of $M'$. $\square$

**4.9 Theorem.** $(V = L)$ *Every hereditarily separable group is free.*

PROOF. Let $A$ be a group which is not free; we shall construct a subgroup $A'$ of $A$ which is not separable. Let $\lambda$ be maximal such that $A$ is $\lambda$-free. If $\lambda = \aleph_0$, then $A$ contains a countable subgroup which is not free, and hence not separable (cf. IV.2.9). So we can assume that $\lambda \geq \aleph_1$; also, $\lambda$ is regular by the Singular Compactness Theorem (IV.3.5). By looking at a subgroup of $A$ of cardinality $\lambda$ which is not free, we can assume without loss of generality that $|A| = \lambda$.

Let $F$ be a free pure subgroup of $A$ of rank $\omega$, and let $F'$ be a subgroup of $F$ such that $F/F' \cong Z(p^\infty)$ and such that there exists $b \in F \setminus F'$ such that $\langle b \rangle$ is pure in $F$ and $pb \in F'$. Let $A = \bigcup_{\alpha < \lambda} B_\alpha$ be a $\lambda$-filtration of $A$ such that $F \subseteq B_0$. Notice that if $A'$ is a subgroup of $A$ such that $A' \cap F = F'$, then $\langle pb \rangle$ is a pure subgroup of $A'$. Our goal is construct such an $A'$ with no projection of $A'$ onto $\langle pb \rangle$. Let $Q$ be a divisible hull of $\langle pb \rangle$.

Starting with this $\lambda$-filtration, we construct, as in 3.4, a $\lambda$-system $\Lambda(A)$ associated with $A$; we shall use the same notation as in 3.4 and 3.5. Our prediction principle will be $\Diamond(\tilde{\Lambda}(A))$, where $\tilde{\Lambda}(A)$ is the $\lambda$-system obtained from $\Lambda(A)$ by replacing $B_\eta$ by $B_\eta \times Q$. This gives us a collection $\{h_\eta : \eta \in S\}$, where each $h_\eta \subseteq \bar{B}_\eta \times Q$. If $h_\eta$ is a function on $\bar{B}_\eta$, it extends in at most one way to a homomorphism on $\langle \bar{B}_\eta \rangle$, the subgroup generated by $\bar{B}_\eta$; denote the extension, if it exists, by $\tilde{h}_\eta$. Note that, since there is a cub of pure subgroups of a group, we can, and do, assume that each $\langle \bar{B}_\eta \rangle$ is pure in $A$.

We are going to define $A'_\eta \subseteq \langle \bar{B}_\eta \rangle$ for all $\eta \in S_f$ by induction with respect to the lexicographic order, $<_\ell$, on $S$. (Here, $<_\ell$ is defined by: $\langle \alpha, \beta, \gamma, \ldots \rangle <_\ell \langle \alpha_1, \beta_1, \gamma_1, \ldots \rangle$ if and only if $\alpha < \alpha_1$, or $\alpha = \alpha_1$ and $\beta < \beta_1$, or $\alpha = \alpha_1$, and $\beta = \beta_1$ and $\gamma < \gamma_1$, etc.) Notice that if $\eta <_\ell \eta_1$ then $A_\eta + \langle \bar{B}_n \rangle \subseteq \langle \bar{B}_{\eta_1} \rangle$.

We are going to require that $A'_\eta$ satisfy:

$$(\#) \qquad\qquad A'_\eta \cap F = F' \quad \text{and} \quad A'_\eta + F = \langle \bar{B}_n \rangle.$$

If $\zeta$ is the least element of $S_f$ (w.r.t. $<_\ell$ ), let $A'_\zeta$ be a subgroup of $\langle \bar{B}_n \rangle$ maximal with respect to the property that $A'_\zeta \cap F = F'$ (cf. 4.8). At limit points of $S_f$ (w.r.t. $<_\ell$ ) we take unions.

Suppose now that $A'_\eta$ has been defined. By construction of the $\lambda$-system, we have a countable group $A_\eta$ such that $(A_\eta + \langle \bar{B}_n \rangle)/\langle \bar{B}_n \rangle$ is not free. Now let $M = A_\eta + \langle \bar{B}_n \rangle$, $N = \langle \bar{B}_n \rangle$, $N' = A'_\eta$; notice that $(\#)$ implies that

$$N/N' \cong F/F' \cong Z(p^\infty).$$

If $\tilde{h}_\eta$ exists and is a projection, apply Lemma 4.7 (with $h = \tilde{h}_\eta {\upharpoonright} A'_\eta$); otherwise apply Lemma 4.8. In either case we get $M' \subseteq A_\eta + \langle \bar{B}_n \rangle$ such that $M' \cap N = N'$ and $M' + N = M$. Notice that $M/M' \cong N/N'$. Let $\eta_1$ be the next element of $S_f$ (w.r.t. $<_\ell$ ); then use Lemma 4.8 to extend $M'$ to $A'_\eta \subseteq \langle \bar{B}_{\eta_1} \rangle$ satisfying $(\#)$. This completes the inductive step of the construction.

We let $A' = \bigcup_{\eta \in S_f} A'_\eta$. Then $A' \cap F = F'$, so $\langle pb \rangle$ is a pure subgroup of $A'$. We claim that $\langle pb \rangle$ is not a direct summand of $A'$, which will show that $A'$ is not separable (cf. IV.2.5). If it were, then there would be a projection $h: A' \to \langle pb \rangle$, which extends to $g: A \to Q$ since $Q$ is injective. But by $\Diamond(\tilde{\Lambda}(A))$, there exists $\eta \in S_f$ such that $g {\upharpoonright} \bar{B}_\eta = h_\eta$. Since we constructed $A'_{\eta_1}$ (for the next element, $\eta_1$, of $S_f$) so that $h_\eta$ does not extend to $A'_{\eta_1}$, we have a contradiction. $\square$

Since every W-group is hereditarily separable, we have the following. (We will give another proof in Chapter XII.)

**4.10 Corollary.** (V = L) *Every W-group is free.* $\square$

Rather than proving 4.7 we will prove a stronger lemma, which we will then use to get a result that follows from CH. (To obtain 4.7 from 4.11, let $K = N$ and $K' = N'$.)

**4.11 Lemma.** *Let $M$ be a free group and $K \subseteq N$ be pure subgroups of $M$ such that $M/N$ is countable and not free. Further suppose there are subgroups $K' \subseteq K$, $N' \subseteq N$ such that $K \cap N' = K'$, $K + N' = N$ and $K/K' \cong Z(p^\infty)$. Let $b \in K \setminus K'$ such that $\langle b \rangle$ is pure in $K$ and $pb \in K'$. Given a homomorphism $h: K' \to \langle pb \rangle$ such that $h(pb) = pb$, there exist $M_0, M_1 \subseteq M$ such that $M_i \cap N = N'$, $M_i + N = M$ ($i = 0, 1$), and whenever $g: N' \to \langle pb \rangle$ is a map extending $h$, then $g$ does not extend to homomorphisms $g_i: M_i \to \langle pb \rangle$, for both $i = 0$ and $i = 1$.*

PROOF. We shall use repeatedly without explicit mention the fact that it is enough to construct $M_0, M_1 \subseteq M$ with all the properties except $M_i + N = M$; for by Lemma 4.8, if we extend these groups to maximal subgroups such that $M_i \cap N = N'$, then $M_0, M_1$ will have all the desired properties.

Since $M/N$ is not free it has a finite rank subgroup which is not free (cf. IV.2.3), so by the preceding observation, it suffices to prove the Lemma when $M/N$ is of finite rank. Moreover, we may suppose that every subgroup of $M/N$ of smaller rank is free. Choose a maximal independent subset $\{a_1 + N, \ldots, a_m + N\}$ of $M/N$; we can suppose that $\{a_2 + N, \ldots, a_m + N\}$ generates a pure (free) subgroup of $M/N$. Let $H = \langle N', pa_1, a_2, \ldots, a_m \rangle$ and

$$\hat{M} \overset{\text{def}}{=} \{x \in M: kx \in H \text{ for some } k \text{ relatively prime to } p\}.$$

Then $\hat{M} \cap N = N'$ because if $x \in \hat{M} \cap N$, then

$$kx = u' + c_1 pa_1 + \Sigma_{i \geq 2} c_i a_i$$

for some $k$ relatively prime to $p$ and some $u' \in N'$ and $c_i \in \mathbf{Z}$; so $c_i = 0$ for all $i$ by the independence of the $a_i$'s, and hence $kx \in N'$; but then $x \in N'$ since $N/N' \cong K/K'$ is a $p$-group. Thus $\hat{M}/N'$ is isomorphic to a subgroup of $M/N$, so every subgroup of $\hat{M}/N'$ of rank $< n$ is free. We now have two cases.

*Case 1*: $\hat{M}/N'$ *is not free*. Then any $g$ has at most one extension to a homomorphism of $\hat{M}$ onto $\langle pb \rangle$, because if $f_1$ and $f_2$ are two different extensions, and $L$ is the kernel of $f_1 - f_2$, then $L/N'$ is a subgroup of $\hat{M}/N'$ of rank $< n$, so it is free, and $\hat{M}/N' \cong L/N' \oplus \mathbb{Z}$ is free. Let $M_0 = \langle \hat{M}, a_1 \rangle$ and let $M_1 = \langle \hat{M}, b + a_1 \rangle$. Suppose some $g$ has extensions $g_0$ and $g_1$ to $M_0$ and $M_1$, respectively. Since the extension to $\hat{M}$ is unique, we will denote it by $g$ as well. Then

$$p(g_1(a_1 + b) - g_0(a_1)) = g(pa_1 + pb) - g(pa_1) = pb.$$

So $pb$ is divisible by $p$ in $\langle pb \rangle$, a contradiction.

*Case 2*: $\hat{M}/N'$ *is free*. It is easy to check that $M/(N + \hat{M})$ is a rank one $p$-group. So, in this case, $M/(N + \hat{M})$ must be infinite, since we have a short exact sequence

$$0 \to (N + \hat{M})/N \to M/N \to M/(N + \hat{M}) \to 0$$

where the first term is isomorphic to $\hat{M}/N'$ and thus is finitely-generated (since it is free), but the second term is not finitely-generated. Hence $M/(N + \hat{M})$ is isomorphic to $Z(p^\infty)$. Now the short exact sequence

$$0 \to (N + \hat{M})/\hat{M} \to M/\hat{M} \to M/(N + \hat{M}) \to 0$$

splits, since $(N + \hat{M})/\hat{M} \cong N/N' \cong Z(p^\infty)$, and so

(†)                    $$M/\hat{M} \cong Z(p^\infty) \oplus Z(p^\infty).$$

Note also that $(N + \hat{M})/\hat{M}$ is the divisible hull of $\langle b + \hat{M} \rangle$.

Fix elements $b_0 = b, b_1, \ldots, b_n, \ldots \in K$ so that $pb_{n+1} = b_n$ (mod $K'$). Fix elements $c_0 = a_1, c_1, \ldots, c_n, \ldots \in M$ so that $pc_{n+1} = c_n$ (mod $\hat{M}$). Note that $b \notin \langle \hat{M} \cup \{c_n : n \in \omega\} \rangle$. Choose a (finite rank) group $L$ so that $\hat{M} = N' \oplus L$.

Let $M_0 = \langle \hat{M} \cup \{c_n : n \in \omega\} \rangle$. We will inductively define a function $\eta : \omega \to 2$; then we will define

$$d_n = c_n + \sum_{k \le n} \eta(k) b_{n-k}.$$

and let $M_1 = \langle \hat{M} \cup \{d_n : n \in \omega\}\rangle$. It is convenient to asssume that our functions are going to $\mathbb{Z}$ rather than $\langle pb \rangle$. Fix an enumeration $\{(f_0^n, f_1^n) : n \in \omega\}$, of all pairs of homomorphisms from $L$ to $\mathbb{Z}$. We will define $\eta$ so that for all $g: N' \to \mathbb{Z}$, there are not both $g_0$ extending $f_0^n$ to $\langle \hat{M} \cup \{c_k : k \leq n\}\rangle$ and $g_1$ extending $f_1^n$ to $\langle \hat{M} \cup \{d_k : k \leq n\}\rangle$. This will complete the proof of the lemma.

Suppose that $\eta \restriction n$ has been defined. Notice that

$$p^{n+1}c_n = e_1 + e_2 \qquad \text{and}$$
$$p^{n+1}\left(c_n + \sum_{k<n} \eta(k)b_{n-k}\right) = e_1 + e_2 + t$$

where $e_1 \in N'$, $e_2 \in L$ and $t \in K'$. Choose $\eta(n)$ so that

$$\eta(n)p^n + f_1^n(e_2) + h(t) - f_0^n(e_2) \not\equiv 0 \pmod{p^{n+1}}.$$

If $g, g_0, g_1$, as above, existed then $p^{n+1}g_1(d_n) - p^{n+1}g_0(c_n)$ would contradict the choice of $\eta(n)$. □

We are able to prove the killing lemma, 4.7, in ZFC only when $M/N$ is countable. Therefore to prove the theorem, 4.9, we needed to reduce the reason for non-freeness of $A$ down to its ultimate countable cause. This is exactly what a $\lambda$-system does, and is the explanation for our use of diamond for $\lambda$-systems. Alternatively — and this is essentially what Shelah did in his original proof — one can use the "ordinary" $\Diamond(E)$ principles and simultaneously prove the killing lemma (for arbitrary cardinality) and the theorem by induction on cardinality.

The stronger version of the lemma can be used to give a proof using weak diamond (and CH):

**4.12 Theorem.** (CH) *If $A$ is an $\aleph_1$-free group of cardinality $\aleph_1$ such that $\Gamma(A) = 1$, then $A$ is not hereditarily separable.*

PROOF. The proof is much the same as that of 4.9. Choose $F$ a free pure subgroup of $A$ of rank $\omega$, and let $F'$ be a subgroup of $F$ such that $F/F' \cong Z(p^\infty)$. Fix $b$ as before. Choose an $\omega_1$-filtration $\{A_\alpha : \alpha < \omega_1\}$ of $A$ with $F = A_0$. We can assume that for all $\alpha$,

$A_{\alpha+1}/A_\alpha$ is not free. Enumerate the homomorphisms from $F'$ to $\langle pb \rangle$ as $\{h_\delta : \delta < \omega_1\}$. Partition $\omega_1$ into $\aleph_1$ disjoint non-small subsets, $\{E_\delta : \delta < \omega_1\}$ (cf. VI.1.11). For $\alpha \in E_\delta$ we are going to define a function $P_\alpha$ on pairs $(A', g)$ where $A' \subseteq A_\alpha$ and $g : A' \to \langle pb \rangle$, and then we will consider the weak diamond function, $\rho_\delta$, relative to $\{P_\alpha : \alpha \in E_\delta\}$ (cf. VI.1.7). If $A'$ is a pure subgroup of $A_\alpha$ so that $A' \cap F = F'$ and $A' + F = A_\alpha$, let $M(A')_{\alpha,0}$, $M(A')_{\alpha,1}$ be as guaranteed by Lemma 4.11, where $h_\delta$, $F'$, $F$, $A'$, $A_\alpha$, and $A_{\alpha+1}$ play the roles of $h$, $K'$, $K$, $N'$, $N$, and $M$ respectively. Define $P_\delta(A', g) = 0$, if $g$ does not extend to a homomorphism of $M(A')_{\alpha,0}$ to $\langle pb \rangle$, and 1 otherwise. (For other pairs, let $P(A', g)$ be arbitrary.)

Now define a subgroup $B_\alpha$ of $A_\alpha$ by induction on $\alpha$. Let $B_0 = F'$. At limit ordinals take unions. If $B_\alpha$ has been defined and $\alpha \in E_\delta$, let $B_{\alpha+1} = M(B_\alpha)_{\alpha, \rho_\delta(\alpha)}$.

Let $A' = \bigcup_{\alpha < \omega_1} B_\alpha$. Notice that for all $\alpha$, $A' \cap A_\alpha = B_\alpha$ and that $A' \cap F = F'$. We claim that $\langle pb \rangle$ is not a summand of $A'$ and hence $A'$ is not separable. Suppose, to the contrary, that there exists a projection $g : A' \to \langle pb \rangle$. Choose $\delta$ so that $g \restriction F' = h_\delta$. By the weak diamond principle there is $\alpha \in E_\delta$ so that $\rho_\delta(\alpha) = P_\delta(B_\alpha, g \restriction B_\alpha)$. So by the choice of $B_{\alpha+1}$, $g \restriction B_\alpha$ does not extend to $B_{\alpha+1}$, a contradiction. $\square$

It is possible to eliminate CH in the theorem above in favor of $2^{\aleph_0} < 2^{\aleph_1}$; also, Theorem 4.9 can be proved under the hypothesis that every stationary subset of a regular cardinal is non-small (see Exercises 25 – 28).

## EXERCISES

1. If $\mathcal{S}$ is a countable family of countable sets and $Z$ is a set such that for every $s \in \mathcal{S}$, $s \setminus Z$ is infinite, then $\mathcal{S}$ has a transversal $T$ such that for all $s \in \mathcal{S}$, $T(s) \notin Z$. [Hint: index the members of $\mathcal{S}$ by $\omega$ and define $T$ by induction on $n \in \omega$.]

2.(i) If $(S, \lambda_\eta, B_\eta : \eta \in S)$ is a $\lambda$-system, and $S_f = \bigcup_{n \in \omega} S_f^n$, let $S^n = \{\eta \in S : \eta \leq \tau \text{ for some } \tau \in S_f^n\}$. Prove that for some $n$, $(S^n, \lambda_\eta, B_\eta : \eta \in S^n)$ contains a $\lambda$-system. [Hint: proof by induction on $\lambda$, using II.4.5.]

(ii) For every $\lambda$-system $(S, \lambda_\eta, B_\eta : \eta \in S)$, there is a subtree $S'$ of $S$ such that $(S', \lambda_\eta, B_\eta : \eta \in S')$ is a $\lambda$-system of height $n$ for some $n$.

The following definitions are used in the exercises below. If $\mathcal{S}$ is a family of countable sets and $\mathcal{S}'$ is a subset of $\mathcal{S}$, say that $\mathcal{S}/\mathcal{S}'$ is *free* if there is a transversal for $\mathcal{S} \setminus \mathcal{S}'$ which takes values in $\bigcup \mathcal{S} \setminus \bigcup \mathcal{S}'$, i.e., a transversal for $\{s \setminus \bigcup \mathcal{S}' : s \in \mathcal{S} \setminus \mathcal{S}'\}$. Similarly $\mathcal{S}/\mathcal{S}'$ is $\lambda$-*free* if $\mathcal{T}/\mathcal{S}'$ is free for every $\mathcal{T} \subseteq \mathcal{S}$ of cardinality $< \lambda$. In analogy with abelian groups, we call $\mathcal{S}' \subseteq \mathcal{S}$ $\lambda$-*pure* if $\mathcal{S}/\mathcal{S}'$ is $\lambda$-free and $\mathcal{S}$ *strongly* $\lambda$-*free* if every subset of $\mathcal{S}$ of cardinality $< \lambda$ is contained in a free $\lambda$-pure subset.

3. (i) Let $\mathcal{S}$ be a family of countable sets of cardinality $\lambda$, a regular uncountable cardinal. Suppose $\mathcal{S}$ is $\lambda$-free. Let $\mathcal{S} = \bigcup_{\nu < \lambda} \mathcal{S}_\nu$ be a $\lambda$-filtration of $\mathcal{S}$. Prove: $\mathcal{S}$ is free if and only if $\{\nu < \lambda : \exists \mu > \nu$ s.t. $\mathcal{S}_\mu / \mathcal{S}_\nu$ is not free$\}$ is not stationary in $\lambda$. [Hint: like the corresponding result for groups: cf. IV.1.7]

(ii) If $\lambda$ is singular and $\mathcal{S}$ is $\lambda$-free of cardinality $\lambda$, then $\mathcal{S}$ is free. [Hint: proceed as in IV.3.3. Define $\mathcal{S}_i^n$ ($n \in \omega$, $i < \mathrm{cf}(\lambda)$) and transversals $T_i^n$ so that for all $n, i$: $\mathcal{S}_i^n$ is $\lambda$-pure; $T_i^{n+1} \supseteq T_i^n$; for all $s \in \mathcal{S}_{i+1}^n$, if $T_{i+1}^n(s) \in \bigcup \mathcal{S}_i^n$ then $s \in \mathcal{S}_i^{n+1}$; and $\{\mathcal{S}_i = \bigcup_{n \in \omega} \mathcal{S}_i^n : i < \mathrm{cf}(\lambda)\}$ is a continuous increasing chain.]

4. If there is a $\lambda$-free $\mathcal{S}$ of cardinality $\lambda$ which is not free, then there is a $\lambda$-free $\mathcal{S}'$ which is based on a $\lambda$-system. [Hint: Build a $\lambda$-system $\Lambda = (S, \lambda_\eta, B_\eta : \eta \in S)$ as in 3.4, using Exercise 3; then let $s_\eta = \bigcup A_\eta \cap \bar{B}_\eta$ — where the $A_\eta$ are the auxiliary sets defined as in 3.4; if necessary, modify the $B_\alpha$, by adding a countably infinite set, so that $s_\eta \neq s_\tau$ if $\eta \neq \tau$. (Handle the case of $\lambda = \omega_1$ separately.)]

The follow exercises construct some $\lambda$-systems and families of countable sets. The notation here is as in sections 3 and 3A.

5. (i) Let $E$ be a non-reflecting stationary subset of $\lambda$ consisting of limit ordinals of cofinality $\omega$. For each $\alpha \in E$, choose a subset $s_\alpha$ of order type $\omega$ in $\alpha$ such that $\sup(s_\alpha) = \alpha$. Let $S = E$, $B_\alpha = \alpha$, $\lambda_\alpha = \omega$ for $\alpha \in E$. Prove that $\mathcal{S} = \{s_\alpha : \alpha \in E\}$ is a $\lambda$-free family based on the $\lambda$-system $(S, \lambda_\eta, B_\eta : \eta \in S)$. [Hint: To prove that $\mathcal{S}$

is $\lambda$-free, prove by induction on $\gamma$ that for all $\mu < \gamma$, if $\mu \notin E$, then $\{s_\alpha : \alpha \in E \cap \gamma\}/\{s_\alpha : \alpha \in E \cap \mu\}$ is free.]

(ii) Compare the $\lambda$-system and family in (i) with that obtained as in 3.4 and 3.7 from a group $A$ constructed as in Theorem VII.1.4.

6. Suppose that $*(\kappa, \lambda)$ holds for some regular $\lambda \le \kappa$ and that $\lambda$-free does not imply $\lambda^+$-free. Assume that $\Lambda = (S, \lambda_\eta, B_\eta : \eta \in S)$ is a $\lambda$-system, $S$ is a family of countable subsets based on $\Lambda$, and $\Lambda$ and $S$ are beautiful. Let $\{f_i : i \in I\}$ be a family of functions from $S_f$ to $\kappa$ so that the ranges of the $f_i$ form an example of $*(\kappa, \lambda)$. (Since $|S_f| = \lambda$, such functions exist.) For $i \in I$ and $\eta \in S_f$, let $t_{i\eta} = \{f_i(\eta)\} \times \{0\} \cup s_\eta \times \{i\} \times \{1\}$. Show that $\{t_{i\eta} : i \in I, \eta \in S_f\}$ is a $\kappa^+$-free family of countable sets which is not strongly $\kappa^+$-free. (This is the set-theoretic version of Theorem 2.13)

7. Use the assumptions and notation of 6. For all stationary $E \subseteq \kappa^+$ such that every element of $E$ has cofinality $\omega$ and is greater than $\kappa$, there is a $\kappa^+$-system $\Lambda'$ and a family $T$ of countable sets based on $\Lambda'$ so that $\Lambda'$ and $T$ are beautiful and $E'_\emptyset = E$. Furthermore for all $\alpha \in E$, $\Lambda'^{(\alpha)}$, $T^{(\alpha)}$ is isomorphic (in the obvious sense) to $\Lambda$, $S$. [Choose a ladder system $\{\eta_\delta : \delta \in E\}$. Identify $E$ with $I$. For $\delta \in E$ and $\nu \in S_f$, let $u_{\delta\nu} = \{\langle\langle \eta_\delta(n), \ldots, \eta_\delta(0)\rangle\langle f_\delta(\nu), 0\rangle\rangle : n \in \omega\} \cup s_\eta \times \{\langle \delta, 1\rangle\}$. Verify that there is an appropriate choice of a $\kappa^+$-system such that it together with a straightforward modification of $\{u_{\delta\alpha} : \delta \in E, \alpha < \lambda\}$ are beautiful.]

8. Use the assumptions and notation of 6. Suppose $\mu$ is an uncountable regular cardinal $\le \lambda$ and that there is some $m$ so that $\lambda_{\eta \restriction m} = \mu$, for all $\eta \in S_f$. Then for all stationary $E \subseteq \kappa^+$ such that every element of $E$ is greater than $\kappa$ and has cofinality $\mu$, there is a $\kappa^+$-system $\Lambda'$ and a family $T$ of countable sets based on $\Lambda'$ so that $\Lambda'$ and $T$ are beautiful and $E'_\emptyset = E$. Furthermore for all $\alpha \in E$, $\Lambda'^{(\alpha)}$, $T^{(\alpha)}$ is isomorphic (in the obvious sense) to $\Lambda$, $S$. [For each $\gamma \in E$, choose a strictly increasing continuous sequence $\{\gamma_i : i < \mu\}$ which is cofinal in $\gamma$. Identify $E$ with $I$. For $\eta \in S_f$, let $u_{\gamma\eta} = \{\langle \gamma_{\eta(m)}, \langle f_\gamma(\eta), n, 0\rangle : n \in \omega\} \cup s_\eta \times \{\langle \gamma, 1\rangle\}.]$

9. For all $m < n \in \omega$ and $E \subseteq \aleph_n$ such that for all $\alpha \in E$,

$\text{cf}(\alpha) = \aleph_m$, there is an $\aleph_n$-system $\Lambda$ and a family of countable sets $\mathcal{S}$ based on $\Lambda$ so that $\Lambda$ and $\mathcal{S}$ are beautiful, $\tilde{E}_\emptyset = \tilde{E}$, and there is $k$ so that for all $\eta \in \mathcal{S}_f$, $\lambda_{\eta \restriction k} = \aleph_m$ and (if $m \neq 0$) for all $\beta \in E_{\eta \restriction k}$, $\text{cf}(\beta) = \omega$. Generalize this result to other cardinals. [Hint: prove the theorem by induction on $n$. For $n > 1$, note that $*(\aleph_{n-1}, \aleph_{n-1})$ holds and apply 7 and 8.]

10. Suppose that $\kappa$-free does not imply $\kappa^+$-free. Then for all $0 < n \in \omega$ and $E \subseteq \kappa^{+n}$ ($n$th successor of $\kappa$) such that there is $\mu \in \{\omega, \kappa, \dots, \kappa^{+(n-1)}\}$ so that for all $\alpha \in E$, $\text{cf}(\alpha) = \mu$ and $\kappa < \alpha$, there is a $\kappa^{+n}$-system $\Lambda$ and a family of countable sets $\mathcal{S}$ based on $\Lambda$ so that $\Lambda$ and $\mathcal{S}$ are beautiful and $E_\emptyset = E$. [Hint: since $\kappa$ is regular, $*(\kappa, \kappa)$ holds. Continue as in 9.]

The following exercises prove the existence of non-free $\kappa^+$-separable groups of cardinality $\kappa^+$ and give information on the $\Gamma$-invariant for strongly $\kappa^+$-free groups for those cardinals $\kappa$ such that $\kappa$-free does not imply $\kappa^+$-free.

11. Suppose $\Lambda$ and $\mathcal{S}$ are beautiful. Let $A(\mathcal{S})$ be defined as in 3.8. Then $\Gamma(A(\mathcal{S})) = \tilde{E}_\emptyset$. [Hint: this is essentially proved in 3.9 and 3.11.]

12. Let $A(\mathcal{S})$ be defined as in 3.8. Show that $A(\mathcal{S})$ is the 2-adic closure of the subgroup generated by $\bigcup_{s \in \mathcal{S}} s$.

13. Suppose $\mathcal{S}$ is a family of countable sets based on a $\Lambda$-system $\Lambda$ such that every small subset can be reshuffled (i.e. has property $(*)$ of 3A.6). Suppose that $\Lambda$ and $\mathcal{S}$ have the following additional property:

$(\dagger)$      any subset of $\mathcal{S}_f$ of size $< \lambda$ is contained in a small subset, $X$, of $\mathcal{S}_f$ so that for all $\eta \in \mathcal{S}_f \setminus X$, $|s_\eta \cap \bigcup_{\nu \in X} s_\nu|$ is finite.

Then $A(\mathcal{S})$ is $\lambda$-separable. [Hint: $A(\mathcal{S})$ is the direct sum of $A(X) \oplus C$, where $C$ is the 2-adic closure of the subgroup generated by $\bigcup_{\eta \in \mathcal{S}_f \setminus X} s_\eta \setminus \bigcup_{\nu \in X} s_\nu$.]

14. Suppose that $\Lambda$ and $\mathcal{S}$ are beautiful. Then $\Lambda$ and $\mathcal{S}$ satisfy (†) if either

(a) every $\gamma \in E_{\emptyset}$ has cofinality $\omega$; or

(b) there is an uncountable $\kappa$ and $m$ so that every $\gamma \in E_{\emptyset}$ has cofinality $\kappa$ and for all $\eta \in S_f$, $\kappa = \lambda_{\eta \restriction m}$ and every $\rho \in E_{\eta \restriction m}$ has cofinality $\omega$; or

(c) every $\gamma \in E_{\emptyset}$ is a regular cardinal and for every $\gamma \in E_{\emptyset}$ and $\rho \in E_{\langle \gamma \rangle}$, $\mathrm{cf}(\rho) = \omega$.

[Hint: (a) for any $\beta$, consider $X = \{\eta : \eta(0) \leq \beta\}$; (b) for any $\beta$, consider $X = \{\eta \in S_f : \eta(m) \leq \rho\}$ where $\gamma = \eta(0)$, $\lambda_{\eta \restriction m} = \mathrm{cf}(\gamma)$, and $\rho = \sup\{\alpha : \gamma_\alpha \leq \beta\}$ (Here $\gamma_\alpha$ is as in 6(b) of the definition of beautiful.). To verify that the definition in (b) works, note that by clause 6(b) in the definition of beautiful if $\rho \in E_{\eta \restriction m}$ then $\gamma_\rho \leq \beta$; (c) is similar.]

15. Suppose $\kappa$-free does not imply $\kappa^+$-free. Then there is a $\kappa^+$-separable group of cardinality $\kappa^+$ which is not free. [Hint: use 14(a), 13 and 10.]

16. Suppose $\kappa$-free does not imply $\kappa^+$-free. Then for all $0 < n \in \omega$ and $E \subseteq \kappa^{+n}$, such that for all $\alpha \in E$, $\mathrm{cf}(\alpha) \in \{\omega, \kappa, \ldots, \kappa^{+(n-1)}\}$, there exists a strongly $\kappa^{+n}$-free group, $A$, so that $\Gamma(A) = \tilde{E}$. (Compare this result with 2.10.) [Hint: it is enough to prove the result where every element of $E$ has the same cofinality. Use 10 and 11.]

17. For $n \in \omega$ if $\kappa = \aleph_0$, then the group in 11 can be taken to be $\aleph_n$-separable. [Hint: use 14(b) and 9.] Generalize this result.

18. For all $0 < n \in \omega$, there is a strongly $\aleph_n$-free group $A$ with an $\aleph_n$-filtration $\{A_\alpha : \alpha < \aleph_n\}$ such that for all limit ordinals $\alpha$, $A/A_\alpha$ is $\aleph_{n-1}$-free but not free. [Hint: use the groups constructed in 16.]

19. If there is a $\lambda$-free abelian group which is not $\lambda^+$-free, then there is a non-commutative group $G$ of cardinality $\lambda$ which is $\lambda$-free in the sense that every subgroup of cardinality $< \lambda$ is a free (non-abelian) group, but $G$ is not a free group. [Hint: construct $G$ analogously to 3.8; show $G$ is not free because $G/G'$ ($G'$ the

commutator subgroup) is not free. Alternatively show that the variety of groups satisfy (CP+) and use 3A.13.]

20. Let $R = \mathbf{Z} \oplus \mathbf{Z}$.
(i) $R$ is not left perfect, and every $R$-module $A$ can be written as $A = A_0 \oplus A_1$ where the $R$-multiplication is given by: $(n,m) \cdot (a,b) = (na, mb)$; if $A$ is a projective $R$-module, then $A_0$ is a free $\mathbf{Z} \times \{0\}$-module. [Hint: $A_0 = (1,0)A$.]
(ii) $\mathrm{Inc}(R) = \mathrm{Inc}(\mathbf{Z}) \cup Succ$. [Hint: use (i) and compare IV.1.15.]
(iii) $\mathrm{Inc}'(R) = \mathrm{Inc}(\mathbf{Z})$

21. Suppose $\kappa < \lambda$, $\lambda$ is a regular cardinal and for all $\kappa \leq \rho < \lambda$, $\rho$-free implies $\rho^+$-free. Let $A$ be a $\lambda$-free group of cardinality $\lambda$ which is not free and $\{A_\alpha : \alpha < \lambda\}$ a $\lambda$-filtration of $A$ such that if $A/A_\alpha$ is not $\lambda$-free then $A_{\alpha+1}/A_\alpha$ is not $\lambda$-free. Show: there is some regular cardinal $\mu < \kappa$ and a stationary subset $S$ of $\lambda$ so that for all $\delta \in S$, $A_{\delta+1}/A_\delta$ is not $\mu^+$-free.

22. In Magidor-Shelah 198?? the following property $\Delta_{\kappa\lambda}$ is defined for $\kappa \leq \lambda$ and $\lambda$ a regular cardinal.

For every $\aleph_0 \leq \mu < \kappa$ and $S \subseteq \lambda$ stationary in $\lambda$, if $\{a_\delta : \delta \in S\}$ is such that for all $\delta$, $a_\delta \subseteq \delta$ and $|a_\delta| \leq \mu$, then there is a regular cardinal $\mu < \lambda' < \kappa$ and an increasing function $f: \lambda' \to \lambda$ such that

$$\{\delta < \lambda': f(\delta) \in S, \ a_{f(\delta)} \subseteq \cup\{a_{f(\gamma)}: \gamma < \delta, \text{ and } \ f(\gamma) \in S\}\}$$

is stationary.
(i) If $\Delta_{\kappa\lambda}$ holds and for all $\kappa \leq \rho < \lambda$, $\rho$-free implies $\rho^+$-free then $\lambda$-free implies $\lambda^+$-free. [Hint: If $A$ is a counterexample, let $\mu$ and $S$ be as in 21; without loss of generality, the underlying set of $A_\delta$ is $\delta$. For each $\delta \in S$, let $C_\delta$ be a subgroup of $A_{\delta+1}$ of cardinality $\leq \mu$ such that $C_\delta + A_\delta/A_\delta$ is not free. Let $a_\delta = C_\delta \cap A_\delta$. Let $\lambda'$ and $f$ be as in $\Delta_{\kappa\lambda}$. Let $B$ be the subgroup of $A$ generated by $\cup\{C_{f(\delta)}: f(\delta) \in S\}$. Show that $B$ is not free.]
(ii) If $\kappa$ is a singular cardinal and $\Delta_{\kappa\kappa^+}$ holds, then $\kappa^+$-free implies $\kappa^{++}$-free.
(iii) If $\kappa$ is a singular cardinal and $\Delta_{\kappa\lambda}$ holds for all regular $\lambda > \kappa$, then $\kappa$-free implies free.

23. If $\Delta_{\aleph_2\aleph_2}$ holds and $A$ is an $\aleph_2$-free group of cardinality $\aleph_2$, then there is an $\aleph_2$-filtration $\{A_\alpha : \alpha < \omega_2\}$ of $A$ so that for all $\alpha$, $A/A_\alpha$ is $\aleph_1$-free.

24. Show that the group $G$ in 4.3 is slender. [Hint: consider $G/H$.]

25. Prove the following variant of Lemma 4.11.
Let $M$ be a free group and $N' \subseteq N$ be subgroups of $M$ such that $M/N$ is countable and not free and $N/N' \cong Z(p^\infty)$. Further assume $N$ is a pure subgroup. Let $b \in N \setminus N'$ such that $\langle b \rangle$ is pure in $N$ and $pb \in N'$. Suppose

$(*_p)$　　there exist $n$ and $\{a_1, \ldots, a_n\} \in M$ independent over $N$ so that any subset of cardinality $n - 1$ is pure independent over $N$ and $\{k : k$ is relatively prime to $p$ and $k$ divides $a_1$ mod $\langle M \cup \{a_2, \ldots, a_n\}\rangle_*\}$ is infinite.

There exist $M_0, M_1 \subseteq M$ such that $M_i \cap N = N'$, $M_i + N = M$ ($i = 0, 1$), and if $g: N' \to \langle pb \rangle$ is a homomorphism such that $g(pb) = pb$ then $g$ does not extend to homomorphisms $g_i: M_i \to \langle pb \rangle$, for both $i = 0$ and $i = 1$. [Hint: this is just the first case of the proof of 4.11.]

26. Suppose $A$ is an $\aleph_1$-free group of cardinality $\aleph_1$ and $\{A_\alpha : \alpha < \omega_1\}$ is an $\omega_1$-filtration of $A$ satisfying $A_{\alpha+1}/A_\alpha$ is free if and only if $A/A_\alpha$ is $\aleph_1$-free. Prove that if $\Gamma(A)$ is non-small (in the sense of VI.1.6), then there is some prime $p$ so that $\{\alpha : A_{\alpha+1}$ and $A_\alpha$ satisfy $(*_p)\}$ is non-small. [Hint: if $A_{\alpha+1}/A_\alpha$ is not free then there is some $p$ so that $A_{\alpha+1}$ and $A_\alpha$ satisfy $(*_p)$. There must be some $p$ which works for a non-small set.]

27. Show Theorem 4.12 is true if we weaken CH to $2^{\aleph_0} < 2^{\aleph_1}$. [Hint: choose a filtration and prime $p$ as in 26. Then argue as in 4.12, using 25.]

28. Suppose $\Lambda$ is a $\lambda$-system associated with a $\lambda$-free group $A$ and $\Diamond(\Lambda)$ holds. Show that $A$ is not hereditarily separable. In

particular, if $A$ is $\aleph_1$-free of cardinality $\aleph_1$ and $\Diamond(\Gamma(A))$ holds, then $A$ is not hereditarily separable.

## NOTES

Section 1 is based on Eklof-Mekler 1988. Prop. 1.1 is due to Bass 1960. The construction of strongly $\kappa$-free abelian groups assuming V = L is first found in Gregory 1973; see also Eklof 1975, where 2.3, 2.4 and 2.5 are proved. The definition of $F(\lambda)$ and 2.2 are from Mekler 1980. Hill 1974a proved that $\aleph_{n+1}$-free does not imply $\aleph_{n+2}$-free for all $n \in \omega$ by a different method, which was used by Mekler 1980 to show, under GCH that $\aleph_{n+1}$-free does not imply strongly $\aleph_{n+1}$-free for all $n \in \omega$.

Theorem 2.6(i) was proved by Shelah 1979b and 2.6(ii) by Magidor-Shelah 19??, both assuming GCH; recently Shelah has been able to eliminate the hypothesis of GCH from these results (personal communication). The implication $(1) \Rightarrow (2)$ in Theorem 2.10 is due to Mekler 1980 as is the implication $(2) \Rightarrow (1)$ for successor cardinals; the rest is from Eklof-Mekler-Shelah 1984. Lemma 2.11 is from Mekler 1980; 2.12 to 2.14 are from Mekler-Shelah 1987.

Section 3 through 3.12 is due to Shelah 1985; the proofs are mainly based on the appendix (by Mekler) to Shelah 1985. For more on transversal theory, see Milner 1974. The equivalence of condition (5) in 3.13 with the other conditions is due to the authors. 3.14 – 3.16 come from Mekler-Shelah 1988. We are grateful to M. Magidor for explaining the proof of the reshuffling theorem to us. The results in 3A on constructions of almost free algebras in varieties are due to Mekler; for related results on almost free groups in varieties see Mekler 19??. The construction principle (CP) is from Eklof-Mekler 1988 and some of the facts mentioned after 3A.16 can be found there or in Mekler-Shelah 19??a.

The results in Exercises 6-17 on the construction of $\lambda$-systems and families of countable sets with the application to abelian groups are new. Some of the group theoretic consequences were known. Mekler 1980 contains the results that there is an $\aleph_n$-separable group of cardinality $\aleph_n$, for all $0 < n < \omega$ and the result in Exercise 16 if $\omega$ is excluded from the list of possibilities.

Section 4 is due to Shelah 1979a; an error there in the proof of Theorem 3.5 (our 4.12) has been corrected by Shelah (personal communication). It is possible to give a simpler proof of 4.9 and 4.12, by using the same trick as in Exercise 27; i.e. we can use an easier version of lemma 4.7; 4.9 and 4.12 are stronger, since the proofs show that (in the terminology of Shelah 1979a) $p$-hereditarily separable implies free.

Exercises. 19:Shelah 1985; 21-22:Magidor-Shelah 19??; 25-27:Shelah (personal communication).

# CHAPTER VIII
# COUNTABLY-SEPARABLE GROUPS

In this chapter we shall consider only groups (**Z**-modules) of cardinality $\leq \aleph_1$. We shall study in particular $\aleph_1$-separable groups; recall that these are the groups such that every countable subset is contained in a countable free summand of the group (cf. IV.2.4). This property is apparently stronger than the property of being strongly $\aleph_1$-free; however, the two properties coincide (for groups of cardinality $\leq \aleph_1$) in models of MA + ¬CH (see XII.1.13).

We are going to investigate the possibility of proving classification and structure results about these groups. In brief, our conclusions will be that we get good positive results assuming MA + ¬CH or the stronger PFA, and we get negative results assuming CH or V = L. We shall begin in section 1 with some constructions that show the variety and abundance of $\aleph_1$-separable groups, and demonstrate the failure of certain attempts to classify them. These examples will motivate the introduction of two important concepts: that of *filtration-equivalence* (see 1.5), a relation between two $\aleph_1$-separable groups, and that of being *in standard form* (see 1.9), a property of an $\aleph_1$-separable group. At the end of section 1, we will survey the four principal methods of constructing $\aleph_1$-free groups. In section 2, we shall investigate these concepts under the hypothesis MA + ¬CH and show, among other things, that filtration-equivalent groups are isomorphic and that groups in standard form have some nice decomposition properties with respect to direct sums. In section 3 we shall show that PFA implies that all $\aleph_1$-separable groups are in standard form.

## §1. Constructions and definitions

In section IV.1 we introduced the $\Gamma$-invariant, $\Gamma(M)$, of a module $M$ (see also the beginning of §VII.2); we showed in VII.1.3 that for every stationary subset $E$ of $\aleph_1$, there is an $\aleph_1$-free group $A$ of cardinality $\aleph_1$ with $\Gamma(A) = \tilde{E}$. We did not, before now, consider the question of whether the $\Gamma$-invariant classifies the group up to isomorphism; here we'll show that, even for $\aleph_1$-separable groups of cardinality $\aleph_1$, the answer is negative.

We'll begin by describing a general construction of an $\aleph_1$-separable group with a given $\Gamma$-invariant. From this we'll obtain a series of examples by varying details of the construction.

First of all, let us establish some conventions that will remain in force throughout this chapter. All groups referred to are of cardinality at most $\aleph_1$. A *filtration* of an $\aleph_1$-separable group $A$ will be a continuous chain $\{A_\nu : \nu < \omega_1\}$ such that $A_0 = 0$, $A = \bigcup_{\nu < \omega_1} A_\nu$, and for all $\nu < \omega_1$, $A_{\nu+1}$ is a countable free direct summand of $A$ (cf. IV.1.11). Thus a filtration is a special kind of $\aleph_1$-filtration, as defined in IV.1.3. Notice that if

$$E \overset{\text{def}}{=} \{\nu \in \lim(\omega_1) : A_{\nu+1}/A_\nu \text{ is not free}\},$$

then $\Gamma(A) = \tilde{E}$ (cf. IV.1.6 and the remark following). Indeed, for any $\mu > \nu$ $A_\mu/A_\nu \cong A_\mu/A_{\nu+1} \oplus A_{\nu+1}/A_\nu$ since $A_\mu/A_{\nu+1}$ is free, so $A_\mu/A_\nu$ is not free if and only if $A_{\nu+1}/A_\nu$ is not free.

Recall that if $B$ is a subgroup of $A$, a *projection*: $A \to B$ is a homomorphism which is the identity on $B$; there is such a projection if and only if $B$ is a direct summand of $A$ (see section I.2).

Most of the examples we construct will be defined as subgroups $A$ of a $\mathbb{Q}$-module $D$ of dimension $\aleph_1$. We shall fix this module $D$ which we assume has as a basis the set $X \cup Y$ where $X = \{x_{\nu,n} : \nu \in \omega_1,\ n \in \omega\}$ and $Y = \{y_{\delta,n} : \delta \in \lim(\omega_1),\ n \in \omega\}$. The *standard filtration* of $A$ will be defined by $A_\nu = $ the pure closure in $A$ of the intersection of $A$ with

$$\{x_{\mu,n} : \mu < \nu, n \in \omega\} \cup \{y_{\delta,n} : \delta \in \nu \cap \lim(\omega_1), n \in \omega\}.$$

(Pure closure is defined in IV.2.1.) Sometimes we shall construct a second group $A'$ inside a $\mathbb{Q}$-module $D'$ which has a basis consisting of elements $x'_{\nu,n}$ and $y'_{\delta,n}$.

**1.1 Example.** Fix a stationary subset $E$ of $\lim(\omega_1)$ and a ladder system $\{\eta_\delta : \delta \in E\}$ on $E$ (cf. II.4.13). We will sometimes write

$\eta(\delta, i)$ for $\eta_\delta(i)$. Also fix a function $\psi: E \times \omega \to \omega$. (Until further notice, we can take $\psi$ to be the constant function 0.) For simplicity of notation let $x_{\eta(\delta,i)} = x_{\eta(\delta,i),\psi(\delta,i)}$ and $y_\delta = y_{\delta,0}$. For each $\delta \in E$, fix a prime $p_\delta$ and for each $n \in \omega$ let

$$z_{\delta,n} = \frac{y_\delta - w_{\delta,n}}{p_\delta^{n+1}}$$

where $w_{\delta,n} = \sum_{i=0}^n p_\delta^i x_{\eta(\delta,i)}$. For convenience, let $z_{\delta,-1} = y_\delta$. Let $A$ be the subgroup of $D$ generated (as **Z**-module) by $\{x_{\nu,n}: \nu \in \omega_1, n \in \omega\} \cup \{z_{\delta,n}: \delta \in E, n \in \omega\}$. Notice that

(1.1.1) $$p_\delta z_{\delta,n} = z_{\delta,n-1} - x_{\eta(\delta,n)}$$

for all $n \geq 0$. These relations generate all the relations among the generators of $A$. Notice that if $\delta \in E$ then $A_{\delta+1}/A_\delta = H_\delta \oplus F$ where $F$ is free with basis $\{x_{\delta,n} + A_\delta: n \in \omega\}$ and $H_\delta$ is the pure closure of $y_\delta + A_\delta$ and is isomorphic to

$$\mathbf{Q}^{(p_\delta)} \stackrel{\text{def}}{=} \{\frac{n}{m} \in \mathbf{Q}: n, m \in \mathbf{Z} \text{ and } m \text{ is a power of } p_\delta\}.$$

If $\mu < \delta$ is such that $y_\delta$ is divisible by $d$ in $A_{\delta+1}/A_\mu$ we say that $y_\delta$ is divisible by $d$ mod $\mu$ (or mod $A_\mu$). Notice that we can recover the ladder $\eta_\delta$ from the fact that $y_\delta$ is divisible by $p_\delta^{n+1}$ mod $\mu$ if and only if $\mu > \eta_\delta(n)$. (To see this, compare coefficients — members of $\mathbf{Q}$ — of the basis $X \cup Y$ of $D$.)

To see that $A$ is $\aleph_1$-free, we use Pontryagin's criterion (IV.2.3). Indeed, if $S$ is any finite subset of $A$, the pure closure of $S$ is contained in the pure closure of a finite subset $S'$ of $X \cup Y$. By enlarging $S'$ we can assume that it has the property that there exists $m$ such that for all $y_\delta \in S'$, $x_{\eta(\delta,n)} \in S'$ if and only if $n \leq m$. Then, using (1.1.1) and comparing coefficients of $X \cup Y$ in $D$, it is not hard to see that the pure closure of $S'$ is generated by the set $(X \cap S') \cup \{z_{\delta,m}: y_\delta \in S'\}$ (and, in fact, this set is even a basis).

To see that $A$ is $\aleph_1$-separable and that $\Gamma(A) = \tilde{E}$, we define for all $\nu \notin E$ a projection $\pi_\nu: A \to A_\nu$. Indeed, for every $\mu \geq \nu$

let $\pi_\nu(x_\mu) = 0$; for $\delta \in E$ with $\delta > \nu$ let $k = k_\delta$ be maximal such that $\eta_\delta(k) < \nu$ and let $\pi_\nu(y_\delta) = w_{\delta,k}$, $\pi_\nu(z_{\delta,n}) = 0$ for $n \geq k$, and $\pi_\nu(z_{\delta,n}) = \sum_{i=n+1}^{k} p_\delta^{i-n-1} x_{\eta(\delta,i)}$ for $n < k$; and, of course, $\pi_\nu \restriction A_\nu$ is the identity. This — by our remarks about the identities (1.1.1) — defines a homomorphism, which is the desired projection.

Obviously, the groups which are constructed as in 1.1 are, in some ways, not typical of $\aleph_1$-separable groups; for example, the quotients $A_{\delta+1}/A_\delta$ have the property that they have rank one modulo a free summand. But they are representative enough to be able to be used to illustrate a number of facts about $\aleph_1$-separable groups. Moreover, later we shall see that, assuming PFA, all $\aleph_1$-separable groups *are* like those constructed in 1.1, in a certain precise way (cf. 1.9 and 3.3).

Now we observe that by varying the choice of $p_\delta$ for fixed $E$, we obtain non-isomorphic groups with the same $\Gamma$-invariant.

**1.2 Example.** Choose two different primes $p$ and $q$. Let $A$ be defined as above with $p_\delta = p$ for all $\delta \in E$. Let $B$ be defined as above with $p_\delta = q$ for all $\delta \in E$. We claim that $A$ and $B$ are not isomorphic. Suppose to the contrary that there is an isomorphism $\varphi: A \to B$. Consider the standard filtrations of $A$ and $B$; then

$$C \overset{\text{def}}{=} \{\nu \in \omega_1 : \varphi[A_\nu] = B_\nu\}$$

is a cub in $\omega_1$ (cf. II.4.12). Since $E$ is stationary, there exists $\delta \in E \cap C$. Let $\nu$ be an element of $C$ which is greater than $\delta$. Then $\varphi$ induces an isomorphism: $A_\nu/A_\delta \to B_\nu/B_\delta$. But by construction $A_\nu/A_\delta \cong A_{\delta+1}/A_\delta \oplus A_\nu/A_{\delta+1} \cong \mathbb{Q}^{(p)} \oplus F$ and $B_\nu/B_\delta \cong \mathbb{Q}^{(q)} \oplus F$, where $F$ is free of rank $\omega$. This is a contradiction since these groups are not isomorphic: one contains a non-zero $p$-divisible (i.e., divisible by all powers of $p$) element, and the other does not.

This suggests defining a new invariant which describes the quotients $A_{\delta+1}/A_\delta$. Rather than define an invariant directly, we define an equivalence relation; the corresponding invariant of the group $A$ is then the equivalence class to which $A$ belongs. We will then show that this invariant does not determine the group up to isomorphism either.

**1.3 Definition.** Two groups $A$ and $B$ of cardinality $\aleph_1$ are called *quotient-equivalent* if they have filtrations $\{A_\nu : \nu \in \omega_1\}$ and $\{B_\nu : \nu \in \omega_1\}$ respectively such that for all $\nu$, $A_{\nu+1}/A_\nu \cong B_{\nu+1}/B_\nu$.

We leave it to the reader to check that this defines an equivalence relation, and to prove that, by varying the choice of the $p_\delta$, one can construct $2^{\aleph_1}$ different quotient-equivalence classes of $\aleph_1$-separable groups for any fixed $\Gamma$-invariant $\tilde{E} \neq 0$ (cf. Exercise 3 and the proof of VII.1.5).

If $A$ has the property that there is a countable non-free group $H$ such that for some filtration of $A$, every quotient $A_{\nu+1}/A_\nu$ is isomorphic to $\mathbf{Z}^{(\omega)}$ or to $H \oplus \mathbf{Z}^{(\omega)}$, we say that $A$ has *quotient-type* $H$. (In this case, the quotient-type together with the $\Gamma$-invariant of $A$ determines the quotient-equivalence class of $A$.) For example, groups constructed as in 1.1 with $p_\delta = $ a fixed prime $p$ are of quotient-type $\mathbf{Q}^{(p)}$.

In order to show that quotient-equivalence does not imply isomorphism for $\aleph_1$-separable groups, we fix $E$ and $p_\delta$ in Example 1.1 and vary $\eta$.

**1.4 Example.** In 1.1, for simplicity, let $p_\delta = p$, for some fixed prime $p$, for all $\delta \in E$. For an arbitrary $\eta$, let $A$ be constructed as in 1.1. Now let a ladder system $\{\eta'_\delta : \delta \in E\}$ be defined by $\eta'_\delta(n) = \eta_\delta(2^n)$ for all $n$, and let $A'$ be constructed as in 1.1 using $\{\eta'_\delta : \delta \in E\}$. Clearly $A$ and $A'$ are quotient-equivalent; we claim that they are not isomorphic. Indeed, suppose to the contrary that $\varphi : A \to A'$ is such an isomorphism. Then there is a cub $C$ such that for all $\mu \in C$, $\varphi[A_\mu] = A'_\mu$. Let $C^*$ be the cub of limit points of $C$ (cf. II.4.1), and let $\delta \in E \cap C^*$; so there is a strictly increasing sequence $\{\mu_n : n \in \omega\}$ of elements of $C$ whose limit is $\delta$. Let $\nu \in C$ such that $\nu > \delta$. Now $\varphi$ induces an isomorphism from $A_\nu/A_\delta$ to $A'_\nu/A'_\delta$ ($\cong \mathbf{Q}^{(p)} \oplus \mathbf{Z}^{(\omega)}$). Then the pure closure of $\varphi(y_\delta) + A'_\delta$ ($\cong \mathbf{Q}^{(p)}$) must equal $\langle z'_{\delta,n} : n \in \omega \rangle + A'_\delta$, so $\varphi(y_\delta)$ must equal $\pm p^t z'_{\delta,k} + u$ for some $t \in \omega$, $k \geq -1$, and $u \in A'_\delta$; by (1.1.1), we can assume that either $k = -1$ (i.e., $z_{\delta,k} = y_\delta$) or $t = 0$. So we have $\varphi(p^m y_\delta) = \pm p^r y'_\delta + b$ for some $m$, $r$ such that either $m$ or $r$ is zero and some $b \in A'_\delta$. Choose $n$ sufficiently large so that $b \in A'_{\mu_n}$ and there exists $j$ so that $\eta_\delta(j+r) \leq \mu_n < \eta'_\delta(j)(= \eta_\delta(2^j))$. Then $p^{j+r+1}$ divides $y_\delta \bmod \mu_n$, so $p^{j+r+1}$ divides $\varphi(y_\delta) \bmod \mu_n$, and hence $p^{j+1}$ divides $y'_\delta \bmod \mu_n$; but this is impossible by construction and the choice of $j$ and $n$.

Having discovered another obstacle to isomorphism we are led to another equivalence relation — one which will insure that, in our examples, the same ladder systems are used in constructing the groups.

**1.5 Definition.** Two groups $A$ and $B$ of cardinality $\aleph_1$ are called *filtration-equivalent* if they have filtrations $\{A_\nu : \nu \in \omega_1\}$ and $\{B_\nu : \nu \in \omega_1\}$ respectively such that for all $\nu \in \omega_1$, there is an isomorphism

$$\theta_\nu \colon A_\nu \to B_\nu$$

where $\theta_\nu[A_\mu] = B_\mu$ for all $\mu \leq \nu$. Such an isomorphism is called a *level-preserving isomorphism*. (Notice that we do *not* require that $\theta_\tau$ extend $\theta_\nu$ when $\tau > \nu$.)

We leave it as Exercise 7 to prove that this relation is an equivalence relation. (It is not quite immediate that the relation is transitive.)

Clearly if $A$ and $B$ are filtration-equivalent then they are quotient-equivalent. The next lemma tells us when our basic example produces filtration-equivalent groups.

**1.6 Lemma.** *Suppose that $A$ and $A'$ are constructed as in Example 1.1 using the same $E$, the same $\eta$, and the same primes $p_\delta$ (so that the only difference is in the functions $\psi$ and $\psi'$ used in constructing $A$ and $A'$ respectively). Then $A$ and $A'$ are filtration-equivalent.*

PROOF. We shall filter $A$ and $A'$ by the standard filtrations. It suffices to prove by induction on $\nu$ the following stronger result:

for all $\mu < \tau \leq \nu < \omega_1$, given a level-preserving iso-morphism $\theta_{\mu+1}: A_{\mu+1} \rightarrow A'_{\mu+1}$ and given a permuta-tion $\sigma$ of $\omega$, there is a level-preserving isomorphism $\theta_{\nu+1}: A_{\nu+1} \rightarrow A'_{\nu+1}$ extending $\theta_{\mu+1}$ and such that for all $i \in \omega$ $\theta_{\nu+1}(x_{\tau,i}) = x'_{\tau,\sigma(i)}$.

There is no problem with the inductive step when $\nu$ is a limit ordinal not in $E$ or $\nu$ is a successor ordinal. So suppose that $\nu = \delta$ where $\delta \in E$; we may suppose by induction that $\tau = \delta$. Suppose first that $\eta_\delta(0) > \mu$. Using the inductive hypothesis we can define by induction on $n$ a chain of level-preserving isomorphisms $\theta_n: A_{\tau_n+1} \rightarrow A'_{\tau_n+1}$ (where $\tau_n = \eta_\delta(n)$ such that $\theta_n(x_{\tau_n,\psi(\delta,n)}) = x'_{\tau_n,\psi'(\delta,n)}$. Then $\bigcup_n \theta_n$ is a level-preserving isomorphism: $A_\delta \rightarrow A'_\delta$ which clearly extends to a level-preserving isomorphism of $A_{\delta+1}$ onto $A'_{\delta+1}$ which takes $y_\delta$ to $y'_\delta$ and maps $x_{\delta,i}$ to $x'_{\delta,\sigma(i)}$. If $\eta_\delta(0) \leq \mu$, let $m$ be maximal such that $\eta_\delta(m) \leq \mu$; then we can easily verify that there is a level-preserving isomorphism of $A_{\delta+1}$ onto $A'_{\delta+1}$ which takes $y_\delta$ to $y'_\delta - w'_{\delta,m} + \theta_{\mu+1}(w_{\delta,m})$ and takes $x_{\tau_n,\psi(\delta,n)}$ to $x'_{\tau_n,\psi'(\delta,n)}$ for $n > m$. $\square$

In section 2, we shall see that under the hypothesis of Martin's Axiom, the notion of filtration-equivalence represents the end of our search; more precisely, assuming MA + ¬CH, filtration-equivalent $\aleph_1$-separable groups are isomorphic. But for now we investigate this concept assuming V = L. First, we shall vary $\psi$ in Example 1.1 to construct non-isomorphic filtration-equivalent groups.

**1.7 Example.** Let $E$, $\eta$, and $p$ be fixed, and assume $\diamondsuit(E)$. Let $A' \subseteq D'$ be constructed as in 1.1 using $p_\delta = p$ and $\psi'(\delta, i) = 0$ for all $\delta$, $i$. By $\diamondsuit(E)$ there is a family of (set) functions $h_\delta: A'_\delta \rightarrow D_\delta$ ($\delta \in E$) such that for any function $\varphi: A' \rightarrow D$, $\{\delta \in E: \varphi{\restriction}A'_\delta = h_\delta\}$ is stationary in $\omega_1$. Now we construct $A \subseteq D$ using the same $E$, $\{\eta_\delta: \delta \in E\}$, and $p$ as for $A'$; we define $\psi(\delta, i)$ by induction on $\delta$ for all $i \in \omega$. Suppose that we have defined $\psi(\gamma, i)$ for all $\gamma \in E \cap \delta$; thus $A_\delta$ is defined. We will define $\psi{\restriction}(\{\delta\} \times \omega)$ to be either the constant function 0 or the constant function 1. Thus we get two possibilities for $A_{\delta+1}$, which we denote respectively $B_0$ and $B_1$; i.e., $B_e$ is generated by $A_\delta$ together with $\{x_{\delta,n} : n \in \omega\}$ and elements

$$z_{\delta,n}^e = \frac{y_\delta - w_{\delta,n}^e}{p^{n+1}}$$

where $w_{\delta,n}^e = \sum_{i=0}^n p^i x_{\eta(\delta,i),e}$ (for $e = 0, 1$). Consider $h = h_\delta : A_\delta' \to D_\delta$. If $h$ is not an isomorphism from $A_\delta'$ onto $A_\delta$, then let $\psi(\delta, i)$ be 0. If it is an isomorphism, we claim that for at least one $e \in \{0, 1\}$, $h$ does not extend to an isomorphism of $A_\delta' + \langle z_{\delta,n}' : n \in \omega \rangle$ onto $A_\delta + \langle z_{\delta,n}^e : n \in \omega \rangle$. Then we will define $A_{\delta+1}$ to be $B_e$ for such an $e$. We will prove the claim shortly, but let us observe now that, constructed in this way, $A$ will not be isomorphic to $A'$. Indeed, for any possible isomorphism $\varphi$, there will be $\delta \in E$ such that $\varphi \upharpoonright A_\delta'$ equals $h_\delta$ and is an isomorphism; but then for all $m$, $\varphi(z_{\delta,m}')$ must be contained in $A_\delta + \langle z_{\delta,n}^e : n \in \omega \rangle (\subseteq A_{\delta+1})$ since it is $p$-divisible mod $A_\delta$; moreover, by considering $\varphi^{-1}$, we see that $\varphi$ must map $A_\delta' + \langle z_{\delta,n}' : n \in \omega \rangle$ onto $A_\delta + \langle z_{\delta,n}^e : n \in \omega \rangle$. However, by choice of $B_e$ we have ruled out that possibility.

It remains to prove the claim. Suppose that it is false, i.e., that there are isomorphisms $f_e : A_\delta' + \langle z_{\delta,n}' : n \in \omega \rangle \to A_\delta + \langle z_{\delta,n}^e : n \in \omega \rangle$ extending $h$. As in Example 1.4, there are $m$, and $r_e$ in $\omega$ and $b_e \in A_\delta$ so that

$$f_e(p^m y_\delta') = \pm p^{r_e} y_\delta + b_e.$$

Say $r_0 \leq r_1$. Choose $k \in \omega$ so that $b_0, b_1 \in A_{\eta(\delta,k)}$ and let $n = k + r_0$. Now $p^{n+1}$ divides $y_\delta' - w_{\delta,n}'$ in $A_\delta + \langle z_{\delta,n}' : n \in \omega \rangle$, so $p^{n+m+1}$ divides

$$\pm p^{r_e} y_\delta + b_e - p^m h(w_{\delta,n}')$$

in $B_e$ for $e = 0, 1$. But also $p^{n+1}$ divides $\pm (y_\delta - w_{\delta,n}^e)$ in $B_e$. Therefore, subtracting, we get that $p^{n+1}$ divides

$$b_e - p^m h(w_{\delta,n}') \pm p^{r_e} w_{\delta,n}^e$$

in $B_e$ and hence in $A_\delta$ since $A_\delta$ is pure in $B_e$. Therefore $p^{n+1}$ divides

$$b_0 - b_1 \pm p^{r_0} w_{\delta,n}^0 \pm p^{r_1} w_{\delta,n}^1$$

in $A_\delta$, so, $p^{n+1}$ divides $p^{r_0} \sum_{i=k}^n p^i (\pm x_{\eta(\delta,i),0} \pm p^{r_1-r_0} x_{\eta(\delta,n),1})$ in $A_\delta/A_{\eta(\delta,k)}$; but by considering the coefficient of $x_{\eta(\delta,k),0}$ we see that this is clearly impossible since by construction the cosets of $x_{\eta(\delta,j),e}$ ($e = 0, 1$; $j \geq k$) are members of a basis of a free summand of $A_\delta/A_{\eta_\delta(k)}$.

In fact, this construction can be done under the hypothesis that $E$ is non-small — see Exercise 8 — so, in particular, assuming $2^{\aleph_0} < 2^{\aleph_1}$ there are filtration-equivalent $\aleph_1$-separable groups that are not isomorphic. Thus, under $V = L$ or even CH, we have again reached the end of our search, but in this case it is a dead end: we are not aware of any invariant which can usefully distinguish between the groups $A$ and $A'$ in 1.7.

We can sum up our constructions so far, by saying that the choice of $E$ determines the $\Gamma$-invariant; the choice of $E$ plus the choice of the $p_\delta$'s determines the quotient-equivalence class; and those choices plus the choice of $\eta$ determines the filtration-equivalence class. However, the last statement should be taken with some caution. For Lemma 1.6, it is crucial that the $w_{\delta,n}$'s are linear combinations of the $x_{\nu,i}$'s and do not involve any $y_{\gamma,n}$'s. In the next example we see how to construct two groups which are not filtration-equivalent but which do use the same ladders. In view of Lemma 1.6, we need to depart somewhat from the type of example described in Example 1.1. The construction will, necessarily, use a principle that is not provable in ZFC.

**1.8 Example.** Fix a stationary set $E \subseteq \lim(\omega_1)$, a decomposition of $E$ as the disjoint union of two stationary sets $E_0$ and $E_1$, and a ladder system $\eta$ on $E$ so that for every cub $C$, there exists $\alpha \in E_0$ such that for all $n$, $\eta_\alpha(n) \in C \cap E_1$. The existence of such sets is implied either by $\Diamond(E_0)$ or NPA (cf. VI.1.5 and VI.4.11). Fix two distinct primes $p$ and $q$; for $\delta \in E_0$, let $p_\delta = p$ and for $\delta \in E_1$, let $p_\delta = q$. Let $\psi =$ the constant function 0. Let $A$ be constructed as in 1.1 using these data. Let $A' \subseteq D'$ be constructed as in 1.1 — using the same $E$, $p_\delta$ and $\eta_\delta$ as for $A$ — *except* that we define

$$(1.8.1) \qquad\qquad z'_{\delta,n} = \frac{y'_{\delta} - w'_{\delta,n}}{p_{\delta}^{n+1}}$$

where $w'_{\delta,n} = \sum_{i=0}^{n} p_{\delta}^{i} a_{\delta,i}$ and

$$(1.8.2) \qquad\qquad a_{\delta,i} = \begin{cases} y'_{\eta_{\delta}(i)} & \text{if } \delta \in E_0 \text{ and } \eta_{\delta}(i) \in E_1 \\ x'_{\eta_{\delta}(i)} & \text{otherwise.} \end{cases}$$

$A'$ is $\aleph_1$-separable since we can define projections as in 1.1 (so $\pi_{\nu}(x'_{\mu}) = 0$ if $\mu \geq \nu$, $\pi_{\nu}(y'_{\delta}) = w'_{\delta,k}$ if $k$ is maximal such that $\eta_{\delta}(k) < \nu$, etc.).

Note that $A$ and $A'$ are built on the same ladder system, $\{\eta_{\delta} \colon \delta \in E\}$, in the sense that for all $\delta$ and all $n$, and all $\mu < \delta$, $p_{\delta}^{n+1}$ divides $y_{\delta}$ mod $A_{\mu}$ if and only if $\mu > \eta_{\delta}(n)$ if and only if $p_{\delta}^{n+1}$ divides $y'_{\delta}$ mod $A'_{\mu}$. (It is here that we need to use two primes in defining $A'$.)

Now suppose, to obtain a contradiction, that $A$ and $A'$ are filtration-equivalent; thus there is a cub $C$ such that for every $\nu \in C$ there is a level-preserving isomorphism $\theta_{\nu} \colon A'_{\nu} \to A_{\nu}$ preserving the levels indexed by elements of $C$. (Here $A_{\nu}$ and $A'_{\nu}$ belong to the standard filtration.) Choose $\delta \in C \cap E_0$ so that for all $n$, $\eta_{\delta}(n) \in C \cap E_1$. Choose $\nu > \delta$ such that $\nu \in C$ and consider $\theta_{\nu}$; as in 1.4, there are $m, r \in \omega$ and $b \in A_{\delta}$ such that $\theta_{\nu}(p^m y'_{\delta}) = \pm p^r y_{\delta} + b$ where either $m$ or $r$ is 0. Moreover since $A$ and $A'$ are built on the same ladder system, it is easy to see that we must have $m = r = 0$. Pick $n$ so that $b \in A_{\eta_{\delta}(n)}$. Let $\gamma = \eta_{\delta}(n) \in E_1$; also $\gamma \in C$, so $\theta_{\nu}[A'_{\gamma}] = A_{\gamma}$. Now by construction $p^{n+1}$ divides $y_{\delta} + b - p^n \theta_{\nu}(y'_{\gamma})$ in $A_{\nu}/A_{\gamma}$, and also $p^{n+1}$ divides $y_{\delta} - p^n x_{\gamma,0}$ in $A_{\nu}/A_{\gamma}$. Therefore $p^{n+1}$ divides $p^n x_{\gamma,0} - p^n \theta_{\nu}(y'_{\gamma})$ mod $A_{\gamma}$. Now $A_{\nu}/A_{\gamma} = H \oplus F$, where $H$ is $q$-divisible and $F$ is free with $\langle x_{\gamma,0} + A_{\gamma} \rangle$ as a cyclic summand. But since $\theta_{\nu}(y'_{\gamma}) + A_{\gamma}$ is $q$-divisible and hence belongs to $H$, we have that $p^{n+1}$ divides $p^n x_{\gamma,0}$ in $A_{\nu}/A_{\gamma}$, which is impossible.

We want to make a definition for general $\aleph_1$-separable groups which captures the distinction between $A$ and $A'$ in the above example, namely, the nature of the elements $w_{\delta,n}$.

**1.9 Definition.** (i) An $\aleph_1$-separable group $A$ of cardinality $\aleph_1$ is said to have a *coherent system of projections* if there is a filtration $\{A_\nu : \nu \in \omega_1\}$ of $A$, and projections $\pi_\nu : A \to A_\nu$ for each $\nu \notin E \overset{\text{def}}{=} \{\nu \in \omega_1 : A_\nu$ is not a direct summand of $A\}$ with the property that for all $\nu < \tau$ in $\omega_1 \setminus E$, $\pi_\nu \circ \pi_\tau = \pi_\nu$.

(ii) $A$ is *in standard form* if it has a coherent system of projections $\{\pi_\nu : \nu \notin E\}$ relative to a filtration $\{A_\nu : \nu \in \omega_1\}$ such that for every $\delta \in E$ there is a subset $Y_\delta$ of $A_{\delta+1}$ such that $A_{\delta+1} = A_\delta + \langle Y_\delta \rangle$ and for all $y \in Y_\delta$ and all $\nu < \delta$ with $\nu \notin E$, there is a finite subset $S$ of $\nu \setminus E$ such that $\pi_\nu(y) = \sum_{\alpha \in S}(\pi_{\alpha+1}(y) - \pi_\alpha(y))$. Let $\pi_{\alpha,\alpha+1}$ denote $\pi_{\alpha+1} - \pi_\alpha$.

**1.10 Example.** Let $A$ be constructed as in Example 1.1. We use the standard filtration except for $A_{\delta+1}$ and $A_{\delta+2}$ when $\delta \in E$: let $A_{\delta+1}$ be the pure closure of

$$\{x_{\mu,n} : \mu < \delta,\, n \in \omega\} \cup \{y_\gamma : \gamma \le \delta,\, \gamma \in E\}.$$

Let $A_{\delta+2} = A_{\delta+1} + \langle\{x_{\nu,n} : \nu = \delta \text{ or } \delta+1,\, n \in \omega\}\rangle$. Note that $A_{\delta+1} = A_\delta + \langle Y_\delta \rangle$ where $Y_\delta = \{z_{\delta,n} : n \in \omega,\, \delta \in E\}$. With appropriate revisions to reflect the changed filtration, the projections defined in 1.1 form a coherent system. Then for $\nu \in \delta \setminus E$, $\pi_\nu(y_\delta) = w_{\delta,k}$ where $k$ is maximal such that $\eta_\delta(k) < \nu$. For $i \le k$, let

$$\alpha_i = \begin{cases} \eta_\delta(i) & \text{if } \eta_\delta(i) \notin E \\ \eta_\delta(i) + 1 & \text{if } \eta_\delta(i) \in E \end{cases}$$

then $\pi_{\alpha_i+1}(y_\delta) - \pi_{\alpha_i}(y_\delta) = p_\delta^i x_{\eta_\delta(i)}$. Hence

$$\pi_\nu(y_\delta) = \sum_{\alpha \in S}(\pi_{\alpha+1}(y) - \pi_\alpha(y))$$

where $S = \{\alpha_i : i \le k\}$. Similar observations apply to $\pi_\nu(z_{\delta,n})$ for all $n$, so $A$ is in standard form. We leave it to the reader to show that the group $A'$ constructed in 1.8 has a coherent system of projections but it is not in standard form — recall the construction required a principle which is not a consequence of ZFC (cf. 2.6 and 3.1).

We should also issue another warning about filtration-equivalence: if $A$ and $B$ are two $\aleph_1$-separable groups in standard form and of the same constant quotient-type $H$ *of rank one*, then they are filtration-equivalent if they are built using the same ladder system $\eta$; however this is not a sufficient condition for filtration-equivalence if $H$ has rank $\geq 2$.

**1.11.** Associated with a coherent system of projections $\{\pi_\nu : \nu \notin E\}$ for $A$ is a *coherent system of complementary summands.* For $\nu \notin E$, let $K_\nu = \ker(\pi_\nu)$; then $A = A_\nu \oplus K_\nu$. For any $\tau > \nu$ let $K_{\nu\tau} = K_\nu \cap A_\tau$; then $A_\tau = A_\nu \oplus K_{\nu\tau}$. Clearly if $\tau$ is a limit ordinal then $K_{\nu\tau} = \cup\{K_{\nu\rho} : \nu < \rho < \tau\}$. Also, if $\nu < \rho$ and $\nu, \rho \in \omega_1 \setminus E$ then by coherency $K_\nu = K_{\nu\rho} \oplus K_\rho$. Moreover, $\pi_\rho - \pi_\nu$ is the projection of $A$ onto $K_{\nu\rho}$ with respect to the decomposition $A = A_\nu \oplus K_{\nu\rho} \oplus K_\rho$; we denote this projection by $\pi_{\nu\rho}$. Using the coherency of the system of projections one can easily verify that whenever $\nu < \rho < \tau$ and $\nu, \rho \in \omega_1 \setminus E$ then $K_{\nu\tau} = K_{\nu\rho} \oplus K_{\rho\tau}$.

Suppose that $A$ is in standard form, and maintain the notation of 1.9 and 1.11. For each $\alpha \notin E$ choose a basis of $K_{\alpha,\alpha+1}$ which we denote $\{x_{\alpha,n} : n \in \omega\}$. If $t$ is a linear combination of elements of $Y_\delta$ for some $\delta$, and $d$ divides $t$ mod $A_\nu$ for some $d \in \mathbb{Z}$ and some $\nu \in \delta \setminus E$, then $d$ divides $t - \pi_\nu(t)$, and $\pi_\nu(t)$ is a linear combination of elements of $\{x_{\alpha,n} : \alpha \in \delta \setminus E, n \in \omega\}$. Thus $A_{\delta+1}$ is generated by $A_\delta$ together with all elements of the form

$$\frac{t - \pi_\nu(t)}{d}$$

It is in this sense that a group in standard form is constructed like those in Example 1.1. For a more precise discussion of this, see Mekler 1987, section 1.

It is an open question whether it is provable in ZFC that every $\aleph_1$-separable group of cardinality $\aleph_1$ has a coherent system of projections.

**1.12.** We have seen four methods for constructing an $\aleph_1$-free group of cardinality $\aleph_1$: as the union of an ascending chain of countable free groups (§VII.1); in terms of generators and relations (§VII.3); as a subgroup of a divisible group (1.1); and as a (pure) subgroup of $\mathbf{Z}^{\omega_1}$. (See also the discussion in VII.4.3.) The approach via generators and relation is perhaps the most general, since it will work in a universal algebraic setting and is not tied to cardinality considerations (i.e., generalizes to the cardinals $\kappa > \aleph_1$). On the other hand, it can be difficult to prove that such constructions define a group with the desired properties. Constructions as unions of ascending chains of free groups are also quite general and are related to the analysis of an almost free group in terms of its $\Gamma$-invariant. The second type of construction has the advantage of giving an explicit definition of the group. Since any torsion-free group is a subgroup of a direct sum of copies of $\mathbf{Q}$, the third method of construction is, in principle, quite general. Similarly any torsionless group can be embedded in a direct product of copies of $\mathbf{Z}$ and any separable group as a pure subgroup of a direct product of copies of $\mathbf{Z}$.

In practice the constructions of subgroups of $\mathbf{Z}^{\aleph_1}$ are clearest when the group is constructed as the pure closure of $\mathbf{Z}^{(\aleph_1)}$ and certain formal sums of the $e_\alpha$. For example, the group constructed in 1.1 is, up to addition of free groups, isomorphic to the pure closure of the subgroup generated by $\mathbf{Z}^{(\aleph_1)} \cup \{\sum_{n \in \omega} p^n e_{\eta(\delta,n)}: \delta \in E\}$.

## §2. $\aleph_1$-separable groups under Martin's Axiom

In this section all our results will assume MA + ¬CH. We will investigate the notions of filtration-equivalence and of standard form under this assumption. By Example 1.7, the following is not a theorem of ZFC.

**2.1 Theorem.** (MA + ¬CH) *Filtration-equivalent* $\aleph_1$*-separable groups of cardinality* $\aleph_1$ *are isomorphic.*

PROOF. Let $A$ and $A'$ be $\aleph_1$-separable groups with filtrations $\{A_\nu: \nu < \omega_1\}$ and $\{A'_\nu: \nu < \omega_1\}$, respectively, such that for every $\nu \in \omega_1$ there is a level-preserving isomorphism $\theta_\nu: A_\nu \to A'_\nu$, i.e., $\theta_\mu[A_\mu] = A'_\mu$ for every $\mu \in \nu$ (cf. Definition 1.5). Let $\mathbf{P}$ be the partial ordering whose elements are all the isomorphisms $\varphi: L \to L'$ where

$\varphi$ is a restriction of a level-preserving isomorphism $\theta_\nu \colon A_\nu \to A'_\nu$ and $L$ (resp. $L'$) is a finitely-generated pure subgroup of $A_\nu$ (resp. $A'_\nu$). Partially order $\mathbf{P}$ by: $\varphi_1 \leq \varphi_2$ if and only if $\varphi_2$ is an extension of $\varphi_1$.

For each $a \in A$ (resp. $a' \in A'$) let $D_a = \{\varphi \in \mathbf{P} \colon a \in \mathrm{dom}(\varphi)\}$ (resp. $D_{a'} = \{\varphi \in \mathbf{P} \colon a' \in \mathrm{rge}(\varphi)\}$). We claim that these sets are dense subsets of $\mathbf{P}$; we will give the argument for $D_a$. If $\psi \in \mathbf{P}$, without loss of generality $\psi$ is a restriction of the level-preserving isomorphism $\theta_{\nu+1} \colon A_{\nu+1} \to A'_{\nu+1}$. Given $a \in A$, there is $\tau > \nu$ such that $a \in A_\tau$. We claim that we can extend $\theta_{\nu+1}$ to a level-preserving isomorphism $\theta_\tau$ on $A_\tau$. If the claim is correct, we let $L$ be a finitely-generated pure subgroup of $A_\tau$ which contains $\mathrm{dom}(\psi) \cup \{a\}$ and let $\varphi = \theta_\tau {\upharpoonright} L$; then $\varphi \in D_a$ and $\psi \leq \varphi$. Now, to prove the claim, choose some level-preserving isomorphism $\tilde{\theta}_\tau$ on $A_\tau$. Then $A_\tau = A_{\nu+1} \oplus H$ for some $H$, and we can define $\theta_\tau$ to be $\theta_{\nu+1}$ on $A_{\nu+1}$ and $\tilde{\theta}_\tau$ on $H$ and easily check that it is level-preserving.

Now suppose for a moment that $\mathbf{P}$ is c.c.c. Then there is a directed subset $\mathcal{G}$ of $\mathbf{P}$ which meets all the $D_x$ ($x \in A$ or $A'$), so the union, $\cup \mathcal{G}$, of the elements of $\mathcal{G}$ is a well-defined function with domain $A$ and range $A'$. Since it is clearly an isomorphism as well, we are finished — except for the proof that $\mathbf{P}$ is c.c.c.

We need to prove that there is a cub $\mathcal{C}$ of countable elementary submodels of $(\mathrm{H}(\kappa), \in)$ — for a large enough $\kappa$ — such that for each $N \in \mathcal{C}$, $0$ is $N$-generic (cf. VI.4.5). Let $\mathcal{C}$ consist of all the countable elementary submodels $N = \bigcup_{i \in \omega} N_i$ where $N_i \prec N_{i+1} \prec N$ and $\omega_1 \cap N_i < \omega_1 \cap N_{i+1}$ for all $i \in \omega$; moreover, we require that $\mathbf{P}$, $A$, $A'$, $\{A_\nu \colon \nu \in \omega_1\}$ and $\{A'_\nu \colon \nu \in \omega_1\}$ belong to $N_0$. Let $\alpha_i = \omega_1 \cap N_i$ and $\alpha = \omega_1 \cap N$. Given $N \in \mathcal{C}$, $\varphi \in \mathbf{P}$ and a dense subset $D$ of $\mathbf{P}$ such that $D \in N$, we must show that $\varphi$ is compatible with an element of $D \cap N$. Say $\varphi \colon L \to L'$ is a restriction of the level-preserving isomorphism $\theta \colon A_\nu \to A'_\nu$ where $\nu > \alpha$. Choose $i$ so that $A_\alpha \cap L = A_{\alpha_i} \cap L$ and $D \in N_i$. For simplicity of notation let $A_i$ denote $A_{\alpha_i}$. Then the canonical map: $(L + A_i)/A_i \to A_\nu/A_{\alpha_i+1}$ is one-one so $(L + A_i)/A_i$ is free; thus $L + A_i = A_i \oplus H$ for some finitely-generated $H \subseteq A_\nu$, and there is a finitely-generated $M \subseteq$

$A_i$ such that $L \subseteq M \oplus H$. Let $\psi = \theta \restriction M$. Then $\psi \in N_i \cap \mathbf{P}$ because $N_i$ is a model of set theory and $\psi$ has an explicit definition involving finitely many elements of $A_\beta \cup A'_\beta$ for some $\beta < \alpha_i$; these elements are elements of $N_i$ because $\beta \in N_i$ so $A_\beta \in N_i$ and hence $A_\beta \subseteq N_i$ — and similarly for $A'_\beta$ (cf. VI.2.3). Therefore since $N_i$ is an elementary submodel and $D, \psi \in N_i$, there exists $\varphi' \in D \cap N_i$ such that $\psi \leq \varphi'$. It remains to show that $\varphi'$ is compatible with $\varphi$. Now $\varphi'$ is the restriction of some level-preserving isomorphism $\theta': A_\sigma \to A'_\sigma$. Choose $\tau + 1 \leq \min(\alpha_i, \sigma)$ such that $\mathrm{dom}(\varphi') \subseteq A_{\tau+1}$. If $B$ is a basis of $H$ then it is pure-independent mod $A_i$ and hence mod $A_{\tau+1}$; so there is a splitting of $A_\nu \to A_\nu/A_{\tau+1}$ which takes $b + A_{\tau+1}$ to $b$ for each $b \in B$. Therefore there is a decomposition $A_\nu = A_{\tau+1} \oplus K$ such that $H \subseteq K$. We can then define a level-preserving isomorphism $\tilde{\theta}: A_\nu \to A'_\nu$ which equals $\theta'$ on $A_{\tau+1}$ and equals $\theta$ on $K$. Then $\tilde{\theta} \restriction (\mathrm{dom}(\varphi') \oplus H)$ belongs to $\mathbf{P}$ and extends both $\varphi'$ and $\varphi$, so we are done. $\square$

**2.2 Example.** Assume MA + ¬CH. Let $A$ and $A'$ be constructed as in Example 1.1 using the same $E$, $p_\delta$ and $\eta$, but possibly different functions $\psi$ and $\psi'$. Then by 1.6, $A$ and $A'$ are filtration-equivalent, and hence isomorphic by 2.1. (Compare this to 1.7.)

It is reasonable to ask whether the classification result given in 2.1 is of any practical use. We will give some evidence for an affirmative answer in the next theorem. (See also Exercise 9.) As we shall see (XIII.4.7), this theorem is not a theorem of ZFC.

**2.3 Theorem.** (MA + ¬CH) *If $A$ is an $\aleph_1$-separable group of cardinality $\aleph_1$, then $A$ is isomorphic to $A \oplus F$, where $F$ is the free group of rank $\aleph_1$.*

PROOF. We can assume that we have chosen a filtration of $A$ so that for all $\nu$, $A_{\nu+1}/A_\nu$ has a free direct summand of rank $\aleph_0$ (cf. Exercise 13). Let $B = A \oplus F$, and write $F = \bigoplus_{\nu < \omega_1} F_\nu$ where each $F_\nu$ is free of rank $\aleph_0$. Filter $B$ by: $B_\nu = A_\nu \oplus \bigoplus_{\mu < \nu} F_\mu$. It suffices to prove that for all $\nu < \omega_1$ there is a level-preserving isomorphism from $A_{\nu+1}$ onto $B_{\nu+1}$. In fact, we prove by induction on $\nu$ the following stronger fact:

(*)

> for all $\mu < \nu < \omega_1$, given a level-preserving isomorphism $\theta_{\mu+1} \colon A_{\mu+1} \to B_{\mu+1}$ and given a decomposition $A_{\nu+1} = A_{\mu+1} \oplus K$ and elements $a_0, \ldots, a_r$ of $K$, there is a level-preserving isomorphism $\theta_{\nu+1} \colon A_{\nu+1} \to B_{\nu+1}$ extending $\theta_{\mu+1}$ such that $\theta_{\nu+1}(a_i) = a_i$ for $i = 0, \ldots, r$.

Suppose first that $A_\nu$ is a summand of $A_{\nu+1}$. Then $K \cap A_\nu$ is a summand of $K$ and $A_{\nu+1} = A_{\mu+1} \oplus (K \cap A_\nu) \oplus H$ where $H \cong A_{\nu+1}/A_\nu \cong \mathbf{Z}^{(\omega)}$. By induction we can assume that $\theta_{\mu+1}$ has been extended to a level-preserving isomorphism $\varphi \colon A_\nu \to B_\nu$, and that $a_0, \ldots, a_r$ belong to $H$. Since $H \cong H \oplus F_\nu$ there is no problem in extending $\varphi$ to an isomorphism of $A_{\nu+1} = A_\nu \oplus H$ with $B_{\nu+1} = A_{\nu+1} \oplus F_\nu = A_\nu \oplus H \oplus F_\nu$ which fixes the $a_i$ ($i = 0, \ldots, r$).

Now suppose that $A_\nu$ is not a summand of $A_{\nu+1}$. Then $\nu$ is a limit ordinal; choose a ladder $\eta$ on $\nu$ such that $\eta(0) = \mu + 1$ and for all $n$, $\eta(n)$ is a successor. For convenience let $A_n$ (resp. $B_n$) denote $A_{\eta(n)}$ (resp. $B_{\eta(n)}$). Since $A_{\nu+1}/A_\nu$ has a free summand of rank $\aleph_0$, we can write $A_{\nu+1} = A'_{\nu+1} \oplus F'$ where $A_\nu \cup \{a_0, \ldots, a_r\} \subseteq A'_{\nu+1}$ and $F' \cong \mathbf{Z}^{(\omega)}$. Then choose $K_n$ and $D_{n+1}$ by induction on $n$ for $n \geq 0$ such that $K_0 = K \cap A'_{\nu+1}$ and

$A_n \oplus K_n = A'_{\nu+1}$

$D_{n+1} \oplus K_{n+1} = K_n$; and

$A_n \oplus D_{n+1} = A_{n+1}$

(cf. Exercise 10). Let $\rho_0$ be the projection of $A'_{\nu+1}$ on $A_0$ relative to the decomposition $A'_{\nu+1} = A_0 \oplus K_0$; and let $\rho_{n+1}$ be the projection of $A'_{\nu+1}$ on $D_{n+1}$ relative to the decomposition $A'_{\nu+1} = A_n \oplus D_{n+1} \oplus K_{n+1}$. Choose a set $Y = \{y_n \colon n \in \omega\}$ such that $A'_{\nu+1} = A_\nu + \langle Y \rangle$, $y_n = a_n$ for $n \leq r$, and $y_n \in K_n$ for $n > r$.

Now define by induction a sequence of level-preserving isomorphisms $\varphi_n \colon A_n \to B_n$ such that $\varphi_0 = \theta_{\mu+1}$ and $\varphi_n(\rho_n(y_k)) = \rho_n(y_k)$ for all $k$. This holds for $n = 0$ because $\rho_0(y_k) = 0$; and the inductive step is possible by (*) because $\{\rho_n(y_k) \colon k \in \omega\}$ is a finite subset of $D_n$. Let $\varphi = \bigcup_n \varphi_n \colon A_\nu \to B_\nu = A_\nu \oplus \bigoplus_{\mu < \nu} F_\mu$. We claim that we can extend $\varphi$ to an isomorphism $\tilde{\theta}$ of $A'_{\nu+1}$ onto

$$B'_{\nu+1} \overset{\text{def}}{=} A'_{\nu+1} \oplus \bigoplus_{\mu<\nu} F_\mu$$

which is the identity on $Y$. Since $A'_{\nu+1}$ is generated by $A \cup Y$, there is at most one choice of $\tilde{\theta}$. To see that $\tilde{\theta}$ is well-defined, suppose $a = \sum_k m_k y_k$ (a finite sum; $m_k \in \mathbf{Z}$) belongs to $A_\nu$. Then $a$ belongs to $A_t$ for some $t$ and so $a = \sum_{n \leq t} \rho_n(a)$. Thus $a = \sum_k \sum_{n \leq t} m_k \rho_n(y_k)$ so

$$\varphi(a) = \varphi_t(a) = \sum_k \sum_{n \leq t} m_k \varphi_t(\rho_n(y_k)) = \sum_k \sum_{n \leq t} m_k \rho_n(y_k) = a.$$

Then we can easily extend $\tilde{\theta}$ to $\theta_{\nu+1}$ since $A_{\nu+1} = A'_{\nu+1} \oplus F'$, $B_{\nu+1} = B'_{\nu+1} \oplus F' \oplus F_\nu$ and $F' \cong F' \oplus F_\nu$. $\square$

It is possible to prove that MA + ¬CH implies that every $\aleph_1$-separable group of cardinality $\aleph_1$ is the direct sum of uncountably many non-free ($\aleph_1$-separable) subgroups (see Eklof 1983), but rather than give that proof — which is technically complicated — we shall assume an additional hypothesis on the group, namely that it is in standard form, and give a simpler proof of a stronger result:

**2.4 Theorem.** (MA + ¬CH) *Let $A$ be an $\aleph_1$-separable group of cardinality $\aleph_1$ in standard form. Suppose $\Gamma(A) = \tilde{E}$ and $E$ is the disjoint union of subsets $\{E_\beta : \beta \in \omega_1\}$ such that for all $\delta \in E_\beta$, $\delta > \beta$. Then $A$ is the direct sum, $A = \bigoplus_{\beta \in \omega_1} A^\beta$, of subgroups $A^\beta$ such that $\Gamma(A^\beta) = \tilde{E}_\beta$.*

Before proving the theorem let us make some remarks. A group which satisfies the conclusion of the theorem is said to have the *decomposition property*. In section 3 we shall prove that, assuming PFA, every $\aleph_1$-separable group (of cardinality $\aleph_1$) is in standard form; hence PFA implies that every $\aleph_1$-separable group has the decomposition property.

Given any stationary $E$, using II.4.9 it is easy to see that there is a partition, $\{E_\beta : \beta < \omega_1\}$, of $E$ as in the theorem such that all the $E_\beta$'s are stationary. The extra hypothesis on the $E_\beta$'s — that $E_\beta \subseteq (\beta, \omega_1)$ — is necessary when we are given uncountably many $E_\beta$'s , but is unnecessary otherwise (cf. Lemma VII.2.8 and the remarks surrounding it). Hence we have the following:

**2.5 Corollary.** (MA + ¬CH)

  (i) *Any $\aleph_1$-separable group of cardinality $\aleph_1$ which is in standard form is the direct sum of $\aleph_1$ non-free subgroups.*

  (ii) *If $A$ is an $\aleph_1$-separable group of cardinality $\aleph_1$ in standard form such that $\Gamma(A) = \tilde{E}$ and $E$ is the disjoint union, $E = \bigcup_{n \in \omega} E_n$, of countably many subsets, then $A = \bigoplus_{n \in \omega} A^n$ where $\Gamma(A^n) = \tilde{E}_n$ for all $n \in \omega$.* □

The following example shows that MA + ¬CH does not imply that *every* $\aleph_1$-separable group has the decomposition property. It follows that MA + ¬CH does not imply that every $\aleph_1$-separable group of cardinality $\aleph_1$ is in standard form.

**2.6 Example.** We will show that in a model of NPA there is an $\aleph_1$-separable group $A'$ such that $\Gamma(A') = \tilde{E}$, $E = E_0 \cup E_1$, $E_0 \cap E_1 = \emptyset$, and $A'$ is *not* a direct sum of subgroups $A^0$ and $A^1$ such that $\Gamma(A^\ell) = \tilde{E}_\ell$ for $\ell = 0, 1$. Let $E = E_0 \amalg E_1$ and let $\eta$ be a ladder system on $E$ such that for every cub $C$ there exists $\delta \in C \cap E_0$ such that for all $n$, $\eta_\delta(n) \in C \cap E_1$ (cf. VI.4.11). Using this ladder system, let $A'$ be the group constructed in Example 1.8, i.e., using equations (1.8.1) and (1.8.2). Suppose, to obtain a contradiction, that $A' = A^0 \oplus A^1$ where $\Gamma(A^\ell) = \tilde{E}_\ell$. Filter $A^\ell$ by: $A^\ell_\nu = A^\ell \cap A'_\nu$. Then it is easy to verify that

$$C \overset{\text{def}}{=} \{\nu \in \omega_1 \colon A'_\nu = A^0_\nu \oplus A^1_\nu\}$$

is a cub. Moreover we can assume that for all $\delta \in C$, $A^\ell/A^\ell_\delta$ is $\aleph_1$-free if and only if $\delta \notin E_\ell$. Then there exists $\delta \in C \cap E_0$ such that for infinitely many $n \in \omega$, $\eta_\delta(n) \in C \cap E_1$. Write $y'_\delta = y_0 + y_1$ where $y_\ell \in A^\ell$. Since $y'_\delta$ is $p$-divisible in $A'/A'_\delta$ and $A^1/A^1_\delta$ is $\aleph_1$-free (because $\delta \in E_0$), it must be that $y_1$ belongs to $A^1_\delta$, and hence to $A^1_\gamma$ for some $\gamma < \delta$; we can suppose that $\gamma = \eta_\delta(n) \in C \cap E_1$. Now by construction $p^{n+1}$ divides $y'_\delta - p^n y'_\gamma$ mod $A'_\gamma$. Moreover, since $y'_\gamma$ is $q$-divisible mod $A'_\gamma$ and $A^0/A^0_\gamma$ is $\aleph_1$-free (because $\gamma \in E_1$), $y'_\gamma = u_0 + u_1$ where $u_0 \in A^0_\gamma$ and $u_1 \in A^1$. Then $p^{n+1}$ divides $(y_0 - p^n u_0) + (y_1 - p^n u_1)$ mod $A'_\gamma$. Because $A' = A^0 \oplus A^1$ it follows that $p^{n+1}$ divides $y_0 - p^n u_0$ mod $A'_\gamma$. Since $u_0$ and $y_1$ belong to $A'_\gamma$, $p^{n+1}$ divides $y'_\delta$ mod $A'_\gamma$. But this contradicts the fact that $p^{n+1}$ divides $y'_\delta$ mod $A'_\mu$ if and only if $\mu > \eta_\delta(n) = \gamma$. (See also Exercise 5(ii).)

**Proof of 2.4.** Given $A$ in standard form, let $E$, $\{\pi_\nu : \nu \notin E\}$ and $\{Y_\delta : \delta \in E\}$ be as in Definition 1.9. We can assume that for all $\nu \in \delta \setminus E$, $\{y \in Y_\delta : \pi_\nu(y) \neq 0\}$ is finite. (Indeed, to achieve the latter, choose an enumeration, $\{y_n : n \in \omega\}$, of $Y_\delta$ and a ladder $\zeta_\delta$ on $\delta$, and replace $y_n$ by $y_n - \pi_{\zeta_\delta(n)+1}(y_n)$.) Then since for all $y \in Y_\delta$ and all $\nu < \delta$, there are only finitely many $\alpha \in \nu \setminus E$ such that $\pi_{\alpha,\alpha+1}(y) \neq 0$, it follows that

(†) $\qquad \{\nu \in \delta \setminus E : \pi_{\nu,\nu+1}(y) \neq 0 \text{ for some } \quad y \in Y_\delta\}$

is of order type $\omega$. (Recall that $\pi_{\alpha,\alpha+1} = \pi_{\alpha+1} - \pi_\alpha$: see 1.11.) Let $\eta_\delta$ be the ladder on $\delta$ which enumerates (†).

Now define a coloring of the ladder system $\{\eta_\delta : \delta \in E\}$ as follows:

$$c_\delta(n) = \begin{cases} \beta & \text{if } \delta \in E_\beta \text{ and } \eta_\delta(n) > \beta \\ 0 & \text{otherwise.} \end{cases}$$

Then by VI.4.6 there is a function $f : \omega_1 \to \omega_1$ which uniformizes $\{c_\delta : \delta \in E\}$, i.e., for every $\delta \in E$ there exists $n_\delta \in \omega$ such that for all $m \geq n_\delta$, $f(\eta_\delta(m)) = c_\delta(m)$. For all $\beta \in \omega_1$ let $K^\beta = \bigoplus \{K_{\alpha,\alpha+1} : \alpha \notin E, f(\alpha) = \beta\}$. (We use the notation of 1.11.) For each $\delta \in E_\beta$, let $\tau_\delta = \max\{\eta_\delta(n_\delta), \beta+1\}$ and let $Y'_\delta = \{y - \pi_{\tau_\delta}(y) : y \in \langle Y_\delta \rangle\}$. Let $A^\beta$ be the subgroup of $A$ generated by $K^\beta \cup \bigcup \{Y'_\delta : \delta \in E_\beta\}$. We claim that these subgroups are the ones that we want. Let $A^\beta_\nu = A^\beta \cap A_\nu$.

First of all, $A$ is generated by the elements of the $K_{\alpha,\alpha+1}$'s and the $Y_\delta$'s; so $A$ is the sum of the $A^\beta$'s since each $K_{\alpha,\alpha+1}$ belongs to some $K^\beta$ and for all $y \in Y_\delta$, $\pi_{\tau_\delta}(y)$ belongs to the sum of the $K_{\alpha,\alpha+1}$'s by definition of standard form.

Secondly, we claim that the sum is direct. Suppose, to the contrary that it is not, and let $\delta$ be minimal such that there exist elements $u^\beta \in A^\beta_{\delta+1}$, not all zero, such that $\sum u^\beta = 0$. Notice that for any $\nu$ there is only one $\beta$ such that $A^\beta_\nu \neq A^\beta_{\nu+1}$. Say $\gamma$ is such that $A^\gamma_\delta \neq A^\gamma_{\delta+1}$; then by the minimality of $\delta$, $u^\gamma \notin A^\gamma_\delta$. Moreover, we claim that $\delta \in E$: if not, then $u^\gamma = k + a$ where $k \in K_{\delta,\delta+1}$ and $a \in A^\gamma_\delta$; but then $k = -a + \sum_{\beta \neq \gamma} u^\beta \in A_\delta$, which implies $k = 0$, a contradiction to the minimality of $\delta$. So $\delta \in E_\gamma$ and we can write $u^\gamma = y' + a$ for some $y' \in \langle Y_\delta' \rangle$ and $a \in A^\gamma_\delta$. Now $y' = y - \pi_{\tau_\delta}(y)$ for some $y \in \langle Y_\delta \rangle$ and there exists $\nu < \delta$ such that $u^\gamma \in (\sum_{\beta \neq \gamma} A^\beta_\nu) \setminus \{0\}$; we can assume that $\tau_\delta < \nu$. Then we have:

$$
\begin{aligned}
y - \pi_{\tau_\delta}(y) + a &= \pi_\nu(y - \pi_{\tau_\delta}(y) + a) \\
&= \pi_\nu(y) - \pi_{\tau_\delta}(y) + \pi_\nu(a) \\
&= \sum_{\alpha \in S} \pi_{\alpha,\alpha+1}(y) + \pi_\nu(a)
\end{aligned}
$$

for some finite subset $S$ of $\nu \setminus (E \cup \tau_\delta)$. By the construction, $f(\alpha) = \gamma$ if $\alpha \in S$ and $\pi_{\alpha,\alpha+1}(y) \neq 0$; so, in that case, $\pi_{\alpha,\alpha+1}(y) \in K_{\alpha,\alpha+1} \subseteq K^\gamma$. Hence $u^\gamma = y - \pi_{\tau_\delta}(y) + a$ belongs to $K^\gamma \cap \sum_{\beta \neq \gamma} A^\beta_\nu$, which is contained in $A^\gamma \cap A_\delta = A^\gamma_\delta$, and we have a contradiction of the minimality of $\delta$.

Finally, since $A_{\delta+1}/A_\delta \cong A^\beta_{\delta+1}/A^\beta_\delta$ if $\delta \in E_\beta$ (because for $\gamma \neq \beta$, $A^\gamma_{\delta+1} = A^\gamma_\delta$), it follows that $\Gamma(A^\beta) = \tilde{E}_\beta \square$

By the way, 2.4 immediately implies 2.3 (since we do not require the $E_\beta$ to be stationary), so this provides another proof of 2.3 assuming PFA (given the main result, 3.3, of the next section).

## §3. $\aleph_1$-separable groups under PFA

We have seen that it is not a theorem of ZFC, or even of ZFC + MA + ¬CH, that every $\aleph_1$-separable group of cardinality $\aleph_1$ is in standard form. (See Example 2.6.) The theorem of this section is that, under the hypothesis of PFA, all $\aleph_1$-separable groups of cardinality $\aleph_1$ *are* in standard form. (Recall that PFA implies MA $+ \neg$ CH.) First we'll look at what happens to the group in Examples 1.8 and 2.6.

**3.1 Example.** Given a stationary set $E \subseteq \lim(\omega_1)$, a decomposition of $E$ as the disjoint union of stationary sets $E_0$ and $E_1$, and a ladder system $\eta$ on $E$, consider $A'$ defined as in Example 1.8, i.e., using equations (1.8.1) and (1.8.2). We saw in 2.6 that, if $E_0$, $E_1$ and $\eta$ are as in the hypothesis of 1.8, then $A'$ is not in standard form. (Such choices were possible assuming NPA or $\diamondsuit(E_0)$.) Now we will assume PFA and show that $A'$ is in standard form, no matter what the choice of $E_0$, $E_1$ and $\eta$.

By VI.4.9, there is a cub $C$ such that: (1) for all $\delta \in E$ there exists $M_\delta$ such that for all $n \geq M_\delta$, $\eta_\delta(n) \notin C$; (2) for all $\beta \in C$, $\beta^+ \notin E$ (where $\beta^+$ denotes the successor of $\beta$ in $C$); and (3) for all $\beta \in C \cap E$, $\beta^+ = \beta + 1$. We will use the filtration $\{A'_\nu : \nu \in C\}$ except that, as in 1.10, we redefine $A'_{\delta+}$ and $A'_{\delta++}$ — and the corresponding projections — when $\delta \in C \cap E$, so that $x'_{\mu,n} \in A'_{\delta++} \setminus A'_{\delta+}$ if $\delta \leq \mu < \delta^{++}$. We claim that $A'$ satisfies the definition of standard form when we take the filtration $\{A'_\nu : \nu \in C\}$, the projections $\{\pi_\nu : \nu \in C \setminus E\}$ (defined as in 1.8), and the sets $Y_\delta = \{z'_{\delta,n} : n \geq M_\delta - 1\}$ (for $\delta \in C \cap E$). Indeed, if $\delta \in C \cap E$, $\nu \in \delta \setminus E$, and $n \geq M_\delta - 1$, then

$$\pi_\nu(z'_{\delta,n}) = \sum_{i=n+1}^{k} p_\delta^{i-n-1} a_{\delta,i}$$

where $k$ is maximal such that $\eta_\delta(k) < \nu$; so it suffices to show that for each $i$ there exists a finite $S_i \subseteq C \setminus E$ such that $a_{\delta,i} = \sum_{\beta \in S_i} \pi_{\beta,\beta^+}(a_{\delta,i})$ (cf. 1.9(ii)). Suppose $\delta \in E_0$ and $\gamma \stackrel{\text{def}}{=} \eta_\delta(i) \in E_1$ (the other cases are even easier); then $a_{\delta,i} = y'_{\gamma,0}$, and by choice of $C$ there exists a maximal $\beta < \gamma$ such that $\beta \in C$; then, by (3), $\beta \notin E$. If $m$ is maximal such that $\eta_\gamma(m) < \beta$, then since $\beta^+ > \gamma$,

$$y'_{\gamma,0} = \pi_{\beta^+}(y'_{\gamma,0}) = \pi_{\beta,\beta^+}(y'_{\gamma,0}) + \pi_\beta(y'_{\gamma,0}) = \pi_{\beta,\beta^+}(y'_{\gamma,0}) + \sum_{j=0}^{m} q^j x'_{\eta_\gamma(j),0}.$$

For each $j \leq m$, let $\tau_j$ be the largest element of $C$ which is $\leq \eta_\gamma(j)$; if $\tau_j \in E$, let $\alpha_j = \tau_j^+$, and otherwise let $\alpha_j = \tau_j$. Then we can take $S_i = \{\alpha_j : j \leq m\} \cup \{\beta\}$.

In the general case we must define the projections $\pi_\nu$ and the sets $Y_\delta$ along with the cub $C$.

**3.2 Definition.** Given a filtration $\{A_\nu : \nu < \omega_1\}$ of an $\aleph_1$-separable group $A$, let $E$ be defined as in 1.9. For any ordinal $\alpha$ and any subset $C$ of $\alpha + 1$, a *coherent system of projections for $A_\alpha$ w.r.t. $C$* is a set $\{\pi_\nu : \nu < \alpha, \nu \in C \setminus E\}$ such that each $\pi_\nu$ is a projection: $A_\alpha \to A_\nu$ and for all $\nu < \tau < \alpha$ in $C \setminus E$, $\pi_\nu \circ \pi_\tau = \pi_\nu$. If $C = \alpha$ or $\alpha + 1$, we omit reference to it.

**3.3 Theorem.** (PFA) *Every strongly $\aleph_1$-free group of cardinality $\aleph_1$ is $\aleph_1$-separable and in standard form.*

PROOF. Let $A$ be an $\aleph_1$-separable group of cardinality $\aleph_1$, and fix a filtration $\{A_\nu : \nu < \omega_1\}$ of $A$. Let $E$ be as in 1.9. Our partial order **P** will consist of all triples $p = (C, \Pi, \mathcal{Y})$ such that:

(a) $C$ is a countable closed set (so, in particular, $\alpha \stackrel{\text{def}}{=} \sup C$ belongs to $C$); $\alpha \notin E$; and for all $\beta \in C \setminus \{\alpha\}$, $\beta^+$ (the successor of $\beta$ in $C$) does not belong to $E$;

(b) $\Pi = \{\pi_\nu : \nu \in C \setminus E\}$ is a coherent system of projections for $A_\alpha$ w.r.t. $C$; and

(c) $\mathcal{Y} = \{Y_\delta : \delta \in C \cap E\}$ where for all $\delta \in C \cap E$, $A_{\delta+1} = A_\delta + Y_\delta$ and for all $y \in Y_\delta$ and $\nu \in (C \setminus E) \cap \delta$, there is a finite set $S \subseteq C \setminus E$ such that $\pi_\nu(y) = \sum_{\beta \in S} \pi_{\beta, \beta^+}(y)$.

In other words, $p$ is a countable approximation to what we are seeking. Define the partial ordering on **P** by: $p = (C, \Pi, \mathcal{Y}) \leq p' = (C', \Pi', \mathcal{Y}')$ if and only if $C = C' \cap \alpha$ (where $\alpha = \sup C$), $\pi_\nu = \pi'_\nu {\restriction} A_\alpha$ for all $\nu \in C \setminus E$, and $Y_\delta = Y'_\delta$ for all $\delta \in C \cap E$.

For each $\mu \in \omega_1$, let $D_\mu = \{(C, \Pi, \mathcal{Y}) \in \mathbf{P} : \mu \leq \sup C\}$. We claim that $D_\mu$ is dense in **P**. Indeed, given $p = (C, \Pi, \mathcal{Y})$, if $\alpha = \sup C$, choose $K$ so that $A_\alpha \oplus K = A_{\mu+1}$ — remember that $\alpha \notin E$. Let $C' = C \cup \{\mu + 1\}$; let $\pi'_\alpha$ be the projection of $A_{\mu+1}$ on $A_\alpha$ along

$K$; let $\pi'_\nu = \pi_\nu \circ \pi'_\alpha$ for $\nu \in C \setminus E$. Then if $\Pi' = \{\pi'_\nu : \nu \in C' \setminus E\}$ and $p' = (C', \Pi', \mathcal{Y})$, $p'$ is an element of $D_\mu$ and $p \leq p'$.

If **P** is proper, then there exists a directed subset $\mathcal{G}$ of **P** which meets all the $D_\mu$. If we let $\tilde{C} = \cup\{C : (C, \Pi, \mathcal{Y}) \in \mathcal{G}\}$, then $\tilde{C}$ will be a cub in $\omega_1$. (It is unbounded because $\mathcal{G}$ meets all the $D_\mu$; it is closed because if $\gamma = \sup X$ for some countable $X \subseteq \tilde{C}$, then there exists $(C, \Pi, \mathcal{Y}) \in \mathcal{G}$ such that $\sup C > \gamma$ and by definition of $\leq$, $X \subseteq C$, so $\gamma \in C$ since $C$ is closed.) For each $\nu \in \tilde{C} \setminus E$, $\mathcal{G}$ determines a well-defined projection $\pi_\nu : A \to A_\nu$. Also for each $\delta \in \tilde{C} \cap E$, we have a uniquely determined $Y_\delta$. It is then clear that, with respect to the filtration $\{A_\nu : \nu \in \tilde{C}\}$, these data show that $A$ is in standard form.

So it remains to show that **P** is proper. Let the cub $\mathcal{C}$ of countable elementary submodels $N$ be defined as in the proof of 2.1, and, as there, let $\alpha_i = \omega_1 \cap N_i$ and $\alpha = \omega_1 \cap N$. We must prove that given $p_0 = (C_0, \Pi_0, \mathcal{Y}_0)$ in $N \cap \mathbf{P}$, there exists $q \geq p_0$ which is $N$-generic (cf. VI.4.8). Enumerate the dense subsets of **P** which belong to $N$ as $\{D_n : n \in \omega\}$. Our procedure will be to define inductively a chain

$$p_0 \leq p_1 \leq \ldots \leq p_n = (C_n, \Pi_n, \mathcal{Y}_n) \leq \ldots$$

of elements of **P** such that $p_{n+1} \in D_n \cap N_n$ (possibly after choosing a subsequence of the $N_n$'s and renumbering). Assuming we have done this, let $p' = (C', \Pi', \mathcal{Y}')$ where $C' = \bigcup_n C_n$, $\alpha = \sup C'$, $\mathcal{Y}' = \bigcup_n \mathcal{Y}_n$, and $\Pi'$ is defined to be the set of projections $\pi'_\nu : A_\alpha \to A_\nu$ (for $\nu \in C' \setminus E$) such that $\pi'_\nu \restriction A_{\sup C_n} = \pi^n_\nu$ if $\nu \leq \sup C_n$. (Here $\pi^n_\nu$ denotes the obvious element of $\Pi_n$.) Now $\alpha \notin C'$, so $C'$ is not closed, and $p' \notin \mathbf{P}$. As long as $q$ is an element of **P** which is $\geq p'$ (in the obvious sense), then $q$ is $N$-generic. If $\alpha \notin E$, we can let $q = (C' \cup \{\alpha\}, \Pi', \mathcal{Y}')$. However, if $\alpha \in E$, we must take more care to be sure that there exists $q \geq p'$ in **P**.

Assume that $\alpha \in E$. Enumerate $A_{\alpha+1}$ as $\{a_n : n \in \omega\}$. By induction, we will define $p_n = (C_n, \Pi_n, \mathcal{Y}_n)$ and a finitely-generated subgroup $G_n$ of $A_{\alpha+1}$ such that if $\beta_n = \sup C_n$, then:

(1) $\beta_n < \beta_{n+1}$;
(2) $G_n \cap A_{\beta_n} = \{0\}$;

(3) $a_n \in A_{\beta_{n+1}} \oplus G_{n+1}$; and

(4) $G_n \subseteq K^{n+1}_{\beta_n, \beta_n^+} + G_{n+1} = K^{n+1}_{\beta_n, \beta_n^+} \oplus G_{n+1}$.

(Here $K^{n+1}_{\beta_n, \beta_n^+} = \ker(\pi^{n+1}_{\beta_n}) \cap A_{\beta_n^+} = \mathrm{rge}(\pi^{n+1}_{\beta_n \beta_n^+})$ : cf. 1.11).

Suppose for a moment that we can do this. Let $p'$ be as above. We will define $q \geq p'$ so that the first coordinate of $q$ is $C' \cup \{\alpha, \alpha+1\}$ and the third coordinate is $\mathcal{Y}' \cup \{Y_\alpha\}$, where $Y_\alpha = \bigcup_n G_n$. (Notice that (3) implies that $A_{\alpha+1} = A_\alpha + Y_\alpha$.) To define the second coordinate of $q$, it will suffice to define $\pi_{\beta_n} : A_{\alpha+1} \to A_{\beta_n}$, for then we will define $\pi_\nu$ to be $(\pi'_\nu \restriction A_{\beta_n}) \circ \pi_{\beta_n}$ if $\nu < \beta_n$; of course, we must do this so that the projections are coherent. By (3) and (4), $A_{\alpha+1} = A_\alpha + \bigcup_{m \geq n} G_m$. Note also that (c) will hold for $y \in Y_\alpha$ because of (4). We define $\pi_{\beta_n}$ to be $\pi'_{\beta_n}$ on $A_\alpha$ and to be identically zero on $\bigcup_{m \geq n} G_m$; To see that it is well-defined, suppose that for some $m \geq n$, $\sum_{i=n}^m g_i \in A_\alpha$ where $g_i \in G_i$. We must show that $\pi'_{\beta_n}(\sum_{i=n}^m g_i) = 0$. Now $\sum_{i=n}^m g_i \in A_{\beta_r}$ for some $r$, and we can assume that $r = m$ and also, by (4), that $\sum_{i=n}^m g_i = g + u$ where $g \in G_m$ and $u \in K^m_{\beta_n \beta_{m-1}^+}$. But then by (2), $g = 0$, so $\sum_{i=n}^m g_i = u$ belongs to $\ker(\pi'_{\beta_n})$. Finally, (4) allows us to verify the coherency condition.

So it remains to do the inductive construction. Let $G_0 = \{0\}$. Now suppose that we have defined $p_i$ and $G_i$ for $i \leq n$ satisfying (1) – (4), and suppose also that

(2′) $G_n \subseteq H_n$, where $A_{\alpha+1} = A_{\beta_n} \oplus H_n$.

Choose a finitely-generated $G \supseteq G_n$ such that $G \subseteq H_n$ and $a_n \in A_{\beta_n} + G$. By choosing a subsequence of the $N_n$'s and renumbering, we can assume that $p_n$ and $D_n$ belong to $N_n$ and that $G \cap A_\alpha = G \cap A_{\alpha_n}$. Let $M$ be the pure closure of $G + (A_{\alpha_n} \cap H_n)$ in $H_n$. Then $M/(A_{\alpha_n} \cap H_n)$ is of finite rank, and the canonical map of $M/(A_{\alpha_n} \cap H_n)$ into $A_{\alpha+1}/A_{\alpha_n+1}$ (which is free) is injective, so $M = (A_{\alpha_n} \cap H_n) \oplus G_{n+1}$ for some finitely-generated $G_{n+1}$. Choose a finitely-generated $L \subseteq (A_{\alpha_n} \cap H_n)$ such that $G \subseteq L \oplus G_{n+1}$ and choose $\gamma$ such that $\beta_n < \gamma < \alpha_n$, $\gamma \notin E$, and $L \subseteq A_\gamma \cap H_n$. Now, in $H(\kappa)$ there is a projection $\pi : A_\gamma \to A_{\beta_n}$ such that $L \subseteq \ker(\pi)$, namely the projection along $A_\gamma \cap H_n$. So since $N_n$ is an elementary submodel of $H(\kappa)$ and $L$ is finitely-generated, there is such a projection, $\pi_{\beta_n}$, which belongs to $N_n$. If we define $\tilde{\Pi} = \{\pi_\nu^n \circ \pi_{\beta_n} : \nu \in C_n \setminus E\} \cup \{\pi_{\beta_n}\}$, and let $\tilde{p} = (C_n \cup \{\gamma\}, \tilde{\Pi}, \mathcal{Y}_n)$, then $\tilde{p} \in N_n \cap \mathbf{P}$. Thus there exists $p_{n+1} \in D_n \cap N_n$ such that $p_{n+1} \geq \tilde{p} \ (\geq p_n)$. It is easy to see that (1) and (3) hold, and (4) holds by the choice of $\gamma$ and since $\gamma = \beta_n^+$, the successor of $\beta_n$ in $C_{n+1}$. To finish the proof we must choose $H_{n+1} \supseteq G_{n+1}$ so that (2') holds. Now $M$ is a summand of $H_n$, since it is a pure subgroup of finite rank, so

$$A_{\alpha+1} = A_{\beta_n} \oplus M \oplus H' = A_{\beta_n} \oplus (A_{\alpha_n} \cap H_n) \oplus G_{n+1} \oplus H'$$

for some $H'$. Also $\beta_{n+1} < \alpha_n$, because $p_{n+1} \in N_n$, so $A_{\beta_{n+1}}$ is contained in (and is therefore a summand of) $A_{\beta_n} \oplus (A_{\alpha_n} \cap H_n)$; hence $G_{n+1}$ is contained in the desired complementary summand of $A_{\beta_{n+1}}$. $\square$

Compare the following with Example 2.6.

**3.4 Corollary.** (PFA) *Every $\aleph_1$-separable group of cardinality $\aleph_1$ satisfies the decomposition property.*

PROOF. This is an immediate consequence of 3.3 and 2.4. $\square$

Finally, we observe that (as in VI.4.13) there is a model of ZFC + GCH in which we get a mix of the phenomena that occur with $\diamondsuit$ and with PFA, depending on the $\Gamma$-invariant involved:

**3.5 Theorem.** $(\mathrm{Ax}(S) + \diamondsuit^*(\omega_1 \setminus S))$ *There is a stationary and co-stationary set $S$ such that*

(1) *if $A$ is strongly $\aleph_1$-free of cardinality $\aleph_1$ and $\Gamma(A) \leq \tilde{S}$, then $A$ is $\aleph_1$-separable and in standard form, and hence has the decomposition property; and*

(2) *for any $E$ such that $E \setminus S$ is stationary, there exists an $\aleph_1$-separable group $A'$ of cardinality $\aleph_1$ such that $\Gamma(A') = \tilde{E}$ and $A'$ is not in standard form and does not have the decomposition property.*

PROOF. (1) We use the same poset $\mathbf{P}$ as in the proof of 3.3; the proof there shows that $\mathbf{P}$ is proper and $(\omega_1 \setminus S)$-complete.

(2) Since $\diamondsuit(E)$ holds, we can use the same construction as in 1.8 and 2.6 and show that $A'$ is not in standard form and does not have the decomposition property. $\square$

EXERCISES

1. (i) A subgroup of a strongly $\aleph_1$-free group is strongly $\aleph_1$-free.
(ii) A direct summand of an $\aleph_1$-separable group is $\aleph_1$-separable. [Hint: use (i).]

2. Let $R$ be a ring which is not left perfect. Let $a_0, a_1, \ldots$ be as in the proof of VII.1.1. Let $E$ be a stationary subset of $\omega_1$ such that $E \subseteq \lim(\omega_1)$; for each $\delta \in E$ choose a ladder system on $\delta$ whose range consists of successor ordinals. Let $P = \prod_{\alpha < \omega_1} Re_\alpha$. Let $A$ be the subgroup of $P$ generated by $\bigoplus_{\alpha < \omega_1} Re_\alpha$ together with

$$\{\sum_{i \geq m} (a_m \cdots a_i)e_{\eta_\delta(i)} : \delta \in E, \, m \in \omega\}.$$

Show that every countable subset of $A$ is contained in a $\leq \aleph_0$-generated direct summand of $A$, but $A$ is not a direct sum of $\leq \aleph_0$-generated modules. [Hint: for any ordinal $\beta$, if $\rho_\beta$ is the projection of $P$ onto $\prod_{\alpha < \beta} Re_\alpha$, then $\rho_\beta[A]$ is a summand of $A$; but for all $\delta \in E$, $A \cap \prod_{\alpha < \delta} Re_\alpha$ is not a summand of $A$ (cf. IV.1.7).]

*In the remainder of these exercises, all groups are $\aleph_1$-separable groups of cardinality $\aleph_1$.*

3. $A$ and $B$ are quotient-equivalent if and only if for any given filtrations $\{A_\nu : \nu \in \omega_1\}$ and $\{B_\nu : \nu \in \omega_1\}$ of $A$ and $B$ respectively, there is a cub $C$ such that for all $\delta \in C$, $(A_{\delta+1}/A_\delta) \oplus \mathbf{Z}^{(\omega)} \cong (B_{\delta+1}/B_\delta) \oplus \mathbf{Z}^{(\omega)}$.

4. For any stationary $E \subseteq \omega_1$ there are $2^{\aleph_1}$ groups with $\Gamma$-invariant $\tilde{E}$ such that any two are not quotient-equivalent. [Hint: use II.4.9 and the idea of the proof of the last part of II.4.9.]

5. (i) Suppose $A$ and $B$ are groups of quotient-type $\mathbf{Q}^{(p)}$ with filtrations $\{A_\nu : \nu \in \omega_1\}$ and $\{B_\nu : \nu \in \omega_1\}$, respectively, such that there is a stationary set $E$ so that $A_{\nu+1}/A_\nu$ is non-free if and only

if $\delta \in E$ if and only if $B_{\nu+1}/B_\nu$ is non-free. Moreover assume that for each $\delta \in E$, there are $y_\delta \in A_\delta$ and $y'_\delta \in B_\delta$ such that $\langle y_\delta, A_\delta \rangle_* / A_\delta \cong \mathbb{Q}^{(p)} \cong \langle y_\delta, B_\delta \rangle_* / B_\delta$ and such that for all $\mu < \delta$ and all $n$, $p^n$ divides $y_\delta$ mod $A_\mu$ if and only if $p^n$ divides $y'_\delta$ mod $B_\mu$. Prove that $A$ is filtration-equivalent to $B$. [Hint: generalize the proof of 1.6; the hypothesis on the quotient-type is used to show that the "witnesses" to divisibility of $y_\delta$ and $y'_\delta$ are like $x_{\nu,i}$'s.] Conclude that, assuming MA $+ \neg$CH, every group of quotient-type $\mathbb{Q}^{(p)}$ is in standard form.

(ii) Use (i) to give an alternate proof that $A'$ in 2.6 has the desired property: let $A$ be as in 1.8; by 2.4, $A = B^0 \oplus B^1$ where $\Gamma(B^\ell) = E_\ell$. Suppose $A' = A^0 \oplus A^1$ where $\Gamma(A^\ell) = E_\ell$; show $A^i$ is filtration-equivalent to $B^i$. So $A$ is filtration-equivalent to $A'$, which contradicts 1.8.

6. Let $R$ be the group of all rationals with square-free denominators. Prove that, assuming NPA, there is a group of quotient-type $R$ which does not have the decomposition property and hence is not in standard form.

7. (i) If $C$ is a cub, say that $A$ and $B$ are *C-filtration-equivalent* if they have filtrations $\{A_\nu : \nu \in \omega_1\}$ and $\{B_\nu : \nu \in \omega_1\}$ such that for all $\nu \in C$, there is an isomorphism $\theta_\nu : A_\nu \to B_\nu$ such that $\theta_\nu[A_\mu] = B_\mu$ for all $\mu \in C$. Prove that if $A$ and $B$ are $C$-filtration-equivalent, then they are filtration-equivalent. [Hint: Note that $\{A_\nu : \nu \in C\}$ may not be a filtration in the sense of this chapter; use Exercise IV.3.]
(ii) Using (i), prove that the relation of filtration-equivalence is transitive.

8. If $E$ is non-small, there are filtration-equivalent groups of type $\mathbb{Q}^{(p)}$ with $\Gamma$-invariant $\tilde{E}$ that are not isomorphic. [Hint: adapt the proof of 1.7, using $\Phi_{\omega_1}(E)$ instead of $\Diamond(E)$.]

9. If $A$ is a subgroup of $B$ such that $B/A$ is a direct sum of a countable group and a free group, then $A$ is filtration-equivalent to $B$ (and hence isomorphic to $B$ assuming MA $+ \neg$ CH). [Hint:

reduce to the case when $B/A$ is countable; choose filtrations such that $B = A + B_0$, $B_\nu = A_\nu + B_0$, and $B_0 \cap A \subseteq A_0$; then mimic the proof of 2.3.]

10. If $A \supseteq \bigcup_{n \in \omega} A_n$ where $A_n \subseteq A_{n+1}$ and $A_n$ is a summand of $A$ for all $n$, then there exist subgroups $C_n$ and $D_{n+1}$ such that for all $n \in \omega$: $A_n \oplus C_n = A$; $D_{n+1} \oplus C_{n+1} = C_n$; and $A_n \oplus D_{n+1} = A_{n+1}$. [Hint: define them by induction: given $C_n$, let $D_{n+1} = C_n \cap A_{n+1}$; then $D_{n+1}$ is a summand of $C_n$ since $A_{n+1}$ is a summand of $A$.]

11. Let $A$ be an $\aleph_1$-separable group with filtration $\{A_\nu : \nu < \omega_1\}$. If $\alpha < \beta$ and $\Pi$ is a coherent system of projections for $A_{\alpha+1}$ (cf. 3.2), then $\Pi$ extends to a coherent system $\Pi'$ of projections for $A_{\beta+1}$. In particular, for all $\alpha$, $A$ has a coherent system of projections for $A_{\alpha+1}$.

12. Formulate and prove analogs of 2.1 and 2.3 for $\aleph_1$-separable $p$-groups.

13. For any filtration $\{A_\nu : \nu \in \omega_1\}$ of $A$, there is a continuous increasing function $f : \omega_1 \to \omega_1$ such that $\{A_{f(\nu)} : \nu \in \omega_1\}$ is a filtration and for all $\nu$, $A_{f(\nu+1)}/A_{f(\nu)}$ has a free summand of rank $\aleph_0$. [Hint: use that fact that $A_\mu/A_{\nu+1}$ is free.]

## NOTES

Fuchs 1960, Problem 29, asked about non-trivial $\kappa$-separable groups. Hill 1969 constructed $\aleph_1$-separable torsion groups which were not the direct sum of cyclic groups; Griffith 1969b extended his construction to torsionless modules over a ring which is not left perfect (see Exercise 2); in particular, this construction yields $\aleph_1$-separable groups of cardinality $\aleph_1$ which are not free.

The notion of quotient-equivalent groups is discussed in Eklof 1980. A construction of families of quotient-equivalent groups which are pairwise strongly non-isomorphic is given in Eklof-Mekler-Shelah 1984 and the logical significance of quotient-equivalence is discussed in Eklof-Mekler 1981. The notion of filtration-equivalence is introduced in Eklof 1983, from which comes also 1.6, 1.7, 1.8, 2.1, 2.3, 2.6, and 3.4. (The proof of 2.3 is based on that in Eklof-Mekler 1983.)

Definitions 1.9 and 1.11 are from Mekler 1983 which also announces 2.4, 2.5 and 3.3; full proofs are given in Mekler 1987. The latter actually deals with the much more general class of "Fuchs 5" groups, that is, groups such that every countable subset is contained in a countable direct summand.

Dugas-Irwin 1989 prove that it is independent of ZFC whether every $\aleph_1$-separable group of cardinality $\aleph_1$ is embeddable in $\mathbf{Z}^\omega$. See also Dugas-Irwin 19??.

$\aleph_1$-separable $p$-groups are studied in Huber 1983b, Eklof-Mekler 1983, and Megibben 1987. In the latter it is proved that a direct summand of an $\aleph_1$-separable $p$-group is $\aleph_1$-separable. $\aleph_1$-separable valued vector spaces are studied in Eklof-Huber 1985 and 1988. The Kaplansky test problems for $\aleph_1$-separable $\aleph_1$-free groups are considered in Thomé 19??a and 19??b. Constructions of $\aleph_1$-separable groups (in the more general sense) of mixed type are given in Mekler 1981. For $\kappa$-separable $p$-groups, see Rychkov 1988.

Exercises. 2:Griffith 1969b; 5,6,8,9:Eklof 1983; 10:Mekler 1982; 12:Eklof-Mekler 1983.

# CHAPTER IX
## QUOTIENTS OF PRODUCTS
## OF THE INTEGERS

We shall say that an abelian group $N$ is a *product* if $N$ is isomorphic to $\mathbb{Z}^\lambda$ for some cardinal $\lambda$. In this chapter we concern ourselves primarily with homomorphic images of products. In the first section we deal with the question of when such a homomorphic image is itself a product. It turns out that a sufficient, and in some cases necessary, condition for $\mathbb{Z}^\lambda/A$ to be a product is that $A$ be a direct summand of $\mathbb{Z}^\lambda$. In section 2 we consider arbitrary quotients of $\mathbb{Z}^\omega$; this leads us to Nunke's characterization of slender groups. In the third section we consider quotients of uncountable products of $\mathbb{Z}$, where current knowledge is more fragmentary. In the last section, we deal with a related subject, that of radicals of groups and how their properties are affected by large cardinal hypotheses.

Throughout this chapter $M^*$ will denote the dual group $\mathrm{Hom}_{\mathbb{Z}}(M, \mathbb{Z})$, and "reflexive" will mean $\mathbb{Z}$-reflexive.

## §1. Perps and products

We begin by investigating necessary and sufficient conditions for a subgroup $A$ of a group $M$ to be a direct summand of $M$ — conditions which arise from the fact that if $A$ is a summand of $M$ then $A^*$ may be regarded as a summand of $M^*$. In this connection we introduce the convenient "perp" notation which we shall use repeatedly later.

If $A$ is a subgroup of $M$ and $\iota: A \to M$ is the inclusion map, denote by $\iota^*$ the induced map: $M^* \to A^*$ which takes $y \in M^*$ to its restriction, $y \circ \iota$, to $A$. Let $A^\perp$ — read "$A$ perp" — denote the kernel of $\iota^*$. Thus

$$A^\perp = \{y \in M^*: \langle y, a \rangle = 0 \text{ for all } a \in A\}.$$

The canonical projection of $M$ onto $M/A$ induces an embedding of $(M/A)^*$ into $M^*$ whose image is $A^\perp$; thus $A^\perp$ is canonically isomorphic to $(M/A)^*$. Hence if $M = A \oplus B$ we can identify $A^\perp$ with $B^*$ via the map which takes $y \in A^\perp$ to its restriction to $B$. In this case it is easy to check that $M^*$ is the internal direct sum of $B^\perp$ and $A^\perp$; indeed, for all $y \in M^*$, $y = y{\restriction}A + y{\restriction}B$, where — in an abuse of notation — $y{\restriction}B$ denotes the element of $A^\perp$ which agrees with $y$ on $B$, and similarly for $y{\restriction}A$. In the same spirit, if $M = A \oplus B$ we shall write $M^* = A^* \oplus B^*$ and regard the elements of $B^*$ $(= A^\perp)$ as either functions on $B$ or as functions on $M$ which are zero on $A$.

Now $A^\perp$ is a subgroup of $M^*$ so we can define $(A^\perp)^\perp$, a subgroup of $M^{**}$. In an abuse of notation we shall denote by $A^{\perp\perp}$ the pre-image of $(A^\perp)^\perp$ under $\sigma_M : M \to M^{**}$. Thus

$$A^{\perp\perp} = \{x \in M : \forall y \in A^\perp (\langle y, x \rangle = 0)\}.$$

Clearly $A \subseteq A^{\perp\perp}$. It is not hard to see that $M/A^{\perp\perp}$ is always torsionless.

**1.1 Lemma.** (i) *If $M/A$ is torsionless, then $A = A^{\perp\perp}$. In particular, if $M$ is torsionless and $A$ is a summand of $M$, then $A = A^{\perp\perp}$.*
   (ii) *If $A^{\perp\perp}$ is a direct summand of $M$, then $(A^{\perp\perp}/A)^* = 0$.*

PROOF. (i) Suppose $x \in M \setminus A$. Since $M/A$ is torsionless there exists $y \in M^*$ such that $y \in A^\perp$ and $\langle y, x \rangle \neq 0$, so $x \notin A^{\perp\perp}$. If $M$ is torsionless and $A$ is a summand of $M$, then $M/A$ is torsionless since it is isomorphic to a summand of $M$.

   (ii) Say $M = A^{\perp\perp} \oplus D$. If $f \in (A^{\perp\perp}/A)^*$ and $\pi$ is the canonical projection: $A^{\perp\perp} \to A^{\perp\perp}/A$, then $f \circ \pi \in (A^{\perp\perp})^* = D^\perp \subseteq M^*$. But $f \circ \pi \in A^\perp$, so by the definition of $A^{\perp\perp}$, $f \circ \pi \equiv 0$; since $\pi$ is surjective, this implies that $f$ is zero. $\square$

In view of this lemma, we seek conditions that imply that $A^{\perp\perp}$ is a direct summand of $M$.

**1.2 Lemma.** *Suppose $M$ is reflexive and $A$ is a subgroup of $M$ such that $M^* = A^\perp \oplus N$. Then $M = A^{\perp\perp} \oplus C$ where $C \cong (A^\perp)^*$, and $A^{\perp\perp} \cong N^*$.*

PROOF. By the remarks at the beginning of this section, $M^{**} = (A^\perp)^\perp \oplus N^\perp$. But then, since $M$ is reflexive, this decomposition of $M^{**}$ pulls back to a decomposition of $M: M = A^{\perp\perp} \oplus C$, where $C$ is isomorphic to $N^\perp$. The final assertions then follow from the identification of $N^\perp$ with $(A^\perp)^*$ and of $(A^\perp)^\perp$ with $N^*$. □

Let us denote the group $\mathbf{Z}^\omega$ by $P$. Say that $A$ is a *countable product* if it is isomorphic to $P$ or to $\mathbf{Z}^n$ for some finite $n$. By III.2.5, $P$ is reflexive.

**1.3 Proposition.** *For any subgroup $A$ of $P$, $A^{\perp\perp}$ is a direct summand of $P$. Moreover, $A^{\perp\perp}$ and $P/A^{\perp\perp}$ are both countable products.*

PROOF. By Lemma 1.2 it's enough to show that $A^\perp$ is a direct summand of $P^*$. By definition of $A^\perp$ we have a short exact sequence

$$0 \to A^\perp \to P^* \xrightarrow{\iota^*} B \to 0$$

where $B$ is the image of $P^*$ under the map $\iota^*: P^* \to A^*$ induced by inclusion. Now $P^*$ is a countable free group (cf. before III.2.5); so $B$ is a countable subgroup of $A^*$ and hence is free, since $A^*$ is $\aleph_1$-free by IV.2.10. Therefore the short exact sequence splits, and $A^\perp$ is a direct summand of $P^* : P^* = A^\perp \oplus N$ where $N \cong B$. By 1.2 $P = A^{\perp\perp} \oplus C$ where $C \cong (A^\perp)^*$ and $A^{\perp\perp} \cong B^*$; hence $A^{\perp\perp}$ and $C$ are countable products since $B$ and $A^\perp$ are countable free groups. □

Proposition 1.3 fails to hold for uncountable products: see Exercise 1.

**1.4 Theorem.** *For any cardinal $\kappa$ which is not $\omega$-measurable and any subgroup $A$ of $\mathbf{Z}^\kappa$, the following are equivalent:*
   (1) *$A$ is a product and $A^{\perp\perp} = A$;*
   (2) *$A$ is a direct summand of $\mathbf{Z}^\kappa$.*
*If these conditions hold, then $\mathbf{Z}^\kappa/A$ is a product.*

PROOF. Assume (1) and consider the short exact sequence

$$0 \to A^\perp \to (\mathbf{Z}^\kappa)^* \xrightarrow{\iota^*} B \to 0$$

where $B$ is the range of the homomorphism $\iota^*\colon (\mathbf{Z}^\kappa)^* \to A^*$ induced by the inclusion of $A$ in $\mathbf{Z}^\kappa$. Since $A^*$ is free by III.3.7, $B$ is free and thus the sequence splits. Since $A^\perp$ is a summand of $(\mathbf{Z}^\kappa)^*$, $A^{\perp\perp}$ is a summand of $\mathbf{Z}^\kappa$ by Lemma 1.2. Thus (2) is proved since $A^{\perp\perp} = A$.

Now assume (2): say $\mathbf{Z}^\kappa = A \oplus D$. Then $(\mathbf{Z}^\kappa)^* = A^\perp \oplus D^\perp$, so by 1.2 $\mathbf{Z}^\kappa = A^{\perp\perp} \oplus C$ where $C \cong (A^\perp)^*$ and $A^{\perp\perp} \cong (D^\perp)^*$. Now $(\mathbf{Z}^\kappa)^*$ is free by III.3.7, so $A^\perp$ and $D^\perp$ are free, and hence $C$ and $A^{\perp\perp}$ are products. By Lemma 1.1(i) $A = A^{\perp\perp}$ so (1) is proved. Finally, notice that $\mathbf{Z}^\kappa/A$ is a product since it is isomorphic to $C$. $\square$

We want to consider when the converse of the last assertion of the theorem is true. Recall that a Whitehead group is a group $G$ such that $\mathrm{Ext}(G, \mathbf{Z}) = 0$.

**1.5 Theorem.** *Let $\kappa$ and $A$ be as in 1.4 and suppose that $\mathbf{Z}^\kappa/A$ is a product. Then $A = A^{\perp\perp}$; moreover, if either*
   (a) *$\kappa = \omega$,*
   (b) *every Whitehead group of cardinality $\leq \kappa$ is free, or*
   (c) *$\mathbf{Z}^\kappa/A$ is a countable product,*
*then $A$ is a direct summand of $\mathbf{Z}^\kappa$.*

PROOF. That $A^{\perp\perp} = A$ follows, by Lemma 1.1(i), from the fact that $\mathbf{Z}^\kappa/A$ is torsionless since it is a product (cf. I.2.1). So it remains to show that $A^{\perp\perp}$ is a direct summand of $\mathbf{Z}^\kappa$. For assumption (a) this is Proposition 1.3. Assume (b); let $M$ denote $\mathbf{Z}^\kappa$. The short exact sequence

$$0 \to A \xrightarrow{\iota} M \xrightarrow{\pi} M/A \to 0$$

induces a short exact sequence

$$0 \to (M/A)^* \xrightarrow{\pi^*} M^* \xrightarrow{\iota^*} B \to 0$$

where $B$ is the image of $\iota\colon M^* \to A^*$. The latter sequence induces the Cartan-Eilenberg sequence

$$0 \to B^* \to M^{**} \xrightarrow{\pi^{**}} (M/A)^{**} \to \mathrm{Ext}(B, \mathbf{Z}) \to \mathrm{Ext}(M^*, \mathbf{Z}) = 0$$

where the last term, $\mathrm{Ext}(M^{**}, \mathbb{Z})$, is zero because $M^{**} \cong \mathbb{Z}^{(\kappa)}$ is free. Now $M/A$ is reflexive because it is a product and $\kappa$ is not $\omega$-measurable (cf. Exercise III.9). Hence $\pi^{**}$ is surjective because $\pi$ is surjective and $\pi^{**} \circ \sigma_M = \sigma_{M/A} \circ \pi$. Therefore $\mathrm{Ext}(B, \mathbb{Z}) = 0$, so by hypothesis, $B$ is free. But then $A^{\perp}$ is a direct summand of $M^*$ since $B \cong M^*/A^{\perp}$. Lemma 1.2 then finishes the proof in this case.

Now assume (c), and retain the notation of case (b). By assumption, $(M/A)^*$ is countable (cf. before III.2.5). Thus since $B \cong M^*/(M/A)^*$ and $M^*$ is free, $B$ is isomorphic to $C \oplus F$ where $C$ is countable and $F$ is free. Hence $\mathrm{Ext}(B, \mathbb{Z}) = \mathrm{Ext}(C, \mathbb{Z})$, so if $\mathrm{Ext}(B, \mathbb{Z}) = 0$, then $C$ is free because every countable Whitehead group is free (cf. XII.1.2). We then finish the proof as in case (b). $\square$

**1.6 Corollary.** (V = L) *For all cardinals $\kappa$ and all subgroups $A$ of $\mathbb{Z}^{\kappa}$, $A$ is a direct summand of $\mathbb{Z}^{\kappa}$ if and only if $\mathbb{Z}^{\kappa}/A$ is a product.*

PROOF. Recall that V = L implies that there are no $\omega$-measurable cardinals (cf. VI.3.15). Then the corollary follows from Theorem 1.5(b) and VII.4.10. $\square$

**1.7 Example.** Suppose there is a Whitehead group $B$ of non-$\omega$-measurable cardinality, $\kappa$, which is not free. There is a short exact sequence

$$0 \to K \to F \to B \to 0$$

for some free group $F \cong \mathbb{Z}^{(\kappa)}$. This induces the Cartan-Eilenberg sequence

$$(*) \qquad 0 \to B^* \overset{\nu}{\longrightarrow} F^* \to K^* \to \mathrm{Ext}(B, \mathbb{Z}) = 0.$$

Without loss of generality we may suppose that $\nu$ is an inclusion map, so $B^*$ is a subgroup of $F^* \cong \mathbb{Z}^{\kappa}$. Thus $F^*/B^*$ is a product because $K$ is free. We claim that $B^*$ is not a direct summand of $F^*$. If it were, then $(*)$ would be a splitting exact sequence and hence so would be

$$0 \to K^{**} \to F^{**} \to B^{**} \to 0.$$

But then since $F$ and $K$ are reflexive, $K$ would be a summand of $F$ and $B$ would be free.

It follows from Example 1.7 and XII.1.11 that Corollary 1.6 is not provable in ZFC. Thus, for example, it is undecidable in ZFC whether for every subgroup $A$ of $\mathbb{Z}^{\omega_1}$, $A$ is a direct summand of $\mathbb{Z}^{\omega_1}$ whenever $\mathbb{Z}^{\omega_1}/A$ is a product.

For future reference we include the following result, which can be proved by a straightforward application of the definitions and remarks at the beginning of this section.

**1.8 Lemma.** (i) *If* $M = B \oplus C \oplus D$, *then* $C^{\perp} = (B \oplus C)^{\perp} \oplus (C \oplus D)^{\perp}$.
   (ii) *If* $M = \bigoplus_{k \in \omega} C_k$, *then* $M^*$ *is canonically isomorphic to* $\prod_{n \in \omega} (\bigoplus_{k \neq n} C_k)^{\perp}$. $\square$

We conclude this section by considering some applications to reflexivity. By I.2.2, there is a homomorphism $\rho: M^{***} \to M^*$ such that $\rho \circ \sigma_{M^*} = 1_{M^*}$. Thus $M^{***} = \operatorname{im}(\sigma_{M^*}) \oplus \ker(\rho)$ and $M^*$ is reflexive if and only if $\ker(\rho) = 0$. From the definition of $\rho$ it is easy to see that $\ker(\rho) = \sigma_M[M]^{\perp}$.

**1.9 Theorem.** *For any group* $A$, $A^*$ *is reflexive if and only if* $A^{**}$ *is reflexive.*

PROOF. Let $\sigma_1: A^* \to A^{***}$ and $\sigma_2: A^{**} \to A^{****}$ be the canonical homomorphisms defined in section I.1. Let $\rho_3: A^{***} \to A^*$ (respectively $\rho_4: A^{****} \to A^{**}$) be the splitting of $\sigma_1$ (respectively, $\sigma_2$) defined in I.2.2. Then $A^{**}$ is reflexive if and only if $\sigma_1[A^*]^{\perp} = 0$ and $A^*$ is reflexive if and only if $\ker(\rho_3) = 0$. But $A^{***} = \sigma_1[A^*] \oplus \ker(\rho_3)$ so $\ker(\rho_3)^* \cong \sigma_1[A^*]^{\perp}$. Since $\ker(\rho_3)$ is torsionless, $\ker(\rho_3) = 0$ if and only if $\ker(\rho_3)^* = 0$, so the theorem is proved. $\square$

**1.10 Theorem.** *A direct summand of a reflexive group is reflexive.*

PROOF. Suppose $M$ is reflexive and $M = A \oplus B$. Then $M^{**} = A^{**} \oplus B^{**}$, with the identifications as described at the beginning of this section. Moreover, $\sigma_M = \sigma_A + \sigma_B$, i.e., for all $a \in A$, $b \in B$, $\sigma_M((a, b)) = (\sigma_A(a), \sigma_B(b)) \in A^{**} \oplus B^{**}$. It follows easily that if $\sigma_M$ is an isomorphism, then so is $\sigma_A$. $\square$

The preceding generalizes in an obvious way to $H$-reflexivity. We will see in X.1.1 that direct sums and products (over non-measurable index sets) of reflexive groups are reflexive. We make use of this and III.2.12 to derive the following.

**1.11 Corollary.** *If $H$ is a free $R$-module of infinite non-$\omega$-measurable rank and $M$ is a projective $R$-module with a generating set which is not $\omega$-measurable, then $M$ is $H$-reflexive.*

PROOF. By hypothesis there is an infinite non-$\omega$-measurable $\kappa$ and a module $P$ such that $M \oplus P \cong R^{(\kappa)} \cong H^{(\kappa)}$. Thus by III.2.12 (with $N = H$), $M \oplus P$ is $H$-reflexive. So by the remark above, $M$ is $H$-reflexive. $\square$

## §2. Countable products of the integers

In this section we will look at quotients of $\mathbb{Z}^\omega$ and use the information gained to characterize slender $\mathbb{Z}$-modules by their subgroups. As before we will denote $\mathbb{Z}^\omega$ by $P$ and say that a group is a countable product if it is isomorphic to $P$ or to $\mathbb{Z}^n$ for some finite $n$.

Let us give $P$ the product topology where $\mathbb{Z}$ is given the discrete topology. Thus $\{U_n : n \in \omega\}$ is a basic system of neighborhoods of $0$, when $U_n = \{x \in P : x(m) = 0$ for all $m < n\}$. For any subset $A$ of $P$, let $\bar{A}$ denote the closure of $A$ in $P$; i.e., $\bar{A} =$ the set of all $x$ in $P$ such that for every $n$, $x$ agrees with some element of $A$ in the first $n$ coordinates.

**2.1 Lemma.** *For any subgroup $A$ of $P$ of infinite rank, $\bar{A}$ is isomorphic to $P$ and $\bar{A}/A$ is a cotorsion group.*

PROOF. For each $n$ let $\pi_n : P \to \mathbb{Z}$ be projection on the $n$th coordinate. Then $\pi_n[A \cap U_n]$ is a cyclic subgroup of $\mathbb{Z}$, say $d_n\mathbb{Z}$; let $a^n$ be an element of $A \cap U_n$ chosen so that $\pi_n(a^n) = d_n$ and $a^n = 0$ if $d_n = 0$. For all $a \in A$ we can choose $k_n \in \mathbb{Z}$ by induction so that

$a = \sum_{n \in \omega} k_n a^n$ (cf. the argument below that $\theta$ is onto $\bar{A}$). Hence since $A$ has infinite rank, infinitely many of the $a^n$'s must be non-zero. Let $\{m(n) : n \in \omega\}$ enumerate in increasing order those $n$ so that $a^n \neq 0$. Define $\theta : P \to P$ by $\theta((k_n)_{n \in \omega}) = \sum_n k_n a^{m(n)}$, which is a well-defined element of $P$ because for all $i \in \omega$, $\sum_n k_n a^{m(n)}(i)$ is a finite sum. Since $\{a^{m(n)} : n \in \omega\}$ is linearly independent, $\theta$ is one-one. Clearly $\theta[\mathbf{Z}^{(\omega)}] \subseteq A$ and $\theta[P] \subseteq \bar{A}$. We claim that $\theta[P] = \bar{A}$. Given $x \in \bar{A}$ we shall define $k_n$ inductively so that for all $i$

$$(*) \qquad x(i) = \sum_{n \leq i} k_n a^{m(n)}(i).$$

Suppose that we have defined $k_n$ for all $n < \ell$ so that $(*)$ holds for all $i < m(\ell)$. Since $x \in \bar{A}$ there exists $y \in A$ such that $y(i) = x(i)$ for $i \leq m(\ell)$. Then by $(*)$, $y - \sum_{n < \ell} k_n a^{m(n)} \in A \cap U_{m(\ell)}$; so by definition of $d_{m(\ell)}$, $(y - \sum_{n < \ell} k_n a^{m(n)})(m(\ell)) = k_\ell d_{m(\ell)} = k_\ell a^{m(\ell)}(m(\ell))$ for some $k_\ell \in \mathbf{Z}$. This defines $k_\ell$ so that $(*)$ holds for all $i \leq m(\ell)$ and hence for all $i < m(\ell + 1)$. This completes the inductive step. But now it is clear that $\theta((k_n)_{n \in \omega}) = x$. Hence $\theta$ is an isomorphism of $P$ onto $\bar{A}$; this isomorphism takes $\mathbf{Z}^{(\omega)}$ into $A$, and therefore induces a homomorphism of $P/\mathbf{Z}^{(\omega)}$ onto $\bar{A}/A$. Since by V.1.16, $P/\mathbf{Z}^{(\omega)}$ is pure-injective, it follows from V.2.2 that $\bar{A}/A$ is cotorsion. $\square$

Since $\mathbf{Z}$ is slender, every element of $P^*$ is continuous when $\mathbf{Z}$ is given the discrete topology and $P$ the product topology as above. Hence $\bar{A}$ is contained in $A^{\perp\perp}$.

**2.2 Lemma.** *For any subgroup $A$ of $P$ of infinite rank, $A^{\perp\perp}/\bar{A}$ is a cotorsion group.*

PROOF. Let $\iota$ be the inclusion of $\bar{A}$ into $A^{\perp\perp}$. Then we have a short exact sequence

$$0 \to (A^{\perp\perp})^* \xrightarrow{\iota^*} \bar{A}^* \to N \to 0$$

because by 1.3 and 1.1(ii) $\ker(\iota^*) = (A^{\perp\perp}/\bar{A})^* = 0$. (Here $N$ is the cokernel of $\iota^*$.) This in turn induces the Cartan-Eilenberg sequence

$$0 \to N^* \to \bar{A}^{**} \xrightarrow{\iota^{**}} (A^{\perp\perp})^{**} \to \mathrm{Ext}(N, \mathbf{Z}) \to \mathrm{Ext}(\bar{A}^*, \mathbf{Z}) = 0$$

where the last term is zero because by 2.1 $\bar{A} \cong P$, so $\bar{A}^*$ is free. Now $\bar{A}$ and $A^{\perp\perp}$ are both reflexive since by 2.1 and 1.3, respectively, they are both countable products (cf. III.2.5). Thus $A^{\perp\perp}/\bar{A}$ is isomorphic to $\mathrm{Ext}(N, \mathbf{Z})$ and hence, by V.2.3, is cotorsion. □

**2.3 Theorem.** *For any subgroup $A$ of $P$, $P/A$ is the direct sum of a cotorsion group and a countable product.*

PROOF. By 1.3 $P = A^{\perp\perp} \oplus C$ where $A^{\perp\perp}$ and $C$ are countable products; so $P/A \cong (A^{\perp\perp}/A) \oplus C$, and it remains to prove that $A^{\perp\perp}/A$ is cotorsion. We first dispose of the case when $A$ has finite rank. In this case the pure closure, $A_*$, of $A$ is a summand of $P$ since $P$ is separable by IV.2.8. Since $A_*$ is clearly contained in $A^{\perp\perp}$, we have $A_* = A_*^{\perp\perp} = A^{\perp\perp}$ by 1.1(i). Then $A^{\perp\perp}/A$ is cotorsion (even pure-injective) since it equals $A_*/A$ and hence is finite.

Now assume $A$ has infinite rank. By Lemmas 2.1 and 2.2, $A^{\perp\perp}/\bar{A}$ and $\bar{A}/A$ are cotorsion; but then by V.2.2 $A^{\perp\perp}/A$ is cotorsion. □

It is not the case that every quotient of $\mathbf{Z}^{\omega_1}$ is of the form given in the previous theorem: see Exercise 2. We shall deal further with quotients of $\mathbf{Z}^\kappa$ when $\kappa$ is uncountable in the next section. Now we derive, as a consequence of 2.3, Nunke's characterization of slender groups. Recall that $Z(p)$ is the cyclic group of order $p$ and $J_p$ is the group of $p$-adic integers.

**2.4 Corollary.** *An abelian group $H$ is slender if and only if it does not contain a copy of $\mathbf{Q}$, $P$, $Z(p)$ or $J_p$ for any prime $p$.*

PROOF. If $H$ is slender, then by V.2.1 it does not contain any pure-injective group and hence does not contain a copy of $\mathbf{Q}$, $Z(p)$ or $J_p$; clearly also it does not contain a copy of $P$. Conversely, suppose $H$ does not contain a copy of any of these groups, and consider a homomorphism $\varphi\colon P \to H$. By 2.3, $\varphi[P]$ is of the form $G \oplus C$ where $C$ is a countable product and $G$ is cotorsion. By V.2.9, $H$ is cotorsion-free, so $G = 0$. Since $H$ does not contain a copy of $P$, $C$ must be isomorphic to $\mathbf{Z}^m$ for some finite $m$. Thus $\varphi[P]$ is a finitely-generated free group, $\mathbf{Z}^m$. By III.1.10, $\mathbf{Z}^m$ is slender, so

$\varphi(e_n) = 0$ for all but finitely many $n$. Since this holds for all $\varphi$, $H$ is slender. $\square$

In other words, 2.4 says that $H$ is slender if and only if it is cotorsion-free and does not contain a copy of $P$.

**2.5 Corollary.** (i) *Every $\aleph_1$-free group which does not contain a copy of $P$ is slender.*

(ii) *Every strongly $\aleph_1$-free group is slender.*

PROOF. Part (i) is immediate from the theorem. For part (ii) we need to prove that a strongly $\aleph_1$-free group $H$ cannot contain a copy of $P$. But, by definition, any countable subgroup $S$ of $H$ is contained in a countable $\aleph_1$-pure subgroup, which must contain the $\mathbf{Z}$-adic closure of $S$ (cf. §I.3). However, $P$ has a countable subset, namely $\mathbf{Z}^{(\omega)}$, whose $\mathbf{Z}$-adic closure in $P$ is uncountable. So $H$ cannot contain a copy of $P$. $\square$

Later (in Chapter XIV), we shall construct non-free strongly $\aleph_1$-free (even $\aleph_1$-separable) groups which are dual groups. Hence we shall have examples of dual groups which are slender but are not free.

We conclude with a characterization, in the same spirit as 2.4, of almost slender groups (see before III.1.11) .

**2.6 Corollary.** *An abelian group $H$ is almost slender if and only if it does not contain a copy of $P$ or of any unbounded cotorsion group.*

PROOF. If $H$ is almost slender, then it clearly does not contain a copy of $P$. Suppose that $H$ contained an unbounded cotorsion group $G$; by V.2.6, $G$ is the homomorphic image of a pure-injective group $A$; say $\varphi: A \to G$ is an epimorphism. It is then easy to construct $\psi: \mathbf{Z}^{(\omega)} \to A$ such that for all $m \neq 0$, $\{m\varphi(\psi(e_n)): n \in \omega\}$ is infinite. Since $A$ is pure-injective, we can extend $\psi$ to $\bar{\psi}: P \to A$; then $\varphi \circ \bar{\psi}$ shows that $H$ is not almost slender. Conversely, if $H$ does not contain a copy of $P$ or of any unbounded cotorsion group, consider a homomorphism $\varphi: P \to H$. By 2.3 $\varphi[P]$ is of the form $G \oplus C$, where $G$ is cotorsion and $C$ is a countable product. By hypothesis, $G$ is bounded and $C$ is a finite product. For some $r \in \mathbf{Z} \setminus \{0\}$, $r\varphi[P] \subseteq C$; we can then finish the proof as in 2.4. $\square$

## §3. Uncountable products of the integers

We are interested in a generalization of Theorem 2.3 to the case where $P$ is replaced by an uncountable product of $\mathbf{Z}$. By Exercise 2, the full analog of 2.3 fails, but we do have the following result.

**3.1 Theorem.** *Let $\kappa$ and $\lambda$ be cardinals which are not $\omega$-measurable. Suppose $A$ is a subgroup of $\mathbf{Z}^\kappa$ which is a homomorphic image of $\mathbf{Z}^\lambda$. Then $\mathbf{Z}^\kappa/A$ is the direct sum of a cotorsion group and a product.*

PROOF. Let us denote $\mathbf{Z}^\lambda$ by $M$ and $\mathbf{Z}^\kappa$ by $N$. Note that $M$ and $N$ are reflexive by Exercise III.9. Let $f \colon M \to N$ be a homomorphism such that $A = f[M]$. Let $\iota \colon A \to N$ be the inclusion map. Then we can factor $f$ as $\iota \circ g$ where $g \colon M \to A$ is a surjection. We have induced maps $\iota^* \colon N^* \to A^*$ and $g^* \colon A^* \to M^*$; the latter is a monomorphism, so $A^*$ is free since $M^* \cong \mathbf{Z}^{(\lambda)}$ is free. The kernel of $\iota^*$ is denoted $A^\perp$ (cf. section 1), and is a (free) direct summand of $N^*(\cong \mathbf{Z}^{(\kappa)})$ because $N^*/A^\perp$ is isomorphic to a subgroup of $A^*$. Therefore by 1.2, $N = A^{\perp\perp} \oplus C$, where $C \cong (A^\perp)^*$; hence $N/A \cong A^{\perp\perp}/A \oplus C$. Now $C$ is a product since $A^\perp$ is free; so it remains to prove that $A^{\perp\perp}/A$ is cotorsion.

Let $B$ denote the image of the map $f^* \colon N^* \to M^*$, i.e., $B = f^*[N^*]$. Now $f^*$ factors as $j \circ h$ where $h \colon N^* \to B$ is a surjection and $j$ is the inclusion of $B$ into $M^*$. Let $D = M^*/B$; so we have a short exact sequence

$$0 \to B \xrightarrow{\ j\ } M^* \to D \to 0$$

which induces the top row of the following commutative diagram:

$$
\begin{array}{ccccccc}
M^* & \xrightarrow{\ j^*\ } & B^* & \to \operatorname{Ext}(D, \mathbf{Z}) \to (M^*, \mathbf{Z}) = 0 \\
\ \downarrow{\scriptstyle \sigma_M^{-1}} & & \ \downarrow{\scriptstyle \theta} & \\
M & \xrightarrow{\ f\ } & N &
\end{array}
$$

where $\theta = \sigma_N^{-1} \circ h^*$. Notice that $\sigma_M$ and $\sigma_N$ are isomorphisms and that the square commutes because $f^{**} = h^* \circ j^*$ and $\sigma_N \circ f =$

$f^{**} \circ \sigma_M$. We claim that the image of $\theta$ is $A^{\perp\perp}$. In fact, this follows from the remarks in the first few paragraphs of section 1 (where we let $N^*$ play the role of $M$ and let $A^{\perp}$ play the role of $A$). The map $h$ may be identified with the canonical projection of $N^*$ onto $N^*/A^{\perp}$ since $\ker(h) = \ker(f^*) = A^{\perp}$. Then $h^*$ in turn may be identified with the induced embedding of $(N^*/A^{\perp})^*$ into $N^{**}$, whose image is $(A^{\perp})^{\perp}$. Hence the image of $h^*$ equals $\sigma_N[A^{\perp\perp}]$.

Now $\theta$ is a monomorphism since it is the composition of two monomorphisms, so

$$\mathrm{Ext}(D, \mathbf{Z}) \cong B^{**}/\mathrm{im}(j^*) \cong \theta[B^*]/\theta[j^*[M^{**}]] \cong \theta[B^*]/f[M]$$

and the last term, as we have just seen, is $A^{\perp\perp}/A$. Thus $A^{\perp\perp}/A$ is isomorphic to $\mathrm{Ext}(D, \mathbf{Z})$, which is cotorsion by V.2.3. $\square$

In fact, the proof of the theorem gives us a little more information, which we state as a corollary.

**3.2 Corollary.** *For any homomorphism $f: \mathbf{Z}^{\lambda} \to \mathbf{Z}^{\kappa}$ where $\kappa$ and $\lambda$ are non-$\omega$-measurable cardinals, there are groups $C$ and $D$ such that $C$ is a product, $\ker(f) \cong D^*$, and $\mathrm{coker}(f) \cong C \oplus \mathrm{Ext}(D, \mathbf{Z})$.*

PROOF. The only new thing that requires checking is the assertion about the kernel of $f$. But it is easily verified from the commutativity of the diagram above that $\sigma_M$ induces an isomorphism between $\ker(f)$ and $D^*$. $\square$

**3.3 Corollary.** *If $\kappa$ is not $\omega$-measurable and $A$ is a subgroup of $\mathbf{Z}^{\kappa}$ which is a product, then $\mathbf{Z}^{\kappa}/A \cong C \oplus \mathrm{Ext}(D, \mathbf{Z})$ where $C$ is a product and $D^* = 0$.* $\square$

We will see in Chapter XII that we have, if we assume $V = L$, a complete structural description of groups of the form $\mathrm{Ext}(D, \mathbf{Z})$ and hence of groups of the form $\mathbf{Z}^{\kappa}/A$, where $A$ is as in the results above. (See also Exercise 8.) We conclude with some notions that will be of use in the next section.

**3.4 Definition.** For any $\kappa \geq \aleph_0$, let $\mathbf{Z}^{<\kappa}$ be the subgroup of $\mathbf{Z}^{\kappa}$ consisting of all $x \in \mathbf{Z}^{\kappa}$ such that $|\mathrm{supp}(x)| < \kappa$. Let $\mathbf{Z}_{\kappa} = \mathbf{Z}^{\kappa}/\mathbf{Z}^{<\kappa}$.

Otherwise said, $\mathbf{Z}_\kappa$ is the reduced power $\mathbf{Z}^\kappa/C_\kappa$ where $C_\kappa$ is the co-$\kappa$ filter on $\kappa$ (cf. II.2.5 and II.3.1).

Say that an abelian group $G$ is *strongly cotorsion-free* if $\mathrm{Hom}(\mathbf{Z}_\kappa, G) = 0$ for all regular $\kappa$ which are not $\omega$-measurable.

The structure of $\mathbf{Z}_\omega$ is discussed in Exercise V.4 and $\mathbf{Z}_{\omega_1}$ is discussed in Exercise IV.11.

**3.5 Lemma.** *For every $\kappa$ of uncountable cofinality, $\mathbf{Z}_\kappa$ is $\aleph_1$-free.*

PROOF. If $A$ is a countable subgroup of $\mathbf{Z}_\kappa = \mathbf{Z}^\kappa/C_\kappa$ then by II.3.4, $A$ is isomorphic to a subgroup of $\mathbf{Z}^\kappa$ since $C_\kappa$ is $\aleph_1$-complete. But $\mathbf{Z}^\kappa$ is $\aleph_1$-free (cf. IV.2.8), so $A$ is free. $\square$

**3.6 Lemma.** *If $\mathrm{Hom}(\mathbf{Z}_{\omega_1}, G) = 0$, then $G$ is cotorsion-free. Hence, a strongly cotorsion-free group is cotorsion-free.*

PROOF. Suppose to the contrary that there is a non-zero cotorsion subgroup $C$ of $G$. Then by V.2.6, $C$ is a homomorphic image of a pure-injective group $H$; say $\psi: H \to C$ is an epimorphism. By 3.5, $\mathbf{Z}_{\omega_1}$ is $\aleph_1$-free, so it contains a countable pure free subgroup $F$. Let $a \in C \setminus \{0\}$. There is a homomorphism $\varphi: F \to H$ such that $a \in (\psi \circ \varphi)[F]$. This homomorphism extends to $\mathbf{Z}_{\omega_1}$ since $H$ is pure-injective. But then it is clear that there is a non-zero homomorphism from $\mathbf{Z}_{\omega_1}$ to $C \subseteq G$, a contradiction. $\square$

In fact, $G$ is cotorsion-free if and only if $\mathrm{Hom}(\mathbf{Z}_\omega, G) = 0$: see Exercise 10. Wald 1983b has proved that, assuming $V = L$, every cotorsion-free group of cardinality $\aleph_1$ is strongly cotorsion-free.

A group $H$ is called *residually slender* if for all $a \in H \setminus \{0\}$, there is a subgroup $N$ of $H$ such that $a \notin N$ and $H/N$ is slender. For example, any separable group is residually slender, but may not be slender (e.g. $\mathbf{Z}^\omega$).

**3.7 Lemma.** *Every residually slender group is strongly cotorsion-free.*

PROOF. Suppose, to the contrary, that there is a residually slender group $H$ and a non-zero homomorphism $\varphi: \mathbf{Z}_\kappa \to H$ for some regular, non-$\omega$-measurable $\kappa$. Say $a \in \mathrm{im}(\varphi) \setminus \{0\}$. By hypothesis there is a subgroup $N$ of $H$ such that $a \notin N$ and $H/N$ is slender. If $\pi$ is the canonical surjection: $H \to H/N$, then $\pi \circ \varphi$ is a non-zero homomorphism: $\mathbf{Z}_\kappa \to H/N$. Hence there is a non-zero homomorphism from $\mathbf{Z}^\kappa$ to $H/N$ which is zero on $\mathbf{Z}^{<\kappa}$; but this is impossible by III.3.4, since $H/N$ is slender and $\kappa$ is not $\omega$-measurable. $\square$

## §4. Radicals and large cardinals

We take up a theme which was first introduced in Chapter II: the effect of large cardinal hypotheses on group-theoretical properties. (The results of this section will not be used later, so the reader not interested in them for their own sake may safely skip this section.)

Let $\mathcal{A}$ denote the class (or category) of abelian groups. A function $T: \mathcal{A} \to \mathcal{A}$ is called a *preradical* (or *subfunctor of the identity*) if for all groups $A$, and all group homomorphisms $h: A \to B$, $TA \subseteq A$ and $h[TA] \subseteq TB$. (In this context we usually write $TA$ instead of $T(A)$.)

If $T$ is a preradical, it is easy to see that $T \circ T \subseteq T$, i.e., for all $A$, $T^2 A \subseteq TA$. We say that $T$ is a *socle* if $T \circ T = T$.

We define a "co-composition", $S{:}T$, of preradicals $S$ and $T$ as follows: for any preradicals $S$ and $T$, if $\rho: A \to A/TA$ is the canonical surjection, $(S{:}T)A = \rho^{-1}[S(A/TA)]$. It is easy to see that $T \subseteq T{:}T$. We say that $T$ is a *radical* if $T = T{:}T$, i.e., $T(A/TA) = 0$ for all $A$.

Using the canonical projections and injections, one can show that for any preradical $T$, and any groups $A_i$, $T(\bigoplus_{i \in I} A_i) = \bigoplus_{i \in I} TA_i$ and $T(\prod_{i \in I} A_i) \subseteq \prod_{i \in I} TA_i$ (see Exercise 9).

**4.1 Examples.** For any class of groups $\mathcal{X}$ and any group $A$ define

$$S_{\mathcal{X}} A = \sum \{im(\varphi) : \varphi \in \mathrm{Hom}(X, A), X \in \mathcal{X}\} \text{ and}$$

$$R_{\mathcal{X}} A = \bigcap \{\ker(\varphi) : \varphi \in \mathrm{Hom}(A, X), X \in \mathcal{X}\}.$$

We will write $S_G$ instead of $S_{\{G\}}$ and $R_G$ instead of $R_{\{G\}}$. It is left to the reader to check that $S_{\mathcal{X}}$ is always a socle and $R_{\mathcal{X}}$ is always a radical.

For example, if $G = \mathbf{Z}/p\mathbf{Z}$, $S_G A = A[p] \overset{\text{def}}{=} \{a \in A : pa = 0\}$. If $G = \bigoplus_{p,n} \mathbf{Z}/p^n\mathbf{Z}$, $S_G A = A_t$, the torsion part of $A$. Also, $R_{\mathbf{Q}} A = A_t$.

For any $A$, $A/R_{\mathbf{Z}} A$ is torsionless, and, in fact, $A$ is torsionless if and only if $R_{\mathbf{Z}} A = 0$. The following is easy to verify.

**4.2 Proposition.** *Suppose $\mathcal{X}$ is a class of groups and $\mathcal{Y}$ is the class of direct sums of elements of $\mathcal{X}$. Then $R_{\mathcal{X}} = R_{\mathcal{Y}}$. In particular, if $F$ is any non-trivial free group then $R_F = R_{\mathbf{Z}}$.* $\square$

Let $\mathcal{X}$ be the class of $\aleph_1$-free groups. The radical $R_{\mathcal{X}}$ is called the *Chase radical* and denoted by $\nu$. In the rest of this section we shall concentrate mainly on the Chase radical and on $R_{\mathbf{Z}}$.

**4.3 Proposition.** (i) *For any countable group $C$, $\nu C = R_{\mathbf{Z}} C$.*
(ii) *If $C$ is countable, $\nu C = C$ if and only if $C^* = 0$.*
(iii) *$\nu$ is a socle.*

PROOF. (i) Let $F$ be the free group of countable rank. If $\varphi$ is any map from $C$ to an $\aleph_1$-free group, then the image of $\varphi$ is a free group. So $\nu C = R_F C$, which, by Proposition 4.2, equals $R_{\mathbf{Z}} C$.

(ii) For any group, $C^* = 0$ if and only if $R_{\mathbf{Z}} C = C$, by definition.

(iii) We must prove that $\nu \subseteq \nu \circ \nu$. For this it suffices to prove that for all groups $A$, $A/\nu\nu A$ is $\aleph_1$-free, for then $a \in \nu A$ implies that the image of $a$ under the canonical surjection of $A$ onto $A/\nu\nu A$ is $0$, which means that $a \in \nu\nu A$. Now we have a short exact sequence

$$0 \to \nu A/\nu\nu A \to A/\nu\nu A \to A/\nu A \to 0$$

and the middle term will be $\aleph_1$-free if $\nu A/\nu\nu A$ and $A/\nu A$ are $\aleph_1$-free (cf. Exercise IV.5). Thus it suffices to show that for all $G$, $G/\nu G$ is $\aleph_1$-free. Let $\mathcal{S} = \{K : K$ is a subgroup of $G$ s.t. $G/K$ is $\aleph_1$-free$\}$. Then $\nu G = \bigcap_{K \in \mathcal{S}} K$. Define

$$\theta : G \to \prod_{K \in \mathcal{S}} G/K$$

by $\theta(g)(K) = g + K$. The kernel of $\theta$ equals $\nu G$, so $\theta$ induces an embedding of $G/\nu G$ into $\prod_{K \in \mathcal{S}} G/K$. Since a product of $\aleph_1$-free groups is $\aleph_1$-free (cf. Exercise IV.5), it follows that $G/\nu G$ is $\aleph_1$-free. $\square$

**4.4 Definition.** Let $T$ be a preradical and $\kappa$ an infinite cardinal. Define

$$T^{[\kappa]}A = \sum \{TB : B \text{ is a subgroup of } A \text{ of cardinality } < \kappa\}.$$

It is easy to check that $T^{[\kappa]}$ is a preradical. We say that $T$ *satisfies the cardinal condition* if there is a $\kappa$ such that $T^{[\kappa]} = T$.

First we will prove that $\nu$ satisfies the cardinal condition. Then we shall show that whether or not $R_\mathbf{Z}$ satisfies the cardinal condition depends on whether a certain large cardinal exists.

**4.5 Theorem.** *The Chase radical satisfies the cardinal condition; in fact, $\nu = \nu^{[\omega_1]}$.*

PROOF. Suppose, to the contrary, that there is a group $A$ and an element $a$ of $\nu A$ such that $a \notin \sum \{\nu C : C \text{ is a countable subgroup of } A\}$. Let $I$ be the set of all $\sigma : C_\sigma \to \mathbf{Z}$ such that $C_\sigma$ is a countable subgroup of $A$, $a \in C_\sigma$, and $\sigma(a) \neq 0$.

For each $x \in A$, let $S_x = \{\sigma \in I : x \in \text{dom } \sigma\}$. We claim that every intersection of countably many of the $S_x$'s is non-empty. Indeed, given $x_n \in A(n \in \omega)$, let $C$ be a countable subgroup of $A$ containing $\{x_n : n \in \omega\}$; since $a \notin R_\mathbf{Z}C$, there exists $\sigma : C \to \mathbf{Z}$ such that $\sigma(a) \neq 0$; but then $\sigma \in I$ and $\sigma \in \bigcap_{n \in \omega} S_{x_n}$. Thus there is an $\omega_1$-complete filter $F$ such that $S_x \in F$ for all $x \in A$.

Define $\psi : A \to \mathbf{Z}^I$ as follows: for all $x \in A$ and $\sigma \in I$,

$$\psi(x)(\sigma) = \begin{cases} \sigma(x) & \text{if } \sigma \in S_x \\ 0 & \text{otherwise} \end{cases}$$

Composing $\psi$ with the canonical surjection onto $\mathbf{Z}^I/F$ we obtain a map $\bar\psi : A \to \mathbf{Z}^I/F$ which is a homomorphism because $S_{x+y} \cap S_x \cap S_y$ belongs to $F$ for all $x, y \in A$. Now $\mathbf{Z}^I/F$ is $\aleph_1$-free since $F$ is $\omega_1$-complete (cf. Exercise II.10), and $\bar\psi(a) \neq 0$ by construction, so $a \notin \nu A$, a contradiction. $\square$

**4.6 Corollary.** $\nu = R_{\mathbf{Z}}^{[\omega_1]}$. □

Now we turn our attention to radicals of the form $R_G$, and consider the question of whether they satisfy the cardinal condition.

**4.7 Theorem.** *Assume there are no measurable cardinals. Then for every non-zero strongly cotorsion-free $G$, $R_G$ does not satisfy the cardinal condition.*

PROOF. Suppose to the contrary that $R_G = R_G^{[\lambda]}$ for some $\lambda$. Choose a regular cardinal $\kappa \geq \lambda$ and let $A = \mathbf{Z}_\kappa$ ($= \mathbf{Z}^\kappa/\mathbf{Z}^{<\kappa}$: see 3.4). Then by hypothesis on $G$, $\text{Hom}(A, G) = 0$, so $R_G A = A$. However, if $B \subseteq A$ and $|B| < \lambda \leq \kappa$, $B$ is isomorphic to a subgroup of $\mathbf{Z}^\kappa$ by II.3.4. Since $R_G \mathbf{Z}^\kappa \subseteq (R_G \mathbf{Z})^\kappa = 0$ (since $G$ is torsion-free), we conclude that $R_G B = 0$. Hence $R_G^{[\lambda]} A = 0 \neq R_G A$, a contradiction. □

This result shows that the absence of certain large cardinals implies the failure of the cardinal condition. Another result of Dugas and Göbel 1985a says that if $0^\#$ does not exist (see VI.3.15) and GCH holds, then for any cotorsion-free $G \neq 0$, $R_G$ does not satisfy the cardinal condition. On the other hand Eda 1989 has proved that Vopenka's Principle — a very large cardinal hypothesis — implies that every preradical satisfies the cardinal condition. In the next results we shall pin down exactly which large cardinal is needed in order for $R_{\mathbf{Z}}$ to satisfy the cardinal condition.

**4.8 Theorem.** *If $\kappa$ is an $L_{\lambda\omega}$-compact cardinal and $G$ is a group of cardinality $< \lambda$, then $R_G = R_G^{[\kappa]}$.*

PROOF. Fix a group $A$ and let $\mathcal{P}_\kappa(A)$, $U_Y$ and $F_\kappa(A)$ be defined as in II.3.9. Let $I = \mathcal{P}_\kappa(A)$. Then $F_\kappa(A)$ is a $\kappa$-complete filter on $I$, so by hypothesis, there is a $\lambda$-complete ultrafilter $D$ on $I$ which contains $F_\kappa(A)$. Define $\psi: A \to \prod_{Y \in I}\langle Y \rangle$ as follows: for all $a \in A$ and $Y \in I$,

$$\psi(a)(Y) = \begin{cases} a & \text{if } a \in Y \\ 0 & \text{otherwise} \end{cases}$$

Then the composition of $\psi$ with the canonical surjection onto the ultraproduct yields a homomorphism $\bar{\psi}: A \to \prod_{Y \in I}\langle Y \rangle/D$.

Now suppose that $a \notin R_G^{[\kappa]} A$; then for all $Y \in I$ such that $a \in \langle Y \rangle$, $a \notin R_G \langle Y \rangle$ so there exists $f_Y : \langle Y \rangle \to G$ such that $f_Y(a) \neq 0$. Note that $f_Y$ is defined for all $Y$ in a member $U_Z$ of $D$: choose $Z$ so that $a \in Z$. Then the $f_Y$'s induce a homomorphism $f : \prod_{Y \in I} \langle Y \rangle / D \to G^I / D$. Now $\delta_G : G \to G^I / D$ is an isomorphism by II.3.2. The composition $\delta_G^{-1} \circ f \circ \bar{\psi}$ is a homomorphism from $A$ to $G$ which is non-zero on $a$. Therefore, $a$ does not belong to $R_G A$. $\square$

**4.9 Theorem.** *Let $G$ be a residually slender group of cardinality $\lambda \geq \aleph_0$ where $\lambda$ is not $\omega$-measurable. Let $\kappa$ be an infinite cardinal. Suppose $R_G = R_G^{[\kappa]}$. Then $\kappa$ is $L_{\lambda^+ \omega}$-compact.*

PROOF. It suffices to show that every $\kappa$-complete filter $F$ on a set $I$ is contained in an $\omega_1$-complete ultrafilter $D$ on $I$. (For, by II.2.11, $D$ is then $\lambda^+$-complete.) Let $A = G^I / F$. We claim that $R_G^{[\kappa]} A = 0$. Indeed, if $B$ is a subgroup of $A$ of cardinality $< \kappa$, then by II.3.4, $B$ is embeddable in $G^I$, so $R_G B = 0$. Therefore $R_G A = 0$. Hence there is a non-zero homomorphism $\varphi : A \to G$; say $x \in G^I$ is such that $\varphi(x_F) \neq 0$. Composing $\varphi$ with the canonical surjection $\rho : G^I \to A$, we obtain a non-zero homomorphism $\theta = \varphi \circ \rho : G^I \to G$. By hypothesis on $G$, there is a subgroup $N$ of $G$ such that $\theta(x) \notin N$ and $G/N$ is slender. Thus we obtain a homomorphism

$$f : G^I \xrightarrow{\;\theta\;} G \to G/N$$

such that $f(x) \neq 0$. Now we apply Theorem III.3.2: there are $\omega_1$-complete ultrafilters $D_1, \ldots, D_n$ on $I$ such that for all $a \in G^I$, if for all $k$, $\text{supp}(a) \notin D_k$ then $f(a) = 0$. If we can prove that $F$ is contained in one of the $D_k$'s, then we will be done. So suppose, to the contrary, that this is not the case, i.e., for all $k = 1, \ldots, n$, there exists $S_k \in F \setminus D_k$. Then $\bigcap_k S_k$ belongs to $F$ but not to any $D_k$. Define

$$x'(i) = \begin{cases} x(i) & \text{if } i \in \bigcap_k S_k \\ 0 & \text{otherwise} \end{cases}$$

Then $\rho(x') = \rho(x)$, so $f(x') = f(x) \neq 0$. But $\text{supp}(x') \notin D_k$ for any $k$, which contradicts Theorem III.3.2. $\square$

**4.10 Corollary.** *$R_{\mathbf{Z}}$ satisfies the cardinal condition if and only if there is an $L_{\omega_1\omega}$-compact cardinal.* $\square$

See Exercises 14 and 17 for more conditions equivalent to the existence of an $L_{\omega_1\omega}$-compact cardinal. We now turn our attention to the question of when a radical commutes with direct products. Recall that always $T\prod_I A_i \subseteq \prod_I TA_i$ (cf. Exercise 9).

**4.11 Proposition.** *If $G$ is cotorsion-free, then $R_G$ commutes with countable products.*

PROOF. We must show that if $a \in \prod_{n\in\omega} R_G A_n$, then we have $a \in R_G \prod_{n\in\omega} A_n$. Given $\varphi: \prod_n A_n \to G$, let $\theta = \varphi \upharpoonright \prod_n \langle a(n)\rangle$. Since each $a(n) \in R_G A_n$, the kernel of $\theta$ contains $\bigoplus_n \langle a(n)\rangle$, and hence $\theta$ induces $\bar{\theta}: \prod_n \langle a(n)\rangle / \bigoplus_n \langle a(n)\rangle \to G$. Now the domain of $\bar{\theta}$ is pure-injective by V.1.16, so the range of $\bar{\theta}$ is cotorsion. But then by hypothesis on $G$, the range of $\bar{\theta}$ is 0, so $\varphi(a) = 0$. $\square$

The converse of 4.11 holds for torsion-free $G$: see Exercise 13. There are cotorsion-free groups which don't commute with uncountable products: see Exercise 15.

**4.12 Theorem.** *If $G$ is strongly cotorsion-free, then $R_G$ commutes with $\prod_\kappa$ whenever $\kappa$ is not $\omega$-measurable.*

PROOF. The proof is by induction on $\kappa$. The case $\kappa = \aleph_0$ is 4.11. The case of a singular $\kappa$ is straightforward, so we assume that $\kappa$ is regular (not $\omega$-measurable) and that for $\lambda < \kappa$ and all groups $A_\mu$, $\prod_{\mu<\lambda} R_G A_\mu \subseteq R_G \prod_{\mu<\lambda} A_\mu$. Suppose, to obtain a contradiction, that there is an element $a$ of $\prod_{\mu<\kappa} R_G A_\mu$ which is not in $R_G \prod_{\mu<\kappa} A_\mu$. Thus there exists $f: \prod_{\mu<\kappa} A_\mu \to G$ such that $f(a) \neq 0$. Let $\psi: \mathbf{Z}^\kappa \to \prod_{\mu<\kappa} A_\mu$ be defined by $\psi((n_\mu)_\mu) = (n_\mu a(\mu))_\mu$. The composition $\theta \overset{\text{def}}{=} f \circ \psi$ is a homomorphism: $\mathbf{Z}^\kappa \to G$ such that $\theta((e_\mu)_\mu) = f(a) \neq 0$. By the inductive hypothesis, $\theta \upharpoonright \mathbf{Z}^{<\kappa} \equiv 0$, so $\theta$ induces a non-zero homomorphism: $\mathbf{Z}_\kappa \to G$, which is a contradiction. $\square$

Measurability is a real barrier, as we see from the next results.

**4.13 Theorem.** *If $\kappa$ is the first measurable cardinal and $G$ is a strongly cotorsion-free group of cardinality $< \kappa$, then $R_G$ does not commute with $\prod_\kappa$.*

PROOF. Let $D$ be a normal $\kappa$-complete ultrafilter on $\kappa$; then $S \overset{\text{def}}{=} \{\alpha < \kappa : \alpha \text{ is a regular cardinal}\}$ belongs to $D$ (cf. II.2.15). Let $I = \bigcup_{\alpha \in S} \alpha \times \{\alpha\}$. For $X \subseteq I$, let

$$W_X \overset{\text{def}}{=} \{\alpha \in S : |\alpha \setminus \{\gamma : (\gamma, \alpha) \in X\}| < \alpha\}$$

and let $F = \{X \subseteq I : W_X \in D\}$. It is routine to check that $F$ is a $\kappa$-complete filter on $I$. We claim that

$(*)$    there is a $\kappa$-complete ultrafilter $U$ on $I$ which contains $F$.

Assuming this for a moment, let us finish the proof. (Note that the claim follows immediately if $\kappa$ is strongly compact.) Now $R_G \mathbf{Z}_\alpha = \mathbf{Z}_\alpha$ for every $\alpha \in S$ because $\text{Hom}(\mathbf{Z}_\alpha, G) = 0$ by hypothesis on $G$. We shall show that $\prod_{\alpha \in S} R_G \mathbf{Z}_\alpha \neq R_G \prod_{\alpha \in S} \mathbf{Z}_\alpha$ by defining a non-zero homomorphism from $\prod_{\alpha \in S} \mathbf{Z}_\alpha$ to $G$.

Since $G$ is torsion-free, we can identify $\mathbf{Z}$ with a subgroup of $G$. Given $z \in \prod_{\alpha \in S} \mathbf{Z}^\alpha$, let $z(\beta, \alpha)$ (for $\beta < \alpha$) denote the $\beta$th-coordinate of $z(\alpha) \in \mathbf{Z}^\alpha$; so $z(\beta, \alpha) \in \mathbf{Z}$. Then since $U$ is $\kappa$-complete, there is a unique $x \in \mathbf{Z} \subseteq G$ such that $\{(\beta, \alpha) \in I : z(\beta, \alpha) = x\}$ belongs to $U$ (cf. proof of II.3.2). Define $\varphi(z) = x$. Since $F \subseteq U$, $\varphi : \prod_{\alpha \in S} \mathbf{Z}^\alpha \to G$ is identically 0 on $\prod_{\alpha \in S} \mathbf{Z}^{<\alpha}$. Therefore $\varphi$ induces $\bar{\varphi} : \prod_{\alpha \in S} \mathbf{Z}_\alpha \to G$ which is clearly non-zero.

So it remains to prove $(*)$. For any $X \subseteq I$ define

$$C_X = \{\gamma \in \kappa : \{\alpha \in S : (\gamma, \alpha) \in X\} \in D\}$$

and let $U = \{X \subseteq I : C_X \in D\}$. We leave it to the reader to verify that $U$ is a $\kappa$-complete ultrafilter on $I$. To show that $F \subseteq U$ we use the normality of $D$. Let $X \in F$. Define $\theta : \kappa \to \kappa$ by:

$$\theta(\alpha) = \begin{cases} \sup(\alpha \setminus \{\gamma \in \alpha : (\gamma, \alpha) \in X\} & \text{if } \alpha \in S \\ 0 & \text{otherwise} \end{cases}$$

Since $X \in F$, $\theta$ is a regressive function, so there exists $\beta \in \kappa$ and $T \subseteq S$ such that $T \in D$ and $\theta(\alpha) = \beta$ for all $\alpha \in T$ (cf. II.2.16). Now one can check easily that $\gamma \in C_X$ if $\gamma > \beta$; hence $\kappa \setminus \beta \subseteq C_X$, so $C_X \in D$, and therefore $X \in U$. $\square$

**4.14 Corollary.** *If $\kappa$ is $\omega$-measurable, then $R_{\mathbf{Z}}$ does not commute with $\prod_\kappa$.* $\square$

In contrast, we have:

**4.15 Theorem.** *For any $\kappa$, if $A$ is a group whose cardinality is not $\omega$-measurable, then $R_{\mathbf{Z}}(A^\kappa) = (R_{\mathbf{Z}}A)^\kappa$.*

PROOF. We must show that $(R_{\mathbf{Z}}A)^\kappa \subseteq R_{\mathbf{Z}}(A^\kappa)$, i.e., if $f\colon A^\kappa \to \mathbf{Z}$ and $x \in (R_{\mathbf{Z}}A)^\kappa$, then $f(x) = 0$. Now $f = \Phi((g_D)_{D \in \mathcal{D}})$ for some $g_D\colon A^\kappa/D \to \mathbf{Z}$ where $D$ is an $\omega_1$-complete ultrafilter on $\kappa$ (cf. III.3.6). But $\delta_A\colon A \to A^\kappa/D$ is an isomorphism by II.3.2, so for all $D$, $g_D(x_D) = (g_D \circ \delta_A)(x(\mu))$ for some $\mu \in \kappa$, and therefore $g_D(x_D) = 0$ since $x(\mu) \in R_{\mathbf{Z}}A$. $\square$

The Chase radical, $\nu$, commutes with countable products: see Exercise 16. Eda 1987 has shown that $\nu$ does not commute with uncountable products.

## EXERCISES

1. If $M = \mathbf{Z}^{\omega_1}$, then there is a subgroup $A$ of $M$ such that $A^{\perp\perp}$ is not a direct summand of $M$. [Hint: start with $K \subseteq M^* = \mathbf{Z}^{(\omega_1)}$ such that $M^*/K$ is torsionless but not free.]

2. (i) $\mathbf{Z}_{\omega_1}$ is not the direct sum of a cotorsion group and a product. [Hint: Assume false. By Exercise IV.11, $(\mathbf{Z}_{\omega_1})^* = 0$, hence $\mathbf{Z}_{\omega_1}$ is cotorsion. Then use the fact that $\mathbf{Z}_{\omega_1}$ is $\aleph_1$-free to obtain from this a contradiction using the results of Chapter V.]
(ii) Prove that $\mathbf{Z}_{\omega_1}$ has no free summand.

3. (i) If $\kappa$ is non-$\omega$-measurable and $A$ is a subgroup of $\mathbf{Z}^\kappa$ such that $A = A^{\perp\perp}$, then $A$ is a dual group, i.e., $A \cong B^*$ for some $B$. [Hint: let $B =$ the image of $\iota^*\colon (\mathbf{Z}^\kappa)^* \to A^*$.]
(ii) For any group $B$, $B^*$ is isomorphic to a subgroup $A$ of $\mathbf{Z}^\kappa$ for some $\kappa$ such that $A = A^{\perp\perp}$.

(iii) Hence, $A$ is a dual group of non-$\omega$-measurable cardinality if and only if it can be embedded in a product $\mathbf{Z}^\kappa$ so that $\kappa$ is not $\omega$-measurable and $\mathbf{Z}^\kappa/A$ is torsionless.

4. (i) For any group $H$ such that $H^* = 0$, there is a subgroup $A$ of $\mathbf{Z}^\kappa$ for some $\kappa$ such that if $L = \mathbf{Z}^\kappa/A$, then $H \cong \ker(\sigma_L)$. [Hint: by III.3.8 there are arbitrarily large free groups which are also dual groups. So by 3(ii), for some $\kappa$ there is a subgroup $F$ of $\mathbf{Z}^\kappa$ such that $F = F^{\perp\perp}$ and $F$ has a subgroup $A$ such that $F/A \cong H$.]
(ii) There is a non-$\omega$-measurable $\kappa$ and a subgroup $A$ of $\mathbf{Z}^\kappa$ such that if $L = \mathbf{Z}^\kappa/A$, $L$ is not the direct sum of a product and a group $H$ such that $H^* = 0$. [Hint: let $H$ be slender such that $H^* = 0$, and use (i).]

5. Use Corollary 2.4 to give another proof that a direct sum of slender groups is slender.

6. (i) A subgroup $A$ of $\mathbf{Z}^\omega$ is a countable product if and only if $A$ is a closed subgroup of $\mathbf{Z}^\omega$ (topology as in section 2) if and only if there exist elements $\{a^n : n \in \omega\}$ of $\mathbf{Z}^\omega$ such that for all $m$, $a^n \in U_m$ for almost all $n$ and $A = \{\sum_n r_n a^n : r_n \in \mathbf{Z}\}$.
(ii) There is a closed subgroup $A$ of $\mathbf{Z}^\omega$ such that $\mathbf{Z}^\omega/A \cong J_p$. [Hint: consider the function $\varphi : \mathbf{Z}^\omega \to J_p$ defined by $\varphi((x_n)_n) = \sum_n x_n p^n$.] This provides a counterexample to Lemma 95.1 of Fuchs 1973.

7. Prove the converse of 3.2: for any product $C$ and group $D$ both of non-$\omega$-measurable cardinality, there exist non-$\omega$-measurable cardinals $\kappa$ and $\lambda$ and a homomorphism $f : \mathbf{Z}^\lambda \to \mathbf{Z}^\kappa$ such that $\ker(f) \cong D^*$ and $\mathrm{coker}(f) \cong C \oplus \mathrm{Ext}(D, \mathbf{Z})$.

8. Prove the following generalization of 3.1: let $R$ be any ring, let $H$ be any $R$-module, let $M$ and $N$ be reflexive $R$-modules and let $f : M \to N$ be a homomorphism such that $f[M]^\perp$ is a direct summand of $N^*$ and $\mathrm{Ext}(M^*, H) = 0$. (Here $M^* = \mathrm{Hom}_R(M, H)$.) Then there is a module $D$, which is a homomorphic image of $M^*$ such that $\ker(f) \cong D^*$ and $\mathrm{coker}(f) \cong C \oplus \mathrm{Ext}(D, H)$, where $C$ is a direct summand of $N$.

9. If $T$ is a preradical, then for any groups $A_i$, $T(\bigoplus_I A_i) = \bigoplus_I T(A_i)$ and $T(\prod_I A_i) \subseteq \prod_I T(A_i)$. [Hint: for $\subseteq$ use $\pi_j: \prod A_i \to A_j$; for $\supseteq$ use $\iota_j: A_j \to \bigoplus A_i$.]

10. $A$ is cotorsion-free if and only if $\text{Hom}(\mathbf{Z}_\omega, A) = 0$. [Hint: for ($\Leftarrow$) use the structural description of $\mathbf{Z}_\omega$ in Exercise V.4]

11. Let $\mathcal{X}$ be a *set* of groups and let $G = \bigoplus_{X \in \mathcal{X}} X$ and $H = \prod_{X \in \mathcal{X}} X$. Show that $R_G = R_{\mathcal{X}} = R_H$ and $S_G = S_{\mathcal{X}}$.

12. (i) An *annihilator class* of groups is a class of groups which is closed under direct products and subgroups. Show that the map which takes a radical $R$ to $\{A: RA = 0\}$ is a one-one correspondence between annihilator classes and radicals.
(ii) Formulate an analogous result for socles.

13. If $G$ is torsion-free and $R_G$ commutes with countable products, then $G$ is cotorsion-free. [Hint: to show that $J_p$ is not contained in $G$, use the fact that $J_p$ is a direct summand of a product of cyclic $p$-groups; to show that $G$ is reduced, use the fact that $R_{\mathbf{Q}}$ is the torsion subgroup functor.]

14. Prove that the following are equivalent. (See Exercise II.8 for the definition of $\kappa$-torsionless.)
(1) $R_\mathbf{Z} = R_\mathbf{Z}^{[\kappa]}$;
(2) for every group $A$, $A$ $\kappa$-torsionless $\Rightarrow$ $A$ torsionless;
(3) for every group $A \neq 0$, $A$ $\kappa$-torsionless $\Rightarrow$ $A^* \neq 0$;
(4) $\kappa$ is $L_{\omega_1 \omega}$-compact.
[Hint: for the proof of (3) $\Rightarrow$ (4), use III.3.2; cf. 4.9.]

15. There is a cotorsion-free group $G$ such that $R_G$ does not commute with $\prod_{\omega_1}$. [Hint: Let $G_\alpha (\alpha \in \omega_1)$ be a rigid system of groups, i.e., $\text{Hom}(G_\alpha, G_\beta) = 0$ if $\alpha \neq \beta$; let $G = \prod_\alpha G_\alpha / \prod_\alpha^{<\omega_1} G_\alpha$. Show that $R_G G_\alpha = G_\alpha$.]

16. The Chase radical $\nu$ commutes with countable products. [Hint: cf. proof of 4.11.]

17. For any radical $R$, define inductively $R^{\alpha+1} = R \circ R^\alpha$ and $R^\mu = \bigcap_{\alpha < \mu} R^\alpha$ for limit $\mu$. Let $R^\infty = \bigcap_\alpha R^\alpha$. Prove:

(i) $R^\infty$ is well-defined and is a radical which is a socle;

(ii) $R_G^\infty(A) = \sum\{B \subseteq A\colon \mathrm{Hom}(B, G) = 0\}$;

(iii) $\mathrm{Hom}(R_G^\infty A, G) = 0$;

(iv) if $\kappa$ is regular, $\kappa$ is $L_{\omega_1\omega}$-compact if and only if $R_{\mathbf{Z}}^\infty = R_{\mathbf{Z}}^{\infty[\kappa]}$. [Hint: show that $(R_{\mathbf{Z}}^\infty)^{[\kappa]} = (R_{\mathbf{Z}}^{[\kappa]})^\infty$.]

## NOTES

1.1 through 1.5 are based on Nunke 1962a. Parts (b) and (c) of Theorem 1.5 come from Dugas-Göbel 1979b. Theorem 1.9 and Corollary 1.11 are due to Huber 1983a. Section 2 is from Nunke 1962b, except for 2.6 which comes from Rychkov 1980 and Göbel-Richkov-Wald 1981.

Theorem 3.1 and Corollaries 3.2 and 3.3 are due to Dugas-Göbel 1979b. For further generalizations see also Huber 1979, Dugas-Göbel 1981, and Huber-Warfield 1981. The notion of strongly cotorsion-free and Lemmas 3.6 and 3.7 are from Dugas-Göbel 1985a. For more on homomorphic images of $\mathbf{Z}^\kappa$ and $\mathbf{Z}^{<\kappa}$, see Wald 1983b, 1984, 1985 and 1987.

Proposition 4.3 is due to Chase 1963a; 4.5 and 4.6 are from Eda 1987. Theorems 4.7, 4.12 and 4.13 are due to Dugas-Göbel 1985a (the last with help from Mekler). Theorem 4.8 is from Dugas 1985, where it is proved for $\kappa$ strongly compact. Theorems 4.9 and 4.15 are due to Eda-Abe 1987. Proposition 4.11 is from Fay-Oxford-Walls 1983b.

Exercises. 3:Nunke 1962a; 4:Dugas-Göbel 1981; 6:Göbel-Wald 1979; 8:Dugas-Göbel 1981; 10:Göbel-Wald 1979; 13:Fay-Oxford-Walls 1983b; 14:Bergman-Solovay 1987/Eda-Abe 1987; 15:Dugas-Göbel 1985a; 17:(i)-(iii)Fay-Oxford-Walls 1983a, (iv)Eda-Abe 1987 and Eda 19??b.

# CHAPTER X
# ITERATED SUMS AND PRODUCTS

In this chapter we shall consider the smallest class of non-zero groups containing $\mathbb{Z}$ and closed under direct sums and products; this class is called the *Reid class* and contains all the "classically" known dual groups. However, Eda showed that there is a dual group, even a reflexive group, which is not in the Reid class: see 2.6.

In the two sections of this chapter we consider two different notions of complexity in forming groups in the Reid class: the size of the index sets used; and the ordinal rank of alternations of $\bigoplus$ and $\prod$ in the construction of the group. The goal is to show that *a priori* different constructions do in fact construct different groups.

## §1. The Reid class

One method of constructing new dual groups from old ones is to take a class of dual groups and close under direct sums and products over index sets of non-$\omega$-measurable cardinality. If each original group is reflexive, then all the groups constructed will be reflexive as well. Until recently the only dual groups known were those constructed in this way, starting with $\mathbb{Z}$.

For a regular infinite cardinal $\kappa$, let $\text{Reid}_\kappa$ denote the least class of non-zero groups containing $\mathbb{Z}$ and closed under direct sums and products over index sets of cardinality $< \kappa$. Throughout this chapter, $\mu$ will denote the first measurable cardinal ($=$ the first $\omega$-measurable cardinal; cf. section II.2). If one assumes that no measurable cardinal exists, then $\text{Reid}_\mu$ is the Reid class.

As we know from III.3.8, $\mathbb{Z}^{(\lambda)}$ — which belongs to $\text{Reid}_{\lambda^+}$ — is not reflexive if $\lambda$ is $\omega$-measurable. However, the following result implies that every group in $\text{Reid}_\mu$ is reflexive.

**1.1 Theorem.** *If $\{A_i : i \in I\}$ is a family of reflexive groups and $I$ is not $\omega$-measurable, then $\bigoplus_I A_i$ and $\prod_I A_i$ are reflexive groups.*

PROOF. We will show that $\prod_I A_i$ is reflexive and leave the analogous proof for $\bigoplus_I A_i$ to the reader. Let $\sigma: \prod_I A_i \to (\prod_I A_i)^{**}$ and $\sigma_i: A_i \to A_i^{**}$ be the canonical maps defined in section I.1. Let $\theta: (\bigoplus_I A_i^*)^* \to \prod_I A_i^{**}$ be the isomorphism which takes $g \in (\bigoplus_I A_i^*)^*$ to $(g \circ \lambda_i)_I$ (cf. section I.1; here $\lambda_i$ is the canonical injection of $A_i^*$ into $\bigoplus_I A_i^*$). Finally, let $\Phi: \bigoplus_I A_i^* \to (\prod_I A_i)^*$ be the map defined in III.1.4. Recall that since $I$ is not $\omega$-measurable, $\Phi$ is an isomorphism and $\langle \Phi((g_i)_I), (a_i)_I \rangle = \sum_I \langle g_i, a_i \rangle$.

Since each $\sigma_i$ is an isomorphism, $(\sigma_i)_I: \prod_I A_i \to \prod_I A_i^*$ is an isomorphism. To show that $\sigma$ is an isomorphism, it suffices to show that $\theta \circ \Phi^* \circ \sigma = (\sigma_i)_I$. So let $a = (a_i)_I \in \prod_I A_i$ and let $\varphi = \langle \theta \circ \Phi^* \circ \sigma, a \rangle \in \prod_I A_i^*$. For any $i \in I$ we must show that the $i$th component, $\varphi_i$, of $\varphi$ equals $\sigma_i(a_i)$. Let $f \in A_i^*$. Then

$$\langle \sigma_i(a_i), f \rangle = f(a_i).$$

On the other hand, $\Phi^*(\sigma(a)) = \sigma(a) \circ \Phi$, so $\varphi_i = \sigma(a) \circ \Phi \circ \lambda_i$ and hence

$$\begin{aligned} \langle \varphi_i, f \rangle &= \langle \sigma(a) \circ \Phi, \lambda_i(f) \rangle = \langle \sigma(a), \Phi(\lambda_i(f)) \rangle \\ &= \langle \Phi(\lambda_i(f)), a \rangle = f(a_i). \end{aligned}$$

Since $\sigma_i(a_i)$ and $\varphi_i$ agree on all elements of $A_i^*$, they are equal and the proof is complete. $\square$

Later (XI.1.12 and XI.4.15) we will show that there is a reflexive group which is not in $\mathrm{Reid}_\mu$. (See also 2.6 in this chapter.)

A major technical tool in the proofs in this chapter is the following consequence of Lemma III.3.9. (See also III.1.9.)

**1.2 Lemma.** *Let $\varphi: \prod_{i \in I} A_i \to \bigoplus_{j \in J} B_j$ be a homomorphism where $B_j$ is reduced and torsion-free for all $j \in J$.*
(i) *If $I$ is not $\omega$-measurable, there is a finite subset $Y$ of $J$ and a finite subset $X$ of $I$ such that $\varphi[K] \subseteq \bigoplus_{j \in Y} B_j$ where $K = \{a: a(i) = 0 \text{ for all } i \in X\}$.*
(ii) *If $I$ is $\omega$-measurable, there is a finite subset $Y$ of $J$ and finitely many $\omega_1$-complete ultrafilters $D_0, \ldots, D_n$ on $I$ such that $\varphi[K] \subseteq \bigoplus_{j \in Y} B_j$ where $K = \{a: \text{for all } k \leq n, \{i \in I: a(i) = 0\} \in D_k\}$.* $\square$

We will show that the Reid class increases as we increase the cardinality of the index sets.

**1.3 Theorem.** *If $\kappa$ is regular, infinite and not $\omega$-measurable, then $Z^{(\kappa)}$ and $Z^\kappa$ do not belong to Reid$_\kappa$. In fact, there is no homomorphism from an element of Reid$_\kappa$ onto either of those groups.*

PROOF. We first establish the theorem for $Z^\kappa$. We show that the class of groups which have no homomorphisms onto $Z^\kappa$ contains $Z$ and is closed under direct sums and products of fewer than $\kappa$ elements; and so this class contains Reid$_\kappa$. This is clear if $A = Z$. Assume for the moment that the class is closed under direct sums of fewer than $\kappa$ elements. With this assumption in hand, suppose that $A = \prod_{\alpha < \lambda} A_\alpha$, where each $A_\alpha$ is a non-zero group in the class and $\lambda$ is a cardinal $< \kappa$. Let $\varphi \colon A \to Z^\kappa$ be a homomorphism. For $\nu < \kappa$, let $\varphi_\nu$ be $\varphi$ composed with the projection of $Z^\kappa$ onto the $\nu$th coordinate. The slenderness of $Z$ implies that for every $\nu$ there is a finite subset $X_\nu \subseteq \lambda$ such that for all $a \in A$, $\varphi_\nu(a) = 0$ if $a{\restriction}X_\nu \equiv 0$. By the pigeon-hole principle there is a subset $I \subseteq \kappa$ of cardinality $\kappa$ and a finite set $X \subseteq \lambda$ so that $X_i = X$ for all $i \in I$. Let $\pi$ be the projection of $Z^\kappa$ onto $Z^I$. Define $\psi \colon \prod_{\alpha \in X} A_\alpha \to Z^I$ by: $\psi(a) = \pi(\varphi(b))$ where $b{\restriction}X = a$; by the choice of $X$, $\psi$ is well defined. Since $X$ is a finite set, $\prod_{\alpha \in X} A_\alpha \cong \bigoplus_{\alpha \in X} A_\alpha$. Since we have assumed that the class is closed under direct sums of cardinality $< \kappa$, $\psi$ is not surjective; hence $\varphi$ is not surjective either. This completes the proof when $A$ is a direct product.

There remains the case when $A = \bigoplus_{\alpha < \lambda} A_\alpha$ $(\lambda < \kappa)$. This case follows immediately from the next lemma, which we state in a more general form for the purposes of an application in section XI.1. Note that $A$ is the direct limit of the direct system $(B_i, \tau_{ij} \colon i < j < \lambda)$ where $B_i = \bigoplus_{\alpha < i} A_\alpha$ and $\tau_{ij}$ is inclusion. (For the immediate purposes, one need only consider the case where $B_i$ is as just defined, and $\tau_i$ is the inclusion of $B_i$ into $A$ — so that $A = \bigcup_{i \in I} B_i$. See the beginning of section XI.1 for a discussion of direct limits including the notation used here.)

**1.4 Lemma.** *Assume $\kappa$ is a cardinal. Suppose $A = \varinjlim B_i$ where $(B_i, \tau_{ij} : i < j \in I)$ is a direct system, $|I| \leq \kappa$, and none of the $B_i$ have $\mathbf{Z}^\kappa$ as a homomorphic image. Then $\mathbf{Z}^\kappa$ is not a homomorphic image of $A$.*

PROOF. Suppose $\varphi : A \to \mathbf{Z}^\kappa$. Partition $\kappa$ into $\{Y_i : i \in I\}$ where each of the $Y_i$ has cardinality $\kappa$. Let $\pi_i$ be the projection of $\mathbf{Z}^\kappa$ onto $\mathbf{Z}^{Y_i}$. For each $i \in I$, choose $a_i \in \mathbf{Z}^{Y_i}$ so that $a_i \notin (\pi_i \circ \varphi \circ \tau_i)[B_i]$. Let $a \in \mathbf{Z}^\kappa$ be such that for all $i \in I$ $a{\restriction}Y_i = a_i$. Then $a \notin \varphi[A]$, since $A$ is the union of the $\tau_i[B_i]$ and for all $i \in I$, $a \notin \varphi[\tau_i[B_i]]$. $\square$

**Proof of 1.3, continued.** We now have to prove the theorem for $\mathbf{Z}^{(\kappa)}$. This time we consider the class of groups $A$ satisfying the property that every homomorphism from $A$ to $\mathbf{Z}^{(\kappa)}$ has a *bounded image,* i.e., the image is contained in $\mathbf{Z}^{(\nu)}$ for some $\nu < \kappa$. Again we need to show that this class contains $\mathbf{Z}$ and is closed under direct sums and products of fewer than $\kappa$ elements. The case when $A$ is $\mathbf{Z}$ or a direct sum is easy, so we suppose that $A = \prod_{\alpha < \lambda} A_\alpha$ where each of the $A_\alpha$ is in the class and $\lambda$ is a cardinal $< \kappa$. Let $\varphi : A \to \mathbf{Z}^{(\kappa)}$ be a homomorphism. By Lemma 1.2 there is a finite subset $X \subseteq \lambda$ and a cofinite subset $I \subseteq \kappa$ so that $\mathbf{Z}^{(I)} \cap \varphi[A] = \mathbf{Z}^{(I)} \cap \varphi[A']$ where $A' = \{a \in \lambda : a(\alpha) = 0, \text{ for all } \alpha \notin X\} \cong \bigoplus_{\alpha \in X} A_\alpha$. By induction, $\varphi {\restriction} A'$ has bounded image and hence so does $\varphi$. $\square$

The second sentence of Theorem 1.3 doesn't always hold for measurable cardinals, as can be seen from the following example.

**1.5 Example.** Let $\lambda$ be a measurable cardinal and let $D$ be a $\lambda$-complete non-principal ultrafilter on $\lambda$; then $D$ is uniform, i.e., every member of $D$ has cardinality $\lambda$. Let $A = (\mathbf{Z}^{(\lambda)})^\lambda$. We will show that there is a homomorphism from $A$ onto $\mathbf{Z}^{(2^\lambda)}$. In fact, there is a homomorphism from $A$ onto the ultrapower $B \overset{\text{def}}{=} (\mathbf{Z}^{(\lambda)})^\lambda / D$. $B$ is a free group by II.3.8. Moreover, by II.3.6, $B$ has cardinality $2^\lambda$. (Note that $\lambda$ is strongly inaccessible by II.2.13.) Let $\kappa = \lambda^+$. We've shown that there is a homomorphism from $A$, a member of $\mathrm{Reid}_\kappa$, onto $\mathbf{Z}^{(\kappa)}$.

Assuming the consistency of the existence of a supercompact

cardinal Silver (cf. Jech 1978, Theorem 88) has shown that it is consistent that there exists a measurable cardinal $\lambda$ such that $2^{\lambda^+} = 2^\lambda$. For such a $\lambda$, $B$ has a basis of size $2^{\lambda^+}$, so there is a homomorphism of $B$ onto $\mathbf{Z}^{\lambda^+}$. Hence there is a homomorphism of $A$, a member of $\mathrm{Reid}_\kappa$, onto $\mathbf{Z}^\kappa$.

## §2. Types in the Reid class

A different way to understand the complexity of groups in the Reid class is in terms of the number of alternations of $\bigoplus$ and $\prod$ in the construction of the group. We will assign to each group in the Reid class a type, which measures this complexity. More exactly, we will say what it means for a group to have a certain type and prove that every group in the Reid class has at least one type and that every group other than $\mathbf{Z}$ in $\mathrm{Reid}_\mu$, where $\mu$ is the first measurable cardinal, has exactly one type.

A *type* of a group is a pair $(\alpha, P)$, $(\alpha, S)$, or $(\alpha, M)$ where $\alpha$ is an ordinal. The letter $P$ is thought of as standing for product, $S$ for sum, and $M$ for mixed. For any group $A$ in the Reid class, we will say that $A$ *has type* $^*\alpha$ if $A$ has type $(\alpha, X)$ for some $X \in \{P, S, M\}$.

We now define by induction on $\alpha$ what it means for $A$ to have type $(\alpha, X)$. To begin, we say that $\mathbf{Z}$ has types $(0, S)$ $(0, P)$ and $(0, M)$ and that any finite direct sum of copies of $\mathbf{Z}$ has type $(0, M)$. If $A = B \oplus C$ where $B$ has type $(\alpha, P)$ and $C$ has type $(\alpha, S)$, then $A$ has type $(\alpha, M)$. Suppose $\alpha = \beta + 1$. A group $A$ in the Reid class has type $(\alpha, P)$ if $A = \prod_{i \in I} A_i$ where each $A_i$ is in the Reid class and has type$^* < \alpha$, and for infinitely many $i$, $A_i$ has type $(\beta, S)$. Similarly, a group $A$ in the Reid class has type $(\alpha, S)$ if $A = \bigoplus_{i \in I} A_i$ where each $A_i$ is in the Reid class and has type$^* < \alpha$, and for infinitely many $i$, $A_i$ has type $(\beta, P)$. Next suppose that $\alpha$ is a limit ordinal. A group $A$ in the Reid class has type $(\alpha, P)$ if $A = \prod_{i \in I} A_i$ where each $A_i$ is in the Reid class and has type$^* < \alpha$, and for all $\beta < \alpha$ there is $i \in I$ so that $A_i$ has type$^* > \beta$. Similarly, a group $A$ in the Reid class has type $(\alpha, S)$ if $A = \bigoplus_{i \in I} A_i$ where each $A_i$ is in the Reid class and has type$^* < \alpha$, and for all $\beta < \alpha$ there is $i \in I$ so that $A_i$ has type$^* > \beta$.

Obviously, all possible types occur in the Reid class. We will first show that every group in the Reid class has at least one type. Later we will show that every group in $\text{Reid}_\mu$ has exactly one type.

**2.1 Proposition.** (i) *Suppose $A$ and $B$ are groups in the Reid class such that $A$ has type $(\alpha, X)$ and $B$ has type $(\beta, Y)$. If $\alpha < \beta$, then $A \oplus B$ has type $(\beta, Y)$;*

(ii) *If $\beta \neq 0$ and $B = \prod_{i \in I} B_i$ ( resp. $B = \bigoplus_{i \in I} B_i$) where each $B_i$ has type $(\beta, P)$ ( resp. $(\beta, S)$), then $B$ has type $(\beta, P)$ ( resp. $(\beta, S)$).*

PROOF. (i) There are three cases to consider ($Y = P, S, M$); since they are all easy we will only do the case where $B$ has type $(\beta, P)$. If $B = \prod_{i \in I} B_i$ attests to the fact that $B$ has type $(\beta, P)$, then $A \times \prod_{i \in I} B_i$ attests to the fact that $A \oplus B$ has type $(\beta, P)$.

(ii) Again we do only one case. If $\prod_{j \in J_i} B_{i,j}$ attests to the fact that $B_i$ has type $(\beta, P)$, then $B = \prod_{i \in I} \prod_{j \in J_i} B_{i,j}$ attests to the fact that $B$ has type $(\beta, P)$. □

**2.2 Theorem.** *Every group in the Reid class has ( at least) one type.*

PROOF. The proof is by induction on the construction of groups in the Reid class. We consider a group $A$ which we can suppose has infinite rank, since otherwise we have defined a type for $A$. Suppose that $A = \prod_{i \in I} A_i$ where each $A_i$ is a group in the Reid class and has a type. (The case of direct sums is entirely analogous and we will not discuss it specifically.) By writing any group of mixed type as the product of two other groups we can assume that each $A_i$ has type $(\alpha_i, X_i)$ where $X_i$ is either $S$ or $P$. Let $F = \{i : A_i$ has finite rank$\}$. If $F$ is infinite, $A = \prod_{i \in I \setminus F} A_i \times \mathbf{Z}^J$. Otherwise, by Exercise 3, $A \cong \prod_{i \in I \setminus F} A_i$. So, using Exercise 4, we can assume that each $\alpha_i \geq 1$. We now proceed to prove by induction on $\alpha \overset{\text{def}}{=} \sup\{\alpha_i + 1 : i \in I\}$ that $A$ has a type with type* $\leq \alpha$. If $\alpha$ is a limit ordinal, then clearly $A$ has type $(\alpha, P)$. Suppose now that $\alpha = \beta + 1$. Let $J = \{i : \alpha_i = \beta$ and $X_i = S\}$. If $J$ is infinite, then by definition $A$ has type $(\alpha, P)$. Otherwise, let $K = \{i : \alpha_i = \beta$ and $X_i = P\}$ and $L = \{i : \alpha_i < \beta\}$. Let $C_1 = \prod_{i \in K} B_i$, $C_2 = \prod_{i \in J} B_i$

and $C_3 = \prod_{i \in L} B_i$. Now $A = C_1 \times C_2 \times C_3$. By 2.1(ii), $C_1$ has type $(\beta, P)$ (or is trivial) and $C_2$ has type $(\beta, S)$ (or is trivial). By the induction hypothesis, $C_3$ has type* at most $\beta$. In this case $A$ has a type, by definition of type and by 2.1(i). $\square$

Our next goal is to show that every group has a unique type. We will only be able to prove this for groups whose type* is not $\omega$-measurable. For now, we will prove it for groups in $\mathrm{Reid}_\mu$ (where $\mu$ is the first measurable cardinal); later we will extend the result to groups constructed using index sets of arbitrary size (see 2.10). Notice that if $A$ belongs to $\mathrm{Reid}_\mu$ and has type $(\alpha, X)$, then $\alpha$ is not $\omega$-measurable.

**2.3 Theorem.** *Let $\alpha$ be a non-zero ordinal. No group in $\mathrm{Reid}_\mu$ of type $(\alpha, P)$ is a direct summand of a group in $\mathrm{Reid}_\mu$ of type $(\alpha, S)$. Also, no group in $\mathrm{Reid}_\mu$ of type $(\alpha, S)$ is a direct summand of a group in $\mathrm{Reid}_\mu$ of type $(\alpha, P)$.*

PROOF. The proof is by induction on $\alpha$. First we do the case $\alpha = 1$. Let $I$ and $J$ be infinite sets. Since $\mathbf{Z}^{(I)}$ is free and $\mathbf{Z}^J$ is not, $\mathbf{Z}^J$ is not a summand of $\mathbf{Z}^{(I)}$. On the other hand, $h \colon \mathbf{Z}^J \to \mathbf{Z}^{(I)}$ cannot be a projection onto a direct summand because by Lemma 1.2(i) the range of $h$ is of finite rank.

Suppose now $\alpha = \beta + 1$; let $A = \prod_{i \in I} A_i$ and $B = \bigoplus_{j \in J} B_j$ be groups of type $(\alpha, P)$ and $(\alpha, S)$ respectively and suppose that the expressions above attest to the fact that $A$ and $B$ have the claimed types. Moreover, we can assume that if type*$(A_i)$ (respectively, type*$(B_j)$) equals $\beta$, then type$(A_i) = (\beta, S)$ (respectively, type$(B_j) = (\beta, P)$). Suppose first that $A$ is a direct summand of $B$. Applying Lemma 1.2(i) to the inclusion of $A$ into $B$, there is a finite subset $X \subseteq I$ and a finite subset $Y \subseteq J$ so that $\{a \in A \colon a(i) = 0$ for all $i \in X\} \subseteq \bigoplus_{j \in Y} B_j$. There must exist $i \notin X$ so that $A_i$ has type $(\beta, S)$. Since $\bigoplus_{j \in Y} B_j$ has type* $\leq \beta$, it is a direct summand of a group of type $(\beta, P)$; so $A_i \cong \{a \in A \colon a(\ell) = 0,$ if $\ell \neq i\}$ is a direct summand of a group of type $(\beta, P)$, which is a contradiction by induction.

Now assume that $B$ is a direct summand of $A$. Let $h \colon A \to B$

be a projection. By Lemma 1.2(i) there is a finite subset $Y \subseteq J$ and a finite subset $X \subseteq I$ such that $h[K] \subseteq \bigoplus_{j \in Y} B_j$, where $K = \{a \in A : a(i) = 0 \text{ for all } i \in X\}$. Let $g$ be a projection of $B$ onto $\bigoplus_{j \notin Y} B_j$. Let $H$ be the kernel of $g \circ h$. So $A \cong \bigoplus_{j \notin Y} B_j \oplus H$ and $H \supseteq K$. Hence $\bigoplus_{j \notin Y} B_j$ is isomorphic to a direct summand of $A/K \cong \bigoplus_{i \in X} A_i$. Now $A/K$ is a direct summand of a group of type $(\beta, S)$ and there is $j \notin Y$ such that $B_j$ is of type $(\beta, P)$; hence the inductive assumption has been contradicted.

The case where $\alpha$ is a limit ordinal is similar but easier. $\square$

**2.4 Corollary.** *Every group in Reid$_\mu$ other than* **Z** *has at most one type.*

PROOF. Suppose that $A$ has two distinct types, $(\alpha, X)$ and $(\beta, Y)$. If $\alpha = \beta = 0$, this is clearly impossible; if $\alpha = \beta > 0$, this is a contradiction of 2.3 since $A$ is a direct summand of itself. So we can assume that $\alpha < \beta$. Suppose now that $Y \neq P$. Since $A$ has type* $\alpha < \beta$, it is a direct summand of a group of type $(\beta, P)$; this contradicts 2.3 because either $A$ has type $(\beta, S)$ or (if $Y = M$) it has a summand of type $(\beta, S)$. Similarly, if $Y = P$ we obtain a contradiction because $A$ is also a summand of a group of type $(\beta, S)$. $\square$

G.A. Reid 1967 posed the problem (Problem 76 in Fuchs 1973, p. 184) whether the groups **Z**$^\omega$, (**Z**$^\omega$)$^{(\omega)}$, ((**Z**$^\omega$)$^{(\omega)}$)$^\omega$, ... are all distinct. The answer to this is now seen to be "yes" because these groups have, respectively, the types $(1, P)$, $(2, S)$, $(3, P)$, ...

It is possible to extend this analysis to wider classes. Namely we can begin with any class $\mathcal{C}$ of non-zero slender groups of non-$\omega$-measurable cardinality, and then consider the smallest class $\mathcal{D}$ (of non-zero groups) containing $\mathcal{C}$ and closed under direct sums and products. In order to analyze the groups in $\mathcal{D}$ it is necessary to change slightly the definition of type: a group has type $(0, S)$ if it is the direct sum of groups in $\mathcal{C}$. Otherwise the definition is as before; in particular, note that for $n < \omega$ there is a group of type $(n, X)$ if and only if either $n$ is even and $X$ is $S$ or $n$ is odd and

$X$ is $P$. We can then prove, very much as before, that every group in $\mathcal{D}$ has one and only one type; the proof of uniqueness of type requires, in addition to 2.3, the following lemma.

**2.5 Lemma.** *With definitions as above and $n < \omega$, no group of type\* $n + 1$ is a direct summand of a group of type $n$.*

PROOF. The proof begins with the observation that a group of type $(1, P)$ is not a summand of a group of type $(0, S)$ because a group of type $(0, S)$ is slender by III.1.10. The proof proceeds by induction on $n$, using 1.2 as in the proof of 2.3. □

**2.6 Example.** The construction of a hierarchy of groups will fail if we choose as a starting point a group which is isomorphic to both the countable direct sum and the countable direct product of copies of itself. There is a very simple example of such a group. Consider $C(\mathbf{Q}, \mathbf{Z})$, the group of continuous functions from the rationals as a topological space to $\mathbf{Z}$. (Here $\mathbf{Q}$ is given the relative topology induced from the usual topology of $\mathbf{R}$, and $\mathbf{Z}$ is given the discrete topology.) We will establish several properties of this group in a series of claims.

**2.6A.** $C(\mathbf{Q},\, \mathbf{Z}) \cong C(\mathbf{Q},\, \mathbf{Z})^{\omega}$.

PROOF. Choose an increasing sequence $0 < a_0 < a_1 < \ldots$ of irrationals. Then $C(\mathbf{Q}, \mathbf{Z})$ is isomorphic to

$$C((-a_0, a_0) \cap \mathbf{Q}, \mathbf{Z}) \times \prod_{n \in \omega} C((a_n, a_{n+1}) \cap \mathbf{Q}, \mathbf{Z}) \times$$
$$\prod_{n \in \omega} C((-a_{n+1},\, -a_n) \cap \mathbf{Q}, \mathbf{Z})$$

which is isomorphic to $C(\mathbf{Q}, \mathbf{Z})^{\omega}$. □

**2.6B.** $C(\mathbf{Q},\, \mathbf{Z}) \cong C(\mathbf{Q},\, \mathbf{Z})^{(\omega)}$.

PROOF. Choose a decreasing sequence $a_n$ $(n \in \omega)$ of irrationals so that $\lim_{n \to \omega} a_n = 0$. Let $\mathbf{1}$ be the function which is constantly 1.

Then $C(\mathbf{Q}, \mathbf{Z})$ is isomorphic to

$$
\begin{aligned}
\langle 1 \rangle \quad &\times \quad C((-\infty, -a_0) \cap \mathbf{Q}, \mathbf{Z}) \times \bigoplus_{n \in \omega} C((-a_n, -a_{n+1}) \cap \mathbf{Q}, \mathbf{Z}) \\
&\times \quad \bigoplus_{n \in \omega} C((a_{n+1}, a_n) \cap \mathbf{Q}, \mathbf{Z}) \times C((a_0, \infty) \cap \mathbf{Q}, \mathbf{Z}) \\
&\cong \quad \mathbf{Z} \oplus C(\mathbf{Q}, \mathbf{Z})^{(\omega)} \\
&\cong \quad \mathbf{Z} \oplus (\mathbf{Z} \oplus C(\mathbf{Q}, \mathbf{Z})^{(\omega)})^{(\omega)} \\
&\cong \quad (\mathbf{Z} \oplus C(\mathbf{Q}, \mathbf{Z})^{(\omega)})^{(\omega)}
\end{aligned}
$$

which is isomorphic to $C(\mathbf{Q}, \mathbf{Z})^{(\omega)}$. $\square$

**2.6C.** $C(\mathbf{Q}, \mathbf{Z})$ *is a separable group.*

PROOF. It is easy to see that $C(\mathbf{Q}, \mathbf{Z})$ is a pure subgroup of $\mathbf{Z}^{\mathbf{Q}}$. Hence it is separable by IV.2.6 and IV.2.8.

By Claims 2.6A and 2.6B, $C(\mathbf{Q}, \mathbf{Z})$ is not in Reid$_\mu$, and hence not in the Reid class, because at least one of the operations of taking an infinite direct sum or an infinite product must increase the type. We shall prove in XI.4.15 that $C(\mathbf{Q}, \mathbf{Z})$ is a reflexive group; then by the claims above it is a non-zero reflexive group which is not in the Reid class. (See also XI.1.12.) For now, we can easily prove the following.

**2.6D.** *The group* $C(\mathbf{Q}, \mathbf{Z})^*$ *is a non-zero dual group which is not in the Reid class.*

PROOF. $C(\mathbf{Q}, \mathbf{Z})^*$ is non-zero by 2.6C, and we have $(C(\mathbf{Q}, \mathbf{Z})^*)^{(\omega)} \cong C(\mathbf{Q}, \mathbf{Z})^* \cong (C(\mathbf{Q}, \mathbf{Z})^*)^\omega$. $\square$

Finally, we show how to extend 2.4 to groups in the Reid class whose type* is not $\omega$-measurable. What we need is a generalization of 2.3. Our proof will proceed by replacing $\omega$-measurable index sets by sets of smaller cardinality. We begin with an *ad hoc* definition.

A group in the Reid class of type* $\alpha$ is called *small* if there is a construction of it from $\mathbf{Z}$ showing that it has type* $\alpha$ in which the index sets have cardinality at most $|\alpha| + \aleph_0$.

**2.7 Lemma.** *Any group in the Reid class has a small direct summand of the same type.*

PROOF. This is a simple induction. $\square$

**2.8 Lemma.** *Suppose $h: A \to B^J$ is a homomorphism. Then $B^J = C \oplus D$ such that $\operatorname{im}(h) \subseteq C$ and $C \cong B^{J'}$ where $|J'| \leq |B|^{|A|}$.*

PROOF. Define an equivalence relation $\equiv$ on $J$ by $i \equiv j$ if and only if for all $a \in A$ $h(a)(i) = h(a)(j)$. Note that $\equiv$ has at most $|B|^{|A|}$ classes. Let $C = \{c \in B^J : \text{if } i \equiv j \text{ then } c(i) = c(j)\}$. Choose a set of equivalence class representatives, $I$, for $\equiv$. Then $B^J = C \oplus B^{J \setminus I}$ and $C \cong B^{J/\equiv}$. $\square$

**2.9 Theorem.** *For any group $A$ there is a cardinal-valued function $\psi$ on $\mathrm{Ord}$ such that:*

*(1) $\psi(\alpha)$ is less than the least strongly inaccessible cardinal greater than $|A|$ and $\alpha$; and*

*(2) if $h: A \to B$ where $B$ is a group in the Reid class of type\* $\alpha$, then $B = C \oplus N$ where $|C| \leq \psi(\alpha)$, $C$ is a group in the Reid class of the same type as $B$ and $\operatorname{im}(h) \subseteq C$.*

PROOF. The proof is by induction on $\alpha$. For $\alpha = 0$, we can let $\psi(\alpha) = \omega$. Now suppose that $\psi(\beta)$ has been defined for all $\beta < \alpha$, and let $\kappa = \sup\{\psi(\beta) : \beta < \alpha\}$. We must consider all homomorphisms $h: A \to B$, where $B$ is in the Reid class and $B$ is of type\* $\alpha$.

Assume first that the type of $B$ is $(\alpha, P)$. Let $B = \prod_{i \in I} B_i$ where each $B_i$ has type\* $< \alpha$. Suppose $h: A \to B$ and let $h_i: B \to B_i$ be $h$ composed with the $i$th projection. Let $C_i$ be a summand of $B_i$ as guaranteed by the induction hypothesis with respect to $h_i$; so $|C_i| \leq \kappa$. Partition $I$ into $\{I_j : j \in J\}$ such that for all $j \in J$ there is $C_j$ such that $C_i \cong C_j$ for all $i \in I_j$. Note that we can choose $J \subseteq I$ so that $|J| \leq 2^\kappa$. Now apply Lemma 2.8 to each $(C_j)^{I_j}$ to get a direct summand $D_j$ of $(C_j)^{I_j}$ such that $D_j \cong (C_j)^{L_j}$ where $|L_j| \leq |C_j|^{|A|} \leq \kappa^{|A|}$ and $\eta_j[A] \subseteq D_j$. Here $\eta_j$ denotes $h$ composed with the projection onto $(C_j)^{I_j}$. So $\prod_{j \in J} D_j$ is as desired and can be increased to get a summand of the same type as $B$ if necessary. For this argument we need only that

$$\psi(\alpha) \geq \left(\kappa^{\left(\kappa^{|A|}\right)}\right)^{2^\kappa}$$

If the type of $B$ is $(\alpha, S)$, then we need only that $\psi(\alpha) \geq |A| \cdot \kappa$.
□

**2.10 Corollary.** *For any non-$\omega$-measurable ordinal $\alpha \neq 0$, no group in the Reid class of type $(\alpha, P)$ is a direct summand of a group in the Reid class of type $(\alpha, S)$. Also, no group in the Reid class of type $(\alpha, S)$ is a direct summand of one of type $(\alpha, P)$.*

PROOF. Suppose $A' \subseteq B'$ is a counterexample. Then there is a small direct summand, $A$, of $B'$ which is of the same type. Applying Theorem 2.9 to the inclusion of $A$ into $A'$, we get $A \subseteq B \subseteq B'$ such that $B$ and $B'$ have the same type and $|B|$ is non-$\omega$-measurable. Since $A$ is a direct summand of $B$, this contradicts Theorem 2.3. □

## EXERCISES

1. Every member of the Reid class is a separable group.

2. Every slender group in the Reid class is free. Hence, strongly $\aleph_1$-free group in the Reid class is free.

3. If $A$ belongs to the Reid class and has infinite rank, then $A \cong A \oplus \mathbb{Z}$.

4. A group in the Reid class has type* 0 if and only if it has finite rank.

5. Suppose $\kappa_i$ $(i \leq 2)$ and $\lambda_j$ $(j \leq 2)$ are non-zero cardinals such that $\kappa_0 > \lambda_j$ for all $j \leq 2$. Prove that $((\mathbb{Z}^{\kappa_0})^{(\kappa_1)})^{\kappa_2}$ is not isomorphic to $((\mathbb{Z}^{\lambda_0})^{(\lambda_1)})^{\lambda_2}$. Generalize this result. (cf. Exercise III.7.)

## NOTES

Reid 1967 conjectured that the groups $Z^\omega$, $\mathbb{Z}^{(\omega)}$, $(\mathbb{Z}^\omega)^{(\omega)}$, $(\mathbb{Z}^{(\omega)})^\omega$, $((\mathbb{Z}^\omega)^{(\omega)})^\omega$, ... are all distinct. He observed that every group in the Reid class is a dual group and asked whether all dual groups belong to the Reid class. Reid called dual groups *kernel groups* and some authors have called the members of the Reid class $\mathbb{Z}$-*kernel groups*.

Theorem 1.1 is found in Heinlein 1971. Theorem 1.3 is new, apparently.

Zimmermann-Huisgen 1979 proved the conjecture of Reid mentioned above. Dugas-Zimmermann-Huisgen 1981 studied (in a more general setting) types for groups in the Reid class.   Eda 1983c showed that every group in $\mathrm{Reid}_\mu$ has a unique type; the results in section 2 through 2.6 are based on this paper; the rest of the section is based on Eda 1984 but the proofs are different.

# CHAPTER XI
# TOPOLOGICAL METHODS

This chapter contains a miscellany of results and methods which are used here and in later chapters to prove results about Ext and about dual groups. Among the results proved here are one (due to Huber) that any $\aleph_1$-coseparable group is reflexive (see 2.8) and one (due to Eda and Ohta) that there is a non-reflexive dual group (see 1.13 and 4.15).

Section 1 discusses duals of inverse and direct systems and introduces the notion of an inverse-direct system; the latter will play an important role in Chapter XIV in the construction of reflexive and non-reflexive groups. Sections 2 and 3 deal with topologies on dual groups; section 2 describes the double dual (with respect to a free module) of a group as a completion; section 3 characterizes pure and dense subgroups of a dual group, and constructs "dual bases" in such a situation. Section 4 studies groups of the form $C(X, \mathbb{Z})$ — i.e., groups of continuous functions from a topological space $X$ to $\mathbb{Z}$ — and their duals.

## §1. Inverse and direct limits

In this section we will review material about inverse and direct systems, as well as introduce the notion of an inverse-direct system. Much of the material will be familiar to most readers, so we will feel free to be somewhat elliptical in some of our proofs, and we will confine ourselves to the setting of groups. We will use inverse-direct systems here and in Chapter XIV to construct some interesting dual groups.

**Definition.** Suppose $I$ is a directed set (i.e., $I$ is a partially ordered set so that for all $i, j \in I$, there is $k$ with $i, j < k$). A set of groups and homomorphisms $(A_i, \tau_{ij}: i < j \in I)$ is a *direct system of groups* if each $\tau_{ij}$ is a homomorphism from $A_i$ to $A_j$ and whenever $i < j < k$, $\tau_{ik} = \tau_{jk} \circ \tau_{ij}$. Similarly a set of groups and homomorphisms $(A_i, \pi_{ji}: i < j \in I)$ is an *inverse system of groups* if each $\pi_{ji}$ is an homomorphism from $A_j$ to $A_i$ and whenever $i < j < k$, $\pi_{ki} = \pi_{ji} \circ \pi_{kj}$. (We will often say "direct system" or "inverse system" in place of "direct system of groups" and "inverse system of groups".)

Given a direct system of groups $(A_i, \tau_{ij} : i < j \in I)$, a group $A$ (together with maps $\tau_i$) is the *direct limit* of the sequence if each $\tau_i$ is a homomorphism from $A_i$ to $A$; for all $i < j$, $\tau_i = \tau_j \circ \tau_{ij}$; and for any $B$ and collection of homomorphisms $\sigma_i$ from $A_i$ to $B$ satisfying $\sigma_i = \sigma_j \circ \tau_{ij}$, there is a unique homomorphism $\varphi \colon A \to B$ such that $\sigma_i = \varphi \circ \tau_i$. Given an inverse system of groups $(A_i, \pi_{ji} : i < j \in I)$, a group $A$ (together with maps $\pi_i$) is the *inverse limit* of the sequence if each $\pi_i$ is a homomorphism from $A$ to $A_i$; for all $i < j$, $\pi_i = \pi_{ji} \circ \pi_j$; and for any $B$ and collection of homomorphisms $\sigma_i$ from $B$ to $A_i$ satisfying $\sigma_i = \pi_{ji} \circ \sigma_j$, there is a unique homomorphism $\varphi \colon B \to A$ such that $\sigma_i = \pi_i \circ \varphi$.

From the definition it is easy to see that the inverse and direct limits, if they exist, are unique. We use, as is standard, $\varinjlim A_i$ and $\varprojlim A_i$ for the direct and inverse limits, respectively (when the maps are clear from context.)

**1.1 Proposition.** *Any direct system of groups has a direct limit. Any inverse system of groups has an inverse limit.*

PROOF. Suppose $(A_i, \tau_{ij} : i < j \in I)$ is a direct system. Define an equivalence relation $\sim$ on the disjoint union $\amalg\{A_i : i \in I\}$ as follows: if $a \in A_i$, $b \in A_j$, $a \sim b$ if and only if there exists $k \geq i, j$ such that $\tau_{ik}(a) = \tau_{jk}(b)$. Let $A$ be the set of equivalence classes, and define $\tau_i \colon A_i \to A$ by $\tau_i(a) = $ the equivalence class of $a$. For any two equivalence classes there is an $A_i$ which has representatives of both; hence there is a well-defined group operation on $A$. Then $A$, with the maps $\tau_i$, is the direct limit of the system.

Consider now an inverse sequence $(A_i, \pi_{ji} : i < j \in I)$. Then the inverse limit of the sequence is the subgroup of $\prod_I A_i$ consisting of those elements $a$ such that for all $i < j$, $a(i) = \pi_{ji}(a(j))$; the maps $\pi_i$ are the restrictions of the projections. $\square$

The reader will recognize that we have given the universal algebraic definition of direct and inverse limits. It is tempting to restrict the definition of direct and inverse systems by requiring that the maps be monomorphisms and epimorphisms respectively. From the proof of the proposition above, it can be seen that any direct limit or inverse limit is the limit of a related system where all the maps are, respectively, monomorphisms or epimorphisms.

We are going to be considering the duals of direct and inverse limits. We begin with the following proposition whose proof is easy. (Recall the definition of $\tau^*$ and the bracket notation, $\langle \, , \, \rangle$, from section I.1.)

**1.2 Proposition.** *Suppose* $(A_i, \tau_{ij} : i < j \in I)$ *is a direct system and* $(A_i, \pi_{ji} : i < j \in I)$ *is an inverse system. Then* $(A_i^*, \tau_{ij}^* : i < j \in I)$ *is an inverse system and* $(A_i^*, \pi_{ji}^* : i < j \in I)$ *is a direct system.* $\square$

Given a direct or inverse system, we call the sequence formed by the duals the *dual system*. We know that if the maps in an inverse system of groups are epimorphisms, then the maps in the dual system are monomorphisms. But even if the maps in a direct system are monomorphisms, the maps in the dual system are not necessarily epimorphisms.

**1.3 Proposition.** *Suppose* $(A_i, \tau_{ij} : i < j \in I)$ *is a direct system and* $A$, *together with maps* $\tau_i$ *from* $A_i$ *to* $A$, *is the direct limit of the system. Then* $A^*$, *together with the maps* $\tau_i^*$, *is the inverse limit of* $(A_i^*, \tau_{ij}^* : i < j \in I)$.

PROOF. Note that $A = \cup \operatorname{rge} \tau_i$. Let $\pi_i$ denote the map from $\varprojlim A_i^*$ to $A_i^*$. Define a map $\varphi$ from $\varprojlim A_i^*$ to $A^*$, by letting $\langle \varphi(f), \tau_i(a) \rangle = \langle \pi_i(f), a \rangle$. To show that $\varphi$ is well-defined, it is enough to prove that if $\tau_i(a) = \tau_j(b)$, then $\langle \pi_i(f), a \rangle = \langle \pi_j(f), b \rangle$. Notice first that since $\pi_i = \tau_{ik}^* \circ \pi_k$, $\langle \pi_i(f), a \rangle = \langle \tau_{ik}^* \circ \pi_k(f), a \rangle = \langle \pi_k(f), \tau_{ik}(a) \rangle$ for all $k > i$. Now if $\tau_i(a) = \tau_j(b)$, then there exists $k > i, j$ such that $\tau_{ik}(a) = \tau_{jk}(b)$; but then $\langle \pi_i(f), a \rangle = \langle \pi_k(f), \tau_{ik}(a) \rangle = \langle \pi_k(f), \tau_{jk}(b) \rangle = \langle \pi_j(f), b \rangle$. So the map is well-defined, and then it is easy to verify that it is one-one and onto and satisfies $\tau_i^* \circ \varphi = \pi_i$ for all $i$. $\square$

The proposition above is a generalization of the fact that $(\bigoplus_{i \in I} A_i)^* = \prod_{i \in I} A_i^*$. Łoś' theorem on duals of direct products does not generalize to inverse limits (see Exercise 1). However we do have the following result.

**1.4 Theorem.** *Suppose $(A_i, \pi_{ji}: i < j \in I)$ is an inverse system such that every $\pi_{ji}$ is an epimorphism. Let $A$ be the inverse limit. Let $B$ be the direct limit of $(A_i^*, \pi_{ji}^*: i < j \in I)$. Then $B$ is a subgroup of $A^*$. More exactly, the map from $B$ to $A^*$ induced by $\{\pi_i^*: i \in I\}$ is an embedding.*

PROOF. This follows easily from the fact that each $\pi_i^*$ is a monomorphism. $\square$

We are going to consider families of groups which simultaneously belong to direct and inverse systems. As a motivating example, consider the direct system $(\mathbb{Z}^n, \tau_{mn}: m < n \in \omega)$ and the inverse system $(\mathbb{Z}^n, \pi_{nm}: m < n \in \omega)$ where the maps are the obvious embeddings and projections. Then $\mathbb{Z}^{(\omega)}$ and $\mathbb{Z}^\omega$ are the direct and inverse limits of the systems respectively. As is usual, we can identify $\mathbb{Z}^{(\omega)}$ with a subgroup of $\mathbb{Z}^\omega$. We generalize this situation by defining the notion of an inverse-direct system.

**1.5 Definition.** Suppose $I$ is a directed set. Then $(A_i, \tau_{ij}, \pi_{ji}: i < j \in I)$ is an *inverse-direct system of groups*, if $(A_i, \tau_{ij}: i < j \in I)$ is a direct system of groups, $(A_i, \pi_{ji}: i < j \in I)$ is an inverse system of groups, and $\pi_{ji} \circ \tau_{ij}$ is the identity map on $A_i$ for all $i < j \in I$.

Note that in an inverse-direct system, the maps in the direct system are monomorphisms and the maps in the inverse system are epimorphisms. In fact, (using the notation of the definition) for $i < j$, $A_i$ is isomorphic to a direct summand of $A_j$: $A_j = \mathrm{im}(\tau_{ij}) \oplus \ker(\pi_{ji})$.

**1.6 Proposition.** *Suppose $(A_i, \tau_{ij}, \pi_{ji}: i < j \in I)$ is an inverse-direct system. Then there is a natural embedding of $\varinjlim A_i$ into $\varprojlim A_i$.*

PROOF. For convenience we will let $\tau_{ii}$ denote the identity map on $A_i$. For each $i$ and $j$, let $\varphi_{ij}$ be $\tau_{ij}$, if $i \leq j$, and let $\varphi_{ij}$ be $\pi_{ij}$, if $i > j$.

It is routine to check that $\pi_{jk} \circ \varphi_{ij} = \varphi_{ik}$ if $j > k$, so by the definition of inverse limits, $(\varphi_{ij}: j \in I)$ determines a homomorphism $\varphi_i$ from $A_i$ to $\varprojlim A_j$. Each map $\varphi_i$ is easily seen to be a monomorphism. Further, by the definition of direct limits, $(\varphi_i: i \in I)$ determines a monomorphism of $\varinjlim A_i$ into $\varprojlim A_i$. □

As the following proposition shows, inverse-direct systems over countable index sets do not represent any real generalization of the notions of direct sum and direct product. In this proposition we only deal with the index set $\omega$, but since any countable directed set contains a cofinal sequence of order type $\omega$, we have not sacrificed any generality.

**1.7 Proposition.** *Suppose $(A_n, \tau_{mn}, \pi_{nm}: m < n \in \omega)$ is an inverse-direct system. Let $B_0 = A_0$ and for $n \geq 1$, let $B_n = $ ker $\pi_{nn-1}$. Then $\varinjlim A_n = \bigoplus_{n \in \omega} B_n$ and $\varprojlim A_n = \prod_{n \in \omega} B_n$.* □

The dual of an inverse-direct system is an inverse-direct system:

**1.8 Proposition.** *Suppose $(A_i, \tau_{ij}, \pi_{ji}: i < j \in I)$ is an inverse-direct system. Then $(A_i^*, \pi_{ji}^*, \tau_{ij}^*: i < j \in I)$ is an inverse-direct system.*

PROOF. It is only necessary to show that for $i < j$, $\tau_{ij}^* \circ \pi_{ji}^*$ is the identity on $A_i^*$. This is true since $\tau_{ij}^* \circ \pi_{ji}^* = (\pi_{ij} \circ \tau_{ij})^* = (id_{A_i})^*$. □

We now investigate double duals of inverse-direct systems. Let $A$ be the direct limit of an inverse-direct system $(A_i, \tau_{ij}, \pi_{ji}: i < j \in I)$. Note that by 1.3 and 1.4 we have an embedding of $\varinjlim A_i^{**}$ into $A^{**}$ induced by the maps $\tau_i^{**}: A_i^{**} \to A^{**}$; we shall refer to this as *the natural embedding*. Recall that $\sigma$, the canonical map from $A$ to $A^{**}$ was defined in section I.1.

**1.9 Proposition.** *Suppose $(A_i, \tau_{ij}, \pi_{ji}: i < j \in I)$ is an inverse-direct system. Let $A$ denote the direct limit of the system. Then the image of $\sigma$, the canonical map from $A$ to $A^{**}$, is contained in the image of $\varinjlim A_i^{**}$ under the natural embedding.*

PROOF. Let $\sigma_i$ denote the canonical map from $A_i$ to $A_i^{**}$. Since

$A$ is generated by the image of the $\tau_i's$, it is enough to check that $\sigma \circ \tau_i = \tau_i^{**} \circ \sigma_i$; this is a routine verification. $\Box$

The last equation of the proof of 1.9 shows that:

**1.10 Corollary.** *Under the assumption of* 1.9, *if each $A_i$ is reflexive, then $\sigma$ maps onto the image of $\varinjlim A_i^{**}$ under the natural embedding.* $\Box$

We conclude this section by discussing in detail two examples which illustrate the power of using inverse-direct systems. These examples and our treatment of them are special cases of the methods we will develop in XIV.4. We hope that the more explicit treatment we give here will help later on.

**1.11 Examples.** We will construct two inverse-direct systems $(A_\alpha, \tau_{\alpha\beta}, \pi_{\beta\alpha}: \alpha < \beta \in \omega_1)$ and $(B_\alpha, \tau_{\alpha\beta}, \pi_{\beta\alpha}: \alpha < \beta \in \omega_1)$. All the groups will be subgroups of $\mathbf{Z}^{\omega_1}$. For $\alpha < \omega_1$, we let $\mathbf{Z}^\alpha$ denote the subgroup of $\mathbf{Z}^{\omega_1}$ consisting of $\{a: a(\beta) = 0 \text{ for all } \beta \geq \alpha\}$. Let $i_\alpha$ denote the inclusion map from $\mathbf{Z}^\alpha$ into $\mathbf{Z}^{\omega_1}$ and $\rho_\alpha$ be the projection from $\mathbf{Z}^{\omega_1}$ onto $\mathbf{Z}^\alpha$. The groups $A_\alpha$ and $B_\alpha$ will be subgroups of $\mathbf{Z}^\alpha$ and the maps $\tau_{\alpha\beta}$ and $\pi_{\beta\alpha}$ will be the restrictions of $i_\alpha$ and $\rho_\alpha$ respectively. (Of course, in the two systems $\tau_{\alpha\beta}$ and $\pi_{\alpha\beta}$ will denote different maps with different domains.) To fix a bit more notation, recall that $e_\alpha$ denotes the element of $\mathbf{Z}^{\omega_1}$ defined by $e_\alpha(\beta) = 1$, if $\alpha = \beta$, and 0 otherwise. Let $\mathbf{1}$ be the function which is constantly 1, and let $\mathbf{1}_\alpha$ be defined by $\mathbf{1}_\alpha(\beta) = 1$, if $\beta < \alpha$ and 0 otherwise.

Fix $E$, a stationary and co-stationary subset of $\omega_1$. We define the groups by induction on $\alpha$. As long as we ensure that for all $\beta < \alpha$, $\rho_\beta[A_\alpha] = A_\beta$ and $\rho_\beta[B_\alpha] = B_\beta$, our commitments will guarantee that we have defined inverse-direct systems. We let, as we must, $A_0 = B_0 = \{0\}$. If $\alpha = \beta + 1$, then $A_\alpha = A_\beta \oplus \langle e_\beta \rangle$ and $B_\alpha = B_\beta \oplus \langle e_\beta \rangle$. It is easy to check that $\rho_\gamma[A_\alpha] = A_\gamma$ and $\rho_\gamma[B_\alpha] = B_\gamma$ for all $\gamma < \alpha$.

For $\alpha$, a limit ordinal, there are two cases. If $\alpha \in E$, let $A_\alpha = \bigcup_{\beta<\alpha} A_\beta$ and let $B_\alpha = (\bigcup_{\beta<\alpha} B_\beta) \oplus \langle \mathbf{1}_\alpha \rangle$. If $\alpha \notin E$, let $A_\alpha = \{a \in \mathbf{Z}^\alpha: \rho_\beta(a) \in A_\beta, \text{ for all } \beta < \alpha\}$ and let $B_\alpha = \{a \in \mathbf{Z}^\alpha: \rho_\beta(a) \in B_\beta,$

for all $\beta < \alpha\}$. It remains to see that $\rho_\beta[A_\alpha] = A_\beta$ and $\rho_\beta[B_\alpha] = B_\beta$. The only case that presents any difficulty is to show that $\rho_\beta[B_\alpha] = B_\beta$, where $\alpha$ is a limit ordinal in $E$; but in this case, we can show by induction that for all $\beta$, $\mathbf{1}_\beta \in B_\beta$.

It is interesting to note that for any limit ordinal $\alpha$, $A_\alpha$ and $B_\alpha$ are (naturally isomorphic to) subgroups of $\varprojlim_{\beta<\alpha} A_\beta$ and $\varprojlim_{\beta<\alpha} B_\beta$, respectively. Also, if $\alpha$ is a limit ordinal not in $E$, then $A_\alpha$ and $B_\alpha$ are (naturally isomorphic to) $\varprojlim_{\beta<\alpha} A_\beta$ and $\varprojlim_{\beta<\alpha} B_\beta$, respectively.

We now want to investigate the direct and inverse limits of the systems. Of course, we view the direct limits of the systems as subgroups of $\mathbf{Z}^{\omega_1}$. In what follows we identify $\varinjlim_{\alpha<\omega_1} A_\alpha$ with its image under the natural embedding given in 1.6.

**1.11A.** $\varinjlim_{\alpha<\omega_1} A_\alpha = \varprojlim_{\alpha<\omega_1} A_\alpha$.

**1.11B.** *Let $\mathbf{1_B}$ be the element of* $\varprojlim_{\alpha<\omega_1} B_\alpha$, *such that for every $\alpha$,* $\pi_\alpha(\mathbf{1_B}) = \mathbf{1}_\alpha$. *Then* $\varinjlim_{\alpha<\omega_1} B_\alpha \oplus \langle\mathbf{1_B}\rangle = \varprojlim_{\alpha<\omega_1} B_\alpha$.

PROOF. Since the proofs of 1.11A and 1.11B are virtually identical we will only prove 1.11B, which is the more difficult of the two. Suppose $b$ is an element of the inverse limit. For each limit ordinal $\alpha \in E$, there is a unique $c_\alpha \in \bigcup_{\beta<\alpha} B_\beta$ and integer $k_\alpha$ so that $\pi_\alpha(b) = c_\alpha + k_\alpha \mathbf{1}_\alpha$. Choose $\beta_\alpha < \alpha$, so that $c_\alpha \in B_{\beta_\alpha}$. By Fodor's lemma (II.4.11), there is $E_1$ a stationary subset of $E$ and $\beta$ so that for all $\alpha \in E_1$, $\beta_\alpha = \beta$. Consider $\alpha, \gamma \in E_1$. Assume $\alpha < \gamma$; then $\pi_\alpha(b) = \pi_\alpha(\pi_\gamma(b)) = \pi_\alpha(c_\gamma + k_\gamma \mathbf{1}_\gamma) = \pi_\alpha(c_\gamma) + k_\gamma \mathbf{1}_\alpha$. Since $c_\gamma \in B_\beta$, $\pi_\alpha(c_\gamma) = c_\gamma$. Hence $c_\gamma = c_\alpha$ and $k_\alpha = k_\gamma$. Let $c$ be the constant value of $c_\alpha$ and $k$ be the constant value of $k_\alpha$, for $\alpha \in E_1$. Then for all $\alpha \in E_1$, $\pi_\alpha(b) = \pi_\alpha(c + k\mathbf{1_B})$. So $b = c + k\mathbf{1_B}$. $\square$

**1.11C.** *For all $\alpha$, $A_\alpha$ and $B_\alpha$ are reflexive.*

PROOF. The proof is by induction on $\alpha$. The only interesting case is that of limit ordinals. Choose $\alpha_n$ $(n < \omega)$ an increasing sequence of ordinals with limit $\alpha$ so that $\alpha_0 = 0$. If $\alpha \notin E$, then $A_\alpha = \prod_{n\in\omega} \ker \pi_{\alpha_{n+1}\alpha_n}$ and $B_\alpha = \prod_{n\in\omega} \ker \pi_{\alpha_{n+1}\alpha_n}$ (cf. Lemma 1.7). If

$\alpha \in E$, then $A_\alpha = \bigoplus_{n \in \omega} \ker \pi_{\alpha_{n+1}\alpha_n}$ and $B_\alpha = \bigoplus_{n \in \omega} \ker \pi_{\alpha_{n+1}\alpha_n} \oplus \langle 1_\alpha \rangle$. Since the class of reflexive groups is closed under taking direct summands, countable direct sums and countable direct products (cf. IX.1.10 and X.1.1), the proof is complete. □

Now we want to consider the dual systems, $(A_\alpha^*, \pi_{\beta\alpha}^*, \tau_{\alpha\beta}^* : \alpha < \beta < \omega_1)$ and $(B_\alpha^*, \pi_{\beta\alpha}^*, \tau_{\alpha\beta}^* : \alpha < \beta < \omega_1)$.

**1.11D.** $\varinjlim_{\alpha < \omega_1} A_\alpha^* = \varprojlim_{\alpha < \omega_1} A_\alpha^*$ and $\varinjlim_{\alpha < \omega_1} B_\alpha^* = \varprojlim_{\alpha < \omega_1} B_\alpha^*$.

PROOF. Since the proofs are the same we will only prove the first assertion. Let $A$ denote $\varinjlim_{\alpha < \omega_1} A_\alpha$. Since $A^*$ is $\varprojlim_{\alpha < \omega_1} A_\alpha^*$ and $A_\beta = A_\alpha \oplus \ker \pi_{\beta\alpha}$ for $\alpha < \beta$, it is enough to show that for any $f \in A^*$, there is an $\alpha$ such that $f \upharpoonright \ker \pi_{\beta\alpha} \equiv 0$ for all $\beta > \alpha$; for then $f$ is the image of $f \upharpoonright A_\alpha$ under the natural embedding given by 1.6. Assume $f$ is a counterexample. Then we can find an increasing sequence $\{\alpha_\beta : \beta < \omega_1\}$ of ordinals so that for all $\beta$, there is $a \in \ker \pi_{\alpha_{\beta+1}\alpha_\beta}$ so that $\langle f, a \rangle \neq 0$. Since the complement of $E$ is stationary, there exists $\alpha \notin E$ and an increasing sequence $\{\beta_n : n \in \omega\}$ so that $\alpha$ is the limit of $\{\alpha_{\beta_n} : n \in \omega\}$. But $A_\alpha = A_{\beta_0} \oplus \prod_{n \in \omega} \ker \pi_{\alpha_{\beta_{n+1}}\alpha_{\beta_n}}$, so $f$ contradicts the slenderness of $\mathbf{Z}$ (cf. III.1.1). □

We can use these examples to give proofs of two interesting results. (Compare Example X.2.6.)

**1.12 Theorem.** *There is a reflexive group which is not in the Reid class.*

PROOF. Let $A$ be the direct limit of the inverse-direct system $(A_\alpha, \tau_{\alpha\beta}, \pi_{\beta\alpha} : \alpha < \beta < \omega_1)$ constructed in 1.11. By 1.11D and 1.3, $A^{**}$ is the inverse limit of the system, $(A_\alpha^{**}, \tau_{\alpha\beta}^{**}, \pi_{\beta\alpha}^{**} : \alpha < \beta < \omega_1)$. Since each $A_\alpha$ is reflexive, the inverse-direct system $(A_\alpha, \tau_{\alpha\beta}, \pi_{\beta\alpha} : \alpha < \beta < \omega_1)$ is isomorphic to $(A_\alpha^{**}, \tau_{\alpha\beta}^{**}, \pi_{\beta\alpha}^{**} : \alpha < \beta < \omega_1)$ via the maps $(\sigma_{A_\alpha} : \alpha < \omega_1)$. So by 1.11A, the inverse limit is the same as the direct limit. By Corollary 1.10 and 1.11C, the map $\sigma$ from $A$ to the direct limit is surjective, so $A$ is reflexive. To complete the proof we must see that $A$ is not in the Reid class. It is easy to see that if $A$ is in the Reid class its type* (see section X.2) is at least $\omega_1$. Hence either $A$ or $A^*$ would have $\mathbb{Z}^{\omega_1}$ as a homomorphic image. But each of these groups is a direct limit of groups in $\mathrm{Reid}_{\omega_1}$, so by X.1.3 and X.1.4, $\mathbb{Z}^{\omega_1}$ is not the homomorphic image of either of these groups. $\square$

**1.13 Theorem.** *There is a non-reflexive dual group.*

PROOF. Let $B$ be the direct limit of the inverse-direct system $(B_\alpha, \tau_{\alpha\beta}, \pi_{\beta\alpha}: \alpha < \beta < \omega_1)$ constructed in 1.11. By 1.11D, $B^{**}$ is the inverse limit of the system, $(B_\alpha^{**}, \tau_{\alpha\beta}^{**}, \pi_{\beta\alpha}^{**}: \alpha < \beta < \omega_1)$. By Corollary 1.10 and 1.11C, the map $\sigma$ from $B$ to $\varinjlim B_\alpha^{**}$ is surjective. But by 1.11B, $\varinjlim B_\alpha^{**}$ is a proper subgroup of $\varprojlim B_\alpha^{**}$; so $B$ is not reflexive. In chapter XIV, we will develop general methods for showing that $B$ is a dual group, but in this special case there is an easy argument: by 1.11B, $B^{**} \cong B \oplus \mathbb{Z}$. Consider the isomorphism $\varphi: \mathbb{Z}^{\omega_1} \oplus \mathbb{Z} \to \mathbb{Z}^{\omega_1}$, defined by $\varphi(a)(0) = 0$, $\varphi(a)(n+1) = a(n)$, for $n \in \omega$ and $\varphi(a)(\alpha) = a(\alpha)$ otherwise, if $a \in \mathbb{Z}^{\omega_1}$ and $\varphi(k) = ke_0$, if $k \in \mathbb{Z}$. Then $\varphi$ induces an isomorphism from $B \oplus \mathbb{Z}$ to $B$. So $B$ is a dual group, since it is isomorphic to the dual group $B \oplus \mathbb{Z}$. $\square$

Theorem 1.13 is due to Eda and Ohta; they used the same group in their proof but viewed it as the group of continuous functions from a topological space to $\mathbb{Z}$ (which is given the discrete topology). In section 4, where we will study groups of continuous functions and their duals, we will discuss this group from a topological point of view.

## §2. Completions

For the first part of this section $A$ will be a (left) $R$-module and $A^*$ will denote the dual of $A$ with respect to a fixed $R$-module $H$, i.e., $A^* = \mathrm{Hom}_R(A, H)$. Recall from section I.1 that $A^*$ is regarded as a right $S$-module where $S = \mathrm{End}_R(H)$, and the double dual $A^{**}$ equals $\mathrm{Hom}_S(A^*, H)$. Recall that linear topologies and their completions were introduced in section I.3.

If $B$ is a submodule of $A^*$ the *B-topology* on $A$ is the linear topology on $A$ obtained by taking as a basis, $\mathcal{U}_B$, of neighborhoods of 0 all sets of the form

$$U_A(F) \stackrel{\text{def}}{=} \bigcap \{\ker(f): f \in F\}$$

where $F$ is a finite subset of $B$. Thus a subset $S$ of $A$ is open if and only if for all $a \in S$, there is a finite subset $F$ of $B$ such that $\{x \in A: f(x) = f(a) \text{ for all } f \in F\} \subseteq S$. Notice that the $B$-topology on $A$ is the weakest topology on $A$ which makes every element of $B$ continuous, if we give $H$ the discrete topology.

The most important examples of $B$-topologies on $A$ are the following.

**2.1 Example.** Let $B = A^*$. The $A^*$-topology on $A$ is also called the $H$-*topology* on $A$, especially in a context where we are considering duals of $A$ with respect to more than one module $H$.

**2.2 Example.** Suppose $G = A^*$, and $G^* = \text{Hom}_S(G, H)$. Let $B = \sigma_A[A] \subseteq A^{**} = G^*$. In this case the $B$-topology on $G$ is called the $A$-*topology*, or *weak topology*, on $A^*$. Notice that in this case a neighborhood basis of 0 in $A^*$ consists of all sets of the form $\{\varphi \in A^*: \varphi(x) = 0 \text{ for all } x \in X\}$, where $X$ ranges over all finite subsets of $A$.

**2.3 Proposition.** $A^*$ *is complete in its $A$-topology.*

PROOF. Let $\{\varphi_X: X \text{ a finite subset of } A\}$ be a Cauchy net in $A^*$. Then for all finite subsets $X$ and $Y$ of $A$, $\varphi_X(a) = \varphi_Y(a)$ for all $a \in X \cap Y$; so define $\psi \in A^*$ by the rule: $\psi(a) = \varphi_X(a)$ if $a \in X$. It is easy to check that $\psi$ is a well-defined homomorphism, and that $\psi$ is the limit of the given Cauchy net. $\square$

For the rest of this section we shall assume that $R$ is a p.i.d., possibly a field. If $H = R^{(\kappa)}$, we shall write $A^{*(\kappa)}$ for $\text{Hom}_R(A, R^{(\kappa)})$. (Note that $A^{*(\kappa)}$ does *not* mean $(A^*)^{(\kappa)}$, the direct sum of $\kappa$ copies of $A^*$.) We shall denote by $\mathcal{U}_\kappa$ the canonical basis of neighborhoods of 0 in the $R^{(\kappa)}$-topology on $A$, i.e., $\mathcal{U}_\kappa = \mathcal{U}_B$, where $B = A^{*(\kappa)}$. Also we shall let $A^{**(\kappa)}$ denote $\text{Hom}_S(A^{*(\kappa)}, R^{(\kappa)})$.

**2.4 Lemma.** *Let $R$ be a p.i.d. For any cardinal $\kappa$ which is not $\omega$-measurable, and any module $A$ which is $R^{(\kappa)}$-torsionless, $A^{**(\kappa)}$*

*is isomorphic to* $\varprojlim(A/U, \pi_{U,V}: U \geq V \in \mathcal{U}_\kappa)$ *by an isomorphism which takes the canonical embedding* $\sigma: A \to A^{**(\kappa)}$ *to the map induced by the canonical maps:* $A \to A/U$.

PROOF. Every $f \in A^{*(\kappa)}$ has kernel $U$ for some $U \in \mathcal{U}_\kappa$ by definition of $\mathcal{U}_\kappa$; so $f \in (A/U)^{*(\kappa)}$. Thus $A^{*(\kappa)} \cong \varinjlim((A/U)^{*(\kappa)}, \pi^*_{U,V}: U \geq V \in \mathcal{U}_\kappa)$. Taking duals (cf. 1.3) we obtain

$$A^{**(\kappa)} \cong \varprojlim((A/U)^{**(\kappa)}, \pi^{**}_{U,V}: U \geq V \in \mathcal{U}_\kappa)$$

Now $A/U$ is free since $R$ is a p.i.d., so $(A/U)^{**(\kappa)}$ is canonically isomorphic to $A/U$ by Theorem IX.1.11, if $\kappa$ is infinite. (If $\kappa$ is finite, $A/U$ is a finite rank free module, and thus reflexive because $R$ is $R^{(\kappa)}$-reflexive.) Thus

(†) $$A^{**(\kappa)} \cong \varprojlim(A/U, \pi_{U,V}: U \geq V \in \mathcal{U}_\kappa)$$

and, moreover, under this isomorphism the canonical embedding $\sigma: A \to A^{**(\kappa)}$ clearly corresponds to the embedding of $A$ into the right-hand side of (†) induced by the canonical maps $A \to A/U$. □

Thus there is a topology on $A^{**(\kappa)}$ — in fact, the weak topology — which makes $\sigma: A \to A^{**(\kappa)}$ the completion of $A$.

**2.5 Proposition.** *Let $R$ be a p.i.d.*
(i) *For any $R$-module $A$ and any infinite non-$\omega$-measurable cardinals $\lambda$ and $\kappa$, $A$ is $R^{(\kappa)}$-reflexive if and only if $A$ is $R^{(\lambda)}$-reflexive.*
(ii) *If $R$ is slender, then for any non-$\omega$-measurable cardinal $\kappa$, $A$ is $R^{(\kappa)}$-reflexive if and only if $A$ is $R$-reflexive.*

PROOF. (i) Say $\lambda < \kappa$. For any $f \in A^{*(\lambda)}$, if $U = \ker(f)$, then $f \in (A/U)^{*(\lambda)}$; moreover, $U \in \mathcal{U}_\kappa$ since $\lambda < \kappa$. Conversely, for any $U \in \mathcal{U}_\kappa$, if $f \in (A/U)^{*(\lambda)}$, then $f \in A^{*(\lambda)}$. Hence

$$A^{*(\lambda)} \cong \varinjlim((A/U)^{*(\lambda)}, \pi_{U,V}: U \geq V \in \mathcal{U}_\kappa).$$

Then just as in the proof of 2.4,

$$A^{**(\lambda)} \cong \varprojlim((A/U)^{**(\lambda)}, \pi^{**}_{U,V} : U \geq V \in \mathcal{U}_\kappa)$$
$$\cong \varprojlim(A/U, \pi_{U,V} : U \geq V \in \mathcal{U}_\kappa).$$

But the latter is canonically isomorphic to $A$ by hypothesis and Lemma 2.4.

(ii) If $R$ is slender, the same proof works because $A/U$ ($U \in \mathcal{U}_\kappa$) is free of non-$\omega$-measurable rank and hence $R$-reflexive by III.3.8. □

Note that part (ii) is false when $R$ is not slender, i.e., when $R$ is a complete discrete valuation ring; for example, in this case $R^{(\omega)}$ is $R^{(\omega)}$-reflexive, but not $R$-reflexive.

Recall the notion of a $\kappa^+$-coseparable group from IV.2.12.

**2.6 Theorem.** *If $A$ is a $\kappa^+$-coseparable $\mathbf{Z}$-module for some $\kappa \geq \aleph_0$, then $A$ is $\mathbf{Z}^{(\kappa)}$-reflexive.*

PROOF. Since $A$ is separable, $A^{**(\kappa)}/\sigma_A[A]$ is torsion-free (cf. Exercise IV.4). So it remains to show that the latter is torsion. Let $D$ be the injective hull of $A$, and $e : A \to D$ the inclusion map; then $D$ is a $\mathbf{Q}$-module, and $D/e[A]$ is torsion. Let $\mathcal{V}_\kappa = \{V \subseteq D : D/V$ is free (as a $\mathbf{Q}$-module) of rank $\leq \kappa\}$.

We are going to define a map $\psi : A^{**(\kappa)} \to \hat{D}$, where we regard $A^{**(\kappa)} \cong \varprojlim(A/U, \pi_{U,V} : U \geq V \in \mathcal{U}_\kappa)$ as in 2.4, and $\hat{D} = \varprojlim(D/V, \pi_{U,V} : U \geq V \in \mathcal{V}_\kappa)$. Note that for all $U \in \mathcal{U}_\kappa$, $\langle e[U] \rangle \in \mathcal{V}_\kappa$. (Here, $\langle e[U] \rangle$ denotes the $\mathbf{Q}$-submodule generated by $e[U] \subseteq D$.) Given $(a_U + U)_{U \in \mathcal{U}_\kappa} \in A^{**(\kappa)}$, for each $U \in \mathcal{U}_\kappa$ let $d_{\langle e[U] \rangle} = e(a_U)$. In general, for any $V \in \mathcal{V}_\kappa$, $e^{-1}[V]$ is such that $A/e^{-1}[V]$ is of cardinality $\leq \kappa$. Hence by hypothesis, there exists $U \in \mathcal{U}_\kappa$ such that $U \subseteq e^{-1}[V]$; then define $d_V = e(a_U)$. We now can define $\psi((a_U + U)_{U \in \mathcal{U}_\kappa})$ to be $(d_V + V)_{V \in \mathcal{V}_\kappa}$, which we can easily check is a well-defined member of $\hat{D}$. Note that $\psi$ is one-one.

By 2.4, $\hat{D}$ is $D^{**(\kappa)}$, where the duals are taken with respect to $\mathbf{Q}^{(\kappa)}$. One can also easily check that $\psi[\sigma_A[A]] = \sigma_D[e[A]]$ But $\sigma_D[D] = \hat{D}$, since $D$ is free (as a $\mathbf{Q}$-module) — see IX.1.11. Hence $\sigma_D^{-1} \circ \psi$ induces an embedding of $A^{**(\kappa)}/\sigma_A[A]$ into $D/e[A]$; and thus $A^{**(\kappa)}/\sigma_A[A]$ is torsion. □

**2.7 Corollary.** *If $A$ is $\aleph_1$-coseparable, then $A$ is $\mathbf{Z}$-reflexive.*

PROOF. This follows immediately from 2.6 and 2.5(ii). □

If MA + ¬CH is assumed, then every $\aleph_1$-separable group of cardinality $\aleph_1$ is $\aleph_1$-coseparable (see XII.1.10 and XII.1.12); hence we have established the following result of Huber.

**2.8 Corollary.** (MA + ¬CH) *Every $\aleph_1$-separable group of cardinality $\aleph_1$ is $\mathbf{Z}$-reflexive.* □

This corollary was the first result which showed that it is consistent that there are reflexive groups which are not in the Reid class. We shall see in section XIV.5 that 2.8 is not a theorem of ZFC. However it is provable in ZFC (cf. XIV.2.4) that there is a non-free $\aleph_1$-separable group of cardinality $\aleph_1$ which is reflexive.

We will finish this section with some material on groups which admit discrete norms.

**Definition.** Let $G$ be a group. A function $\| \ \|: G \to \mathbf{R}$ is said to be a *norm* if

$\| g \| \geq 0$ and $\| g \| = 0$ if and only $g = 0$;

$\| g + h \| \leq \| g \| + \| h \|$;

$\| mg \| = |m| \| g \|$, for all $m \in \mathbf{Z}$.

A norm is said to be *discrete* if there is some $\epsilon > 0$ such that $\| g \| \geq \epsilon$, for all $g \neq 0$.

**2.9 Theorem.** *A group $A$ has a discrete norm if and only if $A$ is free.*

PROOF. ($\Rightarrow$) Suppose the result is false and $A$ is a counterexample of minimum cardinality, $\kappa$. We can assume for all $x$ that $\| x \| \geq 1$. A countable group with a discrete norm is free (see Exercises 11 — 13). By the singular compactness theorem (IV.3.5), $\kappa$ is an uncountable regular cardinal. Let $\{A_\alpha : \alpha < \kappa\}$ be a $\kappa$-filtration of $A$, and let $E = \{\alpha : A_\alpha$ is not $\kappa$-pure in $A\}$. For $\alpha \in E$, choose $\{x_i^\alpha : i \in I_\alpha\}$ so that $\{x_i^\alpha + A_\alpha : i \in I_\alpha\}$ is a maximal independent subset of $A_{\alpha+1}/A_\alpha$. We can view $A_{\alpha+1}/A_\alpha$ as a subgroup of $\bigoplus_{i \in I_\alpha} \mathbf{Q}(x_i^\alpha + A_\alpha)$. Define a norm $\nu$ on $A_{\alpha+1}/A_\alpha$ by $\nu(y + A_\alpha) = \sum_{i \in I_\alpha} |q_i| \| x_i^\alpha \|$, where $y + A_\alpha = \sum_{i \in I_\alpha} q_i(x_i^\alpha + A_\alpha)$.

(Of course all but finitely many of the coefficients in the sum are 0.) Since $A_{\alpha+1}/A_\alpha$ is not free and has cardinality $< \kappa$, $\nu$ is not discrete. Choose $y_\alpha$ so that $\nu(y_\alpha + A_\alpha) < 1/2$. Let $q_i^\alpha$ be such that $y_\alpha + A_\alpha = \sum_{i \in I_\alpha} q_i^\alpha (x_i^\alpha + A_\alpha)$. Choose an integer $k_\alpha > 0$ so that for all $i$, $k_\alpha q_i^\alpha \in \mathbf{Z}$. Let $z_\alpha$ be defined by

$$z_\alpha = k_\alpha y_\alpha - \sum_{i \in I_\alpha} k_\alpha q_i^\alpha x_i^\alpha.$$

Note that $z_\alpha$ is an element of $A_\alpha$. By Fodor's lemma (cf. Exercise II.19), there is $\alpha \neq \beta$ so that $k_\alpha = k_\beta$ ($= k$, say) and $z_\alpha = z_\beta$ ($= z$, say). Calculating we have

$$
\begin{aligned}
0 \;<\; |k|\, \| y_\alpha - y_\beta \| &= \| \textstyle\sum_{i \in I_\alpha} k q_i^\alpha x_i^\alpha + z - \sum_{i \in I_\beta} k q_i^\beta x_i^\beta - z \| \\
&\leq |k| (\textstyle\sum_{i \in I_\alpha} |q_i^\alpha| \, \| x_i^\alpha \| + \sum_{i \in I_\beta} |q_i^\beta| \, \| x_i^\beta \|) \\
&< |k|(1/2 + 1/2) = |k|.
\end{aligned}
$$

Dividing by $|k|$, we have

$$0 < \| y_\alpha - y_\beta \| < 1,$$

which is a contradiction.

($\Leftarrow$) If $A$ is freely generated by $\{x_i : i \in I\}$, define $\| \sum n_i x_i \| = \max\{|n_i| : i \in I\}$. $\square$

Using this theorem it is a possible to give a simple proof of a theorem of Specker and Nöbeling.

**2.10 Theorem.** *Suppose $G$ is a subgroup of $\mathbf{Z}^I$ consisting of bounded functions (i.e., for every $g \in G$, there is $N$ such that for all $i$, $g(i) < N$). Then $G$ is a free group.*

PROOF. Define $\| g \| = \max\{|g(i)| : i \in I\}$. It is easy to verify that $\| \; \|$ is a discrete norm on $G$, so $G$ is free by 2.9. $\square$

## §3. Density and dual bases

In this section we will be considering $\mathbf{Z}$-modules, i.e., groups. Given a group $A$, we will investigate what it means for a subgroup of $A^*$ to be pure and dense in the $A$-topology (cf. Example 2.2). Recall, from the beginning of section I.1, the notation $\langle x, y \rangle$.

**3.1 Lemma.** *A subgroup $B$ of $A^*$ is dense in $A^*$ in the $A$-topology if and only if for every finite set $F = \{a_1, \ldots, a_n\}$ of elements of $A$ and every $y \in A^*$, there exists $b \in B$ such that $\langle b, a_i \rangle = \langle y, a_i \rangle$ for $i = 1, \ldots, n$.*

PROOF. This is an immediate consequence of the definition of the $A$-topology, for every finite subset $F$ of $A$ determines a neighborhood of $y$ equal to $\{z \in A^* : \langle z, a \rangle = \langle y, a \rangle$ for all $a \in F\}$, and conversely there is a basic system of neighborhoods of $y$ of this form. $\square$

When we say $B$ is *dense* in $A^*$ we shall mean that $B$ is dense in $A^*$ in the $A$-topology. As an example, if $A = \mathbf{Z}^{(\omega)}$, and $B$ is the subgroup of $A^*$ ($\cong \mathbf{Z}^\omega$) generated by the canonical projections, then $B$ ($\cong \mathbf{Z}^{(\omega)}$) is pure and dense in $A^*$. We shall see in 3.5 that this example is typical in the countable case.

**3.2 Proposition.** *Let $B$ be a subgroup of $A^*$.*
*(i) $B$ is pure in $A^*$ if and only if whenever $b \in B$ generates a pure subgroup of $B$, there exists $a \in A$ such that $\langle b, a \rangle = 1$.*
*(ii) If $A$ is separable, then $B$ is dense in $A^*$ if and only if whenever $a \in A$ generates a pure subgroup of $A$, there exists $b \in B$ such that $\langle b, a \rangle = 1$.*

PROOF. (i) ($\Leftarrow$) Suppose $y \in mA^* \cap B$; say $y = mz$ for some $z \in A^*$. Now since $B$ is $\aleph_1$-free, $y$ belongs to a pure cyclic subgroup generated by some element $b \in B$; say $y = nb$. By hypothesis there exists $a \in A$ such that $\langle b, a \rangle = 1$. Now calculate:

$$m\langle z, a \rangle = \langle mz, a \rangle = \langle y, a \rangle = \langle nb, a \rangle = n\langle b, a \rangle = n.$$

Hence $n = mr$ for some $r \in \mathbf{Z}$; so $y = m(rb) \in mB$.

(i) ($\Rightarrow$) Suppose that $b$ generates a pure subgroup of $B$. Then by hypothesis, $b$ generates a pure subgroup of $A^*$. Consider $I = \mathrm{rge}(b) \subseteq \mathbf{Z}$; if $I = \mathbf{Z}$ we are done, for then $1 \in \mathrm{rge}(b)$. If not, then

$I = n\mathbf{Z}$ for some $n > 1$. Define $z \in A^*$ by: $\langle z, x \rangle = r$ if $\langle b, x \rangle = nr$; then $nz = b$, which contradicts the fact that $b$ generates a pure subgroup of $A^*$.

(ii) ($\Rightarrow$) Since $A$ is separable, $a$ generates a cyclic summand, $\langle a \rangle$, of $A$: $A = \langle a \rangle \oplus C$; thus it is clear that there is $y \in A^*$ such that $\langle y, a \rangle = 1$. Since $B$ is dense in $A^*$ there exists $b \in B$ such that $\langle b, a \rangle = \langle y, a \rangle = 1$.

($\Leftarrow$) It is enough to show that if $a_1, \ldots, a_n$ freely generate a pure subgroup of $A$, then there are $b_1, \ldots, b_n \in B$ so that $\langle b_j, a_i \rangle = 1$ if $i = j$ and $0$ otherwise. To prove this statement it is enough to show that:

> if $a_1, \ldots, a_n$ freely generate a pure subgroup of $A$, then there is $b \in B$ so that $\langle b, a_n \rangle = 1$ and $\langle b, a_i \rangle = 0$ for $i < n$.

The proof is by induction on $n$. The case $n = 1$ is just the hypothesis of the theorem. Suppose now that $n = k + 1$. By induction, we can choose $c \in B$ so that $\langle c, a_i \rangle = 0$ for $i < k$ and $\langle c, a_{k+1} \rangle = 1$. Suppose $\langle c, a_k \rangle = m$. Also by induction we can choose $d \in B$ so that $\langle d, a_i \rangle = 0$ for $i < k$ and $\langle d, a_k - ma_{k+1} \rangle = 1$. Suppose $\langle d, a_{k+1} \rangle = r$. Note that $\langle d, a_k \rangle = mr + 1$. Let $b = (1 + mr)c - md$. Then $\langle b, a_i \rangle = 0$ for $i < k$,

$$\langle b, a_k \rangle = (1 + mr)m - m(mr + 1) = 0,$$

and

$$\langle b, a_{k+1} \rangle = (1 + mr)1 - mr = 1. \quad \square$$

Notice that if $B$ is dense in $A^*$ and $A$ is separable, then we can identify $A$ with a subgroup of $B^*$ via the map $\tilde{\sigma}: A \to B^*$ which sends $a$ to $\sigma(a){\restriction}B$. There is no ambiguity in the notation $\langle a, b \rangle$ (for $a \in A, b \in B$) since by definition of $\sigma$, $\langle a, b \rangle = \langle \tilde{\sigma}(a), b \rangle$.

**3.3 Lemma.** *Let $A$ and $B$ be free groups such that $B$ is a pure and dense subgroup of $A^*$. Then $A$ is a pure and dense subgroup of $B^*$ (where $A$ is identified with $\tilde{\sigma}[A]$, as above).*

PROOF. We make use of Proposition 3.2. To show that $A$ is pure in $B^*$ consider an element $a \in A$ which generates a pure subgroup of $A$. Since $B$ is dense in $A^*$, Proposition 3.2(ii) implies that there exists $b \in B$ such that $\langle b, a \rangle = 1$. Hence Proposition 3.2(i) implies that $A$ is pure in $B^*$. A symmetric argument shows that $A$ is dense in $B^*$. $\square$

The following lemma is proved by exploiting the symmetry expressed in the above lemma.

**3.4 Lemma.** *Suppose $A$ is a separable group and $B$ is a pure and dense subgroup of $A^*$. Further suppose there are $a_1, \ldots, a_n \in A$ and $b_1, \ldots, b_n \in B$ so that $\langle a_i, b_j \rangle = 1$ if $i = j$, and $0$ otherwise.*
*(i) For any $a$ not in the subgroup generated by $a_1, \ldots, a_n$, there exists $a_{n+1} \in A$ and $b_{n+1} \in B$ so that $a$ is in the subgroup generated by $a_1, \ldots, a_{n+1}$ and $\langle a_i, b_j \rangle = 1$ if $i = j$, and $0$ otherwise.*
*(ii) For any $b$ not in the subgroup generated by $b_1, \ldots, b_n$, there exists $b_{n+1} \in B$ and $a_{n+1} \in A$ so that $b$ is in the subgroup generated by $b_1, \ldots, b_{n+1}$ and $\langle a_i, b_j \rangle = 1$ if $i = j$, and $0$ otherwise.*

PROOF. By 3.3, it is enough to prove (i). First notice that the hypothesis on $a_1, \ldots, a_n$ implies that they freely generate a pure subgroup of $A$; indeed, if $\sum_i k_i a_i$ is divisible by $m$, then so is $k_j = \langle \sum_i k_i a_i, b_j \rangle$ for all $j = 1, \ldots, n$. Now, choose $c$ so that $a_1, \ldots, a_n, c$ freely generate a pure subgroup of $A$ which contains $a$. Let $a_{n+1} = c - \sum_{i=1}^{n} \langle c, b_i \rangle a_i$. Since $B$ is dense in $A^*$, there is $b_{n+1}$, so that $\langle a_i, b_{n+1} \rangle = 0$ for $i \leq n$ and $\langle a_{n+1}, b_{n+1} \rangle = 1$. $\square$

**3.5 Theorem.** *Let $A$ and $B$ be free groups of countably infinite rank such that $B$ is a subgroup of $A^*$. Then $B$ is a pure and dense subgroup of $A^*$ if and only if there are bases $\{a_i : i \in \omega\}$ of $A$ and $\{b_i : i \in \omega\}$ of $B$ such that for every $i, j \in \omega$, $\langle a_i, b_j \rangle = 1$ if $i = j$, and $0$ otherwise.*

PROOF. ($\Leftarrow$) To show that $B$ is pure in $A^*$ consider an element $b$ which generates a pure subgroup of $B$. Then $b = \sum_i n_i b_i$ where $\gcd\{n_i : i \in \omega\} = 1$ (and of course almost all $n_i = 0$). Hence there exist elements $s_i$ in $\mathbf{Z}$ (almost all 0) such that $\sum_i s_i n_i = 1$. If we let $a = \sum_i s_i a_i$, then clearly $\langle b, a \rangle = 1$. Hence by 3.2(i), $B$ is pure in $A^*$. A similar argument using 3.2(ii) shows that $B$ is dense in $A^*$.

($\Rightarrow$) Enumerate $A$ as $\{x_n : n \in \omega\}$ and $B$ as $\{y_n : n \in \omega\}$. By repeatedly applying 3.4, we can choose $\{a_n : n \in \omega\} \subseteq A$ and $\{b_n : n \in \omega\} \subseteq B$ such that for all $i$, $j$ $\langle a_i, b_j \rangle = 1$ if $i = j$ and 0 otherwise. Furthermore, the choice can be made so that $x_n$ is in the subgroup generated $\{a_i : i \leq 2n\}$ and $y_n$ is in the subgroup generated by $\{b_i : i \leq 2n + 1\}$. $\square$

For $A$ and $B$ as above, bases $\{a_i : i \in \omega\}$ and $\{b_i : i \in \omega\}$ with the property given in Theorem 3.5 are called *dual bases* of $A$ and $B$.

For the information of those familiar with the infinitary language $L_{\infty\omega}$, we include the following result which follows easily from Lemma 3.4, since that lemma establishes the necessary back-and-forth criterion for $L_{\infty\omega}$-equivalence. If $A$ is a group and $B$ is a subgroup of $A^*$, let $(A, B, \mathbf{Z}; \langle\ ,\ \rangle)$ denote the 3-sorted structure with universes $A, B, \mathbf{Z}$ — where $A$, $B$ and $\mathbf{Z}$ carry their group structure and $\langle\ ,\ \rangle$ is a function from $A \times B$ to $\mathbf{Z}$.

**3.6 Theorem.** *Suppose $A_1$ and $A_2$ are separable groups of infinite rank and $B_1$, $B_2$ are pure and dense subgroups of $A_1^*$, $A_2^*$ respectively. Then the structures $(A_1, B_1, \mathbf{Z}; \langle\ ,\ \rangle)$ and $(A_2, B_2, \mathbf{Z}; \langle\ ,\ \rangle)$ are $L_{\infty\omega}$-equivalent.* $\square$

Theorem 3.6 can be used to give a proof of direction ($\Rightarrow$) of 3.5. Indeed, if $B$ is a pure and dense subgroup of $A^*$, as in 3.5, then $(A, B, \mathbf{Z}; \langle\ ,\ \rangle)$ is $L_{\infty\omega}$-equivalent to $(\mathbf{Z}^{(\omega)}, \mathbf{Z}^{(\omega)}, \mathbf{Z}; \langle\ ,\ \rangle)$; hence they are isomorphic structures, since they are countable. Thus since the conclusion is true for $(\mathbf{Z}^{(\omega)}, \mathbf{Z}^{(\omega)}, \mathbf{Z}; \langle\ ,\ \rangle)$, it is also true for $(A, B, \mathbf{Z}; \langle\ ,\ \rangle)$.

## §4. Groups of continuous functions

In this section, we study the group of continuous functions from a topological space to the integers. Then we give a characterization of the dual and double dual of a group of continuous functions. We will assume that the reader is familiar with basic ideas of topology. The results in this section are interesting for their own sake and many of the results in Section 1 and Chapter XIV were motivated by the work of Mrówka and Eda and Ohta which we discuss here. Aside from the motivation they provide, the results of this section will not be needed elsewhere.

As in X.2.6, $C(X, \mathbf{Z})$ denotes the group of continuous functions from a topological space, $X$, to $\mathbf{Z}$, where $\mathbf{Z}$ is given the discrete topology. A topological space is called 0-*dimensional* if it has a clopen basis, i.e., a basis of sets which are both closed and open.

For any space $X$, there is an equivalence relation $E$ on $X$ defined by $x \, E \, y$ if and only if $f(x) = f(y)$ for all $f \in C(X, \mathbf{Z})$. We can give $X/E$ a topology whose collection of open sets is $\{f^{-1}[\{n\}]/E : f \in C(X, \mathbf{Z}), n \in \omega\}$. It is easy to see that $X/E$ is a 0-dimensional $T_0$-space and that the natural homomorphism from $C(X/E, \mathbf{Z})$ to $C(X, \mathbf{Z})$ is an isomorphism. So there is no loss of generality in assuming, as we will, that every topological space we deal with is a 0-dimensional $T_0$-space.

**Definition.** Suppose $X$ is a 0-dimensional $T_0$-space. Let $Y$ be a discrete topological space and $C$ be $C(X, Y)$. Let $\varphi : X \to \prod_{f \in C} Y$ be the function defined by $\varphi(x)(f) = f(x)$. Give $\prod_{f \in C} Y$ the product topology. It is easy to verify that $\varphi$ is a topological embedding of $X$ into $\prod_{f \in C} Y$. Define $\beta_Y X$ to be the closure of $\varphi[X]$ in $\prod_{f \in C} Y$. Define $X$ to be $Y$-*compact* if $\varphi[X] = \beta_Y X$, i.e., $\varphi[X]$ is closed in $\prod_{f \in C} Y$.

We are chiefly interested in the case where $Y$ is $\mathbf{Z}$, although we will also use $Y = 2 \, (= \{0, 1\})$. Note that $\beta_2 X$ is compact since it is a closed subspace of a compact space. The definition should be contrasted with the fact that a topological space is compact if and only if it is isomorphic to a closed subspace of a power of the unit interval $[0, 1] \, (\subseteq \mathbf{R})$. It is possible to give a more intrinsic characterization of $\beta_\mathbf{Z}$ and $\mathbf{Z}$-compact.

**Definition.** Suppose $X$ is a 0-dimensional $T_0$-space. An *ultrafilter,*

$D$, on the clopen subsets of $X$ is a subset of the clopen sets which contains $X$, is closed under intersection, and has the property that for every clopen set $U$, exactly one of $U$ and $X \setminus U$ is in $D$. (Note that if $X$ has the discrete topology, then this reduces to the notion of an ultrafilter on $X$ discussed in II.2.) An ultrafilter $D$ on the clopen sets is $\omega_1$-*complete* if and only if for every collection $\{U_n : n \in \omega\}$ of clopen sets such that $X = \bigcup_{n \in \omega} U_n$, there is $n$ such that $U_n \in D$. An ultrafilter on the clopen sets is *principal* if there is a point $x$ so that the ultrafilter is precisely the set of clopen sets which contain $x$.

**4.1 Proposition.** *Suppose $X$ is a 0-dimensional $T_0$-space. $X$ is $\mathbf{Z}$-compact if and only if every $\omega_1$-complete ultrafilter on the clopen subsets of $X$ is principal.*

PROOF. Let $C$ denote $C(X, \mathbf{Z})$. Consider $\varphi \colon X \to \prod_{f \in C} \mathbf{Z}$.

($\Leftarrow$) Suppose first that every $\omega_1$-complete ultrafilter on the clopen subsets of $X$ is principal. Let $y$ be in the closure of $\varphi[X]$; we must show that $y \in \varphi[X]$. Define an ultrafilter $D$ by: for any clopen set $U$, $U \in D$ if and only if $y$ is in the closure of $\varphi[U]$. Let us check that $D$ is closed under intersection. If $U$ and $V$ belong to $D$, we must show that $y$ belongs to the closure of $\varphi[U \cap V]$, i.e., for any $f_1, \ldots, f_n$ in $C$, there exists $x \in U \cap V$ such that $\varphi(x)$ and $y$ agree in their $f_i$-coordinates. Since $U$ and $V$ are clopen there exists $g \in C$ such that $g(x) = 0$ if $x \in U \cap V$; $g(x) = 1$ if $x \in U \setminus V$; $g(x) = 2$ if $x \in V \setminus U$. Without loss of generality we can assume that $f_0 = g$; since $y$ belongs to the closure of $\varphi[U]$ (respectively, $\varphi[V]$) there exists $x$ in $U$ (respectively, in $V$) such that $y(f_i) = \varphi(x)(f_i)\ (= f_i(x))$ for all $i = 0, \ldots, n$; by considering $f_0 = g$ we see that $x$ must belong to $U \cap V$. In a similar manner we can prove that for every clopen $U$, exactly one of $U$ and $X \setminus U$ belongs to $D$.

Suppose now that $X = \bigcup_{n \in \omega} U_n$, where $\{U_n : n \in \omega\}$ is a disjoint collection of clopen sets. Define a function $g$ such that $g(x) = n$ if $x \in U_n$. There is a unique $m$ so that $y(g) = m$; then by an argument like that used above to prove that $D$ closed under intersections, we get that $U_m$ is in $D$. Hence $D$ is $\omega_1$-complete. By hypothesis, there exists $x_0 \in X$ so that $D$ is the ultrafilter determined by $x_0$. For each $f$ in $C$, $x_0 \in f^{-1}[f(x_0)]$, so $f^{-1}[f(x_0)] \in D$. Hence $y$ is in the closure of $\varphi[f^{-1}[f(x_0)]]$ and thus $y(f) = \varphi(x_0)(f)$; therefore $y = \varphi(x_0)$.

($\Rightarrow$) Suppose now that $X$ is $\mathbb{Z}$-compact. If $D$ is an $\omega_1$-complete ultrafilter on the clopen subsets of $X$, then $D$ determines a unique point, $y$, in the closure of $\varphi[X]$ by the rule $y(f) = n$ if and only if $f^{-1}(n) \in D$. Then $y \in \varphi[X]$ implies that $D$ is principal (determined by $x_0$ such that $\varphi(x_0) = y$). $\square$

The proof of this proposition shows that there is a one-one correspondence between the elements of $\beta_{\mathbb{Z}}X$ and the set of $\omega_1$-complete ultrafilters on the clopen subsets of $X$. We will sometimes without further comment identify $\beta_{\mathbb{Z}}X$ with this latter set, in which case $X$ is of course identified with the principal ultrafilters. Every clopen subset, $U$, of $X$ extends to a unique clopen subset $\hat{U}$ of $\beta_{\mathbb{Z}}X$, defined by $D \in \hat{U}$ if and only if $U \in D$. (In the notation of the proof of 4.1, $\hat{U}$ is the closure of $\varphi[U]$.) The sets $\hat{U}$ ($U$ a clopen subset of $X$) form a clopen basis for $\beta_{\mathbb{Z}}X$. Since $U$ is uniquely determined by $\hat{U}$ and vice versa, we will not distinguish between $U$ and $\hat{U}$. Because $X$ and $\beta_{\mathbb{Z}}X$ have the same clopen sets, we have proved the following proposition.

**4.2 Proposition.** *Suppose $X$ is a 0-dimensional $T_0$-space. Then $\beta_{\mathbb{Z}}X = \beta_{\mathbb{Z}}\beta_{\mathbb{Z}}X$, so $\beta_{\mathbb{Z}}X$ is $\mathbb{Z}$-compact.* $\square$

**Definition.** Let $C^b(X, \mathbb{Z})$ denote the bounded homomorphisms from $X$ to $\mathbb{Z}$.

**4.3 Proposition.** *If $X$ is a 0-dimensional $T_0$-space, $C(X, \mathbb{Z}) = C(\beta_{\mathbb{Z}}X, \mathbb{Z})$ and $C^b(X, \mathbb{Z}) = C(\beta_2 X, \mathbb{Z})$. (More precisely, the obvious restriction maps are isomorphisms.)*

PROOF. It suffices to show that the obvious restriction functions map onto $C(X, \mathbb{Z})$ and $C^b(X, \mathbb{Z})$ respectively. In the first case, suppose $f \in C(X, \mathbb{Z})$; for each $n \in \omega$ let $U_n = f^{-1}[\{n\}]$, a clopen subset of $X$. Extend $f$ to $\beta_{\mathbf{Z}} X$ by defining $f(D) = n$ if and only if $D \in \hat{U}_n$, i.e., $U_n \in D$.

In the second case, suppose $f \in C^b(X, \mathbf{Z})$. Without loss of generality we can assume that the range of $f$ is contained in $\omega$. Choose $n$ so that $2^n$ is greater than the maximum of rge $f$. Define $f_0, \ldots, f_{n-1} \in C(X, 2)$, so that $f(x) = f_0(x) + \cdots + f_{n-1}(x)2^{n-1}$. For $y \in \beta_2 X$, let $f(y) = y(f_0) + \cdots + y(f_{n-1})2^{n-1}$. $\square$

**Example.** Let $X$ be a discrete space with $\kappa$ elements. Then $C(X, \mathbf{Z})$ is $\mathbf{Z}^\kappa$. Note that by 4.1, $X$ is $\mathbf{Z}$-compact if and only $\kappa$ is not $\omega$-measurable.

**4.4 Proposition.** *If $X$ is a compact space, then $C(X, \mathbf{Z})$ is free.*

PROOF. $C(X, \mathbf{Z})$ is a subgroup of $\mathbf{Z}^X$ whose elements have bounded range. By Theorem 2.11, the group of bounded (not necessarily continuous) functions from $X$ to $\mathbf{Z}$ is free. So $C(X, \mathbf{Z})$ is free. $\square$

We now begin to study the dual of a group of continuous functions.

**Definition.** Suppose $X$ is a 0-dimensional $T_0$-space and $\varphi \in C(X, \mathbf{Z})^*$. Then a closed subset $A$ of $X$ is a *support* of $\varphi$ if for all $f, g \in C(X, \mathbf{Z})$, $f \upharpoonright A = g \upharpoonright A$ implies $\langle \varphi, f \rangle = \langle \varphi, g \rangle$.

In the language of continuous functions, Corollary III.3.3 says — for $H = \mathbf{Z}$ — that if $X$ is a discrete space of non-$\omega$-measurable cardinality, then any element of $C(X, \mathbf{Z})^*$ has finite support. Since for a discrete space, the compact sets are exactly the finite sets we can restate this result as: "if $X$ is a discrete space of non-$\omega$-measurable cardinality, then any element of $C(X, \mathbf{Z})^*$ has compact support."

**4.5 Lemma.** *If $X$ is a 0-dimensional $T_0$-space and $K$ is a compact subspace, then any continuous function from $K$ to $\mathbf{Z}$ (where $K$ has the relative topology) extends to a continuous function from $X$ to $\mathbf{Z}$.*

PROOF. If $f$ is a continuous function from $K$ to $\mathbf{Z}$, then the range of $f$ is finite, so there exist $m$ and clopen sets $U_0, \ldots, U_m$ covering

$X$ such that $\text{rge}(f) \subseteq \{0, \ldots, m\}$ and for each $i$, $f^{-1}[\{i\}] \subseteq U_i$. Without loss of generality we can assume that the $U_i$'s are pairwise disjoint. Then extend $\varphi$ to $X$ by defining $f(x) = i$ if and only if $x \in U_i$. $\square$

**4.6 Proposition.** *Suppose $X$ is a compact $0$-dimensional $T_0$-space and $\varphi \in C(X, \mathbf{Z})^*$. Then there is a minimum support of $\varphi$, i.e., a support of $\varphi$ which is contained in every other support of $\varphi$.*

PROOF. First we show that the intersection of two supports is a support. Suppose $A = A_1 \cap A_2$, where $A_1$ and $A_2$ are supports of $\varphi$ but $A$ is not. Choose $f, g \in C(X, \mathbf{Z})$ so that $f{\restriction}A = g{\restriction}A$ and $\langle \varphi, f \rangle \neq \langle \varphi, g \rangle$. Since $A$, $A_1$, and $A_2$ are all compact in the relative topology, by 4.5 there are functions $f_1, g_1$ so that $f_1{\restriction}A_1 = g_1{\restriction}A_1$, $f_1{\restriction}A_2 = f{\restriction}A_2$ and $g_1{\restriction}A_2 = g{\restriction}A_2$. But then $\langle \varphi, f_1 \rangle = \langle \varphi, g_1 \rangle = \langle \varphi, g \rangle \neq \langle \varphi, f \rangle = \langle \varphi, f_1 \rangle$. (The first equality holds since $A_1$ is a support and the subsequent equalities hold since $A_2$ is a support.)

To finish the proof it is enough to show, for any infinite cardinal $\lambda$, that if $(A_\alpha : \alpha < \lambda)$ is a decreasing sequence of supports then $A \overset{\text{def}}{=} \bigcap A_\alpha$ is a support. Assume not and choose $f, g \in C(X, \mathbf{Z})$ so that $f{\restriction}A = g{\restriction}A$ and $\langle \varphi, f \rangle \neq \langle \varphi, g \rangle$. Since $X$ is compact the ranges of $f$ and $g$ are finite. So $B \overset{\text{def}}{=} \{x : g(x) = f(x)\}$ is a clopen set containing $A$. Since $B$ is not a support, $A_\alpha \not\subseteq B$ for any $\alpha$. Hence $\{X \setminus A_\alpha : \alpha < \lambda\} \cup \{B\}$ is an open cover of $X$ with no finite subcover, a contradiction. $\square$

**4.7 Proposition.** *Suppose $X$ is a $0$-dimensional $T_0$-space and $\varphi_1$, $\varphi_2 \in C(X, \mathbf{Z})^*$ and agree on $C^b(X, \mathbf{Z})$. Then $\varphi_1 = \varphi_2$.*

PROOF. Suppose not. Let $f \in C(X, \mathbf{Z})$ be such that $\varphi_1(f) \neq \varphi_2(f)$. For $n \in \mathbf{Z}$, let $U_n = f^{-1}[\{n\}]$. Define $f_n \in C^b(X, \mathbf{Z})$ by $f_n(x) = n$, if $x \in U_n$ and $f_n(x) = 0$ otherwise. For any integers $(a_n : n \in \mathbf{Z})$, $\sum a_n f_n$ defines an element of $C(X, \mathbf{Z})$. In this notation, $f = \sum f_n$. We can then define a function $\theta : \mathbf{Z}^\omega \to \mathbf{Z}$ by $\theta((a_n)_n) = (\varphi_1 - \varphi_2)(\sum a_n f_n)$. But this contradicts III.1.3, since $\theta$ is zero on $\mathbf{Z}^{(\omega)}$ but $\theta(f) \neq 0$. $\square$

Now we can state Mrówka's theorem.

**4.8 Theorem.** *If $X$ is a $\mathbf{Z}$-compact $0$-dimensional $T_0$-space, then any $\varphi \in C(X, \mathbf{Z})^*$ has compact support.*

PROOF. If $Y = \beta_2 X$, then $C^b(X, \mathbf{Z}) = C(Y, \mathbf{Z})$ by 4.3. Let $\varphi_Y$ be the restriction of $\varphi$ to $C(Y, \mathbf{Z})$. Since $Y$ is compact, there is a minimum support, $A$, of $\varphi_Y$. It is enough to show that $A \subseteq X$, for then by Proposition 4.7, $A$ is a support of $\varphi$. Suppose, to the contrary that there exists $y \in A \setminus X$. Let $D$ be the ultrafilter on the clopen sets of $X$, defined by $U \in D$ if and only if $y$ is in the closure of $\varphi[U]$. (The verification that $D$ is an ultrafilter is the same as in 4.1.) By 4.1, $D$ is not $\omega_1$-complete so there are disjoint clopen sets $U_n$ ($n \in \omega$), none of which are in $D$, whose union is $X$. Since no finite union of the $U_n$ is a support of $\varphi_Y$, for every $n$ there exists $f_n \in C^b(X, \mathbf{Z})$ such that $f_n \upharpoonright U_m = 0$, for all $m \leq n$, and $\langle \varphi, f_n \rangle \neq 0$. For every $(a_n : n \in \omega)$, $\sum a_n f_n$ defines an element of $C(X, \mathbf{Z})$. If we define $\theta((a_n)_n) = \varphi(\sum a_n f_n)$ we obtain a function such that for all $n$, $\theta(e_n) \neq 0$, contradicting the slenderness of $\mathbf{Z}$. □

Suppose $X$ is a topological space and $K \subseteq X$. Let $\pi_K$ denote the restriction map from $C(X, \mathbf{Z})$ to $C(K, \mathbf{Z})$. The theorem above can be rephrased as follows.

**4.9 Corollary.** *Suppose $X$ is a $\mathbf{Z}$-compact $0$-dimensional $T_0$-space. Then $C(X, \mathbf{Z})^* = \varinjlim C(K, \mathbf{Z})^*$, where the direct limit is taken over the compact subspaces of $X$ and the map from $C(K, \mathbf{Z})^*$ to $C(X, \mathbf{Z})^*$ is $\pi_K^*$.* □

**4.10 Example.** It is possible to deduce the Loś-Eda theorem (III.3.2) in the special case of products of $\mathbf{Z}$ from Mrówka's theorem. Suppose $X$ is a topological space with the discrete topology; so $C(X, \mathbf{Z}) = C(\beta_{\mathbf{Z}} X, \mathbf{Z}) = \mathbf{Z}^X$. As we have seen, $\beta_{\mathbf{Z}} X$ is the set of $\omega_1$-complete ultrafilters on $X$. It is easy to see that any countable subset of $\beta_{\mathbf{Z}} X$ is closed and the relative topology on the set is discrete. So the compact subsets of $\beta_{\mathbf{Z}} X$ are exactly the finite sets. By Theorem 4.8, any element of $C(\beta_{\mathbf{Z}} X, \mathbf{Z})^*$ has finite support (in $\beta_{\mathbf{Z}} X$). If an element has finite support, then it is a linear combination of elements of the form $\varphi_D$, where $D$ is an $\omega_1$-complete ultrafilter on $X$. Here $\varphi_D(f) = f(D)$ and $f(D) = n$ if and only if $\{x \in X: f(x) = n\} \in D$ (cf. proof of 4.5). Thus we have shown that

$$(\mathbf{Z}^X)^* = \bigoplus_{D \in \mathcal{D}} \langle \varphi_D \rangle$$

where $\mathcal{D}$ is the set of $\omega_1$-complete ultrafilters.

In the notation we have developed, Lemma 4.5 says that $\pi_K$ is onto for any compact subspace. We can characterize $C(X, \mathbf{Z})^{**}$ as an inverse limit immediately. But we will see that this inverse limit is actually a group of continuous functions on $X$, if we suitably modify the topology on $X$.

**Definition.** Let $X$ be a 0-dimensional space. A function $f$ from $X$ to $\mathbf{Z}$ is $k_{\mathbf{Z}}$-*continuous* if $f \restriction K$ is continuous (in the relative topology) for every compact subset $K \subseteq X$. Let $k_{\mathbf{Z}} X$ be the set $X$ equipped with the least topology so that every $k_{\mathbf{Z}}$-continuous function is continuous on $k_{\mathbf{Z}} X$.

**4.11 Proposition.** (i) *A subset $U$ of $X$ is clopen in $k_{\mathbf{Z}} X$ if and only if $U \cap K$ is clopen (in the relative topology) for every compact subset $K$ of $X$. Furthermore, these sets form a clopen basis for $k_{\mathbf{Z}} X$.*
(ii) *Every clopen set in $X$ is clopen in $k_{\mathbf{Z}} X$.*
(iii) *$X$ and $k_{\mathbf{Z}} X$ have the same compact sets and induce the same topology on any compact set.*

PROOF. (ii) and (iii) will be obvious once we have proved (i). First note that if $U \cap K$ is clopen for every compact set $K$, then the characteristic function of $U$ is $k_{\mathbf{Z}}$-continuous. So $U$ is a clopen set in $k_{\mathbf{Z}} X$. It is easy to verify that $\{U: U \cap K$ is clopen for every compact set $K\}$ forms a clopen basis for a topology on the set $X$. It remains to check that this topology coincides with $k_{\mathbf{Z}} X$. To see this, it is enough to check that for any $k_{\mathbf{Z}}$-continuous function $f$ and $n \in \mathbf{Z}$, $f^{-1}[\{n\}]$ is clopen in in the new topology. This last statement follows easily from the definition. $\square$

**4.12 Example.** For future use we want to consider two examples. Let $E$ be a stationary and co-stationary subset of $\omega_1$ consisting of limit ordinals. Let $X$ be the topological space $\omega_1 + 1 \setminus E$ in the order topology. Note that the sets of the form $[\alpha + 1, \beta] \cap X$ together with $\{0\}$ and $X$ form a clopen basis of $X$. As well we will consider $\mathbf{Q}$, as a topological space, with the usual topology.

**4.12A.** *Both $X$ and $\mathbf{Q}$ are $\mathbf{Z}$-compact.*

PROOF. We will only give the proof for $X$, since the proof for $\mathbf{Q}$ is similar. Let $D$ be an $\omega_1$-complete ultrafilter on the clopen subsets of $X$. If $\omega_1$ is in every element of $D$, then $D$ is principal. Assume there is $U \in D$ so that $\omega_1 \notin U$. Let $\alpha$ be the smallest ordinal so that $[\alpha + 1, \omega_1] \cap X \notin D$. (There must be such an $\alpha$ since $D$ is $\omega_1$-complete and $\omega_1 \notin U$.) If $\alpha \notin E$, then $D$ is the principal ultrafilter determined by $\alpha$. If $\alpha \in E$, then $\alpha$ is a limit ordinal. Choose $\alpha_n$ $(n \in \omega)$ an increasing sequence of ordinals with limit $\alpha$. Then

$$X = ([0, \alpha_0] \cap X) \cup \bigcup_{n \in \omega} ([\alpha_n, \alpha_{n+1}] \cap X) \cup ([\alpha + 1, \omega_1] \cap X).$$

Each set in the union is a clopen set which is not in $D$. (Note that $[\alpha_n + 1, \omega_1] \cap X$ belongs to $D$ by the minimality of $\alpha$.) So $D$ is not $\omega_1$-complete.

**4.12B.** $\mathbf{Q} = k_{\mathbf{Z}}\mathbf{Q}$.

PROOF. It suffices to show that every closed subset of $k_{\mathbf{Z}}\mathbf{Q}$ is closed in $\mathbf{Q}$. Since $\mathbf{Q}$ is countable, it is enough to show that if $U$ is closed in $k_{\mathbf{Z}}\mathbf{Q}$, then every convergent sequence in $U$ (relative to $\mathbf{Q}$) has its limit in $U$. But any convergent sequence lies in a compact subset of $\mathbf{Q}$, so by 4.11 it has the same limit in $\mathbf{Q}$ as in $k_{\mathbf{Z}}\mathbf{Q}$. $\square$

**4.12C.** $k_{\mathbf{Z}}X = (\omega_1 \setminus E) \cup \{\omega_1\}$, *where $\omega_1$ is an isolated point and the relative topology on $\omega_1 \setminus E$ is the order topology.*

PROOF. To show that $\omega_1$ is an isolated point, we must show that $\{\omega_1\}$ is open in $k_{\mathbf{Z}}X$, i.e., by 4.11, that $\{\omega_1\} \cap K$ is open for all compact subsets $K$ of $X$. It will suffice to show that $K \cap \omega_1$ is bounded. If not, then since $E$ is stationary, there is $\alpha \in E$ so that $\alpha$ is a limit point of $K$. But then $K$ cannot be compact: indeed, as in 4.12A, we can then choose an open cover which does not have a finite subcover.

We now must show that the relative topology on $\omega_1 \setminus E$ in $k_\mathbb{Z} X$ is just the order topology. To prove this it suffices to show that if $U$ is a clopen set in $k_\mathbb{Z} X$ and $U \cap \omega_1$ is unbounded then there is $\alpha < \omega_1$ such that $(\alpha, \omega_1) \setminus E \subseteq U$. Otherwise, as $E$ is co-stationary, there is $\alpha \notin E$ so that $\alpha$ is a limit point of both $\omega_1 \setminus (E \cup U)$ and $U$. Choose $\alpha_n$ $(n \in \omega)$, an increasing sequence of ordinals with limit $\alpha$ so that for all $n$, $\alpha_{2n} \notin E \cup U$ and $\alpha_{2n+1} \in U$. Then $U \cap (\{\alpha_n : n \in \omega\} \cup \{\alpha\})$ is either $\{\alpha_{2n+1} : n \in \omega\} \cup \{\alpha\}$ or $\{\alpha_{2n+1} : n \in \omega\}$. Neither of these sets is a clopen subset of $\{\alpha_n : n \in \omega\} \cup \{\alpha\}$. $\square$

**4.13 Theorem.** *If $X$ is $\mathbb{Z}$-compact and no compact subset of $X$ is of $\omega$-measurable cardinality, then there is an isomorphism, $\varphi$, from $C(X, \mathbb{Z})^{**}$ to $C(k_\mathbb{Z} X, \mathbb{Z})$ and $\varphi \circ \sigma$ is the inclusion map from $C(X, \mathbb{Z})$ into $C(k_\mathbb{Z} X, \mathbb{Z})$.*

PROOF. By Corollary 4.9 and Proposition 1.3, $C(X, \mathbb{Z})^{**} = \varprojlim C(K, \mathbb{Z})^{**}$, where the inverse limit is taken over the compact subspaces of $X$. Since each $C(K, \mathbb{Z})$ is free by 4.4, each one of these groups is reflexive. So we have a natural isomorphism from $C(X, \mathbb{Z})^{**}$ to $\varprojlim C(K, \mathbb{Z})$. For $K_1 \subseteq K_2$, the map in the inverse limit from $C(K_2, \mathbb{Z})$ to $C(K_1, \mathbb{Z})$ is just the restriction map. So to every element of $\varprojlim C(K, \mathbb{Z})$ is associated a function from $X$ to $\mathbb{Z}$, which is $k_\mathbb{Z}$-continuous. As well, any $k_\mathbb{Z}$-continuous function corresponds to an element of $\varprojlim C(K, \mathbb{Z})$. $\square$

**4.14 Corollary.** *Suppose $X$ is a $\mathbb{Z}$-compact space of non-$\omega$-measurable cardinality. Then $C(X, \mathbb{Z})$ is reflexive if and only if $X = k_\mathbb{Z} X$.*

PROOF. By the theorem above, $C(X, \mathbb{Z})$ is reflexive if and only if $C(X, \mathbb{Z}) = C(k_\mathbb{Z} X, \mathbb{Z})$. If $X = k_\mathbb{Z} X$, this is obviously the case. On the other hand, if $X \neq k_\mathbb{Z} X$, there is a clopen subset $U$ of $k_\mathbb{Z} X$, which is not clopen in the topology for $X$, so $C(X, \mathbb{Z}) \neq C(k_\mathbb{Z} X, \mathbb{Z})$. (For example, the characteristic function of $U$ belongs to $C(k_\mathbb{Z} X, \mathbb{Z}) \setminus C(X, \mathbb{Z})$.) $\square$

**4.15 Corollary.** (i) $C(\mathbf{Q}, \mathbf{Z})$ *is a reflexive group* (*which is not in the Reid class*).
(ii) *If $E$ is a stationary and co-stationary subset of $\omega_1$ consisting of limit ordinals, then $C(\omega_1 + 1 \setminus E, \mathbf{Z})$ is a non-reflexive dual group.*

PROOF.   The only assertion that needs to be proved is that $C(\omega_1 + 1 \setminus E, \mathbf{Z})$ is a dual group. If we let $X = \omega_1 + 1 \setminus E$, then $\beta_{\mathbf{Z}} k_{\mathbf{Z}} X \cong X$. So $C(X, \mathbf{Z}) \cong C(\beta_{\mathbf{Z}} k_{\mathbf{Z}} X, \mathbf{Z}) \cong$ (by 4.3) $C(k_{\mathbf{Z}} X, \mathbf{Z}) \cong$ (by 4.13) $C(X, \mathbf{Z})^{**}$. $\square$

Notice that $C(\omega_1 + 1 \setminus E, \mathbf{Z})$ is isomorphic to its double dual, but not by the canonical map $\sigma$; that is, it is non-reflexive, but it is not strongly non-reflexive. In Chapter XIV we will construct dual groups which are strongly non-reflexive, i.e., not isomorphic to their double dual. Part (ii) of the theorem above has already appeared as Theorem 1.13; it was first proved by Eda and Ohta in the manner given here. The reader should verify that the group in part (ii) is the same as the group used in the proof of 1.13.

## EXERCISES

1. Show that there is an inverse-direct system so that the dual of the inverse limit is not the direct limit of the dual system. [Hint: consider the inverse-direct system $(B_\alpha^*, \pi_{\beta\alpha}^*, \tau_{\alpha\beta}^* \colon \alpha < \beta < \omega_1)$ which was constructed in 1.11.]

The following questions concern groups of continuous functions. $X$ will denote a 0-dimensional $T_0$ topological space.

2. $C(X, \mathbf{Z})$ is a separable group. [Hint: show $C(X, \mathbf{Z})$ is a pure subgroup of $\mathbf{Z}^X$.]

3. Let $\mathcal{C}$ denote the class of groups of the form $C(X, \mathbf{Z})$.
(i) $\mathcal{C}$ is closed under direct products. [Hint: consider the disjoint union of the spaces.]
(ii) $\mathcal{C}$ is closed under direct sums. [Hint: for finite direct sums use part (i). Suppose $X_i$ ($i \in I$) is an infinite family of spaces. Let $Y$ be the one point compactification of the disjoint union of the $X_i$. Show $C(Y, \mathbf{Z}) = \bigoplus_{i \in I} C(X_i, \mathbf{Z}) \oplus \mathbf{Z}$. Use 2, to show that this group is isomorphic to $\bigoplus_{i \in I} C(X_i, \mathbf{Z})$.]

4. Show the following are equivalent.
(a) $C(X, \mathbf{Z})$ is free;
(b) $C(X, \mathbf{Z})^*$ is isomorphic to $\mathbf{Z}^I$, for some $I$;
(c) No summand of $C(X, \mathbf{Z})^*$ is isomorphic to $\mathbf{Z}^{(I)}$;
(d) $C(X, \mathbf{Z})$ is slender;
(e) No summand of $C(X, \mathbf{Z})$ is isomorphic to $\mathbf{Z}^\omega$;
(f) Every element of $C(X, \mathbf{Z})$ is bounded.
[Hint: prove (a) $\Rightarrow$ (b), (b) $\Rightarrow$ (c), (c) $\Rightarrow$ (e), (a) $\Rightarrow$ (d), (d) $\Rightarrow$ (e), (e) $\Rightarrow$ (f), and (f) $\Rightarrow$ (a). To show $\neg$ (f) $\Rightarrow$ $\neg$ (e), suppose $\{U_n : n \in \omega\}$ is a partition of $X$ into non-empty clopen sets. For each $n$, choose $x_n \in U_n$. Let $\chi_n$ denote the characteristic function of $U_n$. Then any $g \in C(X, \mathbf{Z})$ can be written uniquely as $\sum g(x_n)\chi_n + f$, where $f(x_n) = 0$ for all $n$. For (f) $\Rightarrow$ (a), use 2.11.]

5. Show the following are equivalent:
(a) $C(X, \mathbf{Z})^*$ is free;
(b) $C(X, \mathbf{Z})^{**}$ is isomorphic to $\mathbf{Z}^I$ for some $I$;
(c) No summand of $C(X, \mathbf{Z})$ is isomorphic to $\mathbf{Z}^{(\omega)}$;
(d) $C(X, \mathbf{Z})^*$ is slender;
(e) No summand of $C(X, \mathbf{Z})^*$ is isomorphic to $\mathbf{Z}^\omega$;
(f) Any compact subset of $\beta_{\mathbf{Z}} X$ is finite.
[Hint: (e) $\Rightarrow$ (f): Suppose $K$ is an infinite compact subset of $\beta_{\mathbf{Z}} X$. Then the restriction map from $C(X, \mathbf{Z})$ onto $C(K, \mathbf{Z})$ is a map onto an infinite rank free group.
(f) $\Rightarrow$ (a): Assume $X$ is $\mathbf{Z}$-compact. Use Mrówka's theorem to show that $C(X, \mathbf{Z})^* = \bigoplus_{x \in X} \langle \varphi_x \rangle$, where $\varphi_x(f) = f(x)$.]

By Exercises 4 and 5, the only slender groups which appear as groups of continuous functions or their duals are free groups.

6. If the cardinality of $X$ is not $\omega$-measurable, $C(X, \mathbf{Z})$ is isomorphic to $\mathbf{Z}^I$ for some $I$ if and only if $X$ is discrete. [Hint: Assume $X = \beta_{\mathbf{Z}} X$. Use 5 to show that $k_{\mathbf{Z}} X = X$ is discrete.]

The next four exercises show that not all groups of continuous functions are dual groups; in these exercises we consider the following topological space. Choose $D$ a non-principal ultrafilter on $\omega$ and let $X = \omega \cup \{\infty\}$. Define an open basis for $X$ to be the power set of $\omega$ together with $\{Y \cup \{\infty\}: Y \in D\}$.

7. $X$ is 0-dimensional and $\mathbf{Z}$-compact.

8. Every compact subset of $X$ is finite.

9. $C(X, \mathbf{Z})$ is the subgroup of $\mathbf{Z}^X$ generated by the function that is constantly 1, and $\{f: \{n: f(n) = 0\} \in D\}$.

10. $C(X, \mathbf{Z})$ is not a dual group. [Hint: by 5, $C(X, \mathbf{Z})^*$ is free. If $C(X, \mathbf{Z})$ were a dual group it would be reflexive, since $C(X, \mathbf{Z})^{**} \cong \mathbf{Z}^I$, for some $I$. By 4.14, $C(X, \mathbf{Z})$ is not reflexive.]

The following exercises show that every countable group with a discrete norm is free.

11. Any group with a norm is torsion-free.

12. Assume that $A$ is a finite rank torsion-free group with a discrete norm $\| \ \|$. Furthermore assume that $\{x_1, \ldots, x_n\}$ is a maximal independent subset of $A$. View $A$ as a subgroup of $\mathbf{Q}x_1 \oplus \ldots \oplus \mathbf{Q}x_n$.
(i) $X \stackrel{\text{def}}{=} \{(q_1, \ldots, q_n): |q_i| \leq 1 \text{ and } \sum q_i x_i \in A\}$ is finite. [Hint: otherwise for every $\epsilon > 0$, there are $(q_1^1, \ldots, q_n^1) \neq (q_1^2, \ldots, q_n^2) \in X$ such that for all $i$, $|q_i^1 - q_i^2| < \epsilon$. So $0 < \| \sum q_i^1 x_i - \sum q_i^2 x_i \| \leq \epsilon \sum \| x_i \|$.]
(ii) $A$ is generated by the finite set $X$ and hence is free.

13. If $A$ is a countable group with a discrete norm, then $A$ is free. [Hint: Pontryagin's Criterion (IV.2.3).]

14. Suppose $(A_\alpha, \tau_{\alpha\beta}, \pi_{\beta\alpha} : \alpha < \beta < \omega_1)$ is an inverse-direct system and there is a stationary set $E$ such that for all $\delta \in E$, $A_\delta = \varinjlim_{\alpha < \delta} A_\alpha$. Then $\varinjlim_{\alpha < \omega_1} A_\alpha = \varprojlim_{\alpha < \omega_1} A_\alpha$.

## NOTES

The material on direct and inverse limits is standard. In this context the notion of an inverse-direct system is due to Mekler. Set theorists used inverse-direct systems to describe iterated forcing. Theorem 1.12 is due to Eda 1983c and 1.13 to Eda-Ohta 1986, but the proofs given here are due to Mekler and Schlitt (cf. Schlitt 1986).

Section 2 through 2.9 is based on Huber 1983a; 2.4 – 2.6 also make use of Mader 1982. Theorem 2.10 is due to Steprāns 1985; the countable case was proved by Lawrence 1984 and Zorzitto 1985. Theorem 2.11 was proved by Specker 1950 assuming CH and by Nöbeling 1968 without assuming CH.

Section 3 is due to Chase 1963b.

4.1 is due to Herrlich 1967. The notion of **Z**-compact is due to Mrówka and 4.5 to 4.9 are from Mrówka 1972; 4.11 to 4.15 are from Eda-Ohta 1986.

Exercises. 2–3:Eda 1983e; 4–8:Eda-Ohta 1986; 10:Reid 1967.

# CHAPTER XII
# THE STRUCTURE OF EXT

Recall that Whitehead's Problem asks if every Whitehead group is free. A group $A$ is called a Whitehead group, or W-group, if $\text{Ext}(A, \mathbb{Z}) = 0$.

In this chapter, in sections 1 and 3, we consider necessary and sufficient conditions on $A$ for $\text{Ext}(A, \mathbb{Z})$ to be zero. In particular, in section 1, we shall show that Whitehead's Problem is undecidable in ZFC. In section 3, we shall give necessary and sufficient conditions, in terms of uniformization properties, for the existence of a non-free Whitehead group of cardinality $\aleph_1$.

In section 2, we consider the structure of Ext. If $A$ is a torsion-free group, then $\text{Ext}(A, \mathbb{Z})$ is a divisible group, so its structure is completely described by certain cardinal invariants. We shall investigate the possible values for these invariants, assuming CH, when $A$ is of cardinality $\aleph_1$.

## §1. The vanishing of Ext

In VII.4.10, it was shown that it is consistent that every Whitehead group is free. We will give another proof of that result here and also show that it is consistent with ZFC that there are non-free W-groups. We will also consider the question of characterizing W-groups (or W-groups of cardinality $\aleph_1$) when there are non-free W-groups. Furthermore, we will deal with the more general problem of characterizing which modules $A$ have the property that $\text{Ext}(A, M) = 0$, for a fixed module $M$.

If $A$ is a group which is not torsion-free, then $A$ has a subgroup isomorphic to $Z(p^n)$; therefore since $\text{Ext}(Z(p^n), \mathbb{Z}) \neq 0$, every W-group is torsion-free. In the following, we will tacitly assume the groups mentioned are torsion-free. Recall, from VII.4, that being a W-group is an hereditary property; i.e., if $B \subseteq A$ and $A$ is a W-group, then so is $B$.

**1.1 Lemma.** *If $A$ is a torsion-free group, then for any group $M$, $\text{Ext}(A, M)$ is divisible.*

PROOF. Consider the short exact sequence $0 \to A \to A$ where the map from $A$ to $A$ is multiplication by $m \in \mathbf{Z}$. This induces the exact sequence

$$\mathrm{Ext}(A, M) \to \mathrm{Ext}(A, M) \to 0$$

where the first map is multiplication by $m$ (cf. I.1.1). Since this map is onto, $\mathrm{Ext}(A, M)$ is divisible by $m$. □

**1.2 Proposition.** *Every Whitehead group is $\aleph_1$-free. In particular, every countable Whitehead group is free.*

PROOF. By Pontryagin's criterion (IV.2.3), it is enough to show that every finite rank W-group is free. The proof is by induction on rank. The result is trivial for the group of rank 0. Assume all W-groups of rank $n$ are free and $A$ is a W-group of rank $n+1$. Let $\hat{\mathbf{Z}}$ denote the $\mathbf{Z}$-adic completion of $\mathbf{Z}$ (cf. I.3.6). Consider the short exact sequence

$$0 \to \mathbf{Z} \to \hat{\mathbf{Z}} \to D \to 0.$$

Note that $\mathbf{Z}$ is pure in $\hat{\mathbf{Z}}$ and $D$ is (torsion-free) divisible and non-zero. This exact sequence induces the exact sequence

$$0 \to \mathrm{Hom}(A, \mathbf{Z}) \to \mathrm{Hom}(A, \hat{\mathbf{Z}}) \to \mathrm{Hom}(A, D) \to \mathrm{Ext}(A, \mathbf{Z}) = 0.$$

The last equality holds because $A$ is a W-group. We claim that $\mathrm{Hom}(A, \mathbf{Z}) \neq 0$. Otherwise, $\mathrm{Hom}(A, \hat{\mathbf{Z}}) \cong \mathrm{Hom}(A, D)$. But $\mathrm{Hom}(A, \hat{\mathbf{Z}})$ is reduced, because $\hat{\mathbf{Z}}$ is torsion-free and reduced; and $\mathrm{Hom}(A, D)$ is a non-zero divisible group, because $A$ is torsion-free and $D$ is a non-zero injective group. Let $g \colon A \to \mathbf{Z}$ be non-zero. Then $A \cong \ker g \oplus \mathbf{Z}$. Since $\ker g$ is a W-group of rank $n$, it is free by induction, and therefore $A$ is free. □

In section 3, we will give a more computational proof of Proposition 1.2. (See the remarks after 3.14.)

**1.3 Proposition.** *Every Whitehead group is separable and slender.*

PROOF. Suppose $A$ is a W-group. The fact that $A$ is separable is a consequence of IV.2.13, but we will give another proof. By IV.2.7 it is enough to show that if $a$ generates a pure subgroup of $A$, then

$\langle a \rangle$ is a summand of $A$. By considering the exact sequence

$$\mathbf{Z} \cong \mathrm{Hom}(\langle a \rangle, \mathbf{Z}) \to \mathrm{Ext}(A/\langle a \rangle, \mathbf{Z}) \to \mathrm{Ext}(A, \mathbf{Z}) = 0$$

we see that $\mathrm{Ext}(A/\langle a \rangle, \mathbf{Z})$ is finitely generated. Since $A/\langle a \rangle$ is torsion-free, $\mathrm{Ext}(A/\langle a \rangle, \mathbf{Z})$ is divisible. But the only finitely generated divisible group is $\{0\}$. So $\mathrm{Ext}(A/\langle a \rangle, \mathbf{Z}) = 0$ and hence the exact sequence

$$0 \to \langle a \rangle \to A \to A/\langle a \rangle \to 0$$

splits. Now to show that $A$ is slender, it suffices, by IX.2.5, to show that $A$ does not contain a copy of $\mathbf{Z}^\omega$. But by Exercise IV.16, $\mathbf{Z}^\omega$ is not a W-group, so it cannot be isomorphic to a subgroup of $A$. □

In order to show (assuming appropriate axioms) that a group is not a W-group, we will need to produce a free resolution of this group.

**1.4 Lemma.** *Suppose $\kappa$ is a regular cardinal and $\{A_\alpha : \alpha < \kappa\}$ is a $\kappa$-filtration of a group $A$. There is a short exact sequence*

$$0 \to K \to F \xrightarrow{\varphi} A \to 0$$

*where $F$, $K$ are free groups such that $F = \bigoplus_{\alpha < \kappa} F_\alpha, K = \bigoplus_{\alpha < \kappa} K_\alpha$ and for all $\alpha < \kappa, |F_\alpha|, |K_\alpha| < \kappa$ and there is a commutative diagram*

$$
\begin{array}{ccccccccc}
0 & \to & \bigoplus_{\beta < \alpha} K_\beta & \to & \bigoplus_{\beta < \alpha} F_\beta & \xrightarrow{\phi_\alpha} & A_\alpha & \to & 0 \\
 & & \downarrow & & \downarrow & & \downarrow & & \\
0 & \to & K & \to & F & \xrightarrow{\phi} & A & \to & 0
\end{array}
$$

*such that the rows are exact and the vertical maps are inclusions.*

PROOF. The groups are defined by induction on $\alpha$. Choose

$$0 \to K_0 \to F_0 \xrightarrow{\varphi} A_1 \to 0$$

a free resolution. Suppose $F_\beta$, $K_\beta$ and $\varphi_\beta$ have been defined for all $\beta < \alpha$ so that $\varphi_\gamma$ is an extension of $\varphi_\beta$ if $\beta < \gamma$. If $\alpha$ is a limit ordinal it is clear how to define $\varphi_\alpha$, so suppose $\alpha = \gamma + 1$

for some $\gamma$. Choose $F_\gamma$ of cardinality $< \kappa$ such that there is a surjective map $\psi_\gamma \colon F_\gamma \to A_{\gamma+1}$. Let $\varphi_\alpha \colon \bigoplus_{\beta < \alpha} F_\beta \to A_{\gamma+1}$ be such that $\varphi_\alpha \upharpoonright \bigoplus_{\beta < \gamma} F_\beta = \varphi_\gamma$ and $\varphi_\alpha \upharpoonright F_\gamma = \psi_\gamma$. Let $\{b_i \colon i \in I\}$ be a set of free generators for $\psi_\gamma^{-1}[A_\gamma]$. Choose $\{f_i \colon i \in I\} \subseteq \bigoplus_{\beta < \gamma} F_\beta$ so that for all $i$, $\varphi_\gamma(b_i) = \psi_\gamma(f_i)$. Let $K_\gamma$ be the group generated by $\{b_i - f_i \colon i \in I\}$. This completes the inductive step in the construction. Finally, define $\varphi$ to be the union of the $\varphi_\beta$'s. $\square$

Given a $\kappa$-filtration we will refer to the resolution given in this lemma as *a free resolution associated with the filtration*. An immediate consequence of Lemma 1.4 is that

$$A_{\alpha+1}/A_\alpha \cong \bigoplus_{\beta \le \alpha} F_\beta / (\bigoplus_{\beta < \alpha} F_\beta + K_\alpha).$$

We are now in a position to give another proof that every W-group is free, assuming $V = L$. First we will state the killing lemma we need. Here, and in what follows, we will use the fact (see section I.2) that if $F$ is free, then $\mathrm{Ext}(F/K, \mathbb{Z}) = 0$ if and only if every homomorphism from $K$ to $\mathbb{Z}$ extends to a homomorphism from $F$ to $\mathbb{Z}$.

**1.5 Lemma.** *Suppose $F$ is a free group and $F/H$ is free and $K$ is such that $\mathrm{Ext}(F/(H + K), \mathbb{Z}) \ne 0$. Then there are homomorphisms $f_0$, $f_1 \colon K \to \mathbb{Z}$ such that no homomorphism $h \colon H \to \mathbb{Z}$ extends to a homomorphism from $F$ to $\mathbb{Z}$ containing $f_0$ and to a homomorphism from $F$ to $\mathbb{Z}$ containing $f_1$.*

PROOF. Let $f_0$ be the zero homomorphism. Since $F/(H + K) \cong (F/H)/(H + K/H)$ there is a homomorphism $g \colon (H + K)/H \to \mathbb{Z}$ which does not extend to a homomorphism from $F/H$ to $\mathbb{Z}$. Let $f_1 = g \circ \rho$ where $\rho \colon K \to (H + K)/H$ is the canonical surjection. Suppose $h_0, h_1 \colon F \to \mathbb{Z}$ are homomorphisms containing $f_0$ and $f_1$, respectively, which agree on $H$. Then $h_1 - h_0$ contradicts the choice of $g$. $\square$

**1.6 Theorem.** (V = L) *Every Whitehead group is free.*

PROOF. Suppose the theorem is false, and let $A$ be a counterexample of minimum cardinality $\kappa$. Since every subgroup of $A$ is a

W-group, $A$ is $\kappa$-free. Since every countable W-group is free, $\kappa$ is uncountable. By the singular compactness theorem (IV.3.5), $\kappa$ is regular. Choose $\{A_\alpha : \alpha < \kappa\}$ a $\kappa$-filtration of $A$ and $\{F_\alpha : \alpha < \kappa\}$, $\{K_\alpha : \alpha < \kappa\}$ an associated free resolution as in 1.4. Let $E = \{\alpha : A_{\alpha+1}/A_\alpha$ is not free$\}$. By the minimality of $\kappa$, for all $\alpha \in E$, $\mathrm{Ext}(A_{\alpha+1}/A_\alpha, \mathbf{Z}) \neq 0$. Let $f_{0\alpha}$, $f_{1\alpha}$ be as in Lemma 1.5, where $\bigoplus_{\beta \le \alpha} F_\beta$ plays the role of $F$, $\bigoplus_{\beta < \alpha} F_\beta$ plays the role of $H$, and $K_\alpha$ plays the role of $K$. (Note that $A_{\alpha+1}/A_\alpha \cong F/(H + K)$.) Define a partition $P$ so that for each $\alpha \in E$ and homomorphism $h : \bigoplus_{\beta < \alpha} F_\beta \to \mathbf{Z}$, $P(h) = 0$ if and only if $h$ does not extend to a homomorphism from $\bigoplus_{\beta \le \alpha} F_\beta$ to $\mathbf{Z}$ which contains $f_{0\alpha}$. Note that if $P(h) = 1$, then $h$ does not extend to a homomorphism which contains $f_{1\alpha}$. Let $\rho$ be a weak diamond function for this partition (cf. VI.1.7(ii)).

Let $f : K \to \mathbf{Z}$ be $\bigoplus_{\alpha < \kappa} f_{\rho(\alpha)\alpha}$. We claim that $f$ does not extend to a homomorphism $h$ from $F$ to $\mathbf{Z}$. Indeed, otherwise, we can choose $\alpha$ so that $P(h \upharpoonright \bigoplus_{\beta < \alpha} F_\beta) = \rho(\alpha)$. But then $h \upharpoonright \bigoplus_{\beta \le \alpha} F_\beta$ contradicts the choice of $P$. $\square$

The proof above doesn't require the full force of V = L, just that the weak diamond principle (cf. VI.1.6) holds for the appropriate stationary set; hence we have the following results.

**1.7 Corollary.** *Suppose $E \subseteq \omega_1$ is a stationary set for which $\Phi_{\omega_1}(E)$ holds. If $A$ is an $\aleph_1$-free group of cardinality $\aleph_1$ and $\Gamma(A) \ge \tilde{E}$, then $A$ is not a W-group.* $\square$

Thus, because $2^{\aleph_0} < 2^{\aleph_1}$ implies $\Phi_{\omega_1}(\omega_1)$, we have:

**1.8 Corollary.** *($2^{\aleph_0} < 2^{\aleph_1}$) If $A$ is a W-group of cardinality $\aleph_1$, then $\Gamma(A) \neq 1$.* $\square$

Moreover, $2^{\aleph_0} < 2^{\aleph_1}$ implies that every W-group (of arbitrary cardinality) is strongly $\aleph_1$-free (cf. Exercise 12).

We cannot improve 1.8 and derive from CH that every W-group of cardinality $\aleph_1$ is free (i.e., has $\Gamma$-invariant $= 0$). In fact, under the assumption of $\mathrm{Ax}(S) + \diamondsuit^*(\omega_1 \setminus S)$ — which is consistent with GCH (cf. VI.4.12) — we can not only show that there are non-free

W-groups, but we can give a good characterization of the W-groups of cardinality $\aleph_1$:

**1.9 Theorem.** $(\mathrm{Ax}(S) + \Diamond^*(\omega_1 \setminus S))$ *There is a stationary co-stationary $S \subseteq \omega_1$ such that an $\aleph_1$-free group $A$ of cardinality $\aleph_1$ is a W-group if and only if $\Gamma(A) \leq \tilde{S}$.*

PROOF. By 1.7, if $A$ is a W-group of cardinality $\aleph_1$ then $\Gamma(A) \leq \tilde{S}$. Assume now that $\Gamma(A) \leq \tilde{S}$. Suppose $0 \to \mathbf{Z} \to B \to A \to 0$ is exact. Then $\Gamma(B) \leq \tilde{S}$ so $B$ is strongly $\aleph_1$-free. By VIII.3.5, $B$ is $\aleph_1$-separable. Hence the sequence splits, so $\mathrm{Ext}(A, \mathbf{Z}) = 0$. $\square$

(It is possible to prove 1.9 without appeal to VIII.3.5. The poset would consist of some approximations to a splitting of the sequence; the proof continues along the same lines as VIII.3.5, but is somewhat easier.)

We can also characterize the W-groups of cardinality $\aleph_1$ if we assume MA + ¬CH. As we shall see, they turn out to be the Shelah groups, a class of groups which properly contains the strongly $\aleph_1$-free groups.

**Definition.** A group $A$ is a *Shelah group* if it is $\aleph_1$-free and for every countable subgroup $B$ there is a countable group $B'$ which has the *Shelah property over $B$*; viz. $B \subseteq B'$ and for any countable $C$ satisfying $C \cap B' = B$, $C/B$ is free.

**1.10 Proposition.** *If $A$ is a strongly $\aleph_1$-free group, then $A$ is a Shelah group.*

PROOF. Given $B$ as in the definition, let $B'$ be a countable $\aleph_1$-pure subgroup containing $B$. Then for $C$ as in the definition, $C/B \cong (C + B')/B'$, which is free because $A/B'$ is free. $\square$

Thus there are many non-free Shelah groups (cf. VII.1.3). There are also Shelah groups of cardinality $\aleph_1$ which are not strongly $\aleph_1$-free; we will not give an example here, but one can be found in Shelah 1979a (or see Eklof 1980, Theorem 8.2).

**1.11 Theorem.** (MA + ¬CH) *If $M$ is a countable group and $A$ is a Shelah group of cardinality $< 2^{\aleph_0}$, then $\mathrm{Ext}(A, M) = 0$. In* *particular, every Shelah group of cardinality $< 2^{\aleph_0}$ is a W-group, so there is a non-free W-group.*

PROOF. Given a short exact sequence

$$0 \to M \to B \xrightarrow{\pi} A \to 0,$$

let **P** be the set of partial splittings whose domains are finitely generated pure subgroups of $A$, i.e., **P** consists of all $g: G \to B$ such that $G$ is a finitely generated pure subgroup of $A$, and $\pi \circ g$ is the identity on $G$. Order **P** by inclusion. Since $A$ is $\aleph_1$-free, it is easy to see that for all $a \in A$, $D_a \overset{\text{def}}{=} \{g \in \textbf{P}: a \in \text{dom } g\}$ is a dense subset of **P**. If **P** is c.c.c., then by MA + ¬CH, there is a directed subset, $\mathcal{G}$, of **P** which meets each $D_a$; then $\cup\mathcal{G}$ is a splitting of the exact sequence. So it remains to show that **P** is c.c.c.

Let $\kappa$ be large enough for **P**. We need to prove that there is a cub $\mathcal{C}$ of countable elementary submodels of $(H(\kappa), \in)$ such that for each $N \in \mathcal{C}$, 0 is $N$-generic (cf. VI.4.5). Let $\mathcal{C}$ consist of all the countable elementary submodels $N = \bigcup_{i \in \omega} N_i$ where $N_i \prec N$ and $N_i \in N_{i+1}$ for all $i \in \omega$; moreover we require **P**, $A$, $B$ and $\pi$ to belong to $N_0$.

Let $N \in \mathcal{C}$ and let $A_i$ denote $A \cap N_i$. We claim that $A_{i+1}$ has the Shelah property over $A_i$. Indeed, since $A_i \in N_{i+1}$,

$N_{i+1} \models$ "∃ a countable $B' \subseteq A$ with the Shelah property over $A_i$".

Hence there is a countable $B'$ which belongs to $N_{i+1}$, and therefore is contained in $N_{i+1}$, which has the Shelah property over $A_i$. But then any extension of $B'$, in particular $A_{i+1}$, has the Shelah property over $A_i$.

Suppose now that $g \in \textbf{P}$ and $D$ is a dense subset which belongs to $N$. Let $G$ denote dom $g$. Choose $i$ so that $G \cap N \subseteq A_i$ and $D \in N_i$. Notice that $\langle G + A_i \rangle_* \cap A_{i+1} = A_i$, where $\langle G + A_i \rangle_*$ denotes the pure closure of $G + A_i$. Since $A_{i+1}$ has the Shelah property over $A_i$, $\langle G + A_i \rangle_*/A_i$ is free and finitely generated. By extending $g$ if necessary we can assume that $G + A_i$ is pure and equals $A_i \oplus K$, for $K \subseteq G$. Let $G_i = G \cap A_i$. So $G = G_i \oplus K$. Since

$M$ is countable, $\pi^{-1}[\{a\}]$ is a countable member of $N_i$ for all $a \in A_i$ and therefore $\pi^{-1}[A_i] \subseteq N_i$. Then since $G_i$ is finitely generated, $g_i \stackrel{\text{def}}{=} g \restriction G_i$ belongs to $N_i$.

Choose $h \in N_i$ so that $g_i \subseteq h$ and $h \in D$. Let $H$ denote the domain of $h$. Since $H$ is a pure subgroup of $A_i$, $H \oplus K$ is pure in $A$. Hence $h \oplus g \restriction K$ belongs to **P** and is a common extension of $h$ and $g$. $\square$

In section 3, we will show that, assuming MA $+$ ¬CH, every W-group of cardinality $\aleph_1$ is a Shelah group. It is tempting to conjecture that a W-group is always a Shelah group, but Shelah has shown that it is consistent that there is a W-group of cardinality $\aleph_1$ which is not a Shelah group (see 3.11). However, every $\aleph_1$-coseparable group *is* a Shelah group (see 3.21). Even for strongly $\aleph_1$-free groups of cardinality $\aleph_1$ there is no nice characterization of W-groups: Shelah has shown it is consistent with GCH that there are two strongly $\aleph_1$-free groups of cardinality $\aleph_1$ with the same $\Gamma$-invariant (even quotient-equivalent), only one of which is a W-group. (See 3.13; compare 1.9.) It is open whether or not it is consistent that the class of W-groups of cardinality $\aleph_1$ is exactly the class of strongly $\aleph_1$-free groups of cardinality $\aleph_1$.

Griffith has shown that a group $A$ is $\aleph_1$-coseparable if and only if $\text{Ext}(A, \mathbf{Z}^{(\omega)}) = 0$ (see Exercises 1 – 4). Thus we have the following:

**1.12 Corollary.** (MA $+$ ¬ CH) *A group of cardinality $\aleph_1$ is $\aleph_1$-coseparable if and only if it is a W-group if and only if it is a Shelah group.* $\square$

Only two of the above implications are consequences of ZFC: the trivial one that every $\aleph_1$-coseparable group is a W-group, and the fact that every $\aleph_1$-coseparable group is a Shelah group (see 3.21). We have shown that it is consistent that there is a Shelah group which is not a W-group (since it is consistent that every W-group is free). We will deal with the other implications in section 3.

**1.13 Corollary.** (MA $+$ ¬CH) *Every strongly $\aleph_1$-free group of cardinality $\aleph_1$ is $\aleph_1$-separable.*

PROOF. It suffices to prove that every countable $\aleph_1$-pure subgroup $M$ of $A$ is a summand of $A$. Given such an $M$, consider the short exact sequence

$$0 \to M \to A \to A/M \to 0.$$

Since $M$ is $\aleph_1$-pure in $A$, $A/M$ is strongly $\aleph_1$-free. But then by 1.10 and 1.11, $\text{Ext}(A, M) = 0$, so the short exact sequence splits, and $M$ is a summand of $A$. $\square$

This last result is not provable in ZFC: in XIII.1.2 we prove that, assuming $V = L$, there is even a strongly $\aleph_1$-free group of cardinality $\aleph_1$ which is indecomposable.

In the remainder of this section we are going to generalize Theorem 1.6 by replacing the ring $\mathbf{Z}$ by a ring $R$, and by replacing the $\mathbf{Z}$-module $\mathbf{Z}$ in the second place of $\text{Ext}(\_, \mathbf{Z})$ by an arbitrary $R$-module $M$. We will need to define an invariant $\Gamma_M$ in the spirit of Definition IV.1.6. We could make the definition and prove a version of 1.15 for arbitrary rings $R$, but to simplify matters, we will confine ourselves to p.i.d.'s. (As usual we will write $\text{Ext}(A, M)$ instead of $\text{Ext}_R^1(A, M)$.)

**Definition.** Suppose $\kappa$ is a regular uncountable cardinal, $R$ is a p.i.d., $M$ and $A$ are $R$-modules and $A$ is $\leq \kappa$-generated. Choose a $\kappa$-filtration $\{A_\alpha : \alpha < \kappa\}$ of $A$ with the property that for all $\alpha$, $\text{Ext}(A_{\alpha+1}/A_\alpha, M) \neq 0$ if for some $\beta > \alpha$ so that $\text{Ext}(A_\beta/A_\alpha, M) \neq 0$. Define $\Gamma_{\kappa,M}(A) = \tilde{E}$, where $E = \{\alpha : \text{Ext}(A_{\alpha+1}/A_\alpha, M) \neq 0\}$. Since $\kappa$ will be clear from context, we will always write $\Gamma_M$ instead of $\Gamma_{\kappa,M}$.

It is easy to show that $\Gamma_M(A)$ is well-defined. Notice that if $R = \mathbf{Z}$, $\Gamma_{\mathbf{Z}}(A) \leq \Gamma(A)$ and, by 1.2, if $A$ has cardinality $\aleph_1$, $\Gamma_{\mathbf{Z}}(A) = \Gamma(A)$; but for larger $A$, this will not be the case in some models of ZFC. (See Example 1.16 below.) We can prove in ZFC the following analog of one direction of IV.1.7.

**1.14 Proposition.** *Let $\kappa$ be a regular uncountable cardinal. If $A$ is a $\leq \kappa$-generated $R$-module such that $\text{Ext}(B, M) = 0$ for every $< \kappa$-generated submodule $B$ of $A$ and if $\Gamma_M(A) = 0$, then $\text{Ext}(A, M) = 0$.*

PROOF. Choose a $\kappa$-filtration $\{A_\alpha : \alpha < \kappa\}$ of $A$ such that for all $\alpha$, $\mathrm{Ext}(A_{\alpha+1}/A_\alpha, M) = 0$. Now consider a short exact sequence

$$0 \to M \overset{\iota}{\longrightarrow} N \overset{\pi}{\longrightarrow} A \to 0.$$

We must show that this sequence splits. To do this, we define by transfinite induction a continuous increasing chain of homomorphisms $\rho_\alpha : A_\alpha \to N$ such that $\pi \circ \rho_\alpha =$ the identity on $A_\alpha$, i.e., $\rho_\alpha$ is a splitting of $\pi \restriction \pi^{-1}[A_\alpha]$. Suppose that $\rho_\alpha$ has been defined for all $\alpha < \beta$. If $\beta$ is a limit ordinal, we let $\rho_\beta$ be the union of the $\rho_\alpha$. If $\beta = \gamma + 1$ for some $\gamma$, let $\sigma : A_\beta \to N$ be some splitting of $\pi \restriction \pi^{-1}[A_\beta]$, which exists since $\mathrm{Ext}(A_\beta, M) = 0$. Since $\rho_\gamma$ and $\sigma \restriction A_\gamma$ are both splittings of $\pi \restriction \pi^{-1}[A_\gamma]$, there is a homomorphism $\theta : A_\gamma \to M$ such that $\iota \circ \theta = \rho_\gamma - \sigma \restriction A_\gamma$. Since $\mathrm{Ext}(A_\beta/A_\gamma, M) = 0$, $\theta$ extends to a homomorphism $\theta' : A_\beta \to M$. If we define $\rho_\beta = \sigma + (\iota \circ \theta')$, then $\rho_\beta$ is a splitting of $\pi \restriction \pi^{-1}[A_\beta]$ which extends $\rho_\gamma$. $\square$

By 1.11 it is consistent with ZFC that there are groups $A$ such that $\mathrm{Ext}(A, \mathbf{Z}) = 0$ but $\Gamma_{\mathbf{Z}}(A) \neq 0$, so the converse of 1.14 is not a theorem of ZFC. But just as in the proof of 1.6, we can show:

**1.15 Theorem.** (V = L) *Suppose that $\kappa$ is a regular uncountable cardinal and $M$ is an $R$-module of cardinality $\leq \kappa$. If $A$ is a $\leq \kappa$-generated $R$-module such that $\mathrm{Ext}(A, M) = 0$, then $\Gamma_M(A) = 0$.*

PROOF. Let $\{A_\alpha : \alpha < \kappa\}$ be a $\kappa$-filtration of $A$. Let $\{F_\alpha : \alpha < \kappa\}$, $\{K_\alpha : \alpha < \kappa\}$ be an associated free resolution, constructed as in 1.4, so that each $F_\alpha$ and $K_\alpha$ is $< \kappa$-generated. Let $X_\alpha$ be a generating set for $F_\alpha$ of cardinality $< \kappa$. Choose also a $\kappa$-filtration of $M$ (as set) so that each $M_\alpha$ is a set of cardinality $< \kappa$. Define a partition $P$ so that for each $\alpha \in E$ and set function $h : \bigcup_{\beta<\alpha} X_\beta \to M_\alpha$, $P(h) = 1$ if and only if $h$ does not extend to a homomorphism from $\bigoplus_{\beta\leq\alpha} F_\beta$ to $M$ which extends $f_{0\alpha}$ (where $f_{0\alpha}$ and $f_{1\alpha}$ are defined as in 1.6). Once we note that for every function $h : \bigcup\{X_\alpha : \alpha < \kappa\} \to M$ there is a cub $C$ so that for $\alpha \in C$, $h[\bigcup_{\beta<\alpha} X_\beta] \subseteq M_\alpha$, the proof continues as in 1.6. $\square$

**1.16 Example.** We consider a model of $\diamondsuit_{\omega_2}(E)$ for every stationary subset, $E$, of $\omega_2$, such that there is a non-free W-group of cardinality $\aleph_1$ (e.g., a model which also satisfies $\mathrm{Ax}(S) + \diamondsuit^*(\omega_1 \setminus S)$ or $\mathrm{MA} + 2^{\aleph_0} = \aleph_2$). By 1.14 and the proof of 1.15, if $A$ is an $\aleph_2$-free group of cardinality $\aleph_2$, then $A$ is a W-group if and only if $\Gamma_{\mathbf{Z}}(A) = 0$. Now we can construct such an $A$ which is the union of an $\aleph_2$-filtration $\{A_\alpha : \alpha < \omega_2\}$ so that $A_{\alpha+1}/A_\alpha$ is not free if and only if $\alpha$ is a limit ordinal of cofinality $\aleph_1$, and in that case $\Gamma(A_{\alpha+1}/A_\alpha) = \tilde{S}$. Then $\Gamma(A) \neq 0$ but, by 1.9, $\Gamma_{\mathbf{Z}}(A) = 0$.

If $A$ is an $R$-module, define $A$ to be a *Whitehead module* if $\mathrm{Ext}^1_R(A, R) = 0$. Recall, from section I.3 that we can form the completion $\hat{R}$ of $R$ in the $R$-topology; $\hat{R}$ is isomorphic to $\Pi_p \hat{R}_p$ where $p$ ranges over a representative set of generators of the prime ideals of $R$, and $\hat{R}_p$ is the completion of $R$ in the $p$-adic topology.

**1.17 Proposition.** (i) *If $R$ is a complete discrete valuation ring, then every torsion-free $R$-module is a Whitehead module.*

(ii) *If $R$ is a p.i.d. which is not a complete discrete valuation ring, then an $R$-module of countable rank is Whitehead if and only if it is free.*

PROOF. (i) By V.1.9, $R$ is pure-injective, and hence cotorsion, i.e., $\mathrm{Ext}(A, R) = 0$ for all torsion-free modules $A$.

(ii) By I.3.5, $\hat{R}$ is torsion-free and reduced and $R$ is pure in $\hat{R}$. Moreover, we claim that $R \neq \hat{R}$. Indeed, if $R$ is not local, then $R \neq \hat{R}$ because $\hat{R}$ is not even an integral domain since it is the product of at least two complete discrete valuation rings of the form $\hat{R}_p$. If $R$ is local, i.e., is a discrete valuation ring, then $R \neq \hat{R}$ by the hypothesis that $R$ is not a complete d.v.r. Note also that Pontryagin's Criterion, IV.2.3, holds for any p.i.d. Therefore the proof of 1.2 applies here. □

Recall, from III.2.9, that a p.i.d. is slender if and only if it is not a complete discrete valuation ring. As a consequence of 1.15 and 1.17 we obtain the following generalization of 1.6.

**1.18 Theorem.** $(V = L)$ *If $R$ is a slender p.i.d. of cardinality $\leq \aleph_1$, then every Whitehead $R$-module is free.* □

This result extends easily to Dedekind domains; that is, if $R$ is a slender Dedekind domain of cardinality at most $\aleph_1$, then every Whitehead $R$-module is projective.

## §2. The rank of Ext

In this section we will study the structure of $\text{Ext}(A, \mathbf{Z})$ for torsion-free $A$. There are few results provable in ZFC when $A$ is uncountable, but we will give an almost complete analysis of the possible structure of Ext under the assumption that $V = L$.

Let $tA$ denote the torsion subgroup of a group $A$. Then $\text{Hom}(tA, \mathbf{Z}) = 0$, so we have a short exact sequence

$$0 \to \text{Ext}(A/tA, \mathbf{Z}) \to \text{Ext}(A, \mathbf{Z}) \to \text{Ext}(tA, \mathbf{Z}) \to 0$$

which splits since $\text{Ext}(A/tA, \mathbf{Z})$ is divisible by Lemma 1.1. Thus

$$\text{Ext}(A, \mathbf{Z}) \cong \text{Ext}(A/tA, \mathbf{Z}) \oplus \text{Ext}(tA, \mathbf{Z}).$$

Now $\text{Ext}(tA, \mathbf{Z}) \cong \text{Hom}(tA, \mathbf{Q}/\mathbf{Z}) \cong \Pi_p \, \text{Hom}(A, Z(p^\infty))$ and its structure is well-known (cf. Fuchs 1970, sections 47 and 52) and determined in ZFC, so we confine ourselves to studying $\text{Ext}(A, \mathbf{Z})$ when $A$ is torsion-free.

A divisible group is a direct sum of copies of $\mathbf{Q}$ and of $Z(p^\infty)$ for each prime $p$, so its structure is completely determined by certain cardinal invariants (cf. I.2.5). For any divisible group $G$, let $r_0 G$ denote the torsion-free rank of $G$ ($= \dim_{\mathbf{Q}}(\mathbf{Q} \otimes G) = $ the number of copies of $\mathbf{Q}$) and for each prime $p$, let $r_p G$ be the $p$-rank of $G$ ($= \dim_{\mathbf{Z}/p\mathbf{Z}}(G[p]) = $ the number of copies of $Z(p^\infty)$). Two divisible groups $G_1$ and $G_2$ are isomorphic if and only if $r_0 G_1 = r_0 G_2$ and $r_p G_1 = r_p G_2$ for all primes $p$. We are going to study these invariants for the divisible group $\text{Ext}(A, \mathbf{Z})$. We begin with the torsion-free rank; for countable $A$ we have a theorem provable in ZFC.

**2.1 Theorem.** *Suppose $A$ is a countable torsion-free group which is not free. Then $r_0 \text{Ext}(A, \mathbf{Z}) = 2^{\aleph_0}$.*

PROOF. By Pontryagin's criterion (IV.2.3), it is enough to show the theorem for groups of finite rank. Suppose for the moment that the theorem holds for groups of rank 1; we claim that it holds for all groups of finite rank. Assume $A$ is a group of finite rank $n$ which is not free. Without loss of generality $A$ is indecompo  _ie; if $B$ is the pure closure of an independent subset of size $n - 1$, then $A/B$ is a non-free group of rank 1. Consider the exact sequence

$$\text{Hom}(B, \mathbb{Z}) \to \text{Ext}(A/B, \mathbb{Z}) \to \text{Ext}(A, \mathbb{Z}).$$

Since $\text{Hom}(B, \mathbb{Z})$ is a countable group and $r_0 \text{Ext}(A/B, \mathbb{Z}) = 2^{\aleph_0}$, $r_0 \text{Ext}(A, \mathbb{Z}) = 2^{\aleph_0}$.

Now suppose that $A$ is a non-free group of rank 1. There are two cases.

*Case 1: For some prime $p$, every element of $A$ is $p$-divisible.* Choose $b \neq 0$ in $A$, and let $B = \{a \in A : ma \in \langle b \rangle$ for some $m \in \mathbb{Z}$ relatively prime to $p\}$; then $A/B \cong Z(p^\infty)$. By considering the exact sequence,

$$\text{Hom}(B, \mathbb{Z}) \to \text{Ext}(Z(p^\infty), \mathbb{Z}) \to \text{Ext}(A, \mathbb{Z}),$$

we see that it suffices to show that the torsion-free rank of $\text{Ext}(Z(p^\infty), \mathbb{Z})$ is $2^{\aleph_0}$. Now consider the sequence

$$0 \to \mathbb{Z} \to \mathbb{Q} \to \mathbb{Q}/\mathbb{Z} \to 0.$$

This induces the exact sequence,

$$
\begin{array}{llll}
0 = & \text{Hom}(Z(p^\infty), \mathbb{Q}) & \to & \text{Hom}(Z(p^\infty), \mathbb{Q}/\mathbb{Z}) \to \\
& \text{Ext}(Z(p^\infty), \mathbb{Z}) & \to & \text{Ext}(Z(p^\infty), \mathbb{Q}) = 0.
\end{array}
$$

Therefore

$$\text{Ext}(Z(p^\infty), \mathbb{Z}) \cong \text{Hom}(Z(p^\infty), \mathbb{Q}/\mathbb{Z}) \cong \text{Hom}(Z(p^\infty), Z(p^\infty)).$$

But this group is isomorphic to the additive group of the $p$-adic integers, and so has torsion-free rank $2^{\aleph_0}$.

*Case 2: Every element of $A$ is divisible by infinitely many primes.* Since this case provides a nice illustration of one of the central methods of this section, we will delay the proof until we prove a couple of general lemmas.

**2.2 Lemma.** *Let $G$ be any group and let $A$ be a torsion-free group. Suppose that*

$$0 \to K \xrightarrow{\varphi} F \to A \to 0$$

*is a free resolution of $A$. If $K = \bigoplus_{\alpha < \kappa} K_\alpha$, then there is a homomorphism from $\mathrm{Ext}(A, G)$ onto $\prod_{\alpha < \kappa} \mathrm{Ext}(F/\varphi[K_\alpha], G)$.*

PROOF. The short exact sequence

$$0 \to K_\alpha \to F \to F/\varphi[K_\alpha] \to 0$$

induces a short sequence sequence

$$\mathrm{Hom}(F, G) \to \mathrm{Hom}(K_\alpha, G) \to \mathrm{Ext}(F/\varphi[K_\alpha], G) \to 0.$$

Since there is a natural isomorphism

$$\psi \colon \mathrm{Hom}(K, G) \to \prod_{\alpha < \kappa} \mathrm{Hom}(K_\alpha, G)$$

we obtain a homomorphism, $\theta$, from $\mathrm{Hom}(K, G)$ onto

$$\prod_{\alpha < \kappa} \mathrm{Ext}(F/\varphi[K_\alpha], G).$$

Now

$$\mathrm{Hom}(F, G) \to \mathrm{Hom}(K, G) \to \mathrm{Ext}(A, G) \to 0$$

is exact, so it suffices to show that the kernel of $\theta$ contains the image of $\mathrm{Hom}(F, G)$. If $f \in \mathrm{Hom}(F, G)$, then $\psi(f \restriction K) = (f \restriction K_\alpha)_{\alpha < \kappa}$. Since $f$ extends each $f \restriction K_\alpha$, $\theta(f \restriction K) = 0$. $\square$

**2.3 Lemma.** *Let $G$ be any group and let $A$ be a torsion-free group of infinite cardinality $\kappa$. Suppose that*

$$0 \to K \xrightarrow{\varphi} F \to A \to 0$$

*is a free resolution of $A$. If $K = \bigoplus_{\alpha < \kappa} K_\alpha$, where for all $\alpha < \kappa$, $\mathrm{Ext}(F/\varphi[K_\alpha], G) \neq 0$, then $r_0 \mathrm{Ext}(A, G) = 2^\kappa$.*

PROOF. Since $|\mathrm{Ext}(A, G)| \leq 2^\kappa$, it is enough to show that $r_0 \mathrm{Ext}(A, G) \geq 2^\kappa$. By Lemma 2.2, there is a map from $\mathrm{Ext}(A, G)$ onto $\prod_{\alpha < \kappa} \mathrm{Ext}(F/\varphi[K_\alpha], G)$. So it suffices to show

$$r_0 \left( \prod_{\alpha < \kappa} \mathrm{Ext}(F/\varphi[K_\alpha], G) \right) \geq 2^\kappa.$$

Partition $\kappa$ into infinite disjoint sets $I_\beta (\beta < \kappa)$. So

$$\prod_{\alpha < \kappa} \operatorname{Ext}(F/\varphi[K_\alpha], G) = \prod_{\beta < \kappa} \prod_{\alpha \in I_\beta} \operatorname{Ext}(F/\varphi[K_\alpha], G).$$

Since $\operatorname{Ext}(F/\varphi[K_\alpha], G)$ is non-zero and divisible, it has elements of arbitrarily large order. Hence for all $\beta$, $\prod_{\alpha \in I_\beta} \operatorname{Ext}(F/\varphi[K_\alpha], G)$ has an element of infinite order. It follows easily that

$$r_0 \left( \prod_{\beta < \kappa} \prod_{\alpha \in I_\beta} \operatorname{Ext}(F/\varphi[K_\alpha], G) \right) \geq 2^\kappa. \quad \square$$

As an example of the use of this lemma, we complete the proof of 2.1.

**Proof of 2.1, case 2.** Fix an element $a \in A$ and let $P$ be the set of primes which divide $a$. For $p \in P$, let $n_p$ be $\omega$ or the maximum power of $p$ which divides $a$, whichever is least. Then $A$ has a free resolution $0 \to K \to F \to A \to 0$ where $F$ is freely generated by $\{x\} \cup \{y_{n,p} : p \in P \text{ and } n < n_p\}$ and $K$ is generated by $\{z_{n,p} : p \in P$ and $n < n_p\}$. Here $z_{0,p} = x - py_{0,p}$ and for $n > 0$, $z_{n,p} = y_{n-1,p} - py_{n,p}$. Partition $P$ into countably many infinite sets $P_m (m < \omega)$ and let $K_m$ be the subgroup of $K$ generated by $\{z_{n,p} : p \in P_m$ and $n < n_p\}$. Then $K = \bigoplus_m K_m$ and $\operatorname{Ext}(F/K_m, G) \neq 0$ because $F/K_m$ is countable and not free. Therefore, by Lemma 2.3, $r_0 \operatorname{Ext}(A, G) = 2^{\aleph_0}$. $\square$

This completes the analysis of torsion-free rank in the case when $A$ is countable. (See Exercise 14 for another proof of Theorem 2.1.) For uncountable $A$, we require a hypothesis that holds in some models of ZFC — for example, L (see 1.6) — but not in others (see 1.11).

**2.4 Theorem.** *Suppose that every W-group is free. Let $A$ be a torsion-free group and $B$ a subgroup of $A$ of minimum cardinality such that $A/B$ is free. Then $r_0 \operatorname{Ext}(A, \mathbb{Z}) = 2^{|B|}$.*

PROOF. Since $A = B \oplus F$, for some free group $F$, $\operatorname{Ext}(A, \mathbb{Z}) = \operatorname{Ext}(B, \mathbb{Z})$. Let $\kappa = |B|$. The proof will be by induction on $\kappa$; since the result is trivial when $\kappa = 0$, we can assume $\kappa > 0$. By 2.1, we can assume that $\kappa > \aleph_0$. Consider first the case when $\kappa$ is regular. Choose a $\kappa$-filtration $\{B_\alpha : \alpha < \kappa\}$ of $B$ so that if $B/B_\alpha$ is not $\kappa$-free then $B_{\alpha+1}/B_\alpha$ is not free. Let

$$0 \to \bigoplus_{\alpha<\kappa} K_\alpha = K \to \bigoplus_{\alpha<\kappa} F_\alpha = F \to B \to 0$$

be a free resolution associated with the $\kappa$-filtration (see 1.4). Since $B/C$ is not free for all $C \subseteq B$ such that $|C| < \kappa$,

$$E \overset{\text{def}}{=} \{\alpha: \bigoplus_{\beta \leq \alpha} F_\beta / (\bigoplus_{\beta < \alpha} F_\beta + K_\alpha) \text{ is not free}\}$$

is stationary. Partition $E$ into $\kappa$ disjoint stationary sets, $E_\beta$ ($\beta < \kappa$). Let $H_\beta = \bigoplus_{\alpha \in E_\beta} K_\alpha$. For all $\beta$, $\Gamma(F/H_\beta) = \tilde{E}_\beta \neq 0$, so $F/H_\beta$ is not free and hence not a W-group. So by Lemma 2.3, $r_0 \operatorname{Ext}(B, \mathbb{Z}) = 2^\kappa = 2^{|B|}$.

Suppose now that $\kappa$ is singular and the theorem is true for cardinals smaller than $\kappa$. For every subgroup $C$ of cardinality $< \kappa$, $B/C$ is not free (by the minimality of $\kappa$), so by the singular compactness theorem (IV.3.5), there exists $D$ such that $C \subseteq D \subseteq B$, $|D| < \kappa$ and $D/C$ is not free. In fact, we claim there exists $D$ so that $C \subseteq D \subseteq B$, $|D|$ is a regular cardinal greater than $|C|$, and if $C \subseteq G \subseteq D$ and $|G| < |D|$, then $D/G$ is not free. Suppose that the claim is false for some $C$ of cardinality $\lambda < \kappa$. We define by induction on $i < \lambda^+$ a continuous chain of groups $C_i$ such that $|C_i| = \lambda$ and $C_{i+1}/C_i$ is not free. Let $C = C_0$. If $C_i$ has been defined, choose $C_{i+1} \supseteq C_i$ of minimum cardinality so that $C_{i+1}/C_i$ is not free. Suppose $|C_{i+1}| > \lambda$. By the inductive hypothesis, $|C_i| = \lambda$. Since the claim is assumed false for $C$ and $C_{i+1} \supseteq C$, there exists $G$ containing $C$ so that $|G| < |C_{i+1}|$ and $C_{i+1}/G$ is free. Since $C_{i+1}/G$ is free there exists $H \supseteq G + C_i$, such that $C_{i+1}/H$ is free and $|H| = |G| + |C_i|$. Since $C_{i+1}/H$ is free, $H/C_i$ is not free, which contradicts the minimality of $|C_{i+1}|$ since $|H| < |C_{i+1}|$. This completes the inductive construction of the $C_i$. Let $D = \bigcup_{i<\lambda^+} C_i$; then $|D| = \lambda^+$ and $D/C$ is not free since

$\Gamma(D/C) = 1$; moreover for any $G \subseteq D$ with $|G| < \lambda^+$, $G \subseteq C_j$ for some $j$, so $D/G = \bigcup_{i \geq j} C_i/G$ is not free. Thus $D$ shows the claim is true for $C$, a contradiction; hence the claim is true.

By the claim there is a continuous chain $\{B_\alpha : \alpha < \text{cf } \kappa\}$ of subgroups of $B$ such that $\bigcup_{\alpha < \text{cf}(\kappa)} B_\alpha = B$ and for all $\alpha$, cf $\kappa < |B_\alpha| < |B_{\alpha+1}|$, $|B_{\alpha+1}|$ is regular, and for all $B_\alpha \subseteq G \subseteq B_{\alpha+1}$, if $|G| < |B_{\alpha+1}|$, then $B_{\alpha+1}/G$ is not free. Let

$$0 \to \bigoplus_{\alpha < \text{cf } \kappa} K_\alpha = K \to \bigoplus_{\alpha < cf\kappa} F_\alpha = F \to B \to 0$$

be a free resolution associated with $\{B_\alpha : \alpha < \text{cf } \kappa\}$ as in Lemma 1.4. We claim that $(\bigoplus_{\beta \leq \alpha} F_\beta)/(K_\alpha + C)$ is not free whenever $|C| < |B_{\alpha+1}|$. If not, and $C$ is a counterexample, there exists $D$ of cardinality $|B_\alpha| + |C|$ such that $\bigoplus_{\beta < \alpha} F_\beta \subseteq D$ and $(\bigoplus_{\beta \leq \alpha} F_\beta)/(K_\alpha + D)$ is free; since $(K_\alpha + D)/(\bigoplus_{\beta < \alpha} F_\beta + K_\alpha)$ is isomorphic to a subgroup of $(\bigoplus_{\beta \leq \alpha} F_\beta)/(\bigoplus_{\beta < \alpha} F_\beta + K_\alpha) \cong B_{\alpha+1}/B_\alpha$ and therefore $(\bigoplus_{\beta \leq \alpha} F_\beta)/(K_\alpha + D)$ is isomorphic to $B_{\alpha+1}/G$ for some $B_\alpha \subseteq G \subseteq B_{\alpha+1}$ with $|G| < |B_{\alpha+1}|$, we arrive at a contradiction of the choice of $B_{\alpha+1}$, and the claim is proved. Now $F/K_\alpha \cong (\bigoplus_{\beta \leq \alpha} F_\beta)/K_\alpha \oplus \bigoplus_{\beta > \alpha} F_\beta$, so by induction, $r_0 \text{Ext}(F/K_\alpha, \mathbb{Z}) \geq 2^{|B_{\alpha+1}|}$. Therefore by Lemma 2.2, $r_0 \text{Ext}(B, \mathbb{Z}) \geq \prod_{\alpha < cf\kappa} 2^{|B_{\alpha+1}|} = 2^\kappa$. $\square$

**2.5 Corollary.** $(V = L)$ *If $A$ is a torsion-free group, then either $\text{Ext}(A, \mathbb{Z}) = 0$ or $r_0 \text{Ext}(A, \mathbb{Z})$ is uncountable.* $\square$

For groups of cardinality $\aleph_1$, there are some other hypotheses under which the conclusion of the Theorem is true. For example, it holds if, for all groups of cardinality $\aleph_1$, the $\Gamma$-invariant of a group determines whether or not it is a W-group. (See Exercise 11.) It also holds under the assumption of MA + ¬CH. (See Exercise 10.) Since we will use it later in this section, we will prove another result along the same lines.

**2.6 Theorem.** $(2^{\aleph_0} < 2^{\aleph_1})$ *Suppose $A$ is an $\aleph_1$-free group of cardinality $\aleph_1$. If $\Gamma(A) = 1$, then $r_0 \text{Ext}(A, \mathbb{Z}) = 2^{\aleph_1}$.*

PROOF. Consider a free resolution

$$0 \to \bigoplus_{\alpha < \omega_1} K_\alpha = K \to \bigoplus_{\alpha < \omega_1} F_\alpha = F \to A \to 0$$

associated with an $\omega_1$-filtration of $A$ (as in 1.4). Since weak diamond holds for $\omega_1$, it is possible to partition $\omega_1$ into disjoint stationary sets $E_\beta$ ($\beta < \omega_1$) such that weak diamond holds for each $E_\beta$ (cf. VI.1.10). Then by Corollary 1.7, $F/(\bigoplus_{\alpha \in E_\beta} K_\alpha)$ is not a W-group since $\Gamma(F/(\bigoplus_{\alpha \in E_\beta} K_\alpha)) \supseteq E_\beta$. The theorem now follows from 2.3. $\square$

From this theorem we can derive Chase's result that $2^{\aleph_0} < 2^{\aleph_1}$ implies that $A$ is strongly $\aleph_1$-free whenever $A$ is torsion-free and $\text{Ext}(A, \mathbf{Z})$ is torsion (cf. Exercise 12). Chase's result and Theorem 2.6 are not provable in ZFC, since by 1.11, in a model of MA + $\neg$CH there are W-groups with $\Gamma$-invariant 1.

It would be tempting to conjecture on the basis of these results that the conclusion of 2.5 is a theorem of ZFC (at least for groups of cardinality $\aleph_1$). However, Shelah 1981b proves that for any countable divisible group $D$, it is consistent that there is a torsion-free group $A$ of cardinality $\aleph_1$ such that $\text{Ext}(A, \mathbf{Z}) \cong D$. So, for example, we can have $\text{Ext}(A, \mathbf{Z}) \cong \mathbf{Q}$ and hence $r_0 \text{Ext}(A, \mathbf{Z}) = 1$. Notice also that it is consistent that $r_p \text{Ext}(A, \mathbf{Z}) > r_0 \text{Ext}(A, \mathbf{Z})$ (if say $D = Z(p^\infty)$).

Now we turn to $p$-ranks. There is a useful characterization of $r_p \text{Ext}(A, \mathbf{Z})$. Fix a prime $p$ and consider the exact sequence

$$\text{Hom}(A, \mathbf{Z}) \xrightarrow{\varphi^p} \text{Hom}(A, \mathbf{Z}/p\mathbf{Z}) \to \text{Ext}(A, \mathbf{Z}) \xrightarrow{p_*} \text{Ext}(A, \mathbf{Z})$$

induced by the short exact sequence $0 \to \mathbf{Z} \xrightarrow{p} \mathbf{Z} \to \mathbf{Z}/p\mathbf{Z} \to 0$. Then $p_*$ is multiplication by $p$, and the kernel of $p_*$ is the $p$-primary subgroup of $\text{Ext}(A, \mathbf{Z})$. Hence $r_p \text{Ext}(A, \mathbf{Z})$ is the dimension of $\text{Hom}(A, \mathbf{Z}/p\mathbf{Z})/\varphi^p[\text{Hom}(A, \mathbf{Z})]$ as a vector space over the field with $p$ elements. Since there is no danger of ambiguity, we will usually write $f/p$ for $\varphi^p(f)$ and $G/p$ for $\varphi^p[G]$. Again, we have a theorem of ZFC when $A$ is countable:

**2.7 Theorem.** *If $A$ is a countable torsion-free group, then for any prime $p$, $r_p \text{Ext}(A, \mathbf{Z}))$ is either finite or $2^{\aleph_0}$.*

PROOF. By Exercise IV.3, $A = B \oplus F$ where $F$ is a free group and $B^* = 0$. So $\text{Ext}(A, \mathbf{Z}) = \text{Ext}(B, \mathbf{Z})$. Since $B^* = 0$, $r_p \text{Ext}(B, \mathbf{Z})$ is the dimension of $\text{Hom}(B, \mathbf{Z}/p\mathbf{Z})$. But this group is just the vector space dual of $B/pB$ so it has finite dimension if $B/pB$ is finite dimensional and dimension $2^{\aleph_0}$ if $B/pB$ has dimension $\aleph_0$. $\square$

The situation for uncountable groups $A$ is quite different: assuming $V = L$, the $p$-ranks can be assigned with only a few restrictions. In this section we will confine ourselves to considering groups of cardinality $\aleph_1$, in which case we need only the assumption CH. The construction of groups with specified $p$-ranks will be done by constructing by induction on $\alpha < \omega_1$ an $\aleph_1$-filtration $\{A_\alpha : \alpha < \omega_1\}$. There are two difficulties which we will have to deal with in the construction. There will be some homomorphisms to $\mathbf{Z}/p\mathbf{Z}$ which we will want to make sure can be lifted to maps to $\mathbf{Z}$ (i.e., belong to the image of $\varphi^p$) and there will be other homomorphisms which we must prevent from having a lifting.

In the filtration we construct, $A_{\alpha+1}/A_\alpha$ will always be isomorphic to $\mathbf{Q}$ (or 0). Suppose that $B$ is a subgroup of $A$ such that $A/B \cong \mathbf{Q}$. By considering the induced exact sequence

$$\text{Hom}(\mathbf{Q}, H) \rightarrow \text{Hom}(A, H) \rightarrow \text{Hom}(B, H) \rightarrow \text{Ext}(\mathbf{Q}, H)$$

for $H = \mathbf{Z}$ or $\mathbf{Z}/p\mathbf{Z}$ we see that each $f \in \text{Hom}(B, H)$ has at most one lifting to an element of $\text{Hom}(A, H)$ because $\text{Hom}(\mathbf{Q}, H) = 0$. Moreover, if $H = \mathbf{Z}/p\mathbf{Z}$, then $f$ has exactly one lifting, since $\text{Ext}(\mathbf{Q}, \mathbf{Z}/p\mathbf{Z}) = 0$

Recall from XI.3 the notion of a dense subgroup of $A^*$. By XI.3.5, if $A$ is a countable free group of infinite rank and $G$ is a countable pure dense subgroup of $A^*$ then there are dual bases $\{a_n : n \in \omega\}$ of $A$ and $\{g_n : n \in \omega\}$ of $G$; i.e., bases of $A$ and $G$ respectively so that for all $n$, $m$, $g_n(a_m) = 1$ if $m = n$ and 0, otherwise. The following lemma will be used in the inductive step of the construction.

**2.8 Lemma.** *Let $B$ be a free group of countably infinite rank, let $G$ be a countable dense pure subgroup of $B^*$, and suppose that $f \in B^*$ but $f \notin G$. Then there exists a short exact sequence*

$$0 \to B \xrightarrow{\beta} A \to \mathbb{Q} \to 0$$

*such that $\beta$ is inclusion and:*
*(a) $A$ is free;*
*(b) each $g \in G$ extends (uniquely) to $g' \in A^*$;*
*(c) $\{g' \in A^*: g \in G\}$ is a pure dense subgroup of $A^*$; and*
*(d) $f$ does not extend to an element of $A^*$.*

PROOF. Choose dual bases $\{b_n : n \in \omega\}$ of $B$ and $\{g_n : n \in \omega\}$ of $G$ (cf. XI.3.5). By induction on $n$ we will choose a basis $\{a_n : n \in \omega\}$ of $B$ and integers $s_n$ so that $n! | s_n$. The group $A$ will be freely generated by elements $\{y_n : n \in \omega\} \cup \{a_{2n+1} : n \in \omega\}$ subject to the relations

$$s_n y_{n+1} = y_n - a_{2n}.$$

Notice that for any value of $f(y_0)$ there is at most one possible value of $f(y_n)$ for each $n$. In fact for all $n$, $f(y_n)$ is determined by $f(y_0)$, $\{f(a_m) : m < 2n\}$ and $\{s_m : m < n\}$. Let $\{r_n : n \in \omega\}$ be an enumeration of $\mathbb{Z}$.

The construction will satisfy various induction hypotheses:

(i) for every $n$, there is a finite set $I_n$ so that $\langle a_m : m < 2n \rangle = \langle b_m : m \in I_n \rangle$;

(ii) $\{0, \ldots, n-1\} \subseteq I_n$;

(iii) for every $n$ and $k \in I_n$, $g_k(a_m) = 0$, if $m \geq 2n$;

(iv) for every $n$, there is no extension of $f$ to $A$ so that $f(y_0) = r_n$.

Suppose $\{a_m : m < 2n\}$ and $\{s_m : m < n\}$ have been defined. Let $r$ be the unique integer, if one exists, so that $f(y_n) = r$ if $f(y_0) = r_n$ and otherwise let $r = 0$. Since $f \notin G$ there are infinitely many $n$ such that $f(b_n) \neq 0$ — for otherwise $f = \sum_n f(b_n) g_n \in G$; hence there exists $t \notin I_n$ such that $f(b_t) \neq 0$. Let $i \in \omega$ be minimal so that $i \notin I_n \cup \{t\}$. Let $a_{2n}$ be either $b_t + b_i$ or $2b_t + b_i$, the choice being made so that $r - f(a_{2n}) \neq 0$. Let $a_{2n+1} = b_t$. Given these

choices, properties (i) – (iii) are satisfied with $I_{n+1} = I_n \cup \{i, t\}$. To
guarantee (iv) it suffices to choose $s_n$ so that $n! | s_n$ but $s_n$ does not
divide $r - f(a_{2n})$. If this choice is made, then there is no extension
of $f$ to $y_{n+1}$ with $f(y_0) = r_n$.

By (i) and (iii) (or directly from the construction), $G$ is freely
generated by the dual basis, $\{h_n : n \in \omega\}$, to $\{a_n : n \in \omega\}$. Each $h_n$
extends to an element $h'_n$ of $A^*$ by defining: $h'_n(y_m) = 0$ for all $m$
if $n$ is odd; and if $n = 2k$, $h'_n(y_k) = 1$, $h'_n(y_m) = 0$ for $m > k$, and

$$h'_n(y_m) = s_{k-1} \cdots s_m$$

for $m < k$. Since $\{h'_n : n \in \omega\}$ is a dual basis to a set of free
generators of $A$, (c) is proved. $\square$

In the proof of Theorem 2.10 we will also need the following
simple lemma.

**2.9 Lemma.** *Suppose $A$ is a free group of infinite rank and $p$ is
a prime. Further assume that $\mathrm{Hom}(A, \mathbb{Z}/p\mathbb{Z}) = H_0 \oplus H_1$ where $H_1$
has uncountable dimension. If $G$ is a countable pure subgroup of $A^*$
such that $G/p \subseteq H_1$ and $h$ is an element of $H_1$ (or of $\mathrm{Hom}(A, \mathbb{Z}/q\mathbb{Z})$
for some prime $q \neq p$), then there is a pure subgroup $G'$ of $A^*$ con-
taining $G$ such that $G'/p \subseteq H_1$ and $h \in G'/p$ ($G'/q$, respectively).*

PROOF. First note that if $f \in A^*$ is such that $f/p \notin G/p$ but $f/p \in$
$H_1$, then if $G'$ is the pure closure of $G \cup \{f\}$, we have $G'/p \subseteq H_1$. If
$h \in G/p$, we can let $G' = G$. If $h \in H_1 \setminus G/p$, choose $f \in A^*$ such
that $f/p = h$. If $h \in \mathrm{Hom}(A, \mathbb{Z}/q\mathbb{Z})$ for some prime $q \neq p$, choose
$f_1$ and $f_2$ in $A^*$ so that $f_1/p \in H_1 \setminus G/p$ and $f_2/q = h$. Choose
$k \in \mathbb{Z}$ so that $kp \equiv 1 \pmod{q}$, and let $f = qf_1 + kpf_2$. Let $G'$ be
the pure closure of $G \cup \{f\}$. $\square$

**2.10 Theorem.** (CH) *Suppose $(\kappa_p : p$ a prime$)$ is a sequence of
cardinals such that for each $p$, $0 \leq \kappa_p \leq \aleph_1$ or $\kappa_p = 2^{\aleph_1}$. Then there
is a group $A$ of cardinality $\aleph_1$ so that $r_0 \mathrm{Ext}(A, \mathbb{Z}) = 2^{\aleph_1}$ and for
all $p$, $r_p \mathrm{Ext}(A, \mathbb{Z}) = \kappa_p$.*

PROOF. Since $\mathrm{Ext}(\bigoplus_{n<\omega} A_n, \mathbb{Z}) = \prod_{n<\omega} \mathrm{Ext}(A_n, \mathbb{Z})$, it is enough to
show that for any prime $p$ and cardinal $\kappa \leq \aleph_1$ there is a group $A$

so that $r_0 \operatorname{Ext}(A, \mathbf{Z}) = 2^{\aleph_1}$, $r_p \operatorname{Ext}(A, \mathbf{Z})) = \kappa$ and $r_q(\operatorname{Ext}(A, \mathbf{Z})) = 0$ for all primes $q \neq p$. Fix now a prime $p$ and a cardinal $\kappa \leq \aleph_1$. We will construct an $\omega_1$-filtration $\{A_\alpha : \alpha < \omega_1\}$ of an $\aleph_1$-free group $A$ by induction on $\alpha < \omega_1$ so that for all $\alpha$, $A_{\alpha+1}/A_\alpha \cong \mathbf{Q}$ or 0. So for any prime $q$, any element of $\operatorname{Hom}(A, \mathbf{Z}/q\mathbf{Z})$ will be determined by its restriction to $A_0$. Also any element of $A^*$ will be determined by its restriction to $A_0$. We will also choose inductively $G_\alpha$ a countable pure dense subgroup of $A_\alpha^*$, such that if $\alpha < \beta$ every element of $G_\alpha$ is the restriction of a (necessarily unique) element of $G_\beta$. To begin the construction let $A_0$ be a free group of countably infinite rank. Let $G_0$ be a countable pure dense subgroup of $A_0^*$. Write $\operatorname{Hom}(A_0, \mathbf{Z}/p\mathbf{Z})$ as $H_0 \oplus H_1$ where the dimension of $H_0$ is $\kappa$, the dimension of $H_1$ is uncountable, and $G_0/p \subseteq H_1$. Since every homomorphism from $A_0$ to $\mathbf{Z}$ or $\mathbf{Z}/q\mathbf{Z}$ will have at most one extension to $A_\alpha$, we can simplify the notation by identifying $A_\alpha^*$ with a subgroup of $A_0^*$ and $\operatorname{Hom}(A_\alpha, \mathbf{Z}/q\mathbf{Z})$ with $\operatorname{Hom}(A_0, \mathbf{Z}/q\mathbf{Z})$. Since we are assuming CH, there is an indexed family $\langle h_\alpha : \alpha < \omega_1 \rangle$ which includes all elements of $A_0^*$ as well as all elements of $\operatorname{Hom}(A_0, \mathbf{Z}/q\mathbf{Z})$, for every prime $q \neq p$. Using this enumeration, we will define an $\omega_1$-filtration $\{A_\alpha : \alpha < \omega_1\}$ and countable pure dense subgroups $G_\alpha$ of $A_\alpha^*$ so that:

(i) for all $\alpha$, $A_{\alpha+1}/A_\alpha \cong \mathbf{Q}$ or 0;

(ii) for all $\alpha$, $G_\alpha/p \subseteq H_1$;

(iii) for all $\alpha < \beta$, every element of $G_\alpha$ extends (uniquely) to an element of $G_\beta$;

(iv) for all $h \in H_1$ (respectively, $h \in \operatorname{Hom}(A_0, \mathbf{Z}/q\mathbf{Z})$ with $q \neq p$), there is $\alpha$ such that $h \in G_\alpha/p$ (resp., $h \in G_\alpha/q$).

(v) for $f \in A_0^*$ if $f/p$ is a non-zero element of $H_0$, then there is $\alpha$ such that $f$ does not extend to an element of $A_\alpha^*$.

Suppose that $A_\alpha$ and $G_\alpha$ have been constructed. If $h_\alpha$ belongs to $H_1$ or to $\operatorname{Hom}(A_0, \mathbf{Z}/q\mathbf{Z})$ for $q \neq p$, we must ensure that condition (iv) holds: let $A_{\alpha+1} = A_\alpha$ and let $G_{\alpha+1}$ be chosen, by Lemma 2.9, so that $h_\alpha$ belongs to $G_{\alpha+1}/p$ or to $G_{\alpha+1}/q$. If $h_\alpha$ belongs to $H_0 \setminus \{0\}$, to ensure that (iii) and (v) hold, we apply Lemma 2.8.

It remains to see that $A \overset{\text{def}}{=} \bigcup_{\alpha < \omega_1} A_\alpha$ is the desired group. By 2.6, since $\Gamma(A) = 1$, $r_0 \operatorname{Ext}(A, \mathbf{Z}) = 2^{\aleph_1}$. Note that by the remarks before Lemma 2.8, $A^*$ is naturally a subgroup of $A_0^*$. Let $G = \{g \in A^* : g{\restriction}A_\alpha \in G_\alpha \text{ for some } \alpha\}$. Then by (iii) and (iv), $G/q = \operatorname{Hom}(A, \mathbf{Z}/q\mathbf{Z})$ for $q \neq p$, and $G/p \supseteq H_1$. Combining this result with (v), we have $A^*/p = H_1$. Hence $\operatorname{Hom}(A, \mathbf{Z}/p\mathbf{Z})/\varphi[\operatorname{Hom}(A, \mathbf{Z})]$ is isomorphic to $H_0$. So $r_p \operatorname{Ext}(A, \mathbf{Z}) = \kappa$. Moreover, for all $q \neq p$, $\varphi^q[\operatorname{Hom}(A, \mathbf{Z})] = \operatorname{Hom}(A, \mathbf{Z}/q\mathbf{Z})$, so $r_q \operatorname{Ext}(A, \mathbf{Z}) = 0$. $\square$

**2.11 Corollary.** (CH) *There is a coseparable group of cardinality $\aleph_1$ which is not free.*

PROOF. By Theorem 2.10 there is a group $A$ which is not free but $r_p \operatorname{Ext}(A, \mathbf{Z}) = 0$ for all primes $p$; i.e., $A$ is coseparable (cf. IV.2.13). $\square$

**2.12 Corollary.** (CH) *There is a coseparable group which is not hereditarily separable.*

PROOF. There is no hereditarily separable group $A$ such that $\Gamma(A) = 1$ by VII.4.12. But by the proof of 2.10, there is such a coseparable group. $\square$

It has been proved by Hulanicki 1957 that a divisible group

$$\mathbf{Q}^{(\alpha)} \oplus \bigoplus_p Z(p^\infty)^{(\beta_p)}$$

admits a compact topology if and only if
    (a) $\alpha$ is of the form $2^\mu$ for some infinite cardinal $\mu$;
    (b) $\beta_p \leq \alpha$ for all primes $p$; and
    (c) $\beta_p$ is finite or of the form $2^{\mu_p}$ ($\mu_p$ infinite) for all primes $p$.
If every W-group is free, then by Theorem 2.4, $\operatorname{Ext}(A, \mathbf{Z})$ satisfies the first two conditions for every torsion-free $A$; the third condition may fail for uncountable $A$ by Theorem 2.10. However, if $A^* = 0$, then (c) holds in ZFC:

**2.13 Proposition.** *If $A$ is a torsion-free group such that $A^* = 0$, then for all primes $p$, $r_p \operatorname{Ext}(A, \mathbf{Z})$ is finite or of the form $2^{\mu_p}$ for some infinite $\mu_p$.*

PROOF. Consider the exact sequence

$$0 = \text{Hom}(A, \mathbf{Z}) \to \text{Ext}(A/pA, \mathbf{Z}) \to \text{Ext}(A, \mathbf{Z}) \xrightarrow{p^*} \text{Ext}(A, \mathbf{Z}) \to 0$$

induced by the short exact sequence $0 \to A \xrightarrow{p} A \to A/pA \to 0$. From this it follows that $r_p \text{Ext}(A, \mathbf{Z})$ equals the dimension of the kernel of $p*$, which in turn equals the dimension of $\text{Ext}(A/pA, \mathbf{Z})$. But $A/pA \cong \mathbf{Z}/p\mathbf{Z}^{(\lambda)}$ for some cardinal $\lambda$ so $\text{Ext}(A/pA, \mathbf{Z}) \cong \Pi_\lambda \mathbf{Z}/p\mathbf{Z}$, from which the result follows. $\square$

**2.14 Corollary.** *Suppose that every W-group is free. If $A$ is an abelian group such that $A^* = 0$, then $A$ admits a compact topology.*

PROOF. By the remarks at the beginning of this section, $\text{Ext}(tA, \mathbf{Z}) \cong \text{Hom}(tA, \mathbf{Q}/\mathbf{Z})$ and this group is known to admit a compact topology. Thus since $A^* \cong (A/tA)^*$, we can assume that $A$ is torsion-free. The conclusion now follows from the result of Hulanicki cited above together with 2.4 and 2.13. $\square$

## §3. Uniformization and W-groups

In the same spirit as section VII.3, we would like to reduce the problem of whether or not there is a non-free W-group, or $\aleph_1$-coseparable group, to a purely set-theoretic problem. This can be done for groups of cardinality $\aleph_1$ using the notion of the uniformization of a coloring of a ladder system (cf. II.4.13):

**3.1 Theorem.** (i) *There is a non-free W-group of cardinality $\aleph_1$ if and only if there is a stationary subset $E$ of $\aleph_1$ and a ladder system $\eta$ on $E$ such that every 2-coloring of $\eta$ can be uniformized.*

(ii) *There is a non-free $\aleph_1$-coseparable group of cardinality $\aleph_1$ if and only if there is a stationary subset $E$ of $\aleph_1$ and a ladder system $\eta$ on $E$ such that every $\aleph_0$-coloring of $\eta$ can be uniformized.*

We will prove this and also prove some other results which show that the characterization of W-groups can be quite complicated: there can be a W-group which is not a Shelah group (3.11), but every $\aleph_1$-coseparable group *is* a Shelah group (3.21); and an $\aleph_1$-separable group quotient-equivalent to an $\aleph_1$-separable W-group may not be a W-group (3.13). Also, we show that, assuming MA + ¬CH, every W-group is a Shelah group (3.20).

To begin, we need to generalize the treatment of uniformization in II.4.13. Suppose $I$ is a set and $\Phi = \{\varphi_\alpha : \alpha < \omega_1\}$ is an indexed family of functions from $\omega$ to $I$. (If $\Phi$ is a ladder system on a stationary set $E$ as in II.4.13, we will index the members of $\Phi$ by $E$ rather than $\omega_1$; this notational change has no real effect.) Suppose $h$ is an ordinal-valued function on $I$. We say that $(\Phi, h)$-*uniformization holds* if every family $\{c_\alpha : \alpha < \omega_1\}$ of ordinal-valued functions on $\omega$ satisfying $c_\alpha(n) < h(\varphi_\alpha(n))$ can be uniformized; i.e., there are functions $(f, f^*)$ such that the domain of $f$ is $I$, $f^* : \omega_1 \to \omega$ and for all $\alpha$, $n \geq f^*(\alpha)$ implies $f(\varphi_\alpha(n)) = c_\alpha(n)$. Often $h$ will be a function with constant value $\lambda$, and we will say that $(\Phi, \lambda)$-*uniformization holds* or, more colloquially, that every $\lambda$-coloring of $\Phi$ can be uniformized.

$\Phi$ is said to be *tree-like* if for all $\alpha$, $\beta$, $n$, $\varphi_\alpha(n) = \varphi_\beta(m)$ implies that $m = n$ and $\varphi_\alpha(k) = \varphi_\beta(k)$ for all $k < n$. For a tree-like $\Phi$, a family $\{c_\alpha : \alpha < \omega_1\}$ can be uniformized if and only if there is a function $f^*$ with domain $\omega_1$ such that for all $\alpha$, $\beta$, if $n \geq f^*(\alpha)$, $f^*(\beta)$ and $\varphi_\alpha(n) = \varphi_\beta(n)$, then $c_\alpha(n) = c_\beta(n)$.

Later we will prove and quote some substantial results about uniformization. Now we will just note some fairly elementary facts.

**3.2 Lemma.** *If $(\Phi, 2)$-uniformization holds, then $(\Phi, \lambda)$-uniformization holds for every finite $\lambda$.*

PROOF. We will prove the stronger fact that for any $k \in \omega$ and any ordinal-valued $h$ and $h'$, if $(\Phi, h)$-uniformization holds, then $(\Phi, h')$-uniformization holds provided that $h'(i) = h(i)^k$ for $i \in I$. Suppose $\{c_\alpha : \alpha \in \omega_1\}$ is a coloring of $\Phi$ where $c_\alpha(n) < h(\varphi_\alpha(n))^k$. We can assume that $c_\alpha(n)$ is a $k$-tuple of natural numbers less than $h(\varphi_\alpha(n))$. For $s < k$, define $c_\alpha^s(n)$ to be the $s$th element of $c_\alpha(n)$. Suppose $(f_s, f_s^*)$ witnesses that $\{c_\alpha^s(n) : \alpha < \omega_1\}$ can be uniformized. Let $f(i) = (f_0(i), \ldots, f_{k-1}(i))$ and $f^*(\alpha) = \max\{f_s^*(\alpha) : s < k\}$. $\square$

Define $\Phi$ to have *disjoint end segments* if there is a function $f^* : \omega_1 \to \omega$ such that $\{\{\varphi_\alpha(n) : n > f^*(\alpha)\} : \alpha < \omega_1\}$ is a disjoint family. Clearly, if $\Phi$ has disjoint end segments, then $(\Phi, \lambda)$-uniformization holds for all cardinals $\lambda$.

**3.3 Proposition.** *Suppose that $(\Phi, \lambda)$-uniformization holds for some $\lambda \leq \aleph_0$. Then either $\Phi$ has disjoint end segments or there is a stationary subset $E$ of $\aleph_1$ and a tree-like ladder system $\eta$ on $E$ so that every $\lambda$-coloring of $\eta$ can be uniformized.*

PROOF. We will first define by induction a one-one function $g: \omega_1 \rightarrow \omega_1$. If $g \restriction \alpha$ has been defined, let $I_\alpha = \bigcup_{\beta < \alpha} \varphi_{g(\beta)}[\omega]$. Next choose $g(\alpha)$ to be the least $\beta$ not in the range of $g \restriction \alpha$ such that $\varphi_\beta[\omega] \cap I_\alpha$ is infinite or, if no such $\beta$ exists, then let $g(\alpha)$ be the least $\beta$ not in the range of $g \restriction \alpha$. Note that $\{I_\alpha : \alpha \in \omega_1\}$ is a continuous chain. By adding extra elements to $I$ we can assume that $I$ has cardinality $\aleph_1$; so we identify $I$ with $\omega_1$. There are two cases to consider.

*Case 1:* $E_1 \overset{\text{def}}{=} \{\alpha \in \omega_1 : \varphi_{g(\alpha)}[\omega] \cap I_\alpha \text{ is infinite}\}$ *is stationary.* Then since $\{\alpha : I_\alpha = \alpha\}$ is a cub in $\omega_1$, $E_2 \overset{\text{def}}{=} E_1 \cap \{\alpha : \alpha = I_\alpha\}$ is stationary in $\omega_1$. For each $\alpha \in E_2$, let $\nu_\alpha$ be a function with domain $\omega$ which enumerates in increasing order a subset of $\varphi_{g(\alpha)}[\omega] \cap \alpha$. Let $\Phi' = \{\nu_\alpha : \alpha \in E_2\}$. Since any $\lambda$-coloring of $\Phi'$ can be extended to a $\lambda$-coloring of $\Phi$, $(\Phi', \lambda)$-uniformization holds. Recall that $^{<\omega}\omega_1$ is the set of finite sequences of elements of $\omega_1$ (cf. II.5.3). Choose a one-one map $\theta$ from $^{<\omega}\omega_1$ to $\omega_1$, such that $\theta(\sigma) \leq \theta(\sigma')$ if $\sigma'$ is a sequence extending $\sigma$ and such that for any $\tau \in^{<\omega} \omega_1$, $\theta(\tau) \geq \tau(n)$ for all $n \in \text{dom}(\tau)$. Let $C$ be a closed unbounded subset of $\omega_1$ consisting of limit ordinals such that for every $\alpha \in C$, $\theta[^{<\omega}\alpha] \subseteq \alpha$. Let $E = E_2 \cap C$. Now choose $\{\eta'_\alpha : \alpha \in E\}$ a ladder system on $E$. For $\alpha \in E$, define $\eta_\alpha(n) = \theta((\eta'_\alpha(0), \nu_\alpha(0), \ldots, \eta'_\alpha(n), \nu_\alpha(n)))$. It is easy to check that $\{\eta_\alpha : \alpha \in E\}$ is the desired ladder system.

*Case 2: not case 1.* Let $C$ be a closed unbounded set which is disjoint from $E_1$. Since every $\gamma \in \omega_1$ is $< \beta$ for some $\beta \in C$, $g$ must be onto $\omega_1$. For $\beta \in C$, let $X_\beta = \{g(\alpha) : \beta \leq \alpha < \beta^+\}$. (Here $\beta^+$ denotes the successor of $\beta$ in $C$.) Each $X_\beta$ is countable and $\omega_1 = \bigcup X_\beta$. Since $(\Phi, 2)$-uniformization holds, $\varphi_\gamma[\omega] \cap \varphi_\delta[\omega]$ is finite for all $\gamma \neq \delta$. (To see that this statement is true consider a coloring so that $c_\gamma$ is constantly 1 and $c_\delta$ is constantly 0.) So there is a function $f_\beta^*: X_\beta \to \omega$ so that the family $\{\varphi_\gamma[\{n: n > f_\beta^*(\gamma)\}]: \gamma \in X_\beta\}$ is disjoint and for all $\gamma \in X_\beta$, $\varphi_\gamma(n) \notin I_\beta$ if $n > f_\beta^*(\gamma)$. Let $f^* = \bigcup_{\beta \in C} f_\beta^*$. Then $\{\{\varphi_\alpha(n): n > f^*(\alpha)\}: \alpha < \omega_1\}\}$ is a disjoint family. $\square$,

Given a family $\Phi$ and an integer-valued function $g$ on $\omega$, we are going to define a group $G(\Phi, g)$ which will be a W-group or an $\aleph_1$-coseparable group if an appropriate uniformization property holds.

**3.4.** Let $g$ be a function from $\omega$ to the integers $\geq 2$. Define $G(\Phi, g)$ to be the group with the free resolution

$$0 \to K \to F \to G(\Phi, g) \to 0$$

where $F$ is freely generated by $\{x_i: i \in I\} \cup \{z_{\alpha n}: \alpha < \omega_1, n < \omega\}$ and $K$ is generated by $\{g(n)z_{\alpha n+1} - z_{\alpha n} - x_{\varphi_\alpha(n)}: \alpha < \omega_1, n < \omega\}$. Let

$$w_{\alpha n} \stackrel{\text{def}}{=} g(n)z_{\alpha n+1} - z_{\alpha n} - x_{\varphi_\alpha(n)}.$$

If $\Phi$ has disjoint end segments, then $G(\Phi, g)$ is free. If $\Phi$ is a ladder system on a stationary subset of $\omega_1$, then $G(\Phi, g)$ will be a non-free $\aleph_1$-separable group (cf. VIII.1.1). But if $I$ is countable, then $G(\Phi, g)$ is not strongly $\aleph_1$-free; in fact it is not even a Shelah group (see Exercise 4 and the proof of Theorem 3.11).

If $g$ is the function with constant value $k$ we will write $G(\Phi, k)$ instead of $G(\Phi, g)$. It is easy to see that $K$ is freely generated by the $w_{\alpha n}$, so $\text{Hom}(K, \mathbf{Z})$ can be identified with the functions from $\{w_{\alpha n}: \alpha < \omega_1, n < \omega\}$ to $\mathbf{Z}$. In order for $G(\Phi, g)$ to be a W-group, every homomorphism $\psi$ from $K$ to $\mathbf{Z}$ must extend to a homomorphism $\theta$ on $F$. What we try to do is to define $\theta$ to be 0 on the $z_{\alpha n}$ and then define $\theta(x_i)$ to be $-\psi(w_{\alpha n})$ if $i = \varphi_\alpha(n)$; but $i$ may belong to the range of many different $\varphi_\alpha$'s, so this is where uniformization comes in.

**3.5 Lemma.** *Use the notation in 3.4. For any $\psi \in \text{Hom}(K, \mathbf{Z})$, there is $\theta \in \text{Hom}(F, \mathbf{Z})$ so that for all $\alpha$ and $n$, $0 \leq (\psi - \theta)(w_{\alpha n}) < g(n)$.*

PROOF. To begin, let $\theta(x_i) = 0$ for all $i$. For each $\alpha$, we will define $\theta(z_{\alpha n})$ by induction on $n$. Let $\theta(z_{\alpha 0}) = 0$. Suppose we have defined $\theta(z_{\alpha k})$ for all $k \leq n$. Let $p = g(n)$. Choose $\psi'(w_{\alpha n}) \in \{0, 1, \ldots, p - 1\}$ so that $\psi'(w_{\alpha n}) \equiv \psi(w_{\alpha n}) + \theta(z_{\alpha n}) \pmod{p}$, i.e.,

$$\psi(w_{\alpha n}) - \psi'(w_{\alpha n}) + \theta(z_{\alpha n} + x_{\varphi_\alpha(n)}) \equiv 0 \pmod{p}.$$

Define $\theta(z_{\alpha n+1}) = (\psi(w_{\alpha n}) - \psi'(w_{\alpha n}) + \theta(z_{\alpha n} + x_{\varphi_\alpha(n)}))/p$. □

The following implies one direction of 3.1.

**3.6 Proposition.** *Use the notation in 3.4.*

(i) *Suppose $h: I \to Ord$ is such that for all $\alpha$ and $n$, $g(n) \leq h(\varphi_\alpha(n))$. If $(\Phi, h)$-uniformization holds, then $G(\Phi, g)$ is a W-group.*

(ii) *If $(\Phi, 2)$-uniformization holds, then $G(\Phi, 2)$ is a W-group.*

(iii) *If $(\Phi, \aleph_0)$-uniformization holds, then $G(\Phi, g)$ is $\aleph_1$-coseparable.*

PROOF. (i) Consider $\psi \in \text{Hom}(K, \mathbf{Z})$; we must show that there exists $\theta \in \text{Hom}(F, \mathbf{Z})$ extending $\psi$. By 3.5, we can assume that $0 \leq \psi(w_{\alpha n}) < g(n)$ for all $\alpha$ and $n$. Let $c_\alpha(n) = \psi(w_{\alpha n})$. By the hypotheses, there is an $(f, f^*)$ which uniformizes this coloring. Define $\theta$ so that $\theta(x_i) = -f(i)$ for all $i \in I$ and $\theta(z_{\alpha n}) = 0$ for $n \geq f^*(\alpha)$. Then for all $n \geq f^*(\alpha)$,

$$\theta(w_{\alpha n}) = -\theta(x_{\varphi_\alpha(n)}) = f(\varphi_\alpha(n)) = c_\alpha(n) = \psi(w_{\alpha n}).$$

For $n < f^*(\alpha)$, we can define $\theta(z_{\alpha n})$ by "downward induction" so that $\theta(w_{\alpha n}) = \psi(w_{\alpha n})$. Hence $\theta$ extends $\psi$.

(ii) is a special case of (i). To prove (iii), enumerate $\mathbf{Z}^{(\omega)}$ by $\omega$. Then as in (i) (without needing 3.5), we can show that $\text{Ext}(G(\Phi, g), \mathbf{Z}^{(\omega)}) = 0$. □

Using 3.6(iii) and VI.4.6, we have another proof that MA + ¬CH implies that there is a non-free $\aleph_1$-coseparable group (cf. 1.12). Next we want to prove converses of 3.6(ii) and (iii) under the assumption that $\Phi$ is tree-like. We'll begin with the simpler case of the converse of 3.6(iii).

**3.7 Proposition.** *Use the notation in 3.4 and suppose as well that $\Phi$ is tree-like. If $G(\Phi, g)$ is $\aleph_1$-coseparable, then $(\Phi, \aleph_0)$-uniformization holds.*

PROOF. Let $\{a_{nm}: n, m \in \omega\}$ enumerate a basis of $\mathbf{Z}^{(\omega)}$. Given an $\aleph_0$-coloring $\{c_\alpha: \alpha < \omega_1\}$, define $\psi \in \mathrm{Hom}(K, \mathbf{Z}^{(\omega)})$ by $\psi(w_{\alpha n}) = a_{n c_\alpha(n)}$. By hypothesis there is an extension of $\psi$ to an element $\theta$ of $\mathrm{Hom}(F, \mathbf{Z}^{(\omega)})$. Define $f^*(\alpha) =$ the least $n$ so that $\theta(z_{\alpha 0}) \in \langle a_{kj}: k < n, j \in \omega \rangle$. It suffices to show that if $\varphi_\alpha(m) = \varphi_\beta(m)$ where $m \geq f^*(\alpha), f^*(\beta)$, then $c_\alpha(m) = c_\beta(m)$. To do this, let $\theta'$ be the composition of $\theta$ with the projection of $\mathbf{Z}^{(\omega)}$ onto $\langle a_{mj}: j \in \omega \rangle$. Then $\theta'(z_{\alpha 0}) = 0 = \theta'(z_{\beta 0})$. Since $\Phi$ is tree-like, $x_{\varphi_\alpha(n)} = x_{\varphi_\beta(n)}$ for all $n \leq m$, and using this we can show by induction that $\theta'(z_{\alpha n}) = \theta'(z_{\beta n})$ for $n \leq m$. Hence $g(m)$ divides $a_{m c_\alpha(m)} - a_{m c_\beta(m)}$, so $a_{m c_\alpha(m)}$ must equal $a_{m c_\beta(m)}$. $\square$

In order to prove the converse of 3.6(ii), we need some definitions and a result about countable groups. By a *presentation* $(x_i (i \in I); w_j (j \in J))$ of a group $A$, we mean that each $w_j$ is in the group freely generated by $\{x_i: i \in I\}$ and $A \cong \langle x_i: i \in I \rangle / \langle w_j: j \in J \rangle$. Using the language of combinatorial group theory, we will call the $x_i$'s *generators* and the $w_j$'s *relations*. We will view the $w_j$ as linear functions of the $x_i$ — or *terms*, in the language of model theory — so we will sometimes write $w_j(x_i: i \in I)$ instead of $w_j$; this point of view will be used later when we will want to consider presentations $(x_i (i \in I); w_j(x_i: i \in I) (j \in J))$ and $(y_i (i \in I); w_j(y_j: j \in J) (j \in J))$ (i.e., ones with different generating sets but the same relations). There is little real difference between a presentation and a free resolution: to every presentation we can associate a free resolution but any free resolution is associated to infinitely many presentations. We call a presentation a *free presentation* if $\{w_j: j \in J\}$ is linearly independent. Given a free presentation of a group $A$, $\mathrm{Ext}(A, G)$ is isomorphic to the group of *functions* from $\{w_j: j \in J\}$ to $G$ modulo the subgroup of restrictions of homomorphisms from $\langle x_i: i \in I \rangle$ to $G$.

**3.8 Lemma.** *Suppose $\{p_m : m \in \omega\}$ is a sequence of integers $\geq 2$ and $G$ is the group with the free presentation $(z_m \ (m \in \omega); \ w_m \ (m \in \omega))$ where $w_m = p_m z_{m+1} - z_m$. Let $X_0$ be the even numbers and $X_1$ the odd numbers. Let $k, r \in \omega$ such that $k \geq 2r$, and let $\ell \in \{0, 1\}$. Suppose $\theta$ is a homomorphism from $\langle z_m : m \in \omega \rangle$ to $\mathbf{Z}$ such that $|\theta(w_m)| < p_m$ for all $m \leq k$. If $|\theta(z_0)| < r$ and $\theta(w_m) = 0$ for all $m \in X_\ell \cap (2r+1)$, then $\theta(w_k) = 0$.*

PROOF. Assume that $\theta$ is as in the statement of the lemma. We claim that for $s < 2r + 1$, the sequence $|\theta(z_s)|$ is non-increasing. Indeed, by the triangle inequality,

$$|\theta(z_{s+1})| \leq p_s^{-1}(|\theta(z_s)| + |\theta(w_s)|) \leq |\theta(z_s)| + \epsilon$$

where $\epsilon < 1$ because $|\theta(w_s)| < p_s$; so $|\theta(z_{s+1})| \leq |\theta(z_s)|$ since $\theta$ is integer-valued. Now $|\theta(z_s)|$ at least halves at every other step, so $\theta(z_{2r}) = 0$. Then $\theta(w_{2r}) = p_{2r}\theta(z_{2r+1})$ and by hypothesis, $|\theta(w_{2r})| < p_{2r}$, so $\theta(z_{2r+1}) = \theta(w_{2r}) = 0$. By induction we can show that $\theta(z_{m+1}) = \theta(w_m) = 0$ for all $2r \leq m \leq k$. $\square$

We can use a strengthening of 3.8 to give a new proof of 2.1: see Exercise 14.

**3.9 Proposition.** *Use the notation in 3.4 and suppose as well that $\Phi$ is tree-like. If $G(\Phi, g)$ is a W-group, then any coloring $\{c_\alpha : \alpha < \omega_1\}$ of $\Phi$ satisfying $c_\alpha(n) < g(n)$ can be uniformized.*

*In particular, if $G(\Phi, 2)$ is a W-group, then $(\Phi, 2)$-uniformization holds.*

PROOF. Suppose we are given a coloring $\{c_\alpha : \alpha < \omega_1\}$ as above. Let $X_0, X_1$ be as in Lemma 3.8. Define two colorings, $\{c_\alpha^\ell : \alpha < \omega_1\}(\ell = 0, 1)$, by $c_\alpha^\ell(n) = c_\alpha(n)$ if $n \notin X_\ell$ and 0 otherwise. It is enough to show that both of these colorings can be uniformized. Fix $\ell$. Define $\psi^\ell \in \mathrm{Hom}(K, \mathbf{Z})$ by $\psi^\ell(w_{\alpha n}) = c_\alpha^\ell(n)$. Let $\theta$ be a homomorphism from $F$ to $\mathbf{Z}$ which extends $\psi^\ell$. Define $f^*(\alpha) = 4|\theta(z_{\alpha 0})| + 2$. To prove that this choice of $f^*$ uniformizes the coloring, it is enough to show that for all $\alpha, \beta$ and $k \geq f^*(\alpha), f^*(\beta)$, if $\varphi_\alpha(k) = \varphi_\beta(k)$ then $\psi^\ell(w_{\alpha k}) = \psi^\ell(w_{\beta k})$. Fix $\alpha$ and $\beta$ and consider $\theta \restriction \langle z_{\alpha m} - z_{\beta m} : m \in \omega \rangle$

and apply 3.8 to the free presentation $(z_{\alpha m} - z_{\beta m}(m \in \omega); w_m(m \in \omega))$ where $w_m = g(m)(z_{\alpha m+1} - z_{\beta m+1}) - (z_{\alpha m} - z_{\beta m})$ and $p_m = g(m)$. Letting $r = 2\max\{|\theta(z_{\alpha 0})|, |\theta(z_{\beta 0})|\} + 1$, we can deduce that $\theta(w_{\alpha k} - w_{\beta k}) = 0$. $\square$

We will use two results, 3.10 and 3.12, due to Shelah, on the consistency of various uniformization properties, which we will state without proof. From each of them we will then derive, as an immediate consequence, an abelian group theory result.

**3.10 Theorem.** *It is consistent that $(\Phi, 2)$-uniformization holds and $I$ is a countable set.* $\square$

**3.11 Corollary.** *It is consistent that there is a W-group which is not a Shelah group.*

PROOF. Let $\Phi$ be as in 3.10. By 3.6(ii), $G(\Phi, 2)$ is a W-group. Let $B$ be the countable subgroup of $G(\Phi, 2)$ generated by the images of the $\{x_i : i \in I\}$. Then $G(\Phi, 2)/B$ is isomorphic to a direct sum of uncountably many copies of the subgroup of $\mathbb{Q}$ generated by $\{1/2^n : n \in \omega\}$. So $G(\Phi, 2)$ is not a Shelah group (see Exercise 4). $\square$

It is *not* consistent with CH that there is a W-group which is not a Shelah group: see Corollary 1.8 and Exercise 12.

Our other consistency results will depend on the following theorem of Shelah.

**3.12 Theorem.** *Let $\eta = \{\eta_\alpha : \alpha \in S\} \in \mathbf{L}$ be a ladder system on $S$, a stationary co-stationary (in $\mathbf{L}$) subset of $\omega_1^{\mathbf{L}}$, and $h \in \mathbf{L}$ a function from $\omega_1^{\mathbf{L}}$ to $\omega$. Then there is a model of $\mathrm{ZFC} + \mathrm{GCH}$ in which all of the following hold:*
(a) $\omega_1^{\mathbf{L}} = \omega_1$;
(b) $(\eta, h)$-*uniformization holds*;
(c) *If $S' \subseteq \omega_1$, $\eta' = \{\eta_\alpha' : \alpha \in S'\}$ is a ladder system on $S'$ and $h' : \omega_1 \to \omega + 1$, then $(\eta', h')$-uniformization does not hold if either*
    (1) $S' \setminus S$ *is stationary; or*

(2) *for every closed unbounded set $C \subseteq \omega_1$, there is $\delta \in S \cap S' \cap C$ such that for every $\alpha < \delta$ and $n < \omega$, there is $\beta \in C$ such that $\alpha < \beta < \delta$ and*

$$(\prod(h(\eta_\delta(m)) : m < \omega, \eta_\delta(m) < \beta))^n <$$
$$\prod(h'(\eta'_\delta(m)) : m < \omega, \eta'_\delta(m) < \beta);$$

(d) *for every regular cardinal $\kappa > \aleph_1$, and stationary subset $E$ of $\kappa$, $\Diamond(E)$ holds.* $\square$

Theorem 3.12 is close to the optimal result. If condition (c) were weakened substantially then $(\eta, h)$-uniformization would imply $(\eta', h')$-uniformization (cf. 3.2).

**3.13 Corollary.** (i) *It is consistent with GCH that there is a stationary subset $S \subseteq \omega_1$ and tree-like ladder systems $\eta = \{\eta_\delta : \delta \in S\}$ and $\eta' = \{\eta'_\delta : \delta \in S\}$ on $S$ so that $(\eta, 2)$-uniformization holds but $(\eta', 2)$-uniformization does not hold.*

(ii) *It is consistent with GCH that there are two quotient-equivalent $\aleph_1$-separable groups of cardinality $\aleph_1$ such that one is a W-group and the other is not.*

PROOF. (i) Let $\eta$ and $\eta'$ be tree-like ladder systems in L such that $\eta_\delta(n) = \eta'_\delta(2^n)$ (cf. Exercise 17). We use the model of 3.12 for this $\eta$. We need only check that $\eta$ and $\eta'$ satisfy (c). Given $C$, let $\delta$ be any element of $S \cap C$. Given $\alpha < \delta$ and $n$, let $\beta > \alpha$ be such that if $k$ is maximal so that $\eta_\delta(k) < \beta$, then $2^k > kn$. Then the inequality in (c) holds since $2^{kn} < 2^{2^k}$.

(ii) Let $\eta$ and $\eta'$ be as in (i). By VIII.1.4, $G(\eta, 2)$ and $G(\eta', 2)$ are quotient-equivalent. By 3.6 and 3.9, respectively, the first group is a W-group but the second group is not. $\square$

Now we turn to the proof of 3.1. One direction of 3.1(i) and (ii) is already implied by 3.6. The proof of the other direction will be a generalization of the proofs of 3.9 and 3.7. First, we need a generalization of Lemma 3.8.

**3.14. Lemma.** *Suppose $G$ is a non-free torsion-free group of finite rank, $n$. There is a free presentation*

$$(y_0, \ldots, y_{n-1}, z_m \ (m \in \omega); w_m \ (m \in \omega))$$

*of $G$ and disjoint infinite subsets $X_0$, $X_1$ of $\omega$ such that for every*
*r there is an integer $m_r$ satisfying*

> *for all $k \geq m_r$ and $\ell \in \{0,1\}$, if $\theta$ is a homomorphism*
> *from $\langle y_0, \ldots, y_{n-1}, z_m : m \in \omega \rangle$ to $\mathbf{Z}$ such that $|\theta(y_i)| <$*
> *$r$ ($i < n$), $|\theta(w_m)| < 2(m \leq k)$, and $\theta(w_m) = 0$ for*
> *$m \in X_\ell \cap (m_r + 1)$, then $\theta(w_k) = 0$.*

PROOF. Since $G$ is a rank $n$ group it has a free presentation of the
following form: it is generated by generators $y_0, \ldots, y_{n-1}$, and $z_{pmj}$
where $p$ ranges over a set of primes $P$; for every $p$ there is some
$d_p \leq \omega$ such that $m < d_p$; and for every $p$ and $m$, there is $n_{pm} \leq n$
so that $j < n_{pm}$. For notational reasons let $z_{p,-1,i}$ denote $y_i$. The
relations are $w_{pmj} = pz_{pmj} - a$ where $a$ is a linear combination of
$\{z_{pdi} : d < m, i < n_{pd}\}$. Let $F$ be the free group on the generators.
   Given a function $f$ from $\{y_0, \ldots, y_{n-1}\}$ to $\mathbf{Z}$, let us say that a
function $g$ to $\{-1, 0, 1\}$ whose domain is contained in the set of
relations is *consistent with* $f$ if there is a homomorphism $\theta : F \to \mathbf{Z}$
so that $\theta$ extends $f \cup g$. For any $f$ and prime $p \neq 2$, there is at most
one $g$ with domain $\{w_{pmj} : m < d_p, j < n_{pm}\}$ which is consistent
with $f$. To see this, suppose $\theta$ extends $f$ and $\theta(w_{pmj}) \in \{-1, 0, 1\}$
for every relation $w_{pmj}$. Note first that $f$ determines $\theta(w_{p0j})$ (mod
$p$); the restriction on the range then determines $\theta(w_{p0j})$. This information lets us calculate $\theta(z_{p0j})$, which in turn determines $\theta(w_{p1j})$
(mod $p$), etc. The same proof shows that for all $d \leq d_p$ there is
at most one $g$ with domain $\{w_{pmj} : m < d, j < n_{pm}\}$ such that $g$ is
consistent with $f$. Choose $d$ maximal such that there is such a $g$
and denote this $g$ by $g_{pf}$.
   For the prime $p = 2$, the situation is somewhat different. We
can assume that $d_2 = \omega$ — otherwise we could change the presentation so that $2 \notin P$. If we try inductively to construct $g$ with domain
$\{w_{2mj} : m < \omega, j < n_{2m}\}$ which is consistent with a given $f$, there
may be two possibilities for $g(w_{2mj})$ for some $m, j$ since $-1 \equiv 1$
(mod 2). To capture the possible extensions we will use a tree of
finite sequences (cf. II.5.3). Let $T_f$ consist of the empty sequence

together with all sequences, $\eta$, of the form $(\theta\restriction\{w_{2,0j}: j < n_{2,0}\}, \dots,$ $\theta\restriction\{w_{2mj}: j < n_{2m}\})$ where $m < \omega$ and $\theta$ ranges over all homomorphisms extending $f$ such that $\theta(w_{2dj}) \in \{-1, 0, 1\}$ for all $d \le m$ and $j < n_{2d}$. Note that $T_f$ is a finitely branching tree; that is, each node has only finitely many immediate successors.

Before continuing the proof we introduce a little more *ad hoc* notation. If $\theta$ is a homomorphism from $F$ to $\mathbf{Z}$, we say $\theta$ is *acceptable* at $w_{pmj}$ if $\theta(w_{pmj}) \in \{-1, 0, 1\}$ and for all $d < m$ and $i < n_{pd}$, $\theta(w_{pdi}) \in \{-1, 0, 1\}$. Let $I$ denote $\{(p, m, j): p \in P, m < d_p, j < n_{pm}\}$. We now make the following claim.

> Suppose $f$ is a function from $\{y_0, \dots, y_{n-1}\}$ to $\mathbf{Z}$ and $W$ is a finite subset of $I$. Then there is a finite subset, $Y$, of $I$ disjoint from $W$ so that whenever $\theta$ is a homomorphism extending $f$ such that $\theta(w_{pmj}) = 0$ and $\theta$ is acceptable at $w_{pmj}$ for all $(p, m, j) \in Y$, then $\theta(w_{pmj}) = 0$ whenever $(p, m, j) \notin W$ and $\theta$ is acceptable at $w_{pmj}$.

Assuming the claim for the moment, let us finish the proof. First, we want to choose disjoint subsets, $X_0'$ and $X_1'$, of $I$ with the following property: for each $\ell \in \{0, 1\}$ and any function $f$ from $\{y_0, \dots, y_{n-1}\}$ to $\mathbf{Z}$, there is a finite subset $X_{f,\ell}$ of $X_\ell'$ and a finite subset $W_{f,\ell}$ of $X_{1-\ell}'$ such that if $\theta$ is any extension of $f$, and for all $(p, m, j) \in X_{f,\ell}$, $\theta(w_{pmj}) = 0$ and $\theta$ is acceptable at $w_{pmj}$, then $\theta(w_{pmj}) = 0$ whenever $(p, m, j) \notin W_{f,\ell}$ and $\theta$ is acceptable at $w_{pmj}$. In order to define $X_0'$ and $X_1'$ we list all pairs $(f, \ell)$ with $f: \{y_0, \dots, y_{n-1}\} \to \mathbf{Z}$ and $\ell \in \{0, 1\}$ in a sequence of order-type $\omega$; we will define by induction disjoint finite subsets $X_{0n}$ and $X_{1n}$ of $I$ such that $X_{\ell n} \subseteq X_{\ell,n+1}$; then we will let $X_\ell' = \bigcup_{n\in\omega} X_{\ell n}$. Let $X_{00}$ and $X_{01}$ be empty. Now suppose that $X_{0n}$ and $X_{1n}$ have been defined and that $(f, \ell)$ is the $n$th element in our list; apply the claim with $W = X_{1-\ell,n}$ to get $Y$ and let $X_{\ell,n+1} = X_{\ell n} \cup Y$ and $X_{1-\ell,n+1} = X_{1-\ell,n}$. (So, for this $(f, \ell)$, $X_{f,\ell}$ will be $X_{\ell,n+1}$ and $W_{f,\ell}$ will be $X_{1-\ell,n}$.)

Re-index the $z_{pmj}$'s as $z_m (m < \omega)$ and the $w_{pmj}$ as $w_m (m < j)$; this enumeration should be chosen so that $w_{pmj}$ is enumerated before $w_{p,m+1,i}$ for all $p$, $m$, $j$, $i$. Let $X_0$ and $X_1$ denote the images of $X_0'$ and $X_1'$ under this renumbering. Choose $m_r$ so that for any $f$ from $\{y_0, \ldots, y_{n-1}\}$ to $\mathbf{Z}$ such that $|f(y_i)| < r$ and for each $\ell \in \{0,1\}$, $X_\ell \cap (m_r + 1) \supseteq X_{f,\ell}$ and $W_{f,\ell} \subseteq m_r$; it is easy to see that $m_r$ has the desired property. So it remains to prove the claim.

First suppose that there is some prime $p \neq 2$ and $m$, $j$ so that $(p, m, j) \notin W$ and $g_{pf}(w_{pmj}) \neq 0$ (or is undefined). In this case we let $Y = \{(p, m, j)\}$. With this choice of $Y$, there is no $\theta$ extending $f$ satisfying the hypothesis of the claim, so the claim is vacuously true. Assume now that for all primes $p \neq 2$ if $(p, m, j) \notin W$ then $g_{pf}(w_{pmj}) = 0$. Consider $T_f$. For $\eta \in T_f$, define $*(\eta)$ to be the property:

> the length of $\eta$ is $m+1$ and there is $j$ so that $(2, m, j) \notin W$ and $\eta(m)(w_{2mj}) \neq 0$.

Let $Y = \{(2, m, j): \exists \eta$ of length $m + 1$ s.t. for all $d \leq m$, $*(\eta \upharpoonright d)$ does not hold, $(2, m, j) \notin W$ and $\eta(m)(w_{2mj}) \neq 0\}$. If $Y$ is finite, it is straightforward to see that $Y$ makes the claim true. So it remains to see that $Y$ is finite.

Let us say that a node, $\eta$, of $T_f$ is a branching node if it has at least two immediate successors; in that case, if the length of $\eta$ is $d - 1$, and $\eta'$ is any immediate successor of $\eta$, there exists $j$ such that $\eta'(w_{2dj}) \neq 0$. Let $n$ be such that $m \leq n$ for all $(p, m, j) \in W$. Let $d$ be such that for each node $\tau$ of length $m$ (there are only finitely many), there is a node $\eta_\tau \geq \tau$ of length $\leq d - 1$ which is a branching node. Then for all $(2, m, j) \in Y$, $m \leq d$, so $Y$ is finite. $\square$

By increasing the sets if necessary, we can assume that $X_0 \cup X_1 = \omega$. Later we will need to refer to the sets $X_\ell$ and the number $m_r$ for more than one presentation. In order to distinguish them, if the presentation in question is called $R$, we will denote these by $X_\ell^R$ and $m_r^R$.

Lemma 3.14 can be used to give a more computational proof that every countable W-group is free. By Pontryagin's criterion we can reduce to considering the case where $G$ is a finite rank non-free group. Let the notation be as in 3.14. Define $f(w_m) = 0$ if $m \in X_0$ and 1 otherwise. By 3.14, $f$ corresponds to a non-trivial element of $\text{Ext}(G, \mathbf{Z})$.

**3.15 Lemma.** *Suppose $H$ is a torsion-free group and $G$ is a pure subgroup of $H$. Then there are free resolutions $0 \to K \to F \to H \to 0$ and $0 \to K_1 \to F_1 \to G \to 0$ so that $K_1$ is a direct summand of $K$ and $F_1$ is a direct summand of $F$. Furthermore, if $G$ is a finite rank non-free group, then the free resolution of $G$ can be taken to be one given by a free presentation satisfying the conclusion of 3.14.*

PROOF. Choose a maximal independent subset $X$ of $G$ and extend the subset to a maximal independent subset $Y$ of $H$. Now view $H$ as a subgroup of $\bigoplus_{y \in Y} \mathbf{Q}y$. Fix a prime $p$. For a subgroup $B$ of $H$, let $p^{-n}B$ denote $\{h \in H : p^n h \in B\}$. Notice that $p^{-n}\langle Y \rangle$ is free and $p^{-n}\langle X \rangle$ is a direct summand of $p^{-n}\langle Y \rangle$. This statement can either be verified directly or by using Theorem XI.2.11. For each $n$, choose a set of free generators, $Z_{pn}$, of $p^{-n}\langle X \rangle$ and extend it to a set of free generators, $W_{pn}$, for $p^{-n}\langle Y \rangle$. Let $F$ be the group freely generated by

$$\bigcup_{p \in \mathcal{P}} \bigcup_{n \in \omega} \{z_h : h \in W_{pn}\}$$

(where $\mathcal{P}$ is the set of primes) and $F_1$ the subgroup generated by

$$\bigcup_{p \in \mathcal{P}} \bigcup_{n \in \omega} \{z_h : h \in Z_{pn}\}.$$

Let $K$ be the kernel of the map from $F$ to $H$ which takes $z_h$ to $h$, and let $K_1$ be the kernel of the restriction of this map to $F_1$. $\square$

**3.16 Lemma.** *Suppose $A$ is a non-free $\aleph_1$-free group of cardinality $\aleph_1$. Then there is a free resolution $0 \to K \to F \xrightarrow{\varphi} A \to 0$, an $\omega_1$-filtration $\{F_\alpha : \alpha < \omega_1\}$ of $F$, a stationary set $E$, and a natural number $n$ so that for all $\alpha \in E$ there are $\{y_0^\alpha, \ldots, y_{n-1}^\alpha, z_m^\alpha : m \in \omega\} \subseteq F$, $\{w_m^\alpha : m \in \omega\}$ a set of linear functions and elements $x_{\alpha m} \in F_\alpha (m \in \omega)$ satisfying:*

(1) $(y_0^\alpha, \ldots, y_{n-1}^\alpha, z_m^\alpha \ (m \in \omega); w_m^\alpha \ (m \in \omega))$ *is a free presentation* $R_\alpha$ *satisfying the conclusion of Lemma 3.14;*

(2) $\{u_m^\alpha : \alpha \in E, m \in \omega\}$ *freely generates a direct summand of* $K$, *where* $u_m^\alpha = w_m^\alpha - x_{\alpha m}$.

PROOF. Choose an $\omega_1$-filtration $\{A_\alpha : \alpha < \omega_1\}$ of $A$. Since $A$ is not free, we can assume that $\{\alpha : A_{\alpha+1}/A_\alpha$ is not free$\}$ is stationary. Choose $n$ so that $E \overset{\text{def}}{=} \{\alpha : A_{\alpha+1}/A_\alpha$ contains a non-free group $G_\alpha$ of rank $n\}$ is stationary (cf. II.4.5). The rest of the proof follows the same lines as 1.4 and uses Lemma 3.15. We define $H_\alpha$ and $K_\alpha$ by induction on $\alpha$, so that $K = \oplus K_\alpha$ and $F_\alpha = \bigoplus_{\beta<\alpha} H_\alpha$. As well, we define maps $\varphi_\alpha : F_\alpha \to A_\alpha$ and let $\varphi = \cup \varphi_\alpha$. The only interesting case occurs when $\alpha \in E$; otherwise we can proceed as in 1.4. Assume $\alpha \in E$ and $F_\alpha$ and $K_\beta$ have been defined for $\beta < \alpha$. Choose a free resolution

$$0 \to K' \to H_\alpha \overset{\pi}{\longrightarrow} A_{\alpha+1}/A_\alpha \to 0$$

guaranteed by Lemma 3.15; that is, there are $\{y_0^\alpha, \ldots, y_{n-1}^\alpha, z_m^\alpha : m \in \omega\} \subseteq H_\alpha$ and $\{w_m^\alpha : m \in \omega\}$ linear functions so that $(y_0^\alpha, \ldots, y_{n-1}^\alpha, z_m^\alpha \ (m \in \omega); w_m^\alpha \ (m \in \omega))$ is a free presentation $R_\alpha$ satisfying the conclusion of Lemma 3.14 and $\{w_m^\alpha : m \in \omega\}$ freely generates a direct summand of $K'$.

Choose $\psi : H_\alpha \to A_{\alpha+1}$ so that the map, $\pi$, from $H_\alpha$ to $A_{\alpha+1}/A_\alpha$ is the composition of $\psi$ with the quotient map. Let $F_{\alpha+1} = F_\alpha \oplus H_\alpha$ and $\varphi_{\alpha+1} = \varphi_\alpha \oplus \psi$. Next choose a set of free generators $\{a_i : i \in I\}$ for $K'$ and $\{b_i : i \in I\} \subseteq F_\alpha$ so that $\psi(a_i) = \varphi_\alpha(b_i)$. Let $K_\alpha = \langle a_i - b_i : i \in I \rangle$. There is a map from $K'$ to $K_\alpha$ defined by taking $a_i$ to $a_i - b_i$. Then $u_m^\alpha$ is the image of $w_m^\alpha$. $\square$

**Proof of 3.1.** (i) If $(\eta, 2)$-uniformization holds for some ladder system $\eta$, then by 3.4 and 3.6(ii), there is a non-free $\aleph_1$-separable W-group.

Assume now that there is a non-free W-group $A$ of cardinality $\aleph_1$. Use the notation of 3.16. Let $X_0^\alpha$, $X_1^\alpha$ denote $X_0^{R_\alpha}$, $X_1^{R_\alpha}$ and $m_r^\alpha$ denote $m_r^{R_\alpha}$. Let $\chi_\alpha$ denote the characteristic function of $X_0^\alpha$.

(Recall that we have assumed that $X_1^\alpha = \omega \setminus X_0^\alpha$.) Define

$$\varphi_\alpha(m) = ((w_0^\alpha, x_{\alpha 0}, \chi_\alpha(0)), \ldots, (w_m^\alpha, x_{\alpha m}, \chi_\alpha(m))).$$

(Here we are viewing $w_i^\alpha$ as a linear function of the variable symbols $y_0, \ldots, y_{n-1}, z_m (m \in \omega)$.) Let $\Phi = \{\varphi_\alpha : \alpha \in E\}$; note that $\Phi$ is tree-like. An application of Fodor's lemma (Exercise II.19) shows that $\Phi$ does not have disjoint end segments. We will prove that $(\Phi, 2)$-uniformization holds. If so, then by 3.3, there is a ladder system $\eta$ on a stationary set such that $(\eta, 2)$-uniformization holds.

Let $\{c_\alpha : \alpha \in E\}$ be a coloring of $\Phi$ by 2 colors. As in the proof of 3.9, define $c_\alpha^\ell(m) = c_\alpha(m)$ if $m \notin X_\ell^\alpha$ and 0 otherwise. It is enough to show each of the colorings $\{c_\alpha^\ell : \alpha \in E\}$ can be uniformized. Fix $\ell \in \{0, 1\}$. Define a function $\psi$ by letting $\psi(u_m^\alpha) = c_\alpha^\ell(m)$. Since $A$ is a W-group, any function from $\{u_{\alpha m} : \alpha < \omega_1, m < \omega\}$ to $\mathbb{Z}$ can be extended to a homomorphism from $F$ to $\mathbb{Z}$. Let $\theta$ be such an extension of $\psi$. Define $f^*(\alpha) = m_r^\alpha$, where $r = 2 \max\{|\theta(y_0^\alpha)|, \ldots, |\theta(y_{n-1}^\alpha)|\} + 1$. To complete the proof we must show that if $k \geq f^*(\alpha), f^*(\beta)$ and $\varphi_\alpha(k) = \varphi_\beta(k)$, then $\psi(u_k^\alpha) = \psi(u_k^\beta)$. Without loss of generality we can assume that $\max\{|\theta(y_0^\alpha)|, \ldots, |\theta(y_{n-1}^\alpha)|\} \geq \max\{|\theta(y_0^\beta)|, \ldots, |\theta(y_{n-1}^\beta)|\}$. Since $\varphi_\alpha(k) = \varphi_\beta(k)$, it is the case that for all $m \leq k$, $w_m^\alpha = w_m^\beta$, $x_{\alpha,m} = x_{\beta,m}$, and $m \in X_\ell^\alpha$ if and only if $m \in X_\ell^\beta$.

Now we consider $\theta \restriction \langle y_0^\alpha - y_0^\beta, \ldots, y_{n-1}^\alpha - y_{n-1}^\beta, z_j^\alpha - z_j^\beta : j \in \omega \rangle$ and apply 3.14 to the presentation

$$(y_0^\alpha - y_0^\beta, \ldots, y_{n-1}^\alpha - y_{n-1}^\beta, z_j^\alpha - z_j^\beta (j \in \omega); w_m(m \in \omega))$$

where

$$w_m = w_m^\alpha(y_0^\alpha - y_0^\beta, \ldots, y_{n-1}^\alpha - y_{n-1}^\beta, z_j^\alpha - z_j^\beta (j \in \omega)).$$

(Note this is the presentation $R_\alpha$ with different names for the generators.) Since $r = 2 \max\{|\theta(y_0^\alpha)|, \ldots, |\theta(y_{n-1}^\alpha)|\} + 1$, $|\theta(y_i^\alpha - y_i^\beta)| < r$, for all $i < n$. Also $w_m = u_m^\alpha - u_m^\beta$, for all $m \leq k$; hence $\theta(w_m) \in \{-1, 0, 1\}$. Since $k \geq m_r^\alpha$, $X_\ell^\alpha \cap (m_r^\alpha + 1) = X_\ell^\beta \cap (m_r^\alpha + 1)$,

$\theta(w_m) = 0$, for all $m \in X_\ell^R \cap (m_r + 1)$. We can now conclude by 3.14 that $\theta(w_k) = 0$, i.e., that $\psi(u_k^\alpha) = \psi(u_k^\beta)$. This completes the proof of (i).

The proof of part (ii) is analogous and uses the method of proof of 3.7. $\square$

**3.17 Corollary.** (i) *If there is a non-free W-group, then there is a strongly $\aleph_1$-free W-group which is not free.*

(ii) *If there is a non-free W-group of cardinality $\aleph_1$, then there is an $\aleph_1$-separable W-group of cardinality $\aleph_1$ which is not free.*

PROOF. Either every W-group is $\aleph_2$-free and hence strongly $\aleph_1$-free, or there is a non-free W-group of cardinality $\aleph_1$. In the latter case, the first line of the proof of 3.1 shows that there is an $\aleph_1$-separable W-group of cardinality $\aleph_1$ which is not free. $\square$

**3.18 Corollary.** *It is consistent that there is a non-free W-group of cardinality $\aleph_1$ and there is no non-free $\aleph_1$-coseparable group (of any cardinality).*

PROOF. We use the model in Theorem 3.12 where the function $h$ is constantly 2. By 3.12(c), there is no ladder system $\eta$ for which $(\eta, \aleph_0)$-uniformization holds. By 3.1(ii), in this model there is no non-free $\aleph_1$-coseparable group of cardinality $\aleph_1$. Using hypothesis (d) we can show by induction (as in 1.6) that there is no non-free $\aleph_1$-coseparable group of any cardinality. $\square$

**3.19 Theorem.** *There is a W-group of cardinality $\aleph_1$ which is not a Shelah group if and only if there is a family, $\Phi$, of functions from $\omega$ to some countable set $I$ such that $(\Phi, 2)$-uniformization holds and $|\Phi| = \aleph_1$.*

PROOF. This theorem is proved in the same way as 3.1. The only difference is that if $A$ is not a Shelah group then in the statement of 3.16 we can demand that $\{x_{\alpha m} : m \in \omega\}$ is countable. $\square$

**3.20 Corollary.** (MA + ¬CH) *Every W-group is a Shelah group.*

PROOF. If there is a W-group which is not a Shelah group, then it has a subgroup of cardinality $\aleph_1$ which is not a Shelah group (cf.

Exercise 4) — but of course is a W-group. However, MA + ¬CH implies that there is no family $\Phi = \{\varphi_\alpha : \alpha \in \omega_1\}$ of functions from $\omega$ to a countable set $I$ such that $(\Phi, 2)$-uniformization holds (see Exercise 9). □

**3.21 Theorem.** *Every $\aleph_1$-coseparable group is a Shelah group.*

PROOF. If there is an $\aleph_1$-coseparable group which is not a Shelah group then there is a family, $\Phi$, of functions from $\omega$ to some countable set $I$ such that $(\Phi, \aleph_0)$-uniformization holds and $|\Phi| = \aleph_1$. But no such family exists (see Exercise 6). □

### EXERCISES
#### A. $\kappa$-coseparable groups.

In the following group of exercises, we outline a proof of Griffith's result that for an uncountable cardinal $\kappa$, a group $A$ is $\kappa$-coseparable if and only if $\text{Ext}(A, \mathbf{Z}^{(\mu)}) = 0$ for all $\mu < \kappa$. (The definition of $\kappa$-coseparable is given in IV.2.12.)

1. If $A$ is $\kappa$-coseparable and $F$ is a free group of rank $\mu < \kappa$, then any exact sequence

$$0 \to F \to B \xrightarrow{\pi} A \to 0$$

splits. [Hint: Let $G$ be a maximal subgroup of $B$ so that $G \cap F = 0$. Hence $\pi \restriction G$ is one-one. By hypothesis, there is $K \subseteq G$ so that $A = \pi[K] \oplus C$ for some free $C$; show that $B = K \oplus \pi^{-1}[C]$ and that $\pi^{-1}[C]/F$ is free.]

2. Suppose $F$ is a free group of rank $\mu$ $(\geq \aleph_0)$ and $\text{Ext}(A, F) = 0$.
(i) $A$ is $\mu^+$-free.
(ii) Suppose $C$ is a subgroup of $A$ of index $\mu$. Let $B \subseteq A$ be a subgroup of cardinality $\mu$ such that $A = B + C$. The exact sequence

$$0 \to K \to B \oplus C \xrightarrow{\pi} A \to 0$$

splits where $\pi(b, c) = b + c$.
(iii) Let $\psi : A \to B \oplus C$ be the splitting map shown to exist in part (ii). Let $\rho : B \oplus C \to B$ be the projection map. If $D$ is $\ker(\rho \circ \psi)$, then $D \subseteq C$ and $D$ is a direct summand of $A$ of index $\mu$.

3. If $\kappa$ is an uncountable cardinal and $\text{Ext}(A, F) = 0$ for all free groups $F$ of rank $< \kappa$, then $A$ is $\kappa$-coseparable.

## B. Shelah Groups and Uniformization

4. Suppose that $A$ is $\aleph_1$-free. The following are equivalent.

(i) $A$ is not a Shelah group

(ii) there is a countable pure subgroup $B$ and countable subgroups $\{A_\alpha : \alpha < \omega_1\}$ such that: for all $\alpha$, $A_\alpha \supseteq B$; for all $\alpha$, $A_\alpha/B$ is not free; and $\langle A_\alpha/B : \alpha < \omega_1 \rangle = \bigoplus_{\alpha < \omega_1} A_\alpha/B$.

(iii) there is a natural number $n$, a countable pure subgroup $B$ and countable subgroups $\{A_\alpha : \alpha < \omega_1\}$ such that: for all $\alpha$, $A_\alpha \supseteq B$; for all $\alpha$, $A_\alpha/B$ is not free and has rank $n$; and $\langle A_\alpha/B : \alpha < \omega_1 \rangle = \bigoplus_{\alpha < \omega_1} A_\alpha/B$.

5. Suppose $I = {}^{<\omega}\omega$ and $\Sigma$ is an uncountable subset of ${}^\omega\omega$. For $\sigma \in \Sigma$ let $\varphi_\sigma(n) = \sigma\lceil n$ and let $\Phi = \{\varphi_\sigma : \sigma \in \Sigma\}$. Then $(\Phi, \aleph_0)$-uniformization does not hold. [Hint: let $c_\sigma(n) = \sigma(n)$; if $(f, f^*)$ uniformizes $\{c_\sigma : \sigma \in \Sigma\}$, then $\{\{\varphi_\sigma(n) : n > f^*(\sigma)\} : \sigma \in \Sigma\}$ is a disjoint family.]

6. If $I$ is a countable set and $\Phi$ is an uncountable family of functions from $\omega$ to $I$, then $(\Phi, \aleph_0)$-uniformization does not hold. [Hint: if $(\Phi, \aleph_0)$-uniformization does hold, then there is $\Phi'$ as in 5 such that $(\Phi', \aleph_0)$-uniformization holds.]

7. Suppose $I = {}^{<\omega}2$ and $\Sigma$ is an uncountable subset of ${}^\omega 2$. For $\sigma \in \Sigma$ let $\varphi_\sigma(n) = \sigma\lceil n$ and let $\Phi = \{\varphi_\sigma : \sigma \in \Sigma\}$. Then $(\Phi, 2)$-uniformization does not hold. [Hint: as in 5.]

8. Suppose $I = {}^{<\omega}\omega$ and $\Sigma$ is an uncountable subset of ${}^\omega\omega$. For $\sigma \in \Sigma$ let $\varphi_\sigma(n) = \sigma\lceil n$ and let $\Phi = \{\varphi_\sigma : \sigma \in \Sigma\}$. Let $\mathbf{P}$ be the following poset. The elements of $\mathbf{P}$ are pairs $(s, X)$ where $s$ is an increasing sequence of natural numbers say $0 = m_0 < \ldots < m_k$ and $X$ is a finite subset of $\Sigma$ satisfying:

    (a) for all $\sigma \neq \tau \in X, \sigma\lceil m_k \neq \tau\lceil m_k$

    (b) for all $\sigma \in X, \{\tau \in \Sigma : \sigma\lceil m_k = \tau\lceil m_k\}$ is uncountable.

    (c) for all $\sigma \in X$ and $j < k$, $|\{\tau\lceil m_{j+1} : \tau \in X$ and $\sigma\lceil m_j = \tau\lceil m_j\}| = 2$

Order $\mathbf{P}$ by $(s, X) \leq (t, Y)$ if $t$ extends $s$, $X \subseteq Y$.

(i) For all $m$, $\{(s, X): \sup s > m\}$ is dense in $\mathbf{P}$.

(ii) For all countable $\Lambda \subseteq \Sigma$, $\{(s, X): X \nsubseteq \Lambda\}$ is dense in $\mathbf{P}$.

(iii) $\mathbf{P}$ is c.c.c. [Hint: begin with an uncountable subset $Z$ of $\mathbf{P}$; find $(s, X)$, $(t, Y) \in Z$ so that $s = t$ and if $m = \sup s$, then $\{\sigma \restriction m: \sigma \in X\} = \{\sigma \restriction m: \sigma \in Y\}$; show these two elements are compatible.]

9. Assume MA + ¬CH. If $I$ is a countable set and $\Phi$ is an uncountable family of functions from $\omega$ to $I$, then $(\Phi, 2)$-uniformization does not hold. [Hint: if $(\Phi, 2)$-uniformization does hold, then there is an uncountable subset $\Sigma$ of $^\omega \omega$ such that if we let $\varphi_\sigma(n) = \sigma \restriction n$ and let $\Phi' = \{\varphi_\sigma: \sigma \in \Sigma\}$, then $(\Phi', 2)$-uniformization holds. Next use $\mathbf{P}$, as in 8, to get a contradiction to 7.]

10. Assume MA + ¬CH. If $A$ is a torsion-free group of cardinality $\aleph_1$, then either $A$ is a W-group or $r_0(\mathrm{Ext}(A, \mathbf{Z})) = 2^{\aleph_1}$ $(= 2^{\aleph_0})$. [Hint: use 2.1, 2.3 and 3.20.]

## C. Miscellaneous

11. For this exercise assume: for $\aleph_1$-free groups of cardinality $\aleph_1$ being a W-group is determined by the $\Gamma$-invariant (i.e., if $\Gamma(A) = \Gamma(B)$, then $A$ is a W-group if and only if $B$ is a W-group).

(i) $\mathcal{I} \overset{\text{def}}{=} \{E \subseteq \omega_1: \text{there is a W-group } A \text{ of cardinality } \aleph_1 \text{ such that } \Gamma(A) = \tilde{E}\}$ is a countably complete normal ideal.

(ii) If $A$ is an $\aleph_1$-free group of cardinality $\aleph_1$ and $A$ is not a W-group, then $r_0(\mathrm{Ext}(A, \mathbf{Z})) = 2^{\aleph_1}$. [Hint: let $\Gamma(A) = \tilde{E}$; using (i), $E$ can be partitioned into $\aleph_1$ subsets none of which are in $\mathcal{I}$; now argue as in 2.4.]

12. (i) Suppose $A$ is $\aleph_1$-free (of arbitrary cardinality) and whenever $B \subseteq A$ and $|B| = \aleph_1$, then $\Gamma(B) \neq 1$. Prove $A$ is strongly $\aleph_1$-free.

(ii) Assuming $2^{\aleph_0} < 2^{\aleph_1}$, show that if $\mathrm{Ext}(A, \mathbf{Z})$ is torsion, then $A$ is strongly $\aleph_1$-free.

13. The group $A$ constructed in Lemma 2.8 is isomorphic to the pure closure in $\mathbf{Z}^\omega$ of $\mathbf{Z}^{(\omega)} \cup \{(1, 0, s_0, 0, s_0 s_1, \ldots)\}$.

14. Use the hypothesis and notation of 3.8.

(i) There is a co-infinite subset $X$ of $\omega$ so that if $\theta$ is a homomorphism of $\langle z_m : m \in \omega \rangle$ to $\mathbf{Z}$ such that $|\{\theta(w_m)| : m \in \omega\}$ is finite and $\theta(w_m) = 0$ for all $m \in X$, then $\theta(w_m) = 0$ for all but finitely many $m$.

(ii) Use (i) to show directly that for any rank one torsion-free non-free group $G$, $r_0 \operatorname{Ext}(G, \mathbf{Z}) = 2^{\aleph_0}$. [Hint: take $Y_\alpha$ ($\alpha < 2^{\aleph_0}$) almost disjoint infinite subsets of $\omega \setminus X$; define $f_\alpha$ by $f_\alpha(w_m) = 1$ if $m \in Y_\alpha$ and 0, otherwise.]

15. Suppose that $F$ is a free group, $H \subseteq F$, and $G \overset{\text{def}}{=} F/H$ is a W-group. In this exercise we will outline a modification of Ehrenfeucht's proof that every W-group is separable.
Consider $H^{\perp\perp}$ (a subgroup of $F$). Define a map $\psi$ from $H^{\perp\perp}$ to $H^{**}$ as follows, if $g \in H^*$, let $g'$ be any extension of $g$ to $F$ (which exists since $G$ is a W-group) and let $\psi(a)(g) = g'(a)$.
(i) The map $\psi$ is well defined and a homomorphism.
(ii) $\psi$ is one-one.
(iii) By Theorem III.3.6, $H^{\perp\perp}/H$ is free.
(iv) Let $\varphi$ be the map from $F$ to $(H^\perp)^*$. Show that the range of $\varphi$ is a pure subgroup of $(H^\perp)^*$. Since $F/\ker\varphi$ is a separable group and the kernel of $\varphi$ is $H^{\perp\perp}$, $F/H^{\perp\perp}$ is separable. As $H^{\perp\perp}/H$ is also separable, so is $F/H$.

16. (i) Assume $\operatorname{Ax}(S) + \diamondsuit^*(\omega_1 \setminus S)$. Under this assumption, every hereditarily separable group of cardinality $\aleph_1$ is a W-group, but not every W-group is free. [Hint: use 1.9, 1.7 and Exercise VII.28.]
(ii) Assume $\operatorname{Ax}(S) + \diamondsuit^*(\omega_1 \setminus S)$ plus $\diamondsuit_\kappa(E)$ for every regular $\kappa > \aleph_1$ and every stationary subset $E$ of $\kappa$. Under this assumption, every hereditarily separable group is a W-group, but not every W-group is free. [Hint: prove by induction that if $\Gamma_{\mathbf{Z}}(A) \neq 0$, then $A$ is not hereditarily separable and, in fact, the killing lemma, VII.4.7, holds when $M/N \cong A$.]

17. For every stationary subset $E$ of $\aleph_1$, there is a tree-like ladder system $\{\eta_\delta : \delta \in E\}$. [Hint: for every $\alpha \in \lim(\omega_1)$, choose a one-one correspondence $\theta_\alpha : {}^{<\omega}\alpha \to \{\alpha + n : n \in \omega, n \text{ odd }\}$; for any $\delta \in E$ of the form $\alpha + \omega(\alpha \in \lim(\omega_1))$, let $\eta_\delta(n) = \alpha + 2n$; otherwise fix a

ladder system on $E$ and define $\eta_\delta(n)$ inductively, using the $\theta_\alpha$ (cf. the proof of 3.3).]

# NOTES

Ehrenfeucht 1955 says that Whitehead asked his question in 1952. Nunke 1977 discusses the topological importance of Whitehead's question and gives a good account of the history of the question. 1.2 is due to Stein 1951 (and independently to Eherenfeucht 1955). 1.3 is due to Rotman 1961 (separable and slender) and Nunke 1962b (slender). 1.4 is from Eklof-Huber 1980 and Mekler 1977. 1.6 is due to Shelah 1974, 1975a. 1.7 and 1.8 are from Devlin-Shelah 1978. Chase 1963a proved that assuming CH, every W-group is strongly $\aleph_1$-free. 1.9 is due to Shelah 1977. 1.11 is from Shelah 1974. 1.14 and 1.15 are from Eklof 1977a. 1.17(ii) is from Gerstner–Kaup-Weidner 1969; 1.18 is from Becker-Fuchs-Shelah 1989. (Becker-Fuchs-Shelah 1989 contains a treatment of Whitehead's problem for modules over domains.)

If $R$ is an integral domain, an $R$-module $B$ is called a Baer module if $\mathrm{Ext}^1_R(B, T) = 0$ for all torsion modules $T$. Griffith 1969a proved (in ZFC) that Baer $\mathbf{Z}$-modules are free. For Baer modules over more general domains, see Eklof-Fuchs 1988 and Eklof-Fuchs-Shelah 19??.

Related to the results on Whitehead groups in section 1 are the results on Crawley's Problem on $\omega$-elongations of $p$-groups: see Megibben 1983 and Mekler-Shelah 1986a and b; see also Eklof-Huber-Mekler 1988.

Section 2 is based on Eklof-Huber 1980 and 1981. Chase 1963a says Lemma 2.1 is well-known. 2.2 and 2.3 are from Eklof-Huber 1980 and distilled from Mekler 1977. 2.4 is due to Eklof-Huber 1980 (to Hiller-Huber-Shelah 1978 assuming V = L). 2.7 is due to C. U. Jensen 1972. 2.8 and 2.9 are from Eklof-Huber 1981. 2.10 is due to Sageev-Shelah 1985, but the proof here is from Eklof-Huber 1981. The special case of 2.10 where $\kappa_p = 0$ for all $p$ (i.e., 2.11) is due to Chase 1963a. In Mekler-Shelah 19??b, it will be shown, assuming V = L, that for every regular cardinal $\kappa$ less than the first weakly compact, there is no restriction on the $p$-rank of Ext for $\kappa$-free groups of cardinality $\kappa$ (other than the obvious one that the $p$-rank is at most $\kappa^+$). However if $\kappa$ is a strong limit singular cardinal of cofinality $\omega$, then Grossberg-Shelah 1989 prove there is no group $A$ of cardinality $\kappa$ such that the $p$-rank of $\mathrm{Ext}(A, \mathbf{Z}) = \kappa$. 2.13 and 2.14 are from Hiller-Huber-Shelah 1978. For the weakly compact case see Sageev-Shelah 1981.

Section 3 is based largely on Shelah 1980. (Not all the results we attribute to that paper are explicitly stated there). Theorem 3.1 is given as an exercise in Shelah 1980; the proof given here is due to Mekler and uses some ideas which do not appear elsewhere. 3.3-3.7 are due to Shelah 1980. 3.8 is due to Mekler. 3.9-3.13 are due to Shelah 1980. 3.14-3.16 are due to Mekler. 3.17 is due to Shelah 1980. 3.20 is due to Shelah 1979a.

It is possible to prove the consistency of uniformization results for cardinals greater than $\aleph_1$. It is also possible to show that it is consistent that for every group $G$, there is a non-free group $A$ such that $\text{Ext}(A, G) = 0$ (see Eklof-Shelah 19??). Also, for any $\kappa$ which is a regular cardinal in L, it is consistent that every cardinal in L is a cardinal in V and $\kappa$ is the first cardinal such that there is a non-free W-group of cardinality $\kappa$.

Exercises.    1-3:Griffith 1970; 5-9:Shelah 1977; 10-11:Eklof-Huber 1980; 14:new; 15:Ehrenfeucht 1955 (Ehrenfeucht only states the result that every countable W-group is free); 16:new.

# CHAPTER XIII
## THE BLACK BOX AND
## ENDOMORPHISM RINGS

In VI.1 we introduced the notion of a prediction principle, exemplified by the diamond and weak diamond principles. Recently, Shelah has discovered a class of $\diamondsuit$-like principles provable in ZFC which go under the generic name of Black Boxes. In this chapter we will explain one of the strongest of these principles and apply it to the construction of endomorphism rings.

In section 1, we show how to construct an endo-rigid group, first using the diamond principle, and then using the Black Box. Section 2 consists of a proof of the Black Box. In section 3, we prove, using the Black Box, a remarkable result (due to Dugas and Göbel) which characterizes the class of endomorphism rings of cotorsion-free groups. In section 4 we use $\diamondsuit$ to construct endomorphism rings of $\aleph_1$-separable groups and we use the Black Box to construct endomorphism rings of separable groups.

One theme of this chapter is the relation between $\diamondsuit$ and the Black Box. Roughly speaking, results that can be proved using $\diamondsuit$ can be proved in ZFC using the Black Box if we are willing to weaken the theorem.

## §1. Black Box

In the prediction principles that we have already met, e.g., weak diamond and diamond for $\lambda$-systems, the strategy in the applications has been that a killing lemma is proved and then the desired object is constructed using the killing lemma and the predictions. The strategy in applications of the Black Box will be the same.

The first example we will consider is the construction, assuming $\diamondsuit$, of a strongly $\aleph_1$-free group $A$ of cardinality $\aleph_1$ which is *endo-rigid*, i.e., $\text{End}(A) \cong \mathbf{Z}$. (Hence $A$ is indecomposable.)

In this section $\hat{G}$ denotes the $p$-adic completion of $G$; so, in particular, $\hat{\mathbf{Z}}$ is the $p$-adic integers (also denoted $J_p$; cf. I.3.4). Suppose

$G$ is contained in the $p$-adic completion of $\bigoplus_{i \in I} \mathbf{Z} a_i$. Since $\prod_{i \in I} \hat{\mathbf{Z}} a_i$ is complete, we can assume $G$ is contained in $\prod_{i \in I} \hat{\mathbf{Z}} a_i$, so any element of $G$ can be written uniquely as $\sum_{i \in I} \xi_i a_i$ where $\xi_i \in \hat{\mathbf{Z}}$. We call $\{i : \xi_i \neq 0\}$ the *support* of $g$, denoted $\operatorname{supp}(g)$. (Note that the support is not, in general, finite!) An easy but important fact is that if $A$ is Hausdorff in the $p$-adic topology, $\{a_n : n \in \omega\}$ is a sequence converging to 0, and $\varphi \in \operatorname{End}(A)$ then $\varphi(\sum_{n \in \omega} a_n) = \sum_{n \in \omega} \varphi(a_n)$, since every endomorphism is continuous in the $p$-adic topology. (Here $\sum_{n \in \omega} a_n$ denotes the limit of the sequence.) The construction will use the following killing lemma.

**1.1 Lemma.** *Suppose $G$ is a pure subgroup of the $p$-adic completion of $\bigoplus_{i \in I} \mathbf{Z} a_i$ containing $\bigoplus_{i \in I} \mathbf{Z} a_i$ such that for all $\xi \in \hat{\mathbf{Z}} \setminus \mathbf{Z}$, $\xi G \cap G = \{0\}$. Further suppose that $\{i_n : n \in \omega\}$ is such that for all $g \in G$, $\{n : i_n \in \operatorname{supp}(g)\}$ is finite. Let $\varphi \in \operatorname{End}(G)$ and $b \in G$ be such that for all $k \in \mathbf{Z}$, $\varphi(b) \neq kb$. Then there is $z_0 \in \hat{G}$ so that $\varphi$ does not extend to an endomorphism of $G_1 \overset{\text{def}}{=} \langle G, z_0 \rangle_*$. In fact, there is an element $y \in \hat{G} \setminus G_1$ so that if $H \supseteq G_1$ and $\psi \in \operatorname{End}(H)$, then $\psi(z_0) = y$.*

*Furthermore, $z_0$ can be chosen to be either*

$$\sum_{n \in \omega} p^n a_{i_n} \quad or \quad \sum_{n \in \omega} p^n (a_{i_n} + b);$$

*and for all $\xi \in \hat{\mathbf{Z}} \setminus \mathbf{Z}$, $\xi G_1 \cap G_1 = \{0\}$.*

PROOF. Let $\zeta = \sum_{n \in \omega} p^n$. For $m \in \omega$, let $x_m = \sum_{n \geq m} p^{n-m} a_{i_n}$ and

(1.1.1)
$$y_m = x_m + \zeta b.$$

Then $\langle G, x_0 \rangle_* = \langle G \cup \{x_m : m \in \omega\} \rangle$ and $\langle G, y_0 \rangle_* = \langle G \cup \{y_m : m \in \omega\} \rangle$. If $z_0$ is either $x_0$ or $y_0$ and $G_1 = \langle G, z_0 \rangle_*$, then $\xi G_1 \cap G_1 = \{0\}$ (by an argument on supports, using the fact that for all $g \in G$, $\{n : i_n \in \operatorname{supp}(g)\}$ is finite).

Let $\varphi$ denote also the unique extension of $\varphi$ to a homomorphism from $\hat{G}$ to $\hat{G}$. It suffices to prove that either $\varphi(x_0) \notin \langle G, x_0 \rangle_*$ or

$\varphi(y_0) \notin \langle G, y_0 \rangle_*$. Suppose that $\varphi(x_0) \in \langle G, x_0 \rangle_*$. Then $\varphi(x_0) = kx_m + g$ for some $m \in \omega$, $k \in \mathbf{Z}$ and $g \in G$. So

$$(1.1.2) \qquad \varphi(y_0) = kx_m + g + \zeta\varphi(b).$$

Subtracting $k$ times (1.1.1) from (1.1.2) we get that

$$\varphi(y_0) - ky_m - g = \zeta(\varphi(b) - kb).$$

If $\varphi(y_0)$ were in $\langle G, y_0 \rangle_*$, then since $\zeta\langle G, y_0 \rangle_* \cap \langle G, y_0 \rangle_* = \{0\}$, we would have $\varphi(b) = kb$, a contradiction. Therefore $\varphi(y_0) \notin \langle G, y_0 \rangle_*$ and we are done. $\square$

If we use this lemma to kill an unwanted endomorphism of $G$, then the endomorphism will not be resurrected in the subsequent extensions so long as $G_1$ remains closed in the $p$-adic topology in the group we are constructing.

**1.2 Theorem.** *Assume that $\Diamond$ holds. Then there is a strongly $\aleph_1$-free group $A$ of cardinality $\aleph_1$ such that $\mathrm{End}(A) \cong \mathbf{Z}$.*

PROOF. We will inductively construct an $\omega_1$-filtration $\{A_\alpha : \alpha < \omega_1\}$ of a strongly $\aleph_1$-free group $A$ so that for each $\alpha$, $A_\alpha$ contains $\bigoplus_{\nu \in \mathrm{succ}(\alpha)} \mathbf{Z}a_\nu$ and is included in the $p$-adic closure of $\bigoplus_{\nu \in \mathrm{succ}(\alpha)} \mathbf{Z}a_\nu$; moreover, for all $\xi \in \hat{\mathbf{Z}} \setminus \mathbf{Z}$, we will have $\xi A_\alpha \cap A_\alpha = \{0\}$. By CH (which is implied by $\Diamond$), we can identify the $p$-adic closure of $\bigoplus_{\alpha \in \mathrm{succ}(\omega_1)} \mathbf{Z}a_\alpha$ with $\omega_1$. By $\Diamond$ there is a sequence of partial functions $\{f_\alpha : \alpha < \omega_1\}$ such that for any function $f : \omega_1 \to \omega_1$, $\{\alpha : f \restriction \alpha = f_\alpha\}$ is stationary. The construction proceeds by induction. For successor ordinals $\alpha$, let $A_{\alpha+1} = A_\alpha \oplus \langle a_\alpha \rangle$. At limit ordinals, we take unions. Consider now a limit ordinal $\delta$; choose an increasing sequence of successor ordinals $\eta_\delta(n)$ $(n < \omega)$ with limit $\delta$. The key case occurs when

(*)     $f_\delta$ is an endomorphism of $A_\delta$ and there is $b \in A_\delta$ so that for all $k \in \mathbf{Z}$, $f_\delta(b) \neq kb$.

If (*) is not satisfied, let $A_{\delta+1} = A_\delta$. Otherwise, we have the hypotheses of Lemma 1.1 with $i_n = \eta_\delta(n)$, $G = A_\delta$, and $\varphi = f_\delta$. In this case, let $A_{\delta+1} = G_1$ as in Lemma 1.1.

It is routine to see that each $A_\alpha$ is free and that $A_\beta/A_{\alpha+1}$ is free whenever $\alpha < \beta$; hence $A = \bigcup_{\alpha < \omega_1} A_\alpha$ is a strongly $\aleph_1$-free group of cardinality $\aleph_1$, and for every successor $\alpha$, $A_\alpha$ is closed in $A$. By $\Diamond$, if $\varphi$ is an endomorphism of $a$ then there is an unbounded set of $\delta$'s such that $\varphi\restriction A_\delta = f_\delta$. If $\varphi$ is not multiplication by an integer, then there is some $b \in A$ so that $\varphi(b) \neq kb$ for any $k \in \mathbb{Z}$ (see Exercise 1). Choose $\delta$ so that $b \in A_\delta$ and $\varphi\restriction A_\delta = f_\delta$. By Lemma 1.1 there is $z \in A_{\delta+1}$ and $y \in \hat{A}_{\delta+1} \setminus A_{\delta+1}$ so that $\varphi(z) = y$. But this contradicts the fact that $A_{\delta+1}$ is closed in $a$. $\square$

We now state the promised combinatorial principle. We shall refer to this principle as the Black Box, but the reader should be warned that the term "Black Box" has also been applied to other, related, principles. (See VI.2 for the model-theoretic terminology used here.)

**1.3 Theorem. (The Black Box)** *Suppose $\kappa$, $\lambda$, $\mu$ are cardinals such that $\lambda = \mu^+$, $\mu^{\aleph_0} = \mu$, $E$ is a stationary subset of $\lambda$ consisting of ordinals of cofinality $\omega$, and $\kappa > \lambda$. Let $N$ be an expansion in a countable language of $(\mathrm{H}(\kappa), \in, <, \lambda)$ where $\mathrm{H}(\kappa)$ is the collection of sets of hereditary cardinality less than $\kappa$ and $<$ is a well ordering of $\mathrm{H}(\kappa)$. There is a family of countable sets $\{(M_i, X_i) : i \in I\}$ such that the following hold.*

*(a) Each $M_i \prec N$.*

*(b) Let $\delta(i)$ denote $\sup(M_i \cap \lambda)$. If $\delta(i) = \delta(j)$ and $i \neq j$ then $(M_i, X_i) \cong (M_j, X_j)$ and $M_i \cap M_j \cap \lambda$ is a proper initial segment of $M_i \cap \lambda$.*

*(c) For all $X \subset \lambda$, $\{\delta \in E: \text{there is } i \text{ such that } \delta(i) = \delta \text{ and } (M_i, X_i) \equiv_{M_i \cap \lambda} (N, X)\}$ is a stationary subset of $\lambda$.*

*(d) There is a ladder system $\{\eta_\delta : \delta \in E\}$ so that for every $i$ with $\delta(i) = \delta$, there is an increasing chain $(M_{in}, X_i)$ of elementary submodels of $(M_i, X_i)$ such that $M_i = \bigcup_{n \in \omega} M_{in}$ and for all $n$, $M_{in} \cap \lambda = M_{in} \cap \eta_\delta(n) = M_i \cap \eta_\delta(n)$ and $\sup(M_{in+1} \cap \lambda) > \eta_\delta(n)$. Further, if $\delta(i) = \delta = \delta(j)$, then for all $n$ the isomorphism from $M_i$ to $M_j$ restricts to an isomorphism between $M_{in}$ and $M_{jn}$.*

Clause (d) of the Black Box expresses a certain uniformity in the $M_i$; in most applications it is not needed. Clause (c) is the prediction clause; notice, in particular, that if $(N, X) \equiv_{M_i \cap \lambda} (M_i, X_i)$ and $X \subseteq \lambda$ then $X \cap M_i = X_i$ because for all $\alpha \in M_i \cap \lambda$,

$$N \models \alpha \in X \text{ if and only if } M_i \models \alpha \in X_i.$$

Moreover, the fact that $M_i \prec N$ and $(N, X) \equiv_{M_i \cap \lambda} (M_i, X_i)$ implies that $X_i$ reflects properties of $X$.

If $i \neq j$ and $\delta(i) = \delta(j)$, then since $N$ and hence $M_i$ and $M_j$ are well-ordered by the relation $<$, the isomorphism guaranteed by clause (b) is unique. (The $\alpha$th element of $M_i$ in the well-ordering $<$ must go to the $\alpha$th element of $M_j$.) The significance of the last part of (b) will become clear below.

We are going to use the Black Box first to give a proof in ZFC of the following (weakened) version of Theorem 1.2.

**1.4 Theorem.** *There is an uncountable $\aleph_1$-free group $A$ so that* $\text{End}(A) \cong \mathbf{Z}$.

The proof will be similar to that of 1.2; in fact, we are giving these two proofs in order to make the similarities plain. The group $A$ that we construct will be a subset of $\lambda$, and will be the union of subgroups $A_\delta$ constructed by induction on $\delta < \lambda$. By a coding argument (identifying $\lambda$ with $\lambda \times \lambda$), the $X_i$ will be used as predictions of endomorphisms that we will kill if they are not multiplication by an integer.

There are two difficulties that did not occur before. First, $X_i$ predicts the restriction of an endomorphism to a *subset* of $A_\delta$ (where $\delta = \delta(i)$). To deal with this, we need the following slightly sharper version of Lemma 1.1, which in fact was proved by the proof of 1.1.

**1.5 Lemma.** *Suppose $G$, $\{a_i : i \in I\}$, $\{i_n : n \in \omega\}$ are as in Lemma 1.1. Let $f$ be a partial function on $G$ so that for all $n$, $f(a_{i_n})$ is defined. If there is $b \in G$ so that for all $k \in \mathbf{Z}$, $f(b) \neq kb$, then there is $z_0 \in \hat{G}$ so that $f$ does not extend to an endomorphism of $G_1 \overset{\text{def}}{=} \langle G, z_0 \rangle_* = \langle G \cup \{z_n : n \in \omega\} \rangle$. Furthermore, $z_m$ can be chosen to be either $\sum_{m \leq n} p^{n-m} a_{i_n}$ or $\sum_{m \leq n} p^{n-m}(a_{i_n} + b)$. In fact, there is* an element $y \in \hat{G} \setminus G_1$ so that if $H \supseteq G_1$ and $\varphi \in \text{End}(H)$ extends $f$, then $\varphi(z_0) = y$. $\square$

The second difficulty is that we will need to construct $A_{\delta+1}$ to kill several endomorphisms at once since there will, in general, be several $i$ with $\delta(i) = \delta$. This difficulty is solved by part (b) of the Black Box which says that the domains of the predicted functions are almost disjoint in the sense that the intersection of any two domains is bounded below $\delta$. In particular, if we choose an increasing sequence contained in $M_i$ of order type $\omega$ with limit $\delta$ and such a sequence contained in $M_j$, then the two sequences have finite intersection. Thus the various strategies for killing the endomorphisms will not interfere with each other.

**Proof of 1.4.** Let $\kappa$, $\lambda$ and $\mu$ be as in 1.3; it should be noted that the hypotheses on $\lambda$ and $\mu$ imply that $\lambda^{\aleph_0} = \lambda$ (cf. Exercise II.26(ii)). Thus if $C = \bigoplus_{\alpha < \lambda} \mathbf{Z}a_\alpha$, $\hat{C}$ has cardinality $\lambda$ and we can identify $\hat{C}$ with $\lambda$ in such a way that if $\alpha < \beta$ then the ordinal representing $a_\alpha$ is less than the ordinal representing $a_\beta$. Let $Y = \{a_\alpha : \alpha \in \lambda\} (\subseteq \lambda)$. Let $\kappa$ be any cardinal $> \lambda^{++}$. The initial structure for the Black Box is $N = (\text{H}(\kappa), \in, <, \lambda, \hat{C}, Y)$. Here $\hat{C}$ denotes the 3-ary relation on $\lambda$ which is the graph of the addition operation on the group $\hat{C}$. Let $\{(M_i, X_i) : i \in I\}$ be as in the statement of the Black Box. We take the first (with respect to $<$) bijection, $g$, of $\lambda \times \lambda$ with $\lambda$ and use it to identify each $X \subseteq \lambda$ with a subset of $\hat{C} \times \hat{C}$. Notice that $g \in M_i$ since $M_i \prec N$.

We will construct a subgroup $A$ of $\hat{C}$ by constructing $A_\alpha$ by induction on $\alpha < \lambda$ and letting $A = \bigcup_{\alpha < \lambda} A_\alpha$. Each $A_\alpha$ will be $\aleph_1$-free and for successor ordinals $\alpha$, $A_\alpha$ will be $\aleph_1$-pure in $A$. Moreover, $A_\alpha$ will contain $\bigoplus_{\nu < \alpha} \mathbf{Z}a_\nu$ and be contained in the $p$-adic completion of the latter group. For successor ordinals $\alpha$, let $A_{\alpha+1} = A_\alpha \oplus \langle a_\alpha \rangle$. At limit ordinals, take unions. Consider now a limit ordinal $\delta$. The key case occurs when

(*)    $\text{cf}(\delta) = \omega$ and for some $i$, $\delta = \delta(i)$, and $X_i$ is the graph of an endomorphism of $M_i \cap A_\delta$ other than multiplication by an integer.

(Otherwise, let $A_{\delta+1} = A_\delta$.) Suppose $i$ satisfies this hypothesis. Let $f_i$ be the function whose graph is $X_i$. Choose $\{\alpha_{ni}: n < \omega\}$ an increasing sequence of successor ordinals in $M_i$ with limit $\delta$ and $b_i \in A_\delta \cap M_i$ so that $f_i(b_i) \neq k b_i$ for all $k \in \mathbb{Z}$. Let $G_i$ be the closure in $A_\delta$ of $M_i \cap A_\delta$; then $G_i$ satisfies the hypotheses of Lemma 1.5 with respect to $f_i$, $a_{\alpha_{ni}}$, and $b_i$. So there exists $z_{0i} \in \hat{G}_i$ and $y_i \in \hat{G}_i \setminus G_{1i}$, where $G_{1i} = \langle G_i, z_{0i} \rangle_*$, such that $f_i(z_{0i}) = y_i$. Let

$$A_{\delta+1} = \langle A_\delta \cup \bigcup \{\hat{G}_{1i}: i \text{ satisfies } (*)\} \rangle_*.$$

By property (b) of the Black Box and an argument on supports, $\hat{G}_i \cap A_{\delta+1} = G_{1i}$, so $f_i$ does not extend to an endomorphism of $A_{\delta+1}$.

If we are able to carry out the construction then — since $A_{\delta+1}$ will be closed in $A$ — $f_i$ will not extend to an endomorphism of $A$. Suppose now that we have been able to carry out the construction and $\varphi$ is an endomorphism of $A$ other than multiplication by an integer. Let $X \subseteq \lambda$ be such that $X$ is the graph of $\varphi$ (under the identification, $g$, of $\lambda$ with $\lambda \times \lambda$). Then by 1.3(c), there will have been some $i$ so that $(M_i, X_i) \equiv_{M_i \cap \lambda} (N, X)$. Let $\delta = \delta(i)$. First note that by construction, $M_i \cap A_\delta = M_i \cap A$ and $X_i = \varphi \restriction M_i \cap A$. Since $(M_i, X_i) \equiv_{M_i \cap \lambda} (N, X)$, $\mathrm{dom}(X_i) = M_i \cap \mathrm{dom}(X) = M_i \cap A$. Furthermore, since

$$(N, X) \models \text{ ``}X \text{ is the graph of a non-trivial}$$
$$\text{endomorphism of } \mathrm{dom}(X)\text{'',}$$

it follows that

$$(M_i, X_i) \models \text{ ``}X_i \text{ is the graph of a non-trivial}$$
$$\text{endomorphism of } \mathrm{dom}(X_i)\text{''.}$$

(Here, "non-trivial" means that it is not multiplication by an integer.) So $\delta$ satisfies $(*)$ and we have already prevented $X_i$ from extending to an endomorphism of $A$ which contradicts the assumption that $\varphi$ exists.

It remains to verify that $A_{\delta+1}$ is $\aleph_1$-free and that for all successor $\alpha < \delta$, $A_\alpha$ is $\aleph_1$-pure in $A_{\delta+1}$. Both of these assertions will be implied by the assertion that for all $\alpha < \beta$, $A_\beta/A_\alpha$ is $\aleph_1$-free whenever $\alpha$ is a successor ordinal or 0. The proof is by induction on $\beta$. Suppose $B$ is a countable subgroup of $A_\beta$. We can assume that $\beta$ is the least ordinal $\geq \alpha$ such that $B \subseteq A_\beta$. We must show that $(B + A_\alpha)/A_\alpha$ is free. If $\beta$ is a limit ordinal, then it has cofinality $\omega$, so we are done by the induction hypothesis and IV.2.3. There is no trouble if $\beta$ is the successor of an ordinal which doesn't satisfy (∗). So we are reduced to considering the case where $\beta = \delta + 1$ and $\delta$ satisfies (∗). We can assume there is $\{i_m : m \in \omega\}$ so that $B$ is the subgroup generated by $D \subseteq A_\delta$ together with $\{z_{n i_m} : n, m \in \omega\}$. Further we can assume that for all $n, m \in \omega$, $a_{\alpha_{n i_m}} \in D$. Choose an increasing sequence of ordinals $\{\alpha_k : k \in \omega\}$ with limit $\delta$ so that $\alpha = \alpha_0$ and for all $m \in \omega$ and all but finitely many $n \in \omega$, $\alpha_{n,i_m} \in \{\alpha_k : k \in \omega\}$. Notice that for any successor ordinal $\gamma < \delta$, $D/(D \cap A_\gamma) \cong (D + A_\gamma)/A_\gamma$ which is free by induction. So for such $\gamma$, $D \cap A_\gamma$ is a direct summand of $D$. Further for all $k$, $D \cap A_{\alpha_k+1} = (D \cap A_{\alpha_k}) \oplus \mathbf{Z}a_{\alpha_k}$. Inductively choose subgroups $D_k$ so that for all $k$, $(D \cap A_{\alpha_k+1}) \oplus D_k = D \cap A_{\alpha_{k+1}}$. Hence $D = \bigoplus_{k \in \omega} D_k \oplus \bigoplus_{k \in \omega} \mathbf{Z}a_{\alpha_k}$.

By property (b) of the Black Box, we can choose $\{n(m) : m \in \omega\}$ so that the collections $\{\alpha_{n,i_m} : n(m) \leq n\}$ are pairwise disjoint. By the choice of the $\alpha_k$'s we can also assume that for every $m$, $\{\alpha_{n,i_m} : n(m) \leq n\} \subseteq \{\alpha_k : k < \omega\}$. Finally we observe that $(B + A_\alpha)/A_\alpha$ is isomorphic to the direct sum of $\bigoplus_{k \neq 0} D_k$ together with the group freely generated by $\{z_{n,i_m} : n(m) \leq n$ and $m \in \omega\} \cup \{a_{\alpha_k} : $ for all $m \in \omega$ and $n(m) \leq n$, $\alpha_k \neq \alpha_{n,i_m}\}$. $\square$

Part of the price we pay for working in ZFC, rather than assuming $\Diamond$, is that the group constructed is not strongly $\aleph_1$-free: in constructing $A_{\delta+1}$ from $A_\delta$, we created a countable set whose $p$-adic closure is uncountable.

## §2. Proof of the Black Box

In this section we give a proof of the Black Box. For convenience we restate the principle.

**2.1 Theorem. (The Black Box)** *Suppose $\kappa$, $\lambda$, $\mu$ are cardinals such that $\lambda = \mu^+$, $\mu^{\aleph_0} = \mu$, $E$ is a stationary subset of $\lambda$ consisting of ordinals of cofinality $\omega$, and $\kappa > \lambda$. Let $N$ be an expansion in a countable language of $(\mathrm{H}(\kappa), \in, <, \lambda)$ where $\mathrm{H}(\kappa)$ is the collection of sets of hereditary cardinality less than $\kappa$ and $<$ is a well ordering of $\mathrm{H}(\kappa)$. There is a family of countable sets $\{(M_i, X_i) : i \in I\}$ such that the following hold.*

(a) *Each $M_i \prec N$.*

(b) *Let $\delta(i)$ denote $\sup(M_i \cap \lambda)$. If $\delta(i) = \delta(j)$ and $i \neq j$ then $(M_i, X_i) \cong (M_j, X_j)$ and $M_i \cap M_j \cap \lambda$ is a proper initial segment of $M_i \cap \lambda$.*

(c) *For all $X \subset \lambda$, $\{\delta \in E : $ there is $i$ such that $\delta(i) = \delta$ and $(M_i, X_i) \equiv_{M_i \cap \lambda} (N, X)\}$ is a stationary subset of $\lambda$.*

(d) *There is a ladder system $\{\eta_\delta : \delta \in E\}$ so that for every $i$ with $\delta(i) = \delta$, there is an increasing chain $(M_{in}, X_i)$ of elementary submodels of $(M_i, X_i)$ such that $M_i = \bigcup_{n \in \omega} M_{in}$ and for all $n$, $M_{in} \cap \lambda = M_{in} \cap \eta_\delta(n) = M_i \cap \eta_\delta(n)$ and $\sup(M_{in+1} \cap \lambda) > \eta_\delta(n)$. Further, if $\delta(i) = \delta = \delta(j)$, then for all $n$ the isomorphism from $M_i$ to $M_j$ restricts to an isomorphism between $M_{in}$ and $M_{jn}$.*

In order to prove Theorem 2.1 we will need two auxiliary results. Before we can state the first result we need some definitions. Assume $\lambda = \mu^+$, $\kappa > \lambda$, and $N$ is an expansion in a countable language of $(\mathrm{H}(\kappa), \in, <, \lambda)$ where $<$ is a well-ordering. An $\omega$-sequence $(M_0, X_0) \prec (M_1, X_1) \prec \ldots$ of countable models is *acceptable* if for all $n$, $M_n \prec N$, $X_n$ is a subset of $M_n \cap \lambda$ and $M_n \cap \mu = M_0 \cap \mu$. Two acceptable sequences $((M_n, X_n) : n \in \omega)$, $((M'_n, X'_n) : n \in \omega)$ have the *same type* if $M_0 \cap \mu = M'_0 \cap \mu$ and there is an isomorphism between the sequences. (The isomorphism is unique because of the well-ordering $<$.) By the *type* of an acceptable sequence we shall mean the set of acceptable sequences having the same type. For technical reasons we can assume that a type is a pair consisting of a countable subset of $\mu$ ($= M_0 \cap \mu$) and a subset of $\omega$ (representing the isomorphism type of the sequence). Note there are at most $\mu^{\aleph_0}$ types.

For $X \subseteq \lambda$, we define a game for $X$, which we will denote as $G(X)$ (cf. section II.5). The game is played as follows. On his first move player I plays a type $s$ and a countable model $(M_0, X_0)$. Thereafter Players I and II alternately choose countable models $(M_n, X_n)$ and ordinals $\beta_n$ — Player I choosing the models and Player II choosing the ordinals. The sequence should satisfy the following conditions:

(I) The sequence $((M_n, X_n): n \in \omega)$ should be an acceptable sequence of type $s$. For all $n$, $(M_n, X_n) \prec (N, X)$. For all $n$, $M_{n+1} \cap \beta_n = M_n \cap \lambda$ and $\sup(M_{n+1} \cap \lambda) > \beta_n$.

(II) For all $n$, $\beta_n > \sup(M_n \cap \lambda)$.

Player I wins a play of the game if it lasts $\omega$ moves, i.e., for each $n$ Player I has a move following the requirements (I); otherwise Player II wins.

**2.2 Theorem.** *Suppose $N$ is as above and $\mu^{\aleph_0} = \mu$. Then for all $X \subseteq \lambda$, Player I has a winning strategy for the game $G(X)$.*

PROOF. Since this is an open game, either Player I or Player II has a winning strategy (cf. II.5.2). Since $N$ is well-ordered by $<$, the intersection of elementary submodels is an elementary submodel. (See Exercise VI.10). Suppose, to obtain a contradiction, that there is a winning strategy $F$ for Player II. Choose an increasing sequence $((N_n, Y_n): n \in \omega)$ of elementary submodels of $(N, X)$ of cardinality $\mu$ so that: for all $n$, $N_n$ is closed under $F$; $\mu \subseteq N_0$; for all $n$, $N_n$ is closed under countable subsets; for all $n$, $N_n \cap \lambda$ is an initial segment of $\lambda$; and $N_n \cap \lambda < N_{n+1} \cap \lambda$. (Since $N_n \cap \lambda$ is an ordinal (cf. VI.2.3(v)), the last clause is a meaningful statement.) Now choose a countable elementary submodel $(M, Y)$ of $\bigcup_{n \in \omega}(N_n, Y_n)$ so that $\sup(M \cap \lambda) = \sup(\bigcup_{n \in \omega} N_n \cap \lambda)$. Let $(M_n, X_n) = (M, Y) \cap (N_n, Y_n)$ and let $s$ be the type of $((M_n, X_n): n \in \omega)$. Then $s$ together with the $(M_n, X_n)$'s give I a winning play against II's strategy, which is a contradiction. To be precise, notice first that since each $N_n$ is closed under countable subsets, $(M_k, X_k) \in N_n$ for all $k \leq n$. Second note that for all $n$, $M_n \cap \mu = M_0 \cap \mu$. Since $((M_k, X_k): k \leq n) \in N_n$ and $N_n$ is closed under $F$, it follows that $\beta_n$ — the move dictated to Player II by $F$ — is also in $N_n$. Thus since $N_n \cap \lambda$ is an initial segment of $\lambda$, $M_{n+1} \cap \beta_n = M \cap \beta_n = M_n \cap \beta_n$. So $(M_{n+1}, X_{n+1})$ is a legal move for player I. $\square$

We also need another result of Shelah.

**2.3 Lemma.** *Suppose $\lambda = \mu^+$ and $\mu = \mu^{\aleph_0}$. For every stationary $S \subseteq \lambda$ consisting entirely of ordinals of cofinality $\omega$, there is a ladder system $\{\eta_\delta : \delta \in S\}$ such that for every closed unbounded set $C \subseteq \lambda \{\delta \in S : \eta_\delta[\omega] \subseteq C\}$ is stationary.*

PROOF. For $\delta \in S$ let $\{\eta_i^\delta : i < \mu\}$ enumerate $\{\eta : \eta$ is an increasing function from $\omega$ to $\delta$ with limit $\delta\}$. For $i < \mu$, let $\bar{\eta}_i = \{\eta_i^\delta : \delta \in S\}$. We claim that one of the $\bar{\eta}_i$ satisfies the conclusion of the theorem. If not, then for each $i < \mu$, there is a cub $C_i \subseteq \lambda$ such that the set $T_i = \{\delta \in S : \eta_i^\delta[\omega] \subseteq C_i\}$ is non-stationary. If $D_i$ is a cub whose intersection with $T_i$ is empty, then by replacing $C_i$ by $C_i \cap D_i$ we can assume for all $\delta \in S$, $\eta_i^\delta[\omega]$ is not contained in $C_i$. Let $C = \bigcap_{i<\mu} C_i$. Choose $\delta \in S$ which is also a limit point of $C$. So for some $i$, $\eta_i^\delta[\omega] \subseteq C \subseteq C_i$. This is a contradiction. $\square$

**Proof of 2.1.** Suppose $N$ is an expansion of $(\mathrm{H}(\kappa), \in, <, \lambda)$ in a countable language. Let $S$ be the set of types for this $N$ (see the definition after the statement of 2.1). Let $h : E \to S$ be a function such that for any type $s$, $h^{-1}[s]$ is stationary. For each type $s$ choose a ladder system $\{\eta_\delta : \delta \in h^{-1}[s]\}$ as guaranteed by Lemma 2.3.

For $\delta \in h^{-1}[s]$, choose a set $\mathcal{M}_\delta$ of $(M, X)$ such that for each $(M, X) \in \mathcal{M}_\delta$ there is an increasing sequence $((M_n, X_n) : n \in \omega)$ of elementary submodels of $(M, X)$ so that $M = \bigcup_{n \in \omega} M_n$; $((M_n, X_n) : n \in \omega)$ is of type $s$; $\sup(M \cap \lambda) = \delta$; and for all $n$, $M_n \prec N$, and $M_n \cap \lambda = M_n \cap \eta_\delta(n) = M \cap \eta_\delta(n)$. Furthermore, make the choice so that for each possible value of $M \cap \lambda$, there is exactly one $(M, X) \in \mathcal{M}_\delta$ so that $M \cap \lambda$ is that value. By the definition of types, if $(M, X)$ and $(M', X') \in \mathcal{M}_\delta$ then $M \cap \mu = M' \cap \mu$ and $(M, X) \cong (M', X')$. Let $\{(M_i, X_i) : i \in I\}$ enumerate $\bigcup_{\delta \in E} \mathcal{M}_\delta$. We claim this is the desired family. Properties (a) and (d) of the Black Box are immediate from the definition. To verify (b) suppose

$(M, X) \neq (M', X') \in \mathcal{M}_\delta$; it is enough to show that if $0 \neq \gamma \in M \cap M' \cap \lambda$, then $M \cap \gamma = M' \cap \gamma$. Let $f$ be the $<$-least surjection of $\mu$ onto $\gamma$. Since $M, M' \prec N$, $f \in M \cap M'$. So $\gamma \cap M = f[M \cap \mu] = f[M' \cap \mu] = \gamma \cap M'$.

To verify (c), consider $X \subseteq \lambda$. Let $F$ be a winning strategy for Player I in the game $G(X)$ and $s$ be the type that Player I plays on his first move. Let $(N_\alpha : \alpha < \lambda)$ be an increasing sequence of elementary substructures of $(\mathrm{H}(\lambda^+), \in)$ such that $C = \{N_\alpha \cap \lambda : \alpha < \lambda\}$ is a closed unbounded set of ordinals and each $N_\alpha$ is closed under $F$. Suppose $\delta \in C \cap h^{-1}[s]$ such that $\eta_\delta[\omega] \subseteq C$. The set of such ordinals is stationary. Let $\alpha_n$ denote an ordinal so that $N_{\alpha_n} \cap \lambda = \eta_\delta(n)$. We choose a sequence of models as follows. Let $(s, (M_0, X_0)) \in N_{\alpha_0}$ be Player I's first move. Since $M_0 \in N_{\alpha_0}$ is countable, $\sup(M_0 \cap \lambda) < \eta_\delta(0)$. So $\eta_\delta(0)$ is a legal play for Player II. In general, we can choose $(M_{n+1}, X_{n+1}) \in N_{\alpha_{n+1}}$ so that

$$(s, (M_0, X_0)), \eta_\delta(0), \ldots, (M_n, X_n), \eta_\delta(n), (M_{n+1}, X_{n+1}))$$

is a play of the game according to the strategy $F$. Let $M = \bigcup_{n \in \omega} M_n$ and $Y = \bigcup_{n \in \omega} X_n$. By construction $(M, Y) \prec (N, X)$. As well, $(M, Y)$ satisfies the requirements for membership in $\mathcal{M}_\delta$. So there is $(M', X')$ in $\mathcal{M}_\delta$ such that $M' \cap \lambda = M \cap \lambda$. Because $(M, Y) \cong (M', X')$ and the isomorphism must be the identity on $M \cap \lambda$, $(M', X') \equiv_{M' \cap \lambda} (N, X)$. Hence we have established (c). $\square$

## §3. Endomorphism rings of cotorsion-free groups

In this section, we show that other rings besides $\mathbf{Z}$, namely all cotorsion-free rings, can be realized as endomorphism rings of groups. Recall, from V.2.9, that a group $A$ is cotorsion-free if and only if it is reduced, torsion-free, and for all primes $p$, $J_p$ is not embeddable in $A$. Call a ring $R$ *cotorsion-free* if $(R, +)$ is a cotorsion-free group. The problem of characterizing the endomorphism rings is put in perspective by the following proposition.

**3.1 Proposition.** *If $A$ is a reduced torsion-free group, then the following are equivalent.*

(1) *A is cotorsion-free.*
(2) *For all primes p, A has no direct summand isomorphic to $J_p$.*
(3) End($A$) *is cotorsion-free.*

PROOF. The equivalence of (1) and (2) has already been proved in Theorem V.2.9. Since $J_p$ carries a ring structure, End($J_p$) $\supseteq J_p$. (In fact End($J_p$) $= J_p$.) So if (2) fails, then End($A$) contains a copy of $J_p$ and hence (3) fails. To complete the proof we will show that (1) implies (3). Assume that $A$ is cotorsion-free. First, End($A$) is torsion-free because for $f \in$ End($A$), if $nf = 0$ where $n \neq 0$ and $f \neq 0$, then there exists $a$ so that $f(a) \neq 0$ but $nf(a) = 0$, a contradiction of the fact that $A$ is torsion-free. To see that End($A$) is reduced, suppose that $0 \neq f \in$ End($A$) is divisible by all integers; then there exists $a$ such that $f(a) \neq 0$ and $f(a)$ is divisible by all integers, a contradiction since $A$ is reduced and torsion-free. Suppose now that End($A$) contains $J_p$ as a subgroup. Regard $A$ as a right End($A$)-module; there must be $a \in A$ so that $aJ_p \neq \{0\}$. But then $aJ_p$ is a non-zero subgroup of $A$ which is a homomorphic image of a pure-injective group and hence cotorsion. So $A$ would not be cotorsion-free. □

For the rest of this chapter, if $A$ is a reduced torsion-free group then $\hat{A}$ denotes the $\mathbb{Z}$-adic closure of $A$. The group $\hat{A}$ can be viewed as a module over $\hat{\mathbb{Z}}$. Recall (see I.3) that any element of $\hat{\mathbb{Z}}$ can be represented as $\sum_{n \in \omega} k_n$ where $n!|k_n$. The representation may not be unique but this is not a problem. Further, $\hat{\mathbb{Z}}$ contains a copy of $J_p$ (cf. I.3.5). As well, since multiplication by an element of $R$ takes convergent sequences to convergent sequences, if $A$ is an $R$-module which is reduced and torsion-free as an abelian group, then $\hat{A}$ has a unique $R$-module structure extending that of $A$.

**3.2 Proposition.** *If $A$ is cotorsion-free, then for any $0 \neq a \in A$ there is $\xi \in \hat{\mathbb{Z}}$ so that $\xi a \notin A$. (Here $\xi a$ is calculated in $\hat{A}$.)*

PROOF. If the assertion were not true then $A$ would contain a non-zero homomorphic image of $\hat{\mathbb{Z}}$ and so wouldn't be cotorsion-free. □

Suppose $R$ is a cotorsion-free ring. Consider $C = \bigoplus_{\alpha \in \text{succ}(\delta)} Ra_\alpha$ where $\delta$ is an ordinal. Since $\prod_{\alpha \in \text{succ}(\delta)} \hat{R}a_\alpha$ is complete (by I.3.2), any element $c$ of $\hat{C}$ can be written uniquely as $\sum_{\alpha \in \text{succ}(\delta)} \rho_\alpha a_\alpha$ where $\rho_\alpha \in \hat{R}$. Call $\rho_\alpha$ the *coefficient* of $a_\alpha$ in $c$. For $c$ in $\hat{C}$ define $\text{supp}(c) = \{\alpha:$ the coefficient of $a_\alpha$ in $c$ is non-zero$\}$. We say $c$ has *bounded support* if $\text{supp}(c)$ is bounded below $\delta$. The proof of the following killing lemma outlines three strategies for killing an unwanted endomorphism.

**3.3 Lemma.** *Let $R$ and $C$ be as above and suppose that $\text{cf}(\delta) = \omega$. Suppose that $G$ is an $R$-submodule of $\hat{C}$ containing $C$ such that $G$ is cotorsion-free as an abelian group and $C$ is dense in $G$ (in the $\mathbf{Z}$-adic topology of $G$). Further suppose that every $a \in G$ has bounded support. Let $\varphi \in \text{End}_{\mathbf{Z}}(G)$ such that $\varphi$ is not multiplication by an element of $R$. Then there exists $G \subseteq G_1 \subseteq \hat{C}$ such that $\varphi$ does not extend to an element of $\text{End}_{\mathbf{Z}}(G_1)$, $C$ is dense in (the restriction of the $\mathbf{Z}$-adic topology to) $G_1$ and $G_1$ is cotorsion-free as abelian group.*

PROOF. To construct $G_1$ we will choose an increasing sequence $\{\alpha_n: n \in \omega\}$ with limit $\delta$, a sequence of natural numbers $\{k_n: n \in \omega\}$ so that for all $n$, $n! | k_n$, and an element $x$ of $G$. We will let

$$z_i = \sum_{n \geq i} \frac{k_n}{d_i}(a_{\alpha_n} + x),$$

where $d_i$ is the greatest common divisor of $\{k_n: i \leq n\}$. Then $G_1$ will be the $R$-module generated by $G \cup \{z_i: i \in \omega\}$. Any element $a$ of $G_1$ can then be written as $rz_n + b$ (for some $n$) where $b \in G$ and $\text{supp}(b)$ is bounded below $\alpha_n$. If $r \neq 0$, then $\text{supp}(a)$ is unbounded. So the only elements of $G_1$ with bounded support are the elements of $G$.

Note that $C$ is dense in $G_1$. We must check that $G_1$ is cotorsion-free. Suppose to the contrary that $f: \hat{\mathbf{Z}} \to G_1$ is a non-zero homomorphism. Say $f(1) = rz_n + b$ where $b \in G$. For any $\zeta \in \hat{\mathbf{Z}}$, calculating in $\hat{C}$, we have that

$$f(\zeta) = (\zeta r)z_n + \zeta b$$

where $\zeta r \in \hat{R}$. We now show that $\zeta r \in R$. There is some $m \geq n$, so that

$$f(\zeta) = sz_m + e$$

where $e \in G$ and $\mathrm{supp}(e)$, $\mathrm{supp}(b)$ are bounded by $\alpha_m$. The coefficient of $a_{\alpha_m}$ in $rz_n + b$ is $k_m/d_n r$ and the coefficient of $a_{\alpha_m}$ in $sz_m + e$ is $k_m/d_m s$; so comparing the coefficients of $a_{\alpha_m}$ in the two expressions for $f(\zeta)$ we have

$$k_m/d_m s = k_m/d_n \zeta r$$

and

$$s = d_m/d_n \zeta r.$$

Thus $s$ is divisible by $d_m/d_n$ in $\hat{R}$ and so in $R$. Hence $\zeta r \in R$. The map which takes $\zeta$ to $\zeta r$ is a homomorphism from $\hat{\mathbf{Z}}$ to $R$. Since $R$ is cotorsion-free, this must be the 0 homomorphism; i.e., $r = 0$. So $f$ is actually a map from $\hat{\mathbf{Z}}$ to $G$, which contradicts the assumption that $G$ is cotorsion-free. Similarly, we can show that for $\zeta \in \hat{\mathbf{Z}}$ if $\zeta r \notin R$, then $\zeta a \notin G_1$.

Next we shall list three hypotheses about $\varphi$ and after each hypothesis a construction of $G_1$ under that hypothesis so that $\varphi$ does not extend to $G_1$. We will show that at least one of the hypotheses must hold, thus proving the lemma. Since any extension of $\varphi$ to $G_1$ is uniquely determined we will also denote the putative extension as $\varphi$; $(\gamma, \delta)$ denotes $\{\nu: \gamma < \nu < \delta\}$.

**Hypothesis (1):** There are disjoint increasing sequences $\{\alpha_n: n \in \omega\}$, $\{\beta_n: n \in \omega\}$ with limit $\delta$ such that for all $n: \beta_n \in \mathrm{supp}(\varphi(a_{\alpha_n}))$; and for all $i < n$, $\mathrm{supp}(\varphi(a_{\alpha_i}))$ is bounded by $\beta_n$.

**Construction (1):** For each $n$, choose $e_n$ which does not divide the coefficient of $a_{\beta_n}$ in $\varphi(a_{\alpha_n})$. Let $k_n = n! \prod_{j<n} e_j$. Define $G_1$ as above with $x = 0$.

Suppose $\varphi$ extends to $G_1$ and consider $\varphi(z_0)$. By construction, for all $n \geq 1$ the coefficient of $a_{\beta_n}$ in $\varphi(z_0)$ is non-zero. Since for any $a \in G_1$ there exists $\gamma < \delta$ such that $\mathrm{supp}(a) \cap (\gamma, \delta) \subseteq \{\alpha_n: n \in \omega\}$ we have a contradiction.

If Hypothesis (1) fails, then there is an ordinal $\beta$ so that for all but a bounded set of $\alpha$, $\mathrm{supp}(\varphi(a_\alpha)) \subseteq \beta \cup \{\alpha\}$.

**Hypothesis (2):** There is an increasing sequence $\{\alpha_n : n \in \omega\}$ with limit $\delta$ and $\beta < \delta$ such that for all $n$, $\mathrm{supp}(\varphi(a_{\alpha_n})) \subseteq \beta \cup \{\alpha_n\}$ and for all $r \in R$, $\{n : \varphi(a_{\alpha_n}) = ra_{\alpha_n} + b_n$ for some $b_n$ with support bounded below $\beta\}$ is finite.

**Construction (2):** Let $k_n = n!$ and $x = 0$. Suppose $\varphi$ extends to $G_1$. Then in $G_1 \varphi(z_0) = rz_i + b$ for some $i$, $r$, and some $b$ of bounded support. Now $\varphi(z_0)$ also equals $\sum_{n \geq 0} k_n \varphi(a_{\alpha_n})$, so by comparing coefficients of $a_{\alpha_n}$ we see that for all but finitely many $n$, $\varphi(a_{\alpha_n}) = \frac{r}{d_i} a_{\alpha_n} + b_n$ where $\mathrm{supp}(b_n) \subseteq \beta$. This contradicts hypothesis (2).

**Hypothesis (3):** There is an increasing sequence $\{\alpha_n : n \in \omega\}$ with limit $\delta$, $\beta < \delta$, $r \in R$, and $a \in G$ such that for all $n$, $\mathrm{supp}(\varphi(a_{\alpha_n})) \subseteq \beta \cup \{\alpha_n\}$ and $\varphi(a_{\alpha_n}) = ra_{\alpha_n} + b_n$; and $ra \neq \varphi(a)$.

Note that if hypotheses (1) and (2) fail, then there must exist such a sequence of ordinals and such an $r$. Since $\varphi$ is not multiplication by $r$, such an $a$ must exist as well.

**Construction (3):** Choose $\xi = \sum_{n \in \omega} k_n$ so that for all $n$, $n! | k_n$ and $\xi(\varphi(a) - ra) \notin G$ (cf. 3.2). We try two possibilities for $G_1$. First let $x = 0$; call the resulting group $G_1$. Now try $x = a$ and call the resulting group $G_1'$ and let $z_i'$ correspond to $z_i$. (So for example, $z_0' = \sum_{n \geq 0} k_n(a_{\alpha_n} + a) = z_0 + \xi a$.)

Suppose $\varphi$ extends in both cases; then $\varphi(z_0) = rz_0 + b$ for some $b \in G$. Also $\varphi(z_0') = rz_0 + \xi\varphi(a) + b$. Then $\varphi(z_0') - rz_0' - b = \xi(\varphi(a) - ra)$, which contradicts the fact that $\xi(\varphi(a) - ra) \notin G$, and hence does not belong to $G_1'$. $\square$

It is useful to note the information that we need to construct $G_1$ as above. We need to know $\varphi(a_\alpha)$ for an unbounded set of $\alpha$ and if $r$ is as in construction (3), we need to know some $a$ and $\xi \in \hat{Z}$ so that $\xi(\varphi(a) - ra) \notin G$. To be precise, the proof of 3.3 proves the following. (Here, $\langle X \rangle$ denotes the $R$-*module* generated by $X$.)

**3.4 Corollary.** *Suppose $G$ and $C$ are as in the statement of Lemma 3.3. Further suppose that $f$ is a partial function from $G$ to $G$ sat-*

*isfying one of hypotheses* (1), (2) *or* (3). *Then there are elements* $z_n \in \hat{C}$ *such that if* $G_1 = \langle G \cup \{z_n : n \in \omega\} \rangle$ *then:* $G_1$ *is cotorsion-free as an abelian group;* $G = \{b \in G_1 : \text{supp}(b) \text{ is bounded}\};$ $C$ *is dense in* $G_1$; *and there is an element* $y \in \hat{G}_1 \setminus G_1$ *so that if* $H \supseteq G_1$ *is reduced and torsion-free and* $\varphi \in \text{End}_{\mathbf{Z}}(H)$ *extends* $f$ *then* $\varphi(z_0) = y$. $\square$

It is actually this Corollary which serves as the killing lemma which makes it possible to prove the following theorem about realization of rings as endomorphism rings.

**3.5 Theorem.** *Suppose* $R$ *is a cotorsion-free ring. Suppose* $\mu^{\aleph_0} = \mu$ *and* $\lambda = \mu^+ \geq |R|$. *Then there is a reduced torsion-free cotorsion-free group* $A$ *of cardinality* $\lambda$ *so that* $\text{End}(A) \cong R$.

PROOF. To begin, we let $R$ be a ring structure on some cardinal $\leq \lambda$. Let $C = \bigoplus_{\alpha \in \text{succ}(\lambda)} Ra_\alpha$ and identify $\hat{C}$ with $\lambda$ in such a way that if $\alpha < \beta$ then the ordinal representing $a_\alpha$ is less than the ordinal representing $a_\beta$. Let $Y = \{a_\alpha : \alpha < \lambda\} (\subseteq \lambda)$. Let $\kappa$ be any cardinal $> \lambda^{++}$. The initial structure for the Black Box is $N = (\text{H}(\kappa), \in, <, \lambda, R, \hat{C}, Y)$. Here $R$ denotes the ring structure and $\hat{C}$ the $R$-module structure. Note that $C$ is definable in $N$. We identify $\lambda \times \lambda$ with $\lambda$ (as in the proof of 1.4) and so each $X \subseteq \lambda$ is identified with a subset of $\hat{C} \times \hat{C}$. Suppose $M \prec N$. For all $\alpha < \lambda$, $\alpha \in M$ if and only if $a_\alpha \in M$. Furthermore, $R_M = R \cap M$ is a ring structure, which we denote $R_M$. (This notation has nothing to do with localization!). Also, $\hat{C} \cap M$ is an $R_M$-module contained in the $\mathbf{Z}$-adic completion of $\bigoplus_{\alpha \in M \cap \text{succ}(\lambda)} R_M a_\alpha$ and $\hat{C} \cap M$ contains $\bigoplus_{\alpha \in M \cap \text{succ}(\lambda)} R_M a_\alpha$.

We will construct $A_\alpha \supseteq \bigoplus_{\beta \in \text{succ}(\alpha)} Ra_\beta$ by induction on $\alpha < \lambda$. For all $\alpha$, $A_\alpha$ will be contained in the completion of $\bigoplus_{\beta \in \text{succ}(\alpha)} Ra_\beta$. We will verify inductively that for all $\alpha < \beta$ and $a \in A_\beta$, if $\text{supp}(a) \subseteq \alpha$ then $a \in A_{\alpha+1}$. In particular this will imply that $A_{\alpha+1}$ is closed in $A$. Let $(M_i, X_i)$ be as in the statement of the Black Box with respect to $N$ and the stationary set $\{\delta < \lambda : \text{cf}(\delta) = \omega\}$. The key case in the construction occurs in the construction of $A_{\delta+1}$ when

for some $i$, $\delta = \delta(i) = \sup M_i \cap \lambda$, $X_i$ is the graph of an
(*) endomorphism of $A_\delta \cap M_i$ which is not multiplication
by an element of $R_{M_i}$.

In the other cases we let $A_{\delta+1} = A_\delta \oplus Ra_\delta$, for successor ordinals $\delta$, $A_{\delta+1} = A_\delta$, for limit $\delta$ not satisfying (*) and take unions at limit ordinals.

We turn to the key case. Let $M_i$ be as in (*). Let $\varphi_i$ denote the endomorphism coded by $X_i$. Since $\varphi_i$ and $A_\delta \cap M_i$ satisfy the hypotheses of Lemma 3.3 (as an $R_{M_i}$-module), there exists an increasing sequence $\{\alpha_{ni}: n \in \omega\} \subseteq M_i$ with limit $\delta$ and — in case (3) — $a$ satisfying one of the hypotheses (1), (2), or (3) as an $R_{M_i}$-module. To apply lemma 3.4, we need to know that the sequence satisfies (1), (2), or (3) as $R$-modules. If the sequence and — in case (3) — the element $a$ satisfy (1) or (3) with respect to $R_{M_i}$ then it is clear from the definition of the cases that they also satisfy (1) or (3) with respect to $R$. Suppose now that $\{\alpha_{ni}: n \in \omega\}$ and $\beta$ satisfy (2) with respect to $R_{M_i}$. To prove the sequence satisfies (2) with respect to $R$, note first that:

for all $\alpha \in M_i \cap \lambda$ and $r \in R$, if the coefficient of $a_\alpha$ in $\varphi_i(a_\alpha) = r$ then $r \in M_i$.

To see this, let $b = \varphi_i(a_\alpha)$. The binary function, $g$, which associates to each $a \in \hat{C}$ and $\alpha < \lambda$ the (unique) $r$ so that the coefficient of $a_\alpha$ in $a$ is $r$ is a definable element of $N$. Since $M_i \prec N$, $g \in M_i$. As $\alpha$, $b \in M_i$, $g(b, \alpha) = r \in M_i$.

Suppose that there is some $r \in R$ which is a counterexample to the statement that the sequence satisfies (2), i.e., for infinitely many $n$, $\varphi_i(a_{a_{ni}}) = ra_{a_{ni}} + b_n$ for some $b_n$ with support bounded below $\beta$. For all but finitely many $n$, $\alpha_{ni} > \beta$. Hence there is $n$, so that $r$ is the coefficient of $a_{\alpha_{ni}}$ in $\varphi_i(a_{\alpha_{ni}})$. But then $r \in M_i$.

Let $G_i$ be the closure in $A_\delta$ of $\bigoplus_{\alpha \in M_i \cap \lambda} R_{M_i} a_\alpha$. Then by 3.4 (with $G = G_i$) there exists $G_{1,i} = \langle G_i \cup \{z_{n,i}: n \in \omega\}\rangle$ such that $\varphi(z_{0,i}) \in \hat{G}_{1,i} \setminus G_{i,i}$ for any homomorphism $\varphi$ extending $\varphi_i$. Let $A_{\delta+1} = (A_\delta + \langle G_{1,i}: i \text{ satisfies } (*)\rangle) \oplus Ra_\delta$. As in the proof of 3.3

we can show that $A_{\delta+1}$ is cotorsion-free, and hence for all $\alpha$, $A_\alpha$ is cotorsion-free.

Let $A = \bigcup_{\alpha<\lambda} A_\alpha$. Suppose now that $\varphi \in \mathrm{End}(A)$ and $\varphi$ is not multiplication by an element of $R$. Let $X \subseteq \lambda$ be such that $X$, under the identification of $\lambda$ with $\lambda \times \lambda$, is the graph of $\varphi$. Consider $(M_i, X_i) \equiv_{M_i \cap \lambda} (N, X)$, where $\delta = \delta(i)$ is such that $\varphi[A_\delta] \subseteq A_\delta$. (Since the set of $\delta$ with this last property is closed and unbounded, the Black Box implies that there is such an $i$.) Since $A_{\delta+1}$ is closed in $A$ and contains the closure of $A_\delta$, $\varphi$ maps the closure of $A_\delta$ into $A_{\delta+1}$. Now by property (b) of the Black Box and an argument on supports, $\hat{G}_{1,i} \cap A_{\delta+1} = G_{1,i}$. Hence, by construction, $\varphi(z_{0,i}) \notin A_{\delta+1}$, which is a contradiction. $\square$

In particular, there are arbitrarily large reduced torsion-free cotorsion-free groups $A$ so that $\mathrm{End}(A) \cong R$. The groups constructed are in fact $\aleph_1$-free as $R$-modules. Not surprisingly, if we assume $V = L$, then the groups can be constructed so that they are strongly $\lambda$-free as $R$-modules.

## §4. Endomorphism rings of separable groups

In this section we deal with the realization of rings as endomorphism rings of separable or $\aleph_1$-separable groups. Since a separable group has many summands, and consequently there are many projections which belong to the endomorphism ring, we cannot realize $\mathbb{Z}$ as the endomorphism ring of a separable group. However, we can realize $\mathbb{Z}$, and other rings, as the quotient of the endomorphism ring by an ideal of "inevitable" endomorphisms.

**4.1 Definition.** For any group $A$, let $E_f(A)$ (respectively $E_c(A)$) denote the ideal of $\mathrm{End}(A)$ consisting of all elements of $\mathrm{End}(A)$ whose image is finitely-generated (resp. countably generated).

In fact, these ideals determine the group: if $A$ and $B$ are separable groups such that $E_f(A) \cong E_f(B)$ (as rings) then $A \cong B$ (see Exercise 3).

When we concern ourselves with separable groups, we shall realize the given ring $R$ as $\mathrm{End}(A)/E_f(A)$; in fact, we get something stronger, namely that $\mathrm{End}(A)$ is a *split extension* of $R$ by $E_f(A)$ which means that there are ring homomorphisms

$$R \xrightarrow{\iota} \mathrm{End}(A) \xrightarrow{\rho} R$$

such that $\rho \circ \iota = 1_R$ and $\ker(\rho) = E_f(A)$. Notice that then

$$\mathrm{End}(A) \cong E_f(A) \oplus R$$

*as abelian groups.* Similarly, when we deal with $\aleph_1$-separable groups, we construct $A$ so that $\mathrm{End}(A)$ is a split extension of $R$ by $E_c(A)$. For the former case we use the Black Box, but for the latter we will need the diamond principle.

**4.2 Definition.** If $C = \bigoplus_{i \in I} \mathbf{Z}a_i$ is a free group let $\bar{C}$ be the intersection of $\hat{C}$ (the $\mathbf{Z}$-adic completion of $C$) with $\prod_{i \in I} \mathbf{Z}a_i$; i.e., $\bar{C} = \{x \in \prod_{i \in I} \mathbf{Z}a_i \colon \text{for all } n > 0 \ \ n! \text{ divides } x(i) \text{ for all but finitely many } i\}$. (The definition of $\bar{C}$ depends on the choice of a basis for $C$, but this choice will always be clear from the context.) $\bar{C}$ is separable since it is a pure subgroup of $\prod_{i \in I} \mathbf{Z}a_i$ (cf. IV.2.6).

As an *ad hoc* definition, we shall say that a ring $R$ is *separably realizable* if:

$R$ is a ring with free additive group $R^+ = \bigoplus_{j \in J} \mathbf{Z}e_j$ such that $\bar{R}$ is an $R$-submodule of $\hat{R}$.

This definition of separably realizable arises from the facts that, first, we want to construct a group $A$ (realizing $R$) to be an $R$-module, and, second, we want to construct $A$ as a pure subgroup of a product of copies of $\mathbf{Z}$, so that it will be separable. Note that there are endomorphism rings of separable groups which are not separably realizable; for example $\mathrm{End}(\mathbf{Z}^\omega)$ contains the ring $\mathbf{Z}^\omega$. However, the endomorphism ring of a separable group is always separable as a group. In particular, a countable endomorphism ring of a separable group is free. The definition of separably realizable was given by Dugas and Göbel, who proved the following theorem using a different version of the Black Box.

**4.3 Theorem.** *Suppose $R$ is a separably realizable ring. Suppose $\mu^{\aleph_0} = \aleph_0$ and $\lambda = \mu^+ \geq |R|$. Then there is a separable group $A$ of cardinality $\lambda$ so that $\text{End}(A)$ is a split extension of $R$ by $E_f(A)$.*

We will delay the proof of this theorem and first prove an easier realization theorem. Note however the following consequence.

**4.4 Corollary.** *There are arbitrarily large separable groups $A$ with the property that $A$ is not the direct sum of two non-finitely-generated subgroups.*

PROOF. If $\text{End}(A)/E_f(A) \cong \mathbf{Z}$ then $A$ has the desired property: indeed otherwise, if $A = B \oplus C$ where $B$ and $C$ are not finitely-generated, then the projection on $B$ along $C$ is not equivalent mod $E_f(A)$ to multiplication by an element of $\mathbf{Z}$. $\square$

Other special separable groups can be constructed using Theorem 4.3 and the methods of Corner: see, for example, Corner-Göbel 1985.

It follows from VIII.2.3 that, assuming MA $+ \neg$ CH, if $A$ is $\aleph_1$-separable of cardinality $\aleph_1$, then $\text{End}(A)/E_c(A)$ cannot be isomorphic to $\mathbf{Z}$ (for example). So we cannot hope to prove in ZFC a theorem analogous to 4.3 about $\aleph_1$-separable groups. But we can do so, assuming V = L, at least for countable rings:

**4.5 Theorem.** ($\diamondsuit$) *Suppose $R$ is a ring whose additive group is a countable free group. Then there is an $\aleph_1$-separable group $A$ of cardinality $\aleph_1$ such that $\text{End}(A)$ is a split extension of $R$ by $E_c(A)$.*

PROOF. Choose free generators $e_i (i \in I)$ for $R^+$ (as a $\mathbf{Z}$-module). The group will be constructed between

$$C \stackrel{\text{def}}{=} \bigoplus_{\alpha \in \text{succ}(\omega_1)} Ra_\alpha = \bigoplus_{\alpha \in \text{succ}(\omega_1)} \bigoplus_{i \in I} \mathbf{Z}e_{i\alpha} \quad \text{and}$$
$$\hat{C} \cap \prod_{\alpha \in \text{succ}(\omega_1)} Ra_\alpha.$$

(Of course $Ra_\alpha = \bigoplus_{i \in I} \mathbf{Z}e_{i\alpha}$ and we identify $a_\alpha$ with $e_{0\alpha}$.) We give $\hat{C}$ the obvious structure as an $R$-module. For limit ordinals $\delta$ we will choose a ladder $\eta_\delta: \omega \to \delta$ and a sequence of natural numbers

$d_n$ such that $n! | d_n$. We will then let $z_\delta = \sum_{n \in \omega} d_n b_n$, where $b_n \in Ra_{\eta_\delta(n)}$. The group $A$ will be the pure closure (in $\hat{C}$) of the $R$-module generated by $C \cup \{z_\delta : \delta \in \lim(\omega_1)\}$. If we do this, then, since $R$ is countable, $A$ will be $\aleph_1$-separable.

We will define $z_\delta$ by induction and let $A_\delta$ be the pure closure of the $R$-module generated by $\bigoplus_{\alpha \in \mathrm{succ}(\delta)} \bigoplus_{n \in \omega} \mathbf{Z} e_{n\alpha} \cup \{z_\beta : \beta < \delta\}$. There is no harm in assuming that as a set $A_\delta = \delta \times \omega$. More exactly we will inductively define a group structure on $\delta \times \omega$ by choosing a one-one map to a subgroup of $\hat{C}$. $\Diamond$ gives us a family of functions $f_\delta : A_\delta \to A_\delta$ with the usual prediction property (cf. VI.1.2). The key case occurs when $f_\delta$ is an endomorphism from $A_\delta$ to $A_\delta$ so that $f_\delta$ is not multiplication by $r \bmod A_{\alpha+1}$ for any $\alpha < \delta$. By the killing lemma, 4.6, which follows this proof, there is $\eta_\delta$ and sequences $d_n$ $(n \in \omega)$, $b_n$ $(n \in \omega)$ as above so that $f_\delta$ does not extend to an endomorphism of the pure closure of the $R$-module generated by $A_\delta \cup \{\sum_{n \in \omega} d_n b_n\}$.

Otherwise, i.e., if we are not in the key case, let $\eta_\delta$, $d_n(n \in \omega)$, $b_n$ $(n \in \omega)$ be arbitrary.

This completes the inductive construction. Suppose that there exists $\varphi \in \mathrm{End}(A)$ such that for every $r \in R$, the image of $r - \varphi$ is not countably generated.

Since $R$ is countable, the set

$$C \overset{\text{def}}{=} \{\delta \in \omega_1 : \varphi[A_\delta] \subseteq A_\delta \text{ and for all } \beta < \delta \text{ and } r \in R,$$
$$\varphi[A_\beta] \subseteq A_\alpha \text{ for some } \alpha < \delta \text{ and there exists}$$
$$a \in A_\delta \setminus A_\beta \text{ such that } (r - \varphi)(a) \notin A_\beta\}$$

is easily seen to be a cub. By the properties of the $\Diamond$ sequence, there is $\delta \in C$ so that $\varphi {\restriction} A_\delta = f_\delta$. Since the $\mathbf{Z}$-adic closure of $A_\delta$ (in $A$) is $A_{\delta+1}$, $\varphi {\restriction} A_{\delta+1}$ is an endomorphism of $A_{\delta+1}$. But this contradicts the construction. Thus we are done with the proof of 4.5, except for the killing lemma. $\square$

We will continue to use the notation above. By the *support* of an element of $\hat{C}$ we mean the set of pairs $(i, \beta)$ such that the projection of the element onto $\hat{\mathbf{Z}} e_{i,\beta}$ is non-zero. When we say that $\mathrm{supp}(g)$

is bounded by $\beta$, we will mean that if $(i, \alpha)$ is in the support of $g$, then $\alpha < \beta$. If $G$ is a subgroup of $\hat{C}$, let $G_\beta$ denote the subgroup of $G$ consisting of elements whose support is bounded by $\beta$. In the following, "bounded" means bounded by $\beta$ for some $\beta < \delta$.

**4.6 Lemma.** *Let $R$ be as in 4.5. Suppose $\delta$ is a countable limit ordinal, $C_\delta = \bigoplus_{\alpha \in \text{succ}(\delta)} Ra_\alpha = \bigoplus_{\alpha \in \text{succ}(\delta)} \bigoplus_{n \in \omega} \mathbf{Z}e_{n\alpha} \subseteq G \subseteq \bar{C}_\delta$, and $G$ is a pure subgroup of $\bar{C}_\delta$ such that every element of $G$ has bounded support. Suppose $\varphi \in \text{End}_{\mathbf{Z}}(G)$ such that for all $\beta < \delta$, $\varphi[G_\beta]$ is bounded, and for all $r \in R, (r - \varphi)[G]$ is unbounded. Then there is a ladder $\eta_\delta$ on $\delta$, elements $b_n \in Ra_{\eta_\delta(n)}$, and a sequence $d_n$ $(n \in \omega)$ of non-zero integers so that $n!|d_n$ and $\varphi$ does not extend to an endomorphism of $\langle G + Rz_0 \rangle_*$, where $z_0 = \sum_{n \in \omega} d_n b_n$.*

PROOF. The proof of this lemma is much like the proof of 3.3, but simpler. We will list three hypotheses and under each hypothesis give a construction of $G_1$.

**Hypothesis (1):** For all $\beta < \delta$ there is $\alpha > \beta$ and $b_\alpha \in Ra_\alpha$ so that $\varphi(b_\alpha) \notin Ra_\alpha + G_\beta$.

**Construction (1):** Let $\alpha_n$ $(n \in \omega)$ be an increasing sequence of ordinals cofinal in $\delta$. Let $\beta_0 = \alpha_0$. Suppose $b_{n-1}$ and $\beta_n$ have been defined; choose $\eta_\delta(n) > \beta_n$ and $b_n \in Ra_{\eta_\delta(n)}$ so that the support of $\varphi(b_n)$ contains some $(j, \gamma)$ where $\gamma \geq \beta_n$ and $\gamma \neq \eta_\delta(n)$.

Next choose a successor ordinal $\beta_{n+1} > \alpha_{n+1}$ so that the support of $\varphi(b_n)$ is bounded by $\beta_{n+1}$. For simplicity let $G_n$ denote $G_{\beta_n}$.

Let $d_0 = 1$. In general, if $d_n$ has been found choose $d_n|d_{n+1}$ so that $(n + 1)!|d_{n+1}$ and $d_{n+1}$ does not divide $n!d_n\varphi(b_n)$ (mod $G_n + Ra_{\eta_\delta(n)}$).

It remains to verify that the sequence has the desired property. Suppose not; then for some $m > 0, r \in R$ and $\beta_k, m\varphi(z_0) = rz_0 + b$ where $\text{supp}(b)$ is bounded by $\beta_k$. Notice that

$$m\varphi(z_0) = \sum_{n \in \omega} md_n\varphi(b_n).$$

Choose $n > k, m$. Working mod $G_n + Ra_{\eta_\delta(n)}$ and modulo $d_{n+1}$, we have $md_n\varphi(b_n)$ is equivalent to 0, which contradicts the choice of $d_{n+1}$.

**Hypothesis (2):** there is an ordinal $\beta$ and a ladder $\eta_\delta$ on $\delta$ so that for all $n$, $\varphi(a_{\eta_\delta(n)}) \in Ra_{\eta_\delta(n)} + G_\beta$, but for all $r \in R$, $\{n: \varphi(a_{\eta_\delta(n)}) \neq ra_{\eta_\delta(n)} \bmod G_\beta\}$ is infinite.

**Construction (2):** Let $b_n = a_{\eta_\delta(n)}$ and $d_n = n!$. If $\varphi$ extends to an endomorphism of $\langle G + Rz_0 \rangle_*$, then arguing as in Case 1 we find $m \neq 0$, $r$ and $\gamma$ so that for all $n$ such that $\eta_\delta(n) > \gamma$, $m\varphi(a_{\eta_\delta(n)}) = ra_{\eta_\delta(n)} \bmod G_\beta$. But then $m$ divides $r$ and we have contradicted the choice of $\eta_\delta$.

**Hypothesis (3):** there is an ordinal $\beta$ so that there exists $r_0 \in R$ such that for all $\alpha > \beta$ and all $b \in Ra_\alpha$, $\varphi(b) \in Ra_\alpha + G_\beta$, and $\varphi(a_\alpha) = r_0 a_\alpha \bmod G_\beta$.

**Construction (3):** In this case choose a ladder $\eta_\delta$ with $\eta_\delta(0) > \beta$ so that we can choose $b_n = a_{\eta_\delta(n)}$ for $n$ even, and $b_n \in Ra_{\eta_\delta(n)}$ with $\varphi(b_n) \neq r_0 b_n \bmod G_\beta$ for $n$ odd. (Such a choice is possible by the hypothesis on $\varphi$.) Now if $\varphi$ extends to an endomorphism and $m$ and $r$ are as above, then looking at the even terms we can deduce that $mr_0 = r$, i.e., $\varphi(z_0) = r_0 z_0 \bmod G_\gamma$ for some $\gamma < \delta$. But then for some odd $n$ we have $\varphi(b_n) = r_0 b_n + G_\beta$, a contradiction. $\square$

In the proof of 4.6, in the definition of $z_0$, we could have chosen $b_{2n} = a_{\eta_\delta(2n)}$ in all three cases. This refinement will be used in the proof of Lemma 4.9.

The following corollary to 4.5, whose proof is analogous to that of 4.4, shows that Theorem VIII.2.3 is not a theorem of ZFC.

**4.7 Corollary.** ($\Diamond$) *There is an $\aleph_1$-separable group $A$ of cardinality $\aleph_1$ with the property that $A$ is not the direct sum of two uncountable subgroups.* $\square$

We now turn to the proof of Theorem 4.3. Before stating the killing lemma we will state an easy characterization of the $\aleph_1$-free reduced slender groups. The proof, which we omit, is just an appeal to Nunke's characterization of the slender groups (IX.2.4); the hypothesis that the group be $\aleph_1$-free is much stronger than needed for the proposition.

**4.8 Proposition.** *Suppose $G$ is an $\aleph_1$-free reduced group and $I$ is a coinfinite subset of $\omega$. $G$ is slender if and only if for any sequence of non-zero elements $\{a_n : n \in \omega\} \subseteq G$ there are integers $\{d_n : n \in \omega\}$ such that: for all $n$, $n! | d_n$; if $n \in I$, $n! = d_n$; and $\sum_{n \in \omega} d_n a_n \notin G$ (the sum is taken in $\hat{G}$).* □*

The key to the proof of Theorem 4.3 is the following killing lemma. The rather involved hypotheses summarize the situation that is met in the proof using the Black Box.

**4.9 Lemma.** *Let $R$ be a separably realizable ring with $R^+ = \bigoplus_{j \in J} \mathbf{Z} e_j$. Let $C = \bigoplus_{\alpha < \delta} R a_\alpha = \bigoplus_{\alpha < \delta} \bigoplus_j \mathbf{Z} e_{j,\alpha}$ where $\delta$ is an ordinal of cofinality $\omega$ and let $G$ be a pure slender subgroup of $\bar{C}$ which is an $R$-submodule of $\bar{C}$ and contains $C$. Suppose that every element of $G$ has bounded support. Suppose $G'$ is a subgroup of $G$ so that:*
*(1) $G'$ is generated as a group by $\{e_{j\alpha} : e_{j\alpha} \in G'\}$;*
*(2) $\{\alpha : a_\alpha \in G'\}$ is unbounded in $\delta$;*
*(3) for all $\alpha$ and $j$, if $e_{j\alpha} \in G'$, then $a_\alpha \in G'$.*
*Suppose $f$ is a function from $G'$ to $G$ such that for all $r \in R$ there are infinitely many $e_{j\alpha} \in G'$ such that $f(e_{j\alpha}) \neq r e_{j\alpha}$. Then there exists $z_0 \in \bar{C}$ such that $f$ does not extend to an endomorphism of $G_1 \stackrel{\text{def}}{=} \langle G, R z_0 \rangle_*$. Further, $G_1$ is slender and has no new elements of bounded support.*

PROOF. Note first that $G_1$ will be an $R$-submodule of $\bar{C}$ since $\bar{C} \subseteq \prod_{\alpha < \delta} \bar{R}_\alpha$ and $R$ is separably realizable. We will construct a ladder $\eta_\delta$ and choose a sequence of elements $b_n \in G$ and natural numbers $d_n$ ($n \in \omega$) so that $z_0 = \sum_{n \in \omega} d_n b_n$ and for all even $n$, the projection of $z_0$ on $\bar{R} a_{\eta_\delta(n)} = d_n a_{\eta_\delta(n)}$ and $d_n \neq 0$. There are two cases to consider.

*Case 1: there is some $\beta < \delta$ and $r \in R$ so that for all $\alpha > \beta$, $f(e_{j\alpha}) = r e_{j\alpha} \mod G_\beta$ for all $j$ such that $e_{j\alpha} \in G'$.* We define the ladder and the sequence $b_n$ inductively. If there is a bound $\gamma$ on $\{\alpha : $ for some $j$, $f(e_{j\alpha}) \neq r e_{j\alpha}\}$ then let $\eta_\delta(0) > \gamma$. Otherwise let $\eta_\delta(0)$ be arbitrary. Let $b_0 = a_{\eta_\delta(0)}$. For odd $n$, let $\eta_\delta(n) = \eta_\delta(n-1)+1$ and inductively choose $b_n = e_{j\alpha}$ so that $b_n \notin \{b_m : m < n\}$ and $f(b_n) \neq r b_n$ and $\alpha \neq \eta_\delta(m)$ for $m \leq n$. For even $n$, choose $\eta_\delta(n)$ larger than any of the ordinals used so far (and greater than the $n$th element in some fixed sequence cofinal in $\delta$); let $b_n = a_{\eta_\delta(n)}$. Now choose $d_n$ $(n \in \omega)$ so that $n! | d_n$, $(2n)! = d_{2n}$ and $\sum_{n \in \omega} d_n(f(b_n) - rb_n) \notin G$. (This is possible by 4.8.) Notice that the support of this infinite sum is bounded by $\beta$. Let $z_0 = \sum_{n \in \omega} d_n b_n$. Since any $e_{j\alpha}$ appears in the support of at most one of the $b_n$, $z_0 \in \bar{C}$. To verify that no new elements of bounded support are added, notice that for any $r' \neq 0$, the support of $r'z_0$ contains $\eta_\delta[\{2n : n \in \omega\}]$. Since no elements of bounded support are added, $f$ cannot be extended to an endomorphism.

*Case 2: not Case 1.* The construction can be done as in Lemma 4.6, except that we take care to make sure that for every even $n$, $b_n = a_\alpha$, for some $\alpha$.

It remains to prove that $G_1$ is slender. Note first that $G$ is $\aleph_1$-free. Suppose that $\{c_n : n \in \omega\}$ is a subset of $G_1$ such that for all $k_n$ $(n \in \omega)$ so that $n! | k_n$, $\sum_{n \in \omega} k_n c_n \in G_1$. Since $G_1$ adds no elements of bounded support and $G$ is slender, there is no bound on the supports of any infinite subset of the $c_n$'s (see 4.8). Recall that any element of $G_1$ of unbounded support has a non-zero projection on all but finitely many of the $\bar{R}a_{\eta_\delta(2n)}$; in fact the projection will be $ra_{\eta_\delta(2n)}$ for some $r \in R$. So there are increasing sequences of natural numbers $n_p$ and even natural numbers $m_p (p \in \omega)$ so that the projection of $c_{n_p}$ on $\bar{R}a_{\eta_\delta(m_p)}$ is non-zero. Let $s_{np} \in \bar{R}$ be such that the projection of $c_n$ on $\bar{R}a_{\eta_\delta(m_p)}$ equals $s_{np}a_{\eta_\delta(m_p)}$.

Consider now any sequence $k_n$ $(n \in \omega)$ so that for all $n$, $n! | k_n$. Choose $r$, $b$ and $m \neq 0$, so that

$$m \sum_{n \in \omega} k_n c_n = rz_0 + b, \text{ where } b \in G.$$

Since $b$ is of bounded support, we have for almost all $p$,

$$d_{m_p}r = m \sum_{n \in \omega} k_n s_{np}.$$

Note that $r$ is uniquely determined by $m$ and the sequence of $k_n$'s. By varying the sequences of $k_n$'s we obtain $2^{\aleph_0}$ $r$'s (see Exercises 9 – 11). Hence there is a countable subset of $R$ — namely a set whose **Z**-adic closure contains $\{s_{np} : n \in \omega\}$ — whose **Z**-adic closure in $R$ is uncountable. This contradicts the fact that $R^+$ is free. $\square$

**Proof of 4.3.** Once we have the killing lemma, the proof follows closely along the lines of the proofs of 1.4 and 3.5, so we shall give only a brief sketch. Let $C = \bigoplus_{\alpha \in \mathrm{succ}(\lambda)} Ra_\alpha$ and make the identification of $\bar{C}$ with $\lambda$ as in the proof of 3.5. As in that proof, let $Y = \{a_\alpha : \alpha < \lambda\}$ and let $Z = \{e_{j\alpha} : j \in J, \alpha < \lambda\}$. Let $N = (\mathrm{H}(\kappa), \in, <, \lambda, R, \bar{C}, Y, Z)$, where $\kappa > \lambda^{++}$. We construct $A_\alpha \supseteq \bigoplus_{\beta \in \mathrm{succ}(\alpha)} Ra_\beta$ by induction on $\alpha$ so that it is a pure slender subgroup of

$$\overline{\bigoplus_{\beta \in \mathrm{succ}(\alpha)} Ra_\alpha}.$$

In the key case of the construction, if we are looking at $M_i$ and the endomorphism guessed is not equal to multiplication by a fixed $r$ for all but finitely many of the $e_{j\alpha} \in M_i$ then we can kill it using Lemma 4.8.

Let $A = \bigcup_{\alpha < \lambda} A_\alpha$. Since $A$ is a pure subgroup of $\bar{C}$, it is separable. By construction $A$ is an $R$-module, so multiplication by $r \in R$ determines an endomorphism of $A$ and therefore we have a canonical embedding, $\iota$, of $R$ into $\mathrm{End}(A)$. Suppose there is some $\varphi \in \mathrm{End}(A)$ such that for every $r \in R$, $\varphi(e_{j\alpha}) \neq re_{j\alpha}$ for infinitely many $e_{j\alpha}$. Then we can find $M_i$ where $A_{\delta(i)} \cap M_i = A \cap M_i$ and the restriction of $\varphi$ are correctly predicted. Furthermore

$$M_i \models \text{``for every } r, \varphi(e_{j\alpha}) \neq re_{j\alpha}, \text{ for infinitely many } e_{j\alpha}\text{''}.$$

To see that $\varphi$ was killed (i.e., that we could have applied Lemma 4.8), we must show that for every $r$, $\varphi(e_{j\alpha}) \neq re_{j\alpha}$, for infinitely many $e_{j\alpha} \in M_i$. The elementariness only gives us this for every $r \in M_i$. Suppose there is $r$ so that $\varphi(e_{j\alpha}) = re_{j\alpha}$ for all but finitely many $e_{j\alpha} \in M_i$. Then there is $a_\alpha \in M_i$, so that $\varphi(a_\alpha) = ra_\alpha$. Since $\varphi(a_\alpha) \in M_i$ and $r$ is uniquely determined by the equation, $r \in M_i$. Hence $\varphi$ was killed. $\square$

## EXERCISES

1. If $\varphi \in \mathrm{End}(A) \setminus Z$ and $A$ is torsion-free, then there exists $b \in A$ so that $\varphi(b) \neq kb$ for any $k \in Z$. [Hint: if $b, c \in A$ and $b$ and $c$ are linearly independent, consider $\varphi(b), \varphi(c)$ and $\varphi(b - c)$.]

2.  Prove that the following groups are *not* the additive groups of endomorphism rings of abelian groups: $\mathbf{Q} \oplus \mathbf{Q}$; $J_p \oplus J_p$; and $\mathbf{Z}/p\mathbf{Z} \oplus \mathbf{Z}/p\mathbf{Z}$. (Note that all these groups are the additive group of some ring.) [Hint: consider $\mathbf{Q} \oplus \mathbf{Q}$. The group would be a $\mathbf{Q}$ vector space. The endomorphism ring of an infinite rank vector space has infinite rank, as a vector space; while the endomorphism ring of a vector space of dimension $n$ — i.e., the $n \times n$ matrices — has dimension $n^2$.]

3. Suppose $A$ is a separable group and $g: A \to A$.
(a) $g$ is a projection if and only if $g^2 = g$ and $(1 - g)^2 = 1 - g$.
(b) $g$ is a projection onto a rank 1 summand if and only if it is a non-zero projection and if $g = h + f$ where $h$, $f$ are projections and $hf = fh = 0$ then either $g = f$ or $g = h$.
(c) Define an equivalence relation $\equiv$ on the rank 1 projections by $f \equiv g$ if $fg = g$. Show that this relation says $f$ and $g$ project onto the same rank 1 summand.
(d) Suppose for separable groups $A$ and $B$, $\varphi: \text{End}(A) \to \text{End}(B)$ is an isomorphism. Show that the following scheme defines an isomorphism $\psi$ from $A$ to $B$. Let $a$ be any element which generates a pure subgroup. Let $g$ be a projection onto $\langle a \rangle$. Choose $b$ a generator of the image of $\varphi(g)$. Let $\psi(a) = b$. For any $c \in A$, choose $h$ so that $h(a) = c$ and define $\psi(c) = \varphi(h)(b)$.

The next sequence of exercises characterizes in $L$ those rings which are endomorphism rings for $\kappa$-free groups with $\kappa$ arbitrarily large.

4. $\text{End}(\mathbf{Z}^{(\kappa)}, \mathbf{Z}^{(\kappa)}) \cong \text{End}(\mathbf{Z}, \mathbf{Z}^{(\kappa)})^{\kappa} \cong (\mathbf{Z}^{(\kappa)})^{\kappa}$. So if $F$ is a free group of infinite rank, then $(\text{End}(F), +)$ is not free.

5. If $R = \text{End}(A)$, $A$ is $\kappa^+$-free and $|R| \leq \kappa$, then $R$ is a subring of $\text{End}(\mathbf{Z}^{(\kappa)})$. (Hint: find a subgroup $B \subseteq A$ of cardinality $\kappa$ which is an $R$-module such that for all $r \in R$, if $r \neq 0$, then there is $b \in B$ so that $rb \neq 0$.)

6. Suppose $|R| \leq \kappa$ and $R$ is a subring of $\text{End}(\mathbf{Z}^{(\kappa)})$. Show there is an $R$-module $M = \bigoplus_{a \in A} Ra$ which is free as an abelian group (in particular there is no assumption that $Ra$ is a free $R$-module)

so that $R$ acts faithfully on $M$; $|A| \leq \kappa$; and for all $b \in M$ there is $a \in A$ so that $Rb \cong Ra$ and the isomophism is induced by taking $a$ to $b$. [Hint: To begin let $A_0 = \mathbf{Z}^{(\kappa)}$ and $M_0 = \bigoplus_{a \in A_0} Ra$. Define $A_n$ inductively so that the last clause will be satisfied.]

7. Suppose $M_n = \bigoplus_{a_n \in A_n} Ra_n$ is a copy of $M$ (i.e. $Ra_n$ is a copy of $Ra$ with $a_n$ corresponding to $a$). Suppose $N = (\bigoplus_{n \in \omega} M_n)$ is a free abelian group and $\varphi \in \mathrm{End}_{\mathbf{Z}}(N)$ is an endomorphism other than multiplication by an element of $R$. Show there is $a \in A$, $b \in \bigcup_{n \in \omega} M_n$, and an increasing sequence $k_n$ of natural numbers so that if $b \neq 0$ then $Ra \cong Rb$ via the map which takes $a$ to $b$ and $\varphi$ does not extend to the $R$-submodule $N_1$ of $(\widehat{\bigoplus_{n \in \omega} M_n})$ generated by $N$ together with $\{y_n : n \in \omega\}$ where $y_n = \sum_{n \leq m} \frac{m!}{n!}(a_{k_m} + b)$. Furthermore, if $b \in \bigcup_{n < k_0} M_n$, then $N_1$ is free. In fact $N_1$ is generated as module by $\{c \in \bigcup_{n \in \omega} A_n : \text{for all } n, c \neq a_{k_n}\} \cup \{y_n : n \in \omega\}$ and is isomorphic to $N$ via the map which takes $y_n$ to $a_{k_n}$ and is the identity on $\{c \in \bigcup_{n \in \omega} A_n : \text{for all } n, c \neq a_{k_n}\}$.

8. Assume V = L. For all subrings $R \subseteq \mathrm{End}(\mathbf{Z}^{(\kappa)})$ and regular non-weakly compact cardinals $\lambda > \kappa$, there is a strongly $\lambda$-free group $A$ so that $\mathrm{End}(A) \cong R$.

In the following exercises we use the notation of 4.9.

9. Suppose $k_j^0$, $k_j^1$ ($j < n_p$) are sequences of integers so that $j! | k_j^i$, $m^0$, $m^1$ are non-zero integers, and $\ell$ is a non-zero natural number. There are $k_p^0$, $k_p^1$ and a non-zero natural number $t$ so that $\ell$, $n_p! | k_{n_p}^i$, and if $k_j^i$ ($j > n_p$) are such that $t$, $j! | k_j^i$, then

$$m^0 \sum_{j \in \omega} k_j^0 s_{jp} \neq m^1 \sum_{j \in \omega} k_j^1 s_{jp}.$$

[Hint: since $s_{n_p p} \neq 0$, it is possible to choose $k_{n_p}^0$, $k_{n_p}^1$ and $t$ so that $t$ does not divide

$$m^0 \sum_{j \leq n_p} k_j^0 s_{jp} - m^1 \sum_{j \leq n_p} k_j^1 s_{jp};$$

in fact one of $k_{n_p}^0$, $k_{n_p}^1$ can be taken to be 0.]

10. Show that there are sequences $\{k_n^\sigma : n \in \omega, \sigma \in \omega^2\}$ so that for all $\sigma \neq \nu$, and $m^0$, $m^1 \neq 0$, there are infinitely many $p$ so that

$$m^0 \sum_{n \in \omega} k_n^\sigma s_{np} \neq m^1 \sum_{n \in \omega} k_n^\nu s_{np}.$$

[Hint: define the $k_n^\sigma$ so that there is a strictly increasing $f : \omega \to \omega$ such that if $\sigma \restriction n = \nu \restriction n$ then $k_m^\sigma = k_m^\nu$ for all $m \leq f(n)$; this value is denoted $k_m^{\sigma \restriction n}$ and is defined by induction on $n$.]

11. As is claimed in 4.9, by varying the $k_n$'s, we can obtain $2^{\aleph_0}$ $r$'s.

## NOTES

The Black Box is due to Shelah. The earliest forms can be found in Shelah 1978 Chapter VIII and later developments in Shelah 1984b, among other places. The most complete account of Black Boxes will appear in Shelah 19??a. The Black Box given here, Theorem 2.1 (also called 1.3), is due to Shelah and proved for the first time here.

Fuchs 1973, with help from Corner 1969b, proved the existence of endo-rigid groups in all cardinalities less than the first strongly inaccessible cardinal. Shelah 1974 constructed such groups in all cardinalities. Eklof-Mekler 1977 showed that V = L implies that that for all regular non-weakly compact $\kappa$, there is a strongly $\kappa$-free group which is indecomposable. This was generalized in Dugas 1981 to show that the groups could be chosen to be endo-rigid. Shelah 1981a showed that for $\kappa = \aleph_1$, the hypothesis could be weakened to $2^{\aleph_0} < 2^{\aleph_1}$. Theorem 1.4, or at least the existence of an $\aleph_1$-free indecomposable group, is attributed in Dugas-Göbel 1982a to an unpublished manuscript of Corner.

The study of which rings can be realized as endomorphism rings of abelian groups goes back to Corner 1963. In the present terminology, he showed that every countable cotorsion-free ring is the endomorphism ring of a countable torsion-free abelian group. Dugas-Göbel 1982a showed, assuming a consequence of V = L, that every cotorsion-free ring is the endomorphism ring of a cotorsion-free group. In Dugas-Göbel 1982b, the theorem was proved in ZFC, but the cardinalities of the groups constructed were certain strong limit cardinals. The construction in Dugas-Göbel 1982b uses a prediction principle from Shelah 1975b. The prediction principle used is closer to the Dushnik-Miller one than to $\Diamond$. In Shelah 1984b it is shown, using a version of the Black Box, that for every cotorsion-free ring, $R$, if $\mu > |R|$ and $\mu^{\aleph_0} = \mu$ then there is a cotorsion-free group $A$ so that $\text{End}(A) \cong R$. This result contains most of Theorem 3.5 and information about other cardinals.

This chapter is only an introduction to the extensive subject of endomorphism rings of modules. Corner-Göbel 1985 contains a comprehensive account of the use of the Black Box (a different version) to obtain results about realizing endomorphism rings in many settings.

Theorem 4.3 is due to Dugas-Göbel 1985b. Theorem 4.5 is from Thomé 1988. Corollary 4.7 is from Eklof 1983.

Exercises. 2:Corner 1963; 3:Wolfson 1963

# CHAPTER XIV
# DUAL GROUPS

In this chapter we investigate dual groups in depth; by a *dual group* we mean a group of the form $\text{Hom}(A, \mathbb{Z})$, denoted $A^*$. Among the dual groups are, of course, the reflexive groups. In chapters III, $X$ and XI we have encountered some dual groups, notably those in the Reid class, and $C(\mathbb{Q}, \mathbb{Z})$ and its dual. We have seen that there are dual groups which are not in the Reid class, and that there are non-reflexive dual groups. Here we will give many more examples of dual groups and give some general classes of dual groups.

We begin in section 1 with some basic concepts and problems. In section 2 we define a general class of groups, the tree groups, such that every group in the class is reflexive; we show that this class includes many $\aleph_1$-separable groups in standard form. In section 3 we give some sufficient conditions for a group to be a dual group. Sections 4 and 5 construct many non-reflexive dual groups; the constructions in section 4 are done in ZFC; those in section 5 require some additional set-theoretic assumptions. Sections 2, 3 and 4 may be read independently, but section 5 uses sections 2 and 3.

In this chapter we will often assume during the discussion, without repeatedly stating it, that all the cardinals considered are less than the first measurable cardinal (if it exists). However the theorem statements will include, where necessary, the hypothesis that the cardinals are not $\omega$-measurable.

## §1. Invariants of dual groups

Much of the work on the structure of dual groups till now has been an investigation of the extent to which the structure of the class of dual groups resembles the Reid class. Thus the question of whether or not every dual group is reflexive has been central to the study of dual groups. In later sections we will see that there are many non-reflexive dual groups. Once we know that there are dual groups which are non-reflexive, we are free to ask more insightful questions. In this section we will make some elementary

observations about dual groups and ask many questions. Much of
the rest of this chapter is devoted to answering those questions to
which the answer is known. However, many questions remain open.

Given a dual group $A$, there may be many groups $B$ such that
$B^* \cong A$. For example, both the $\mathbb{Z}$-adic closure of $\mathbb{Z}^{(\omega)}$ in $\mathbb{Z}^\omega$ and $\mathbb{Z}^\omega$
have $\mathbb{Z}^{(\omega)}$ as their duals. Indeed, there will be many subgroups of
$\mathbb{Z}^\omega$ with the same dual as $\mathbb{Z}^\omega$.

**1.1 Definition.** A group $B \subseteq A^*$ is a *predual* of $A$ if $A = B^*$, or,
more precisely, the map $\varphi_B$ from $A$ to $B^*$ is an isomorphism, where
$\varphi_B$ is defined by $\varphi_B(a) = \sigma(a) \restriction B$, i.e., for all $b \in B$, $\varphi_B(a)(b) =$
$\langle a, b \rangle$. We say that $B$ is a *maximal predual* of $A$ if it is maximal
under containment among the preduals of $A$. If $B$ is a torsionless
group then we will say that $B$ is a maximal predual if its image in
$B^{**}$ is a maximal predual of $B^*$.

Note that if $A$ is identified with $B^*$ via $\varphi_B$, then $\sigma_B[B]$ is identified with $B$.

**1.2 Proposition.** *If $A$ is a dual group, then any predual of $A$ is
contained in a maximal predual.*

PROOF. In order to apply Zorn's Lemma it enough to verify that
the union of an increasing chain of preduals is also a predual. Sup-
pose $B = \bigcup_{i \in I} B_i$ where each $B_i$ is a predual of $A$. Clearly $\varphi_B$ is
one-one. To see that $\varphi_B$ is onto consider $f \in B^*$. Fix $i_0 \in I$ and
choose $a$ so that $f \restriction B_{i_0} = \sigma(a) \restriction B_{i_0}$; this is possible since $B_{i_0}$ is a
predual of $A$. It suffices to show that for all $j > i_0$, $f \restriction B_j = \sigma(a) \restriction B_j$.
Otherwise, there is $a \neq c \in A$, so that $f \restriction B_j = \sigma(c) \restriction B_j$. But then
$\sigma(c - a) \restriction B_{i_0} = 0$. This last statement contradicts the fact that $B_{i_0}$
is a predual of A. □

This proposition leads to some natural questions. Are all maximal preduals of a dual group isomorphic? If one predual is a dual
group must all maximal preduals be dual groups? Does every dual
group possess a predual which is itself a dual group? At the moment we don't know the answer to the second of these questions; we
will answer the other two in the negative in section 4. It is worth

noting that all these questions are trivial if we add the assumption that the group is reflexive: the only maximal predual of a reflexive group $A$ is the group $A^*$. In fact, $A^*$ is a predual of $A$ if and only if $A$ is reflexive.

Note that a maximal predual of $A$ is a pure subgroup of $A^*$. The converse is false, as seen by the example of the $\mathbf{Z}$-adic closure of $\mathbf{Z}^{(\omega)}$. However, if a predual of $A$ is a direct summand of $A^*$, then it is a maximal predual, as the next proposition shows. Also, if a predual is a dual group in its own right, then it will be a maximal predual.

**1.3 Proposition.** (i) *If $B$ is a predual of $A$ which is a direct summand of $A^*$, then $B$ is a maximal predual of $A$.*

(ii) *Any predual of $A$ which is itself a dual group is a maximal predual of $A$.*

PROOF. (i) Suppose $A^* = B \oplus D$. It suffices to prove that for any $C$ such that $B \subseteq C \subseteq A^*$, $\varphi_C : A \to C^*$ is not onto if $C \neq B$. But there clearly exists $y \in A^{**}$ such that $y{\upharpoonright}B \equiv 0$ and $y{\upharpoonright}C \not\equiv 0$; then $y{\upharpoonright}C$ does not belong to the image of $\varphi_C$ since $\varphi_B$ is one-one.

(ii) It suffices to show that if $B$ is a predual of $A$ and $B = G^*$ then $B$ is a direct summand of $A^*$. Consider the map $\rho$ from $A^*$ to $B$ defined by: $\rho(y)(g) = <y, a>$ if $y \in A^*$ and $a$ is such that $\varphi_B(a) = \sigma_G(g) \in B^* = G^{**}$. It is easy to check that $\rho$ is a splitting of the inclusion of $B$ into $A^*$. $\square$

**1.4 Proposition.** *If $A$ is a dual group, then $A$ has a unique maximal predual if and only if $A$ is reflexive.*

PROOF. If $A$ is reflexive then $A^*$ is a predual of $A$ and hence the unique maximal predual. Assume $A$ is not reflexive and choose $B \subseteq A^*$ a maximal predual of $A$. Choose $f \in A^* \backslash B$ which generates a pure subgroup of $A^*$. Let $A_0 \subseteq A$ be the kernel of $f$. Since $f$ maps $A$ onto $\mathbf{Z}$, there is an element $a \in A$ so that $A = A_0 \oplus \langle a \rangle$ and $\langle a, f \rangle = 1$. As $\varphi_B(a)$ generates a pure subgroup of $B^*$, there is $g \in B$ so that $\langle a, g \rangle = 1$ and $B = B_0 \oplus \langle g \rangle$ where $B_0$ is the kernel of $\varphi_B(a)$. Let $B' = B_0 \oplus \langle f \rangle$.

We claim that $B'$ is a predual of $A$. First we must check that $A$ acts faithfully on $B'$, i.e., $\varphi_{B'}$ is one-one. Note that for any $0 \neq x \in A_0$ there is $y \in B_0$ so that $\langle x, y \rangle \neq 0$; otherwise, $x - \langle x, g \rangle a$ would be a non-zero element of $A$ which acts trivially on $B$. So for all $x \in A_0$ and $k \in \mathbf{Z}$, if $\varphi_{B'}(x + ka) = 0$, then $x = 0$ and $k = 0$. To see that $\varphi_{B'}$ is surjective, consider $\psi \colon B' \to \mathbf{Z}$. Choose $x \in A$ so that $\varphi_B(x) \restriction B_0 = \psi \restriction B_0$. Then

$$\psi = \varphi_{B'}(x - \langle x, f \rangle a + \psi(f)a).$$

□

Given a predual to a dual group, we may be able to find a predual to the predual, and so on. One way to measure how often we can repeat this process is through the notion of the *foundation rank* of a dual group. A dual group is defined to have foundation rank 0 if none of its preduals is a dual group. In general we say a dual group $A$ has foundation rank $\geq \alpha$ if for all $\beta < \alpha$ there is a predual of $A$ with rank at least $\beta$. Let the foundation rank of a dual group be the least ordinal it can be, or $\infty$ if there is no such ordinal. We do not know if all foundation ranks are either $\infty$ or a finite ordinal.

A more tractable, and we think more interesting, invariant of a dual group is its length-rank. Before we define this invariant we need to make some observations about dual groups.

**1.5.** Consider a sequence $\{A_i \colon i \in \omega\}$ where $A_{i+1}$ is the dual of $A_i$. Let $\sigma_i \colon A_i \to A_{i+2}$ be the natural map and for $i \geq 3$ let $\rho_i \colon A_i \to A_{i-2}$ be the restriction map, i.e., $\rho_i(a)$ is the unique element of $A_{i-2}$ so that for all $b$ in $A_{i-3}$, $\langle \rho_i(a), b \rangle = \langle a, \sigma_{i-3}(b) \rangle$. Note that, by I.2.2, for $i \geq 3$, $A_i = \mathrm{im}(\sigma_{i-2}) \oplus \ker(\rho_i)$.

**1.6 Proposition.** *Let* $A_i$, $\sigma_i$ ($i \leq 3$) *and* $\rho_3$ *be as in 1.5. Then* $\ker(\rho_3)$ *is a dual group. In fact it is isomorphic to the dual of* $A_2/\sigma_0[A_0]$.

PROOF. Since every element of $\ker(\rho_3)$ is 0 on $\sigma_0[A_0]$, any such element induces a unique homomorphism from $A_2/\sigma[A_0]$ to $\mathbf{Z}$. Furthermore, this representation must be faithful: otherwise there would be a non-zero element of $A_3$ which induces the trivial map on $A_2$; since $A_3$ is the dual of $A_2$, no such element can exist. To complete the proof of the proposition we must show that if $\psi: A_2/\sigma_0[A_0] \to \mathbf{Z}$ is a homomorphism then there is an element of $\ker(\rho_3)$ which induces $\psi$. Since $A_3$ is the dual of $A_2$, there must be an element $a \in A_3$ so that for all $b \in A_2$, $\langle a, b \rangle = \psi(b + \sigma_0[A_0])$. By definition, this element is in $\ker(\rho_3)$. $\square$

As stated, the proposition above appears to depend on the choice of the predual. However since $\sigma_1\rho_3$ is the identity on $\sigma_1[A_1]$, $A_3 = \sigma_1[A_1] \oplus \ker(\rho_3)$. So $\ker(\rho_3) \cong A_3/\sigma_1[A_1]$. We can restate 1.6 in the following form.

**1.7 Proposition.** *Suppose $A$ is a dual group and $\sigma: A \to A^{**}$ is the natural embedding. Then $A^{**}/\sigma[A]$ is a dual group.* $\square$

In the proof of 1.6 we could have chosen to look at $A_2/B$ where $B$ is the subgroup annihilated by $\ker(\rho_3)$. In general, $\sigma_0[A_0] \subseteq B$ and no proper extension of $B$ inside of $A_2$ can be a predual of $A_1$. However we don't know if $B$ must be a (maximal) predual of $A_1$.

An important question is which groups can appear as $A^{**}/\sigma[A]$. We conjecture that any dual group is possible. Although we will get quite a bit of information in later sections, this is not settled.

**1.8 Definition.** Suppose $A$ is a dual group. We will define an invariant of $A$ called the *length-rank* of $A$. A length rank is an ordered pair $(n, m)$, where $n$, $m$ take values in $\omega \cup \{\infty\}$ with the stipulations that if $n = \infty$ then $m = \infty$ and if $n > 0$ then $m > 0$. If $A$ is a reflexive group then the length-rank of $A$ is the ordered pair $(0, m)$ where $m$ is the rank of $A$ as an abelian group if the rank is finite and $\infty$ if the rank is infinite. If $A$ is not reflexive and the length-rank of $A^{**}/\sigma[A]$ is $(n, m)$, then the length-rank of $A$ is $(n + 1, m)$. (Here we conventionally let $\infty + 1 = \infty$.)

In view of Proposition 1.7, length-rank is a well defined invariant of dual groups. In section 4, we will show that all length-ranks can be realized.

## §2. Tree groups

The first result which showed that there may be dual groups outside the Reid class was Huber's result that, assuming MA + ¬CH, every $\aleph_1$-separable group of cardinality $\aleph_1$ is reflexive (see XI.2.8). In section 5 we will see that it is consistent that there is an $\aleph_1$-separable group of cardinality $\aleph_1$ which is not reflexive but is a dual group. In this section we describe a certain class of groups called the tree groups all of which are reflexive. We shall show that this class includes many $\aleph_1$-separable groups.

**2.1 Definition.** A group $A$ is called a *tree group* if the following situation holds. We are given a tree $T$ of height $\omega$ (see II.5) and a collection $B$ of branches through $T$, and for each $\nu \in T$ a free group $F_\nu$. Also, for each $\eta \in B$, we are given a free group $A_\eta$ containing $\bigoplus_{n\in\omega} F_{\eta \restriction n}$ and groups $A_{\eta n}$ so that $A_{\eta 0} = A_\eta$ and for all $n$, $A_{\eta n+1} \oplus F_{\eta \restriction n} = A_{\eta n}$. The groups $A_\eta (\eta \in B)$ generate $A$ and for all $\eta \neq \zeta \in B$, $A_\eta \cap A_\zeta = \bigoplus_{m<n} F_{\eta \restriction m}$ where $n$ is the least natural number so that $\zeta \restriction n \neq \eta \restriction n$.

It should not be hard to see that any free group is a tree group and that a direct sum of tree groups is a tree group. Less trivial examples of tree groups are provided by the following result.

**2.2 Theorem.** (i) *There is a non-free $\aleph_1$-separable group of cardinality $\aleph_1$ which is a tree group.*

(ii) *(MA + ¬CH) Every $\aleph_1$-separable group of cardinality $\aleph_1$ in standard form is a tree group.*

PROOF. (i) Fix a stationary subset $E$ of $\omega_1$ and choose a ladder system $\{\eta_\alpha : \alpha \in E\}$ such that for all $\alpha \in E$ and and $n \in \omega$, $\eta_\alpha(n) \in \omega_1 \setminus E$, and for any $\gamma \in E$ and $m \in \omega$, if $\eta_\gamma(m) = \eta_\alpha(n)$, then $m = n$ and for all $k < n$, $\eta_\gamma(k) = \eta_\alpha(k)$ (cf. Exercise XII.17). Let $A$ be an $\aleph_1$-separable group in standard form, say of type $\mathbf{Q}^{(p)}$, built on this ladder system, as in VIII.1.1. So $A$ is a subgroup of

$$\bigoplus_{\alpha\in E} \mathbf{Q}y_\alpha \oplus \bigoplus_{\beta\in\omega_1\setminus E} \mathbf{Q}x_\beta.$$

Let the tree $T$ be the set of all rungs of the ladders, partially ordered so that $\beta <_T \delta$ if and only if for some $\alpha \in E$ and $n < m$, $\eta_\alpha(n) = \beta$ and $\eta_\alpha(m) = \delta$. For each $\alpha \in E$, by a harmless abuse of notation we can regard $\eta_\alpha$ as a branch of $T$. Let $B$ be the set of all the $\eta_\alpha$'s. For $\beta$ in $T$, let $F_\beta$ be $\mathbf{Z}x_\beta$. For each $\alpha \in E$ and $n \in \omega$, let $A_{\eta_\alpha n}$ be the pure closure in $A$ of

$$\{p^{-n}(y_\alpha - \sum_{\ell < n} p^\ell x_{\eta_\alpha(\ell)})\} \cup \{x_{\eta_\alpha}(m) : m \geq n\}.$$

If $\tilde{A}$ is the subgroup of $A$ generated by the $A_{\eta_\alpha}$'s, then it is straight-forward to verify that $\tilde{A}$ is a tree group and that $A = \tilde{A} \oplus F$ for some free group $F$. ($F$ has a basis consisting of those $x_\beta$ such that $\beta \notin T$.) Thus $A$ is a tree group.

(ii) Suppose $A$ is an $\aleph_1$-separable group in standard form; we will use the notation of VIII.1.9 and VIII.1.11. In effect, we will show that $A$ is constructed like the group in (i). Without loss of generality we can assume that for each $\alpha \in E$

$$\{\beta \in \alpha \setminus E: \text{for some } \ y \in Y_\alpha, \ \pi_{\beta\beta+1}(y) \neq 0\}$$

is of order type $\omega$ (cf. the proof of VIII.2.4); let $\eta_\alpha : \omega \to \alpha$ be an enumeration in increasing order of this set. For each $\beta < \omega_1$, let $g_\beta$ be a map from $\omega$ onto the set of finite increasing sequences of ordinals less than $\beta$. Define a coloring $\{c_\alpha : \alpha \in E\}$ of the ladders $\eta_\alpha$ with $\omega$ colors by choosing $c_\alpha(n)$ so that $g_{\eta_\delta(n)}(c_\alpha(n)) = (\eta_\alpha(m) : m < n)$, a finite sequence of ordinals $< \eta_\alpha(n)$. Since we have assumed $MA + \neg CH$, there is $(f, f^*)$ which uniformizes it (cf. VI.4.6). We can assume every value of $f$ is a finite sequence of ordinals, that is we replace $f(\beta)$ by $g_\beta(f(\beta))$. Let $T = \omega_1 \setminus E$. Define a partial order on $T$ by setting $\beta_1 <_T \beta_2$ if and only if $f(\beta_1)^\frown(\beta_1)$ is an initial segment of $f(\beta_2)$. Since for any $\beta$, $f(\beta)$ is a finite sequence, $\beta$ can have at most finitely many predecessors in $<_T$; so $T$ is a tree of height at most $\omega$. For $\beta$ in $T$, let $F_\beta$ be $K_{\beta\beta+1}$. For each $\alpha \in E$ and $n \geq f^*(\alpha)$, $c_\alpha(n) = f(\eta_\alpha(n))$; let $b_\alpha$ be the branch through $T$ which is cofinal with $\{\eta_\alpha(n): n \geq f^*(\alpha)\}$. Define $A_{\eta\alpha}$ to be the subgroup of $A$ generated by

$$\{y - \pi_{0\eta_\alpha(n_\alpha)}(y): y \in Y_\alpha\} \cup \bigcup_{\beta \in b_\alpha} K_{\beta\beta+1}.$$

Let $\tilde{A}$ be the group generated by the $A_{\eta\alpha}$'s. It should be clear that $\tilde{A}$ is a tree group and that $A = \tilde{A} \oplus \bigoplus \{K_{\beta\beta+1}: \beta \text{ is not in any } b_\alpha\}$. Since $A$ is the sum of a free group and a tree group, $A$ is a tree group itself. $\square$

Another source of examples of tree groups is found by considering subgroups of direct products. Let $T$ be a tree of height $\omega$ and $B$ a collection of branches through $T$. For $\nu \in T$ let $x_\nu$ be the function on $T$ defined by

$$x_\nu(\mu) = \begin{cases} 1 & \text{if } \nu = \mu \\ 0 & \text{otherwise.} \end{cases}$$

Suppose for each $\eta \in B$, $z_\eta$ is an element of the **Z**-adic completion of $\langle x_{\eta\restriction n}: n < \omega \rangle$ inside $\mathbf{Z}^T$. If $A$ is the pure closure in $\mathbf{Z}^T$ of the subgroup generated by $\{x_\nu: \nu \in T\} \cup \{z_\eta: \eta \in B\}$, then $A$ is a tree group.

Our interest in tree groups is justified by the following theorem.

**2.3 Theorem.** *Every tree group whose cardinality is not $\omega$-measurable is reflexive.*

PROOF. Assume $A$ is a tree group and $T$, $B$, $A_{\eta n}$, $F_\nu$ are as in 2.1. We will use certain direct sum decompositions of $A$. Let $T_n$ denote the elements of $T$ of length $n$. For $\nu \in T_n$, let $D_\nu = \sum\{A_{\eta n}: \eta \in B, \eta \restriction n = \nu\}$. Notice that for $\nu \in T_{n+1}$, $D_{\nu\restriction n} = D_\nu \oplus F_{\nu\restriction n}$. Also, for any $n$,

$$A = \bigoplus_{\nu \in T_n} D_\nu \oplus \bigoplus_{m < n} \bigoplus_{\mu \in T_m} F_\mu.$$

To simplify notation, let us denote $\bigoplus_{m<n} \bigoplus_{\mu \in T_m} F_\mu$ as $F_n$.

Now we will fix $\varphi \in A^{**}$ and show that $\varphi$ belongs to the image of $A$ under $\sigma_A$. Since for any $n$, $A^{**} = \bigoplus_{\nu \in T_n} D_\nu^{**} \oplus F_n^{**}$, there is a smallest finite set $X_n \subseteq T_n$ so that $\varphi \in \bigoplus_{\nu \in X_n} D_\nu^{**} \oplus F_n^{**}$. It is easy to see that for all $n < m$ if $\nu \in X_m$ then $\nu {\restriction} n \in X_n$. If for some $n$, $X_n$ is empty then $\varphi \in F_n^{**}$, and since $F_n$ is reflexive, in this case we would have $\varphi$ in the image of $A$.

Suppose now that for all $n$, $X_n$ is non-empty. Let $X = \cup X_n$. Call an element $\nu$ of $X$ a *splitting node* if there are incomparable $\mu, \tau \in X$ each of which is greater than $\nu$. We claim that there are only finitely many splitting nodes. Otherwise by König's lemma (II.5.4) there would be an infinite increasing sequence $\{\nu_n : n \in \omega\}$ of splitting nodes. Then if we let $\mu_n \in X$ be a successor of $\nu_n$ which is not a $\nu_k$, $\{\mu_n : n \in \omega\}$ is a set of pairwise incomparable elements of $X$; let $m_n$ be the length of $\mu_n$. Let us consider what it means for $\mu_n$ to be an element of $X$. Since we identify $D_{\mu_n}^*$ with $(\bigoplus_{\mu_n \neq \nu \in T_{m_n}} D_\nu \oplus F_{m_n})^\perp$ — see section IX.1 — it means that there is a function $f_n \in A^*$ which is constantly 0 on $\bigoplus_{\mu_n \neq \nu \in T_{m_n}} D_\nu \oplus F_{m_n}$ such that $\langle \varphi, f_n \rangle \neq 0$. Now for any sequence $(k_n : n \in \omega)$ of integers, $\sum_{n \in \omega} k_n f_n$ defines an element of $A^*$ because for all $a \in A$, $f_n(a) = 0$ for all but finitely many $n$. Hence $\psi : \mathbf{Z}^\omega \to \mathbf{Z}$ defined by $\psi((k_n : n \in \omega)) = \langle \sum_{n \in \omega} k_n f_n, \varphi \rangle$ contradicts the slenderness of $\mathbf{Z}$. Thus we have shown that the set of splitting nodes is finite.

Choose $n$ so that every splitting node has length less than $n$ and so that every element of $X_n$ has extensions in $X$ of all lengths. So each element $\nu$ of $X_n$ determines a unique branch, $\{\mu \in \bigcup_{k \geq n} X_k : \mu {\restriction} n = \nu\}$, through $T$. Enumerate these branches as $\eta_0, \ldots, \eta_m$. (A simple argument would show that each of these branches must be in $B$. However for simplicity if some $\eta_i$ is not in $B$, let $A_{\eta_i k} = \bigoplus_{k < \ell} F_{\eta_i {\restriction} \ell}$.) Let

$$Y = \bigcup_{k \geq n} \{\nu \in T_k : \nu \neq \eta_i {\restriction} k \text{ for all } i; \text{ and}$$
$$k > n \text{ implies } \nu {\restriction} (k-1) = \eta_i {\restriction} (k-1) \text{ for some } i\}.$$

It is easy to see that

$$A = \bigoplus_{i \leq m} A_{\eta_i n} \oplus \bigoplus_{\nu \in Y} D_\nu \oplus F_n.$$

View $A^{**}$ as $\bigoplus_{i \leq m} A_{\eta_i n}^{**} \oplus \bigoplus_{\nu \in Y} D_\nu^{**} \oplus F_n^{**}$. We claim that $\varphi \in \bigoplus_{i \leq m} A_{\eta_i n}^{**} \oplus F_n^{**}$; indeed, otherwise there would be $\nu \in Y \cap X$ and this would contradict our assumptions. Since each of the $A_{\eta_i}$ and $F_n$ are free and hence reflexive, $\varphi$ is in the image of $A$.

Using direct sum decompositions similar to the ones in the paragraph above, it is not hard to see that $A$ is separable. So the map $\sigma$ from $A$ to $A^{**}$ is one-one. $\square$

From 2.2(i), 2.3, and IX.2.5 we can conclude immediately:

**2.4 Corollary.** *There is a non-free $\aleph_1$-separable group of cardinality $\aleph_1$ which is reflexive. In particular, there is a non-free dual group which is slender.* $\square$

As an immediate consequence of 2.2(ii), 2.3, and VIII.3.3, we also obtain Huber's result, albeit under the hypothesis of PFA rather than MA + ¬CH:

**2.5 Corollary.** (PFA) *Every $\aleph_1$-separable group of cardinality $\aleph_1$ is reflexive.* $\square$

There is a more general approach to tree groups which makes it clear that Theorem 2.3 is a generalization of the theorem of Łoś on the dual of a product of groups. We will outline this approach. Since the proofs are virtually the same as the ones above we will leave them as exercises for the reader.

**2.6 Definition.** Suppose $T$ is a tree of height $\omega+1$. A *tree of groups* is a collection $\{A_\nu : \nu \in T\}$ of groups indexed by $T$ and homomorphisms $\{\iota_{\nu\eta}, \pi_{\nu\eta} : \nu < \eta \in T\}$ satisfying the following properties, for all $\nu < \eta < \tau : \iota_{\nu\eta} : A_\nu \to A_\eta$ is an injection; $\pi_{\eta\nu} : A_\eta \to A_\nu$; $\pi_{\eta\nu} \circ \iota_{\nu\eta}$ is the identity on $A_\nu$; $\iota_{\nu\tau} = \iota_{\eta\tau} \circ \iota_{\nu\eta}$; and $\pi_{\tau\nu} = \pi_{\eta\nu} \circ \pi_{\tau\eta}$. Let $T_n$ denote the elements of $T$ of height $n$. The *tree product* of $\{A_\nu : \nu \in T\}$ is the subgroup of $\Pi_{\nu\in T_\omega} A_\nu$ composed of those elements $a$ such that for all $\nu, \eta \in T_\omega$ if $\nu\restriction n = \eta\restriction n$ then $\pi_{\nu\nu\restriction n}(a(\nu)) = \pi_{\eta\eta\restriction n}(a(\eta))$. The *tree sum* of $\{A_\nu : \nu \in T\}$ is the subgroup of the tree product consisting of those elements $a$ of the tree product which have finite support in the following sense: there is a finite set $\nu_1, \ldots, \nu_n$ of elements of $T_\omega$ such that for any other element $\eta$ of $T_\omega$, if we choose $m$ maximal so that there is $i$ such that $\eta\restriction m = \nu_i\restriction m \ (= \zeta)$, then $a(\eta) = \iota_{\zeta\eta}(\pi_{\nu_i\zeta}(a(\nu_i)))$.

The usual notions of direct sums and products are examples of tree products where for all $\nu \in T \setminus T_\omega$, $A_\nu = \{0\}$. It is also possible to give other descriptions of tree sums and tree products. The tree sum is just the direct sum modulo the identification of the common subgroups. What we called a tree group in 2.1 is the tree sum of free groups.

Suppose $T$ is a tree of height $\omega + 1$ and $\{A_\nu : \nu \in T\}$ is a tree of groups relative to the maps $\{\iota_{\nu\eta}, \pi_{\eta\nu} : \nu < \eta \in T\}$. Let the *dual tree of groups* be defined as $\{A_\nu^* : \nu \in T\}$ relative to the maps $\{\varphi_{\nu\eta}, \psi_{\nu\eta} : \nu < \eta \in T\}$ where $\varphi_{\nu\eta} = \pi_{\eta\nu}^*$ and $\psi_{\eta\nu} = \iota_{\nu\eta}^*$.

**2.7 Theorem.** (i) *The dual of the tree sum of a tree of groups is the tree product of the dual tree of groups.*
(ii) *The dual of the tree product of a tree of non-$\omega$-measurable cardinality of reflexive groups is the tree sum of the dual tree of groups.*

PROOF. Part 1 is easy to prove and the proof of part 2 uses the ideas in the proof of Theorem 2.3. □

In this approach Theorem 2.3 becomes a special case of the following corollary.

**2.8 Corollary.** *A tree sum or tree product of non-$\omega$-measurable cardinality of reflexive groups is reflexive.* □

## §3. Criteria for being a dual group

When we come to construct dual groups with desired properties, we will normally be able to construct the group, $A$, while maintaining control over its dual, $A^*$. Our other concern will be to insure that the group $A$ being constructed is in fact a dual group. In this section we will give some criteria for being a dual group.

**3.1 Theorem.** *Suppose $A$ is an $\aleph_1$-separable group of cardinality $\aleph_1$ which has a coherent system of complementary summands and $\Gamma(A) \neq 1$. Then $A$ is a dual group.*

PROOF. Choose a filtration $\{A_\alpha : \alpha < \omega_1\}$ of $A$ as in §VIII.1. Let $E = \{\alpha : A_\alpha$ is not a direct summand of $A\}$; then $E$ is co-stationary.

Let $\{K_{\alpha\beta}: \alpha < \beta, \alpha \notin E\}$ be a system of complementary summands for $A$ (cf. VIII.1.11); i.e., for all appropriate $\alpha < \beta < \gamma: A_\alpha \oplus K_{\alpha\beta} = A_\beta$; $K_{\alpha\beta} \subseteq K_{\alpha\gamma}$; and $K_{\alpha\beta} \oplus K_{\beta\gamma} = K_{\alpha\gamma}$. Let $K_\alpha$ denote $\bigcup_{\beta>\alpha} K_{\alpha\beta}$; so $A = A_\alpha \oplus K_\alpha$. For $\alpha \notin E$, let $G_\alpha = \{f \in A^*: f{\restriction}K_\alpha = 0\}(= K_\alpha^\perp)$. We will show that $G = \bigcup_{\alpha \notin E} G_\alpha$ is a predual of $A$.

It is necessary to show that the map $\varphi_G: A \to G^*$, in definition 1.1, is one-one and onto. It is easy to see that $\varphi_G$ is one-one: consider $a \in A \setminus \{0\}$ and choose $\alpha \notin E$ so that $a \in A_\alpha$; then there is $f \in G_\alpha$ so that $\langle f, a \rangle \neq 0$.

If $\alpha, \beta \notin E$ and $\alpha < \beta$, let $H_{\alpha\beta} = \{f \in A^*: f{\restriction}(A_\alpha \oplus K_\beta) = 0\} = (A_\alpha \oplus K_\beta)^\perp$. For any $\alpha \notin E$, $G = G_\alpha \oplus \bigcup_{\beta>\alpha} H_{\alpha\beta}$ (cf. IX.1.8(i)). Further, if $\alpha$ is a limit ordinal not in $E$ and $\{\alpha_n: n \in \omega\}$ is an increasing sequence of ordinals outside $E$ with limit $\alpha$, then $G_\alpha = G_{\alpha_0} \oplus \prod_{n\in\omega} H_{\alpha_n\alpha_{n+1}}$ (cf. IX.1.8(ii)). We now show that $\varphi_G$ is onto. Suppose that $\psi \in G^*$. We claim that there is some $\alpha \notin E$ so that $\psi{\restriction}\bigcup_{\beta>\alpha} H_{\alpha\beta} = 0$. Otherwise, we can choose an increasing sequence of ordinals $\{\alpha_i: i \in \omega_1\}$ disjoint from $E$ such that for all $i < \omega_1$, $\psi{\restriction}H_{\alpha_i\alpha_{i+1}} \neq 0$. Since $E$ is co-stationary, we can choose $\alpha \notin E$ and an increasing sequence $\{i_n: n \in \omega\}$ so that $\alpha$ is the limit of $\{\alpha_{i_n}: n \in \omega\}$. Then $\psi{\restriction}\prod_{n\in\omega} H_{\alpha_{i_n}\alpha_{i_{n+1}}}$ contradicts the slenderness of $\mathbf{Z}$. This proves the claim.

Now, $A^* = A_\alpha^\perp \oplus G_\alpha$; since $A_\alpha$ is reflexive (in fact free) there exists $a \in A_\alpha$ such that $\sigma_A(a){\restriction}G_\alpha = \psi{\restriction}G_\alpha$. Since $G = (A_\alpha^\perp \cap G) \oplus G_\alpha$ and both $\sigma_A(a)$ and $\psi$ are constantly 0 on $A_\alpha^\perp \cap G$ $(= \bigcup_{\beta>\alpha} H_{\alpha\beta})$, we can conclude that $\varphi_G(a) = \psi$. $\square$

We don't know if it is provable in ZFC that every $\aleph_1$-separable group of cardinality $\aleph_1$ is a dual group. If we assume MA + ¬CH, then they are all reflexive and so dual groups (cf. XI.2.9). The next class of groups which we will consider may seem artificial, but will be of importance in section 5. The groups we will consider all lie between $\mathbf{Z}^{(\lambda)}$ and $\mathbf{Z}^\lambda$. For a set $X$, let $\langle X \rangle$ denote the subgroup generated by $X$. Recall that two sets are said to be almost disjoint if their intersection is finite, and if $z \in \mathbf{Z}^\lambda$, supp$(z)$ is defined to be $\{\alpha \in \lambda: z(\alpha) \neq 0\}$.

**3.2 Theorem.** *Let $\lambda$ be a non-$\omega$-measurable cardinal. Suppose $\mathbf{Z}^{(\lambda)} \subseteq A \subseteq \mathbf{Z}^{\lambda}$ and $A$ is the pure closure of $\mathbf{Z}^{(\lambda)} \cup \{z_i : i \in I\}$ where $\operatorname{supp}(z_i) = t_i$, a countable set. If the $t_i$'s are pairwise almost disjoint and satisfy condition (∗) below, then $A$ is a dual group.*

(∗)
> *Suppose $X \subseteq \lambda$ and $X$ is not contained in the union of finitely many of the $t_i$ together with a finite set; then there exists an infinite subset $Y$ of $X$ such that $Y$ is almost disjoint from each of the $t_i$.*

PROOF. $A^*$ contains for each $\alpha < \lambda$ the projection $e_\alpha$ defined by $e_\alpha(a) = a(\alpha)$. Since $\mathbf{Z}^{(\lambda)} \subseteq A$, $\{e_\alpha : \alpha < \lambda\}$ is a free basis of the subgroup $S$ of $A^*$ that it generates. Let $B$ be the $\mathbf{Z}$-adic closure of $S$ in $A^*$. We claim that $B$ is a predual of $A$. Let $\varphi_B : A \to B^*$ be as in Definition 1.1. Clearly $\varphi_B$ is one-one since $B$ contains $S$. Let $f \in B^*$; note that $f$ is determined by $f{\upharpoonright}S$. Define $\operatorname{supp}(f) = \{\alpha < \lambda : f(e_\alpha) \neq 0\}$. We claim that $\operatorname{supp}(f)$ is contained in the union of finitely many of the $t_i$ and a finite set. If not, then condition (∗) implies we can choose $Y = \{\alpha_n : n \in \omega\} \subseteq \operatorname{supp}(f)$ so that $Y$ is almost disjoint from the $t_i$'s. Notice then that for all $(r_n : n \in \omega) \in \mathbf{Z}^\omega$, $b = \sum_{n \in \omega} n! r_n e_{\alpha_n}$ is defined on $A$ and hence belongs to $B$. But then we obtain a contradiction of the slenderness of $\mathbf{Z}$ since we can define $\theta : \mathbf{Z}^\omega \to \mathbf{Z}$ by $\theta((r_n)) = f(\sum_{n \in \omega} n! r_n e_{\alpha_n})$.

Let $y_\alpha \in \mathbf{Z}^{(\lambda)}$ be defined by: $y_\alpha(\beta) = 1$ if $\beta = \alpha$ and 0 otherwise. By subtracting from $f$ a finite sum of $\varphi_B(y_\alpha)$'s, we can assume $\operatorname{supp}(f) \subseteq t_{i_0} \cup \ldots \cup t_{i_k} = X$. Let $G$ be the pure closure of $\langle y_\alpha, z_{i_m} : \alpha \in X, m \leq k \rangle$. Since the $t_i$'s are pairwise almost disjoint, $G$ is a direct summand of $A$; let $H$ be a complementary summand of $G$ in $A$. As discussed at the beginning of section IX.1, we can identify $G^*$ with $H^\perp$. Let $L$ be the subgroup of $B$ generated by $\{e_\alpha : \alpha \in X\}$. It is an easy calculation to see that $L$ is pure and dense in $H^\perp$. Both $L$ and $G$ are countable free groups; hence by Chase's Theorem on dual bases (XI.3.5), there is a basis $\{u_n : n < \omega\}$ of $L$ which is dual to a basis $\{g_n : n < \omega\}$ of $G$. So for all $(r_n : n \in \omega) \in \mathbf{Z}^\omega$, $\sum_{n \in \omega} n! r_n u_n$ is an element of $G^* = H^\perp$ which belongs to $B$. Unless $f$ vanishes on all but finitely many of the $u_n$, then just as in the previous paragraph, we can arrive at a contradiction. But if $f$ vanishes on all but finitely many of the $u_n$, then $f \in \varphi_B[A]$; indeed $f = \varphi_B(\sum_n f(g_n)u_n)$ since these two functions agree on $L$ by construction and are both identically zero on $\{e_\alpha : \alpha \notin X\}$. $\square$

Groups which satisfy the hypothesis of Theorem 3.2 arise naturally in constructions using the Black Box. In the following we use the notation of Theorem XIII.2.1.

**3.3 Proposition.** *Suppose $\{(M_i, X_i) : i \in I\}$ is as in the statement of the Black Box relative to a cardinal $\lambda$ and a stationary set $E$. For each $i \in I$, suppose $t_i \subseteq M_i \cap \lambda$ is an increasing sequence of order type $\omega$ and $\sup(t_i) = \delta(i)$. Further suppose for all $\delta \in E$, if $\delta(i) = \delta = \delta(j)$, then $t_i$ is carried to $t_j$ by the isomorphism between $M_i$ and $M_j$. Then the $t_i$'s are pairwise almost disjoint and satisfy condition $(*)$ in Theorem 3.2.*

PROOF. The $t_i$'s are almost disjoint because of XIII.2.1(b). In order to verify $(*)$, first we attach to each ordinal $\delta \in E$ a tree of ordinals. Define $T_\delta$ to have universe $\bigcup_{\delta(i)=\delta} t_i$. For convenience we can assume $0 \in t_i$, for all $i \in I$. Define $\alpha <_\delta \beta$ if $\alpha < \beta$ and for some $i$ with $\delta(i) = \delta$, $\alpha, \beta \in t_i$. To see that $T_\delta$ is a tree of height $\omega$ when $T_\delta$ is non-empty, note that if $\beta \in t_i \cap t_j$ and $\delta(i) = \delta = \delta(j)$, then because the isomorphism between $M_i$ and $M_j$ is the identity on $M_i \cap \beta$, we have that for all $\alpha < \beta$, $\alpha \in t_i$ if and only if $\alpha \in t_j$.

Suppose now that $X$ is a set which is not contained in the union of finitely many of the $t_i$ together with a finite set. We need to find an infinite subset $Y$ of $X$ which is almost disjoint from all the $t_i$'s. Let $L = \{\delta : \delta$ is a limit point of $X$ and $cf(\delta) = \omega\}$. There are three cases to consider.

*Case 1: there is a unique element $\delta \in L$.* If $T_\delta$ is empty then we can take $Y = X$. So we may assume $T_\delta$ is non-empty. Furthermore, we can assume $X \subseteq T_\delta$ since if $X \cap T_\delta$ is finite, we can take $Y = X \setminus T_\delta$. Give $X$ the ordering induced by $<_\delta$. So $X$ is a tree. If some level of $X$ is infinite, then the elements at this level give the desired subset $Y$ of $X$ (because if $\delta(i) = \delta$, then $t_i$ intersects a level of $X$ in at most one point.) So we can assume that $X$ is finitely branching. Furthermore we can assume that any branch through $X$ is contained in some $t_i$: otherwise some branch through $X$ would provide the desired set. Let $[X]$ be the set of branches through $X$. The branches of $X$ can be viewed as a closed subset of the product space $\prod_{n \in \omega} X_n$, where $X_n$ is the set of elements of $X$ of height $n$ and $X_n$ is given the discrete topology — this topology is sometimes called the *tree topology*. Since $X$ was assumed to be finitely branching each $X_n$ is finite and hence compact. So, by the Tychonoff theorem, $[X]$ is a compact topological space. Hence either $[X]$ is finite or there is a non-isolated branch $B$ through $X$. König's Lemma (II.5.4) says that an infinite finitely branching tree has a branch of length $\omega$. Since we have assumed that every branch is contained in some $t_i$, if $[X]$ is finite then $X$ is contained in finitely many of the $t_i$ together with a finite set. Hence, there is a non-isolated branch $B$. So we can find $\beta_0 <_\delta \beta_1 <_\delta \ldots$ an increasing sequence in $B$ so that for all $n$, there is $\gamma_n \in X \setminus B$ such that $\beta_n <_\delta \gamma_n$. Then $Y = \{\gamma_n : n \in \omega\}$ is the desired set.

*Case 2: L is finite.* The proof is by induction on the cardinality of $L$. Let $\delta_0 < \ldots < \delta_m$ enumerate $L$. If $X \cap \delta_0$ is not contained in the union of finitely many of the $t_i$ together with a finite set, then we can apply case 1 to $X \cap \delta_0$. So we can assume that $X \cap \delta_0$ is contained in $\bigcup_{k<n} t_{i_k}$ together with a finite set. Next we can apply the induction hypothesis to $X \setminus \bigcup_{k<n} t_{i_k}$.

*Case 3: case 2 doesn't hold.* Let $\{\delta_n : n \in \omega\}$ enumerate the first $\omega$ elements of $L$ and let $\delta$ be the limit of the $\delta_n$. Note that $X \cap \delta$ is not contained in the union of finitely many of the $t_i$ together with a finite set, since the order type of any finite union of the $t_i$ is less than $\omega^2$ but the order type of $X \cap \delta$ is at least $\omega^2$. So we can assume that $X \subseteq \delta$. Again we can assume $T_\delta$ is non-empty. Consider $X \cap T_\delta$. If this set were contained in finitely many of the $t_i$ with $\delta(i) = \delta$ together with a finite set, then for all $n \in \omega$, $\delta_n$ would be a limit point of $X \setminus (X \cap T_\delta)$. Hence there would exist an increasing sequence of order type $\omega$ in $X \setminus (X \cap T_\delta)$ — namely any sequence $\{\alpha_n : n \in \omega\}$ such that for all $n$, $\delta_n < \alpha_n < \delta_{n+1}$ —

with limit point $\delta$. This sequence would be the desired set. So we can assume that $X = X \cap T_\delta$. It is conceivable that, in the course of making our simplifying assumptions, we reverted to a previous case, but such an eventuality is only for the better. Of course the ordinals which we now denote as $\delta_n$ may not be the same as those we started with.

Give $X$ the order induced by $<_\delta$. For any $n$, $X \cap (\delta_n, \delta_{n+1}) = X_n$ is infinite and any $<_\delta$-sequence in $X_n$ is finite. If we color the pairs in $X_n$ according to whether or not they are comparable, then by Ramsey's theorem (II.5.6), there is an infinite subset $B_n \subseteq X_n$ such that either every pair of elements in $B_n$ is comparable or every pair of elements in $B_n$ is incomparable. Since there is no infinite set of pairwise comparable elements contained in $X_n$, the elements of $B_n$ must be pairwise incomparable. Let $C_n$ be a subset of $B_n$ of size $n$. Since each $C_n$ consists of elements which are pairwise incomparable in $<_\delta$, if $\delta(i) = \delta$ then $|t_i \cap C_n| \leq 1$. If $\delta(i) \neq \delta$, then $t_i$ can intersect only finitely many of the $C_n$ non-trivially. Hence $\bigcup_{n \in \omega} C_n = C$ is not contained in any finite union of the $t_i$ together with a finite set. So we can use $C$ in place of $X$, and then we are in Case 1. $\square$

## §4. Some non-reflexive dual groups

As we have seen in section 1, a key to understanding the structure of dual groups is the question of which groups can appear as the kernel of the restriction map from a triple dual onto a dual group. In this section we shall introduce a class of groups, called $\omega_1$-dual groups, which do appear as such a kernel. (For inverse and inverse-direct systems, see section XI.1.)

**4.1 Definition.** Suppose $(A_\alpha, \pi_{\beta\alpha} : \alpha < \beta < \omega_1)$ is an inverse system of reflexive groups and that each $\pi_{\beta\alpha} : A_\beta \to A_\alpha$ is onto. Also suppose that for limit ordinals $\delta$, $A_\delta$ is a subgroup of $\varprojlim_{\alpha < \delta} A_\alpha$ and for each $\alpha < \delta$, $\pi_{\delta\alpha}$ is the restriction of the map from the inverse limit. Then $(A_\alpha, \pi_{\beta\alpha} : \alpha < \beta < \omega_1)$ is called an $\omega_1$-*inverse system* and the inverse limit of the system, $A = \varprojlim_{\alpha < \omega_1} A_\alpha$, is called an $\omega_1$-*dual*.

Every $\omega_1$-dual group is a dual group. Indeed, since each $A_\alpha$ is reflexive, $A$ is isomorphic to $\varprojlim A_\alpha^{**}$; but the latter (by XI.1.3) is the dual of $\varinjlim A_\alpha^*$.

It is often more convenient to work with a concrete description of the $\omega_1$-inverse system and its inverse limit $A$. Consider $P \overset{\text{def}}{=} \prod_{\alpha<\omega_1} A_\alpha$. We identify $\prod_{\alpha<\beta} A_\alpha$ with $\{c \in P : c(\nu) = 0$ for all $\nu \geq \beta\}$ and $\prod_{\gamma \leq \alpha < \beta} A_\alpha$ with $\{c \in P : c(\nu) = 0$ if $\nu < \gamma$ or $\nu \geq \beta\}$. Let $\rho_\beta$ and $\rho_{\gamma\beta}$ denote the projections on $\prod_{\alpha<\beta} A_\alpha$ and $\prod_{\gamma \leq \alpha < \beta} A_\alpha$ respectively. If we apply the usual construction of an inverse limit as a subgroup of a direct product (cf. XI.1.1), we obtain the following proposition.

**4.2 Proposition.** *Suppose* $(A_\alpha, \pi_{\beta\alpha} : \alpha < \beta < \omega_1)$ *is an* $\omega_1$-*inverse system and* $A = \varprojlim_{\alpha<\omega_1} A_\alpha$. *Let* $\pi_\alpha : A \to A_\alpha$ *be the canonical map associated with the inverse limit. Then with the notation above:*

(i) *there is a monomorphism* $\iota$ *from* $A$ *into* $\prod_{\alpha<\omega_1} A_\alpha$ *defined by* $\iota(a)(\alpha) = \pi_\alpha(a)$;

(ii) *for all* $\alpha$, $\rho_{\alpha\alpha+1}[\iota[A]] = A_\alpha$ *and if* $\alpha$ *is a limit ordinal* $\rho_\alpha[\iota[A]] \cong A_\alpha$;

(iii) *for all limit ordinals* $\delta$, *if* $\rho_\delta(\iota(a)) \neq 0$ *then* $\{\alpha < \delta : \iota(a)(\alpha) \neq 0\}$ *is unbounded in* $\delta$.

PROOF. Most of this is obvious from the definitions. For the second part of (ii), note that for limit $\alpha$, since $A_\alpha$ is contained in $\varprojlim_{\gamma<\alpha} A_\gamma$, the map from $\rho_\alpha[\iota[A]]$ to $A_\alpha$ defined by sending $\rho_\alpha(\iota(a))$ to $\pi_\alpha(a)$ is an isomorphism. Also, (iii) follows from the hypothesis that for limit $\delta$, $A_\delta$ is a subgroup of $\varprojlim_{\alpha<\delta} A_\alpha$. $\square$

**4.3 Example.** Of course, any reflexive group $A$ is the inverse limit of $\omega_1$ copies of $A$ where the maps are all the identity function. However there are other ways in which a group can be realized as an $\omega_1$-dual:

Let $\kappa$ be a (finite or infinite) cardinal $\leq \omega_1$. For each $\alpha < \kappa$, let $A_\alpha$ be the group freely generated by $\{x_{\alpha\gamma} : \gamma < \min(\kappa,\alpha)\}$. For $\alpha < \beta$, let $\pi_{\beta\alpha}$ be the homomorphism from $A_\beta$ to $A_\alpha$ defined by

$$\pi_{\beta\alpha}(x_{\beta\gamma}) = \begin{cases} x_{\alpha\gamma} & \text{if } \gamma < \alpha \\ 0 & \text{otherwise.} \end{cases}$$

Then $(A_\alpha, \pi_{\beta\alpha} : \alpha < \beta < \omega_1)$ is an $\omega_1$-inverse system whose inverse limit is $\mathbf{Z}^{(\kappa)}$; in fact, a basis for the inverse limit consists of $\{x_\alpha : \alpha < \kappa\}$ where

$$\pi_\beta(x_\alpha) = \begin{cases} x_{\beta\alpha} & \text{if } \alpha < \beta \\ 0 & \text{otherwise.} \end{cases}$$

If $\kappa = 1$, then $\iota$, in Proposition 4.2, embeds $A$ as the subgroup of $\mathbf{Z}^{\omega_1}$ generated by $\mathbf{1}$, the function which is constantly 1, and $\rho_\alpha(\mathbf{1}) = \mathbf{1}_\alpha$, which is defined as in Example XI.1.11.

The key construction in this section is given in the following definition. (Inverse-direct systems are defined in XI.1.5.)

**4.4 Definition.** Let $E$ be a stationary and co-stationary subset of $\omega_1$. Suppose $(A_\alpha, \pi_{\beta\alpha} : \alpha < \beta < \omega_1)$ is an $\omega_1$-inverse system with $A = \varprojlim A_\alpha$. We will construct an inverse-direct system $(B_\alpha, \tau_{\alpha\beta}, p_{\beta\alpha} : \alpha < \beta < \omega_1)$. Each $B_\alpha$ will be a subgroup of $\prod_{\gamma < \alpha} A_\gamma$, and $\tau_{\alpha\beta}$ will be a restriction of the inclusion map from $\prod_{\gamma < \alpha} A_\gamma$ into $\prod_{\alpha < \omega_1} A_\alpha$. Each $p_{\beta\alpha}$ will be a restriction of $\rho_\alpha$. (We maintain the notation of the paragraph preceding 4.2, as well as of 4.2.)

To begin, we let $B_0 = \{0\}$. If $B_\alpha$ has been defined, then let $B_{\alpha+1} = B_\alpha \oplus A_\alpha$. Suppose that $\delta$ is a limit ordinal. If $\delta \in E$, let $B_\delta = \bigcup_{\alpha < \delta} B_\alpha \oplus \rho_\delta[\iota[A]]$; this sum is direct because of 4.2(iii). If $\delta \notin E$, let $B_\delta = \varprojlim_{\alpha < \delta} B_\alpha$; more exactly, let $B_\delta = \{b \in \prod_{\beta < \delta} A_\beta :$ for all $\alpha < \delta$, $\rho_\alpha(b) \in B_\alpha\}$. Define $B = \varinjlim B_\alpha (= \bigcup_{\alpha < \omega_1} B_\alpha)$.

By induction we can show that for all $\alpha < \beta$, $B_\alpha \subseteq B_\beta$ and $\rho_\alpha[B_\beta] = B_\alpha$, and so we do indeed have an inverse-direct system. We can also show by induction that for all $\alpha$, $\rho_\alpha[\iota[A]] \subseteq B_\alpha$.

In the case when we start with the $\omega_1$-inverse system defined in 4.3 for $\kappa = 1$, then we obtain the second inverse-direct system — $(B_\alpha, \tau_{\alpha\beta}, \pi_{\beta\alpha} : \alpha < \beta \in \omega_1)$ — of Example XI.1.11

In the following lemmas, $(A_\alpha, \pi_{\beta\alpha} : \alpha < \beta < \omega_1)$ forms an $\omega_1$-inverse system with inverse limit $A$, and $B$ and the $B_\alpha$'s are defined as in 4.4; we shall identify $A$ with $\iota[A]$. (These lemmas should be compared with XI.1.11B, C, and D.)

**4.5 Lemma.** $\varprojlim_{\alpha<\omega_1} B_\alpha = B \oplus A$.

PROOF. By the construction, $\varprojlim_{\alpha<\omega_1} B_\alpha$ can be identified with the subgroup of $\prod_{\alpha<\omega_1} A_\alpha$ consisting of those elements $c$ such that for all $\alpha$, $\rho_\alpha(c) \in B_\alpha$. Under this identification, $B$ and $A$ are subgroups of $\varprojlim_{\alpha<\omega_1} B_\alpha$. Since any element of $B$ is bounded (i.e., contained in $\prod_{\gamma<\alpha} A_\gamma$ for some $\alpha$) and any non-zero element of $A$ is unbounded, $B \cap A = \{0\}$. So $B + A = B \oplus A$, and what we need to prove is that this subgroup of $\varprojlim_{\alpha<\omega_1} B_\alpha$ is the whole group.

Suppose $b$ is an element of $\varprojlim_{\alpha<\omega_1} B_\alpha$. For each limit ordinal $\alpha \in E$, there is $\beta_\alpha < \alpha$ and $c_\alpha \in B_{\beta_\alpha}$ and $a_\alpha \in A$ so that $\rho_\alpha(b) = c_\alpha + \rho_\alpha(a_\alpha)$; furthermore, $c_\alpha$ and $\rho_\alpha(a_\alpha)$ are uniquely determined by $b$. By Fodor's lemma (II.4.11), there is $E_1$, a stationary subset of $E$, and $\beta$ so that for all $\alpha \in E_1$, $\beta_\alpha = \beta$. Consider $\alpha < \gamma$ in $E_1$; then $\rho_\alpha(b) = \rho_\alpha(\rho_\gamma(b)) = \rho_\alpha(c_\gamma + \rho_\gamma(a_\gamma)) = \rho_\alpha(c_\gamma) + \rho_\alpha(a_\gamma)$. Since $c_\gamma \in B_\beta$, $\rho_\alpha(c_\gamma) = c_\gamma$. Hence $c_\gamma = c_\alpha$ and $\rho_\alpha(a_\alpha) = \rho_\alpha(a_\gamma)$. Let $c$ be the common value of $c_\alpha$ for $\alpha \in E_1$. Then $b = c + a$ where $a$ is the element of $A$ such that for all $\alpha \in E_1$, $\rho_\alpha(a) = \rho_\alpha(a_\alpha)$. $\square$

**4.6 Lemma** *For all $\alpha$, $B_\alpha$ is reflexive.*

PROOF. The proof is by induction on $\alpha$. The only interesting case is that of limit ordinals. Choose an increasing sequence $\{\alpha_n : n < \omega\}$ with limit $\alpha$ so that $\alpha_0 = 0$. If $\alpha \notin E$, then $B_\alpha = \prod_{n\in\omega} \ker p_{\alpha_{n+1}\alpha_n}$ (cf. Lemma XI.1.7). If $\alpha \in E$, then $B_\alpha = \bigoplus_{n\in\omega} \ker p_{\alpha_{n+1}\alpha_n} \oplus \rho_\alpha[A]$. Since the class of reflexive groups is closed under taking direct summands, countable direct sums and countable direct products (cf. IX.1.10 and X.1.1), we are done. $\square$

Now we want to consider the dual system $(B_\alpha^*, p_{\beta\alpha}^*, \tau_{\alpha\beta}^* : \alpha < \beta < \omega_1)$.

**4.7 Lemma** $\varinjlim_{\alpha<\omega_1} B_\alpha^* = B^* = \varprojlim_{\alpha<\omega_1} B_\alpha^*$.

PROOF. The second equality is by Proposition XI.1.3. As for the first equality, since $B_\beta = B_\alpha \oplus \ker p_{\beta\alpha}$ for $\alpha < \beta$, it is enough to show that for any $f \in B^*$, there is an $\alpha$ such that $f \restriction \ker p_{\beta\alpha} \equiv 0$ for all $\beta > \alpha$; for then $f$ is the image of $f \restriction B_\alpha$ under the natural embedding of $\varinjlim_{\alpha<\omega_1} B_\alpha^*$ into $\varprojlim_{\alpha<\omega_1} B_\alpha^*$ given in Proposition XI.1.6. Assume $f$ is a counterexample. Then we can find an increasing sequence $\{\alpha_\beta : \beta < \omega_1\}$ of ordinals so that for all $\beta$, there is $b_\beta \in \ker p_{\alpha_{\beta+1}\alpha_\beta}$ so that $\langle f, b_\beta \rangle \neq 0$. Since the complement of $E$ is stationary, there exists $\alpha \notin E$ and an increasing sequence $\{\beta_n : n \in \omega\}$ so that $\alpha$ is the limit of $\{\alpha_{\beta_n} : n \in \omega\}$. But $B_\alpha = B_{\alpha_{\beta_0}} \oplus \prod_{n \in \omega} \ker p_{\alpha_{\beta_{n+1}}\alpha_{\beta_n}}$, so $f \restriction B_\alpha$ contradicts the slenderness of $\mathbf{Z}$. $\square$

**4.8 Theorem.** *For every $\omega_1$-dual group $A$ there is a dual group $B$ so that $A \cong B^{**}/\sigma[B]$. Furthermore, $B \oplus A$ is an $\omega_1$-dual group.*

PROOF. Let $B$ and $(B_\alpha, \tau_{\beta\alpha}, p_{\alpha\beta} : \alpha < \beta < \omega_1)$ be as in Definition 4.4. By 4.7 and XI.1.3, $B^{**}$ is the inverse limit of the system, $(B_\alpha^{**}, \tau_{\beta\alpha}^{**}, p_{\alpha\beta}^{**} : \alpha < \beta < \omega_1)$. By 4.6 and XI.1.10, the map $\sigma$ from $B$ to $\varinjlim B_\alpha^{**}$ is onto. Now since each $B_\alpha$ is reflexive, the inverse-direct system $(B_\alpha, \tau_{\alpha\beta}, p_{\beta\alpha} : \alpha < \beta < \omega_1)$ is isomorphic to $(B_\alpha^{**}, \tau_{\alpha\beta}^{**}, p_{\beta\alpha}^{**} : \alpha < \beta < \omega_1)$, so by 4.5 $B^{**} = \varinjlim B_\alpha^{**} \oplus A = \sigma[B] \oplus A$. Again by 4.5, $B \oplus A$ is an $\omega_1$-dual since it is the inverse limit of the $\omega_1$-inverse system $(B_\alpha, p_{\beta\alpha} : \alpha < \beta < \omega_1)$.

So it remains only to prove that $B$ is a dual group. In order to do that, let us first consider how $B^{**}$ acts on $B^*$, i.e., if we identify $B^{**}$ with $B \oplus A$ as above, and if $x \in B \oplus A$, $f \in B^*$, what is $\langle x, f \rangle$? If $x = b \in B$, then by the definition of $\sigma$, $\langle b, f \rangle = f(b)$, the value of $f$ at $b$. If $x = a \in A$, and $f$ is identified with $f \restriction B_\alpha \in B_\alpha^*$, as in the proof of 4.7 (i.e., $f \restriction \ker p_{\beta\alpha} \equiv 0$ for all $\beta > \alpha$), then $\langle a, f \rangle = f(\rho_\alpha(a))$.

Let $C$ be the subgroup of $B^*$ which is annihilated by $A$, i.e., $C = \bigcup_{\alpha<\omega_1} \{c \in B_\alpha^* : c(\rho_\alpha(a)) = 0 \text{ for all } a \in A\}$. We will show that $C$ is a predual of $B$, i.e., the map $\varphi_C : B \to C^*$ defined by $\varphi_C(b)(c) = \langle c, b \rangle$ is an isomorphism (cf. Definition 1.1).

First note that $B$ acts faithfully on $C$, i.e., $\varphi_C$ is one-one. Consider $0 \neq b \in B_\alpha$. Choose $c_1 \in B_\alpha^*$ so that $\langle c_1, b \rangle \neq 0$. Next define $c_2 \in B_{\alpha+1}^* = (B_\alpha \oplus A_\alpha)^*$ by $c_2 \restriction B_\alpha \equiv 0$ and for all $x \in A_\alpha$, $\langle c_2, x \rangle = -\langle c_1, \rho_\alpha(a) \rangle$, where $\pi_\alpha(a) = x$. Then $c_1 + c_2 (\in B_{\alpha+1}^*)$ is an element of $C$ because for all $a \in A$,

$$\langle c_1 + c_2, \rho_{\alpha+1}(a) \rangle = \langle c_1 + c_2, \rho_\alpha(a) + \rho_{\alpha\alpha+1}(a) \rangle = 0.$$

But $c_1 + c_2$ is non-trivial on $b$, so $\varphi_C$ is one-one.

It remains to show that for any $f \in C^*$ there is $b \in B$ so that $\varphi_C(b) = f$. First we show that there is an ordinal $\alpha$ so that for all $c \in C$, if $c \restriction B_\alpha \equiv 0$ then $f(c) = 0$. Suppose not. Then we can find increasing sequences $\alpha_i < \beta_i < \alpha_{i+1}$ $(i \in \omega_1)$ and a set of elements $\{c_i : i \in \omega_1\}$ of $C$ so that for all $i$, $f(c_i) \neq 0$, $c_i \in B_{\beta_i}^*$ (i.e., $c_i \restriction \ker p_{\gamma\beta_i} \equiv 0$ for all $\gamma > \beta_i$) and $c_i \restriction B_{\alpha_i} \equiv 0$. Since $E$ is stationary, there exists $\delta \in E$ and an increasing sequence $\{i_n : n \in \omega\}$ so that $\delta$ is the limit of $\{\alpha_{i_n} : n \in \omega\}$. We claim that for any sequence of integers $(k_n : n \in \omega)$, the formal sum $\sum_{n \in \omega} k_n c_{i_n}$ defines an element of $C$. If so, we obtain a contradiction of the slenderness of $\mathbf{Z}$ by considering the map which takes $(k_n : n \in \omega)$ to $f(\sum_{n \in \omega} k_n c_{i_n})$.

For any $b \in B$, there is $\gamma < \delta$, $b_1 \in B_\gamma$, and $a \in A$ so that $\rho_\delta(b) = b_1 + \rho_\delta(a)$; so $b = b_1 + \rho_\delta(a) + \rho_{\delta\omega_1}(b)$. Now for all $n$, $c_{i_n}(\rho_\delta(a)) = 0 = c_{i_n}(\rho_{\delta\omega_1}(b))$. Moreover by the choice of the $c_i$, for all but finitely many $n$, $c_{i_n}(b_1) = 0$. Therefore, for all but finitely many $n$, $c_{i_n}(b) = 0$, so $\sum_{n \in \omega} k_n c_{i_n}(b)$ is defined. Thus $\sum_{n \in \omega} k_n c_{i_n}$ belongs to $B^*$; finally, one may check that it belongs to $C$.

Now let $f$ and $\alpha$ be as in the first two sentences of the paragraph before last. Consider the map $\varphi : B_\alpha^* \to B_{\alpha+1}^* (= B_\alpha^* \oplus A_\alpha^*)$ defined by $\varphi(g) \restriction B_\alpha = g$ and for all $a \in A$, $\varphi(g)(\rho_{\alpha\alpha+1}(a)) = -g(\rho_\alpha(a))$. This map is the identity on $C \cap B_\alpha^*$ and $\varphi[B_\alpha^*] \subseteq C$. Since $B_\alpha$ is reflexive, there exists $b \in B_\alpha$ so that for all $g \in B_\alpha^*$, $\langle b, g \rangle = f(\varphi(g))$. It remains to see that for all $c \in C$, $\langle b, c \rangle = f(c)$. Since $c - \varphi(c \restriction B_\alpha)$ is 0 on $B_\alpha$, $f(c - \varphi(c \restriction B_\alpha)) = 0$. Thus $f(c) = f(\varphi(c \restriction B_\alpha)) = \langle b, c \restriction B_\alpha \rangle = \langle b, c \rangle$. $\square$

**4.9 Corollary.** *For every pair $(n, m)$ which is a length-rank, there is a dual group whose length-rank is $(n, m)$.*

PROOF. Notice first that length-ranks are additive in the following sense: suppose $\{A_i : i \in I\}$ is a set of dual groups and the length-rank of $A_i$ is $(n_i, m_i)$; then the length-rank of $\bigoplus_{i \in I} A_i$ is $(n, m)$ where $n = \sup\{n_i : i \in I\}$ and $m = \sum m_i (i \in I$ and $n_i = n)$ if $n$ is finite and $m = \infty$ if $n$ is $\infty$. To prove the Corollary, it suffices to show that for every $n \in \omega$ and $m \in (\omega \cup \{\infty\}) \setminus \{0\}$, there is an $\omega_1$-dual group with length-rank $(n, m)$. The proof is by induction on $n$. Since every free group is also an $\omega_1$-dual, the case $n = 0$ is known. In general, if $A$ is an $\omega_1$-dual group with length-rank $(n, m)$, then there is a dual group $B$ so that $B^{**}/\sigma[B] \cong A$. So the length-rank of $B$ is $(n + 1, m)$. Also $B \oplus A$ is an $\omega_1$-dual and its length-rank is $(n + 1, m)$ by the first observation. $\square$

**4.10 Corollary.** *There is a dual group $B$ such that $B$ is not of the form $G^{**}$ for any group $G$.*

PROOF. Let $A$ be the $\omega_1$-dual constructed in 4.3 with $\kappa = \aleph_1$. Let $B$ be constructed from $A$ as in 4.4. Notice that for each $\alpha$, $\rho_\alpha[A]$ is a countable free group. In fact it is freely generated by $\{\rho_\alpha(x_\beta) : \beta < \alpha\}$. We can prove by induction that for $\alpha < \beta$, ker $p_{\beta\alpha}$ is in the least class containing $\mathbb{Z}$ and closed under *countable* direct sums and products. (Recall from section X.1 that this class is denoted as $\text{Reid}_{\omega_1}$). Since each $B_\alpha$ is in $\text{Reid}_{\omega_1}$, so is each $B_\alpha^*$. Hence by Lemma 4.7, $B^*$ is a direct limit of $\omega_1$ elements of $\text{Reid}_{\omega_1}$. By X.1.3 and X.1.4, $\mathbb{Z}^{\omega_1}$ is not isomorphic to a direct summand of $B^*$. Suppose that $B = H^*$ and $H = G^*$. Let $\sigma_H$ and $\sigma_B$ denote the canonical maps from $H$ and $B$ to $H^{**}$ and $B^{**}$ respectively. Then by 1.6 $B^* = \sigma_H[H] \oplus K$, where $K$ is a dual group. Now $B^{**} = \sigma_B[B] \oplus A$, where $\sigma_B[B] = K^\perp$ and $A \cong \sigma_H[H]^\perp \cong K^*$. So $K^* \cong \mathbb{Z}^{(\omega_1)}$. Since $\mathbb{Z}^{(\omega_1)}$ is reflexive, $K$ is reflexive (cf. IX.1.9) and so $K \cong \mathbb{Z}^{\omega_1}$. Thus we have arrived at a contradiction. $\square$

Recall from the preface that a group, $A$, is *strongly non-reflexive* if $A$ is not isomorphic to $A^{**}$ under any isomorphism.

**4.11 Corollary.** *There is a strongly non-reflexive dual group, i.e., there is a dual group $B$ which is not isomorphic to $B^{**}$.* $\square$

Notice that with $B$ as in 4.10, $B^{**} \cong B \oplus \mathbf{Z}^{(\omega_1)} \cong B^{****}$. So $B^{**}$ is an example of a dual group with non-isomorphic maximal preduals (even ones which are dual groups themselves). This example also indicates the difficulties inherent in trying to understand the foundation-rank of a dual group. Another curious feature of our examples is that all $\omega_1$-duals which we have constructed are weakly reflexive. We don't know if all $\omega_1$-duals are weakly reflexive.

It is possible to generalize the notion of an $\omega_1$-dual to the notion of a $\lambda$-dual. Let $\lambda$ be a regular cardinal. Suppose $(A_\alpha, \pi_{\beta\alpha}: \alpha < \beta < \lambda)$ is an inverse system of reflexive groups and that each $\pi_{\beta\alpha}$ is onto. Also suppose that for limit ordinals $\delta$, $A_\delta$ is a subgroup of $\varprojlim_{\alpha < \delta} A_\alpha$ and for each $\alpha < \delta$ $\pi_{\delta\alpha}$ is the restriction of the map from the inverse limit. Then $(A_\alpha, \pi_{\beta\alpha}: \alpha < \beta < \lambda)$ is called a $\lambda$-*inverse system* and the inverse limit of the system, $A = \varprojlim_{\alpha < \lambda} A_\alpha$, is called a $\lambda$-*dual*.

The analog of Theorem 4.8 is true for $\lambda$-dual groups with almost the same proof, except we need to make the additional set theoretic assumption that $\mathrm{E}(\lambda)$ holds. (See VI.3.1.)

**4.12 Theorem.** *Assume that* $\mathrm{E}(\lambda)$ *holds and* $\lambda$ *is non-$\omega$-measurable. Then for any* $\lambda$-*dual group* $A$ *there is a dual group* $B$ *so that* $A \cong B^{**}/\sigma[B]$. *Furthermore,* $B \oplus A$ *is a* $\lambda$-*dual group.*

PROOF. $\mathrm{E}(\lambda)$ says that there is a non-reflecting stationary subset, $E$, of $\lambda$ consisting of ordinals of cofinality $\omega$. The construction of $B$ is the same as in 4.4 except that the induction takes $\lambda$ steps. The only change in the proof comes in the verification of the analog of 4.6; it is here that we use the hypothesis on $E$. In the case where $\delta$ is a limit ordinal outside $E$, choose a continuous increasing sequence $\{\alpha_\beta: \beta < \gamma\}$ outside $E$ with limit $\delta$ such that $\alpha_0 = 0$. Then $B_\alpha = \prod_{\beta < \gamma} \ker p_{\alpha_\beta \alpha_{\beta+1}}$. $\square$

As a consequence of 4.12, we can prove, assuming $\mathrm{E}(\lambda)$ for arbitrarily large non-$\omega$-measurable $\lambda$, that various duals groups, e.g., every free group of non-$\omega$-measurable cardinality, appear as the kernel of the restriction map from a triple dual onto a dual group. In the next section we will prove, assuming some set-theoretic hypothesis, that any double dual can appear in this way.

## §5. More dual groups

In this section we shall show that for any dual group $G$ (of non-$\omega$-measurable cardinality) there is a dual group $A$ such that $G^* \cong A^{**}/\sigma[A]$. Unfortunately, our proof will require some modest set-theoretical assumptions. We will need to assume that there are arbitrarily large cardinals $\lambda$ (less than the first measurable cardinal, if any) such that $\lambda = \mu^+$ where $\mu^{\aleph_0} = \mu$ and $E(\lambda)$ holds. Although it is consistent, assuming the consistency of certain large cardinal axioms, that $E(\lambda)$ fails to hold, the failure of this principle implies the consistency of very strong large cardinal principles: see VI.3.17. Of course this principle is true in L.

Inside L, it is possible to prove stronger results which are not consequences of ZFC. Since the proof in L is a model of the proof of the more general result and the stronger version is of independent interest, we will first give that proof.

**5.1 Theorem.** $(V = L)$ *Suppose $G = H^*$. Then for every regular non-weakly compact cardinal $\lambda \geq |H|$ there is a $\lambda$-separable dual group $A$ of cardinality $\lambda$ so that $A^{**}/\sigma[A] \cong G^*$. Furthermore, for any stationary set $E$ of ordinals of cofinality $\omega$, $A$ can be chosen so that $\Gamma(A) = \tilde{E}$.*

PROOF. It is enough to prove the theorem for a non-reflecting stationary set $E$ consisting of ordinals of cofinality $\omega$. For then, if $E \subseteq \lambda$ is any stationary set consisting of ordinals of cofinality $\omega$, by VI.3.5, $E = \amalg_{\alpha<\lambda} E_\alpha$ where each $E_\alpha$ is a non-reflecting subset of $\lambda$ and $E_\alpha \subseteq \{\beta : \alpha < \beta\}$; we can choose $\{A_\alpha : \alpha < \lambda\}$ a set of $\lambda$-separable dual groups of cardinality $\lambda$ so that: for all $\alpha$, $\Gamma(A_\alpha) = \tilde{E}_\alpha$; $A_0^{**}/\sigma[A_0] \cong G^*$; and for all $\alpha \neq 0$, $A_\alpha^{**}/\sigma[A_\alpha] \cong 0$; then $\bigoplus_{\alpha<\lambda} A_\alpha$ is the desired group (cf. VII.2.8).

So we can assume that $E \cap \delta$ is not stationary in $\delta$ for any limit $\delta < \lambda$. As well, we can assume that $E$ is co-stationary. Fix $\theta : \lambda \to H$ so that for all $h \in H$, $\theta^{-1}[h]$ is cofinal in $\lambda$. The group

$A$ will be a subgroup of $\mathbf{Z}^\lambda$ and will contain $\mathbf{Z}^{(\lambda)}$. Moreover $A$ will be contained in the $\mathbf{Z}$-adic closure of $\mathbf{Z}^{(\lambda)}$ in $\mathbf{Z}^\lambda$. (See I.3.1 for the definition of the $\mathbf{Z}$-adic topology.) Let $x_\alpha$ be the element of $\mathbf{Z}^\lambda$ defined by

$$x_\alpha(\beta) = \begin{cases} 1 & \text{if } \alpha = \beta \\ 0 & \text{otherwise.} \end{cases}$$

Let $\psi$ be the embedding of $G$ into $(\mathbf{Z}^{(\lambda)})^*$ defined by letting $\langle \psi(g), x_\alpha \rangle = g(\theta(\alpha))$. By $\Diamond(E)$ we have a sequence $\{f_\delta : \delta \in E\}$ which predicts functions from $\mathbf{Z}^{(\lambda)}$ to $\mathbf{Z}$. More exactly, each $f_\delta$ is a homomorphism from $\mathbf{Z}^{(\delta)}$, the subgroup generated by $\{x_\alpha : \alpha < \delta\}$, to $\mathbf{Z}$ and for any homomorphism $f$ from $\mathbf{Z}^{(\lambda)}$ to $\mathbf{Z}$, $\{\delta \in E : f \restriction \mathbf{Z}^{(\delta)} = f_\delta\}$ is stationary. Since $A$ will be a subgroup of the $\mathbf{Z}$-adic closure of $\mathbf{Z}^{(\lambda)}$, every element of $A^*$ is uniquely determined by its restriction to $\mathbf{Z}^{(\lambda)}$ (cf. I.3.6(ii)). So the family $\{f_\delta : \delta \in E\}$ can be viewed as predicting functions from $A$ to $\mathbf{Z}$. Our goal is to keep unwanted elements out of $A^*$ while ensuring that $\psi[G] \subseteq A^*$ (cf. (5.1.3) below).

Consider $\delta \in E$. If possible choose an increasing sequence of ordinals $\{\alpha_n : n \in \omega\}$ with limit $\delta$ and elements $\{z_{n\delta} : n \in \omega\}$ such that for all $n$: $z_{n\delta} \in \langle x_\alpha : \alpha_n < \alpha < \delta \rangle$; $f_\delta(z_{n\delta}) \neq 0$; and $\langle \psi(g), z_{n\delta} \rangle = 0$, for all $g \in G$. If such a choice is possible, let $y_\delta = \sum_{n<\omega} k_n z_{n\delta}$, where $n!$ divides $k_n$ and $f_\delta$ does not extend to a homomorphism from the pure closure of $\mathbf{Z}^{(\lambda)} \cup \{y_\delta\}$ to $\mathbf{Z}$. (Here the infinite sum denotes the coordinate-wise sum; the existence of the $k_n$ follows from the fact that $\mathbf{Z}$ is slender: see XIII.4.8.) Otherwise, let $y_\delta = 0$. Let $A$ be the pure closure of $\mathbf{Z}^{(\lambda)} \cup \{y_\delta : \delta \in E\}$. It is straightforward to verify that $A$ is a $\lambda$-separable group. In the case where $\lambda = \aleph_1$, $A$ is in standard form. It remains to verify that $A$ is a dual group and that $A^{**}/\sigma[A] \cong G^*$.

First we fix some notation. Let $\mathbf{Z}^\alpha$ denote the subgroup of $\mathbf{Z}^\lambda$ consisting of $\{x : x(\gamma) = 0$ for all $\alpha \leq \gamma\}$. Let $\rho_\alpha$ denote the projection of $\mathbf{Z}^\lambda$ on $\mathbf{Z}^\alpha$. Let $A_\alpha \overset{\text{def}}{=} \rho_\alpha[A]$. Notice that $A_\alpha$ is the subgroup of $A$ generated by $\{x_\beta : \beta < \alpha\} \cup \{y_\delta : \delta \leq \alpha\}$, in fact $A_\alpha$

is a direct summand of $A$. Since each $A_\alpha$ has cardinality $< \lambda$ and so is free,

(5.1.1)                    for all $\alpha$, $A_\alpha$ is reflexive.

(In L, there are no measurable cardinals so every free group is reflexive: see III.3.8.)

Next we verify that

(5.1.2)                    $\psi[G]$ is a subgroup of $A^*$.

Consider $g \in G$. Let $\psi(g)$ also denote the unique extension of $\psi(g)$ to a map from $A$ to $\hat{\mathbf{Z}}$. But since $\langle \psi(g), y_\delta \rangle = 0$ for all $\delta \in E$, $\mathrm{rge}(\psi(g)) \subseteq \mathbf{Z}$.

An element of $f \in A^*$ is said to be *bounded* if there is some $\alpha$ so that $f{\restriction}(\ker \rho_\alpha {\restriction} A) = 0$. Let $B$ denote the bounded elements of $A^*$. It is enough to show that

(5.1.3)                    $$A^* = B \oplus \psi[G]$$

and (5.1.4)

(5.1.4)                    $B$ is a predual of A.

For then we can apply 1.6 (with $A_1 = A$), to get

$$A^{**}/\sigma[A] \cong \ker \rho_3 \cong (A_2/\sigma[A_0])^* \cong (A^*/B)^* \cong \psi[G]^* \cong G^*.$$

We first prove (5.1.3). Suppose that $f \in A^*$. We claim that there is $\alpha < \lambda$, so that for all $\gamma$, $\beta > \alpha$, if $\theta(\beta) = \theta(\gamma)$, then $f(x_\beta) = f(x_\gamma)$. Assume not. Then, by an argument which is by now familiar, we can find $\delta \in E$ so that $f{\restriction}\mathbf{Z}^{(\delta)} = f_\delta$ and there is an increasing sequence of ordinals $\{\alpha_n : n \in \omega\}$ with limit $\delta$ such that for each $n$, there is $\alpha_n \le \beta_n < \gamma_n < \alpha_{n+1}$ such that $\theta(\beta_n) = \theta(\gamma_n)$ but $f(x_{\beta_n}) \ne f(x_{\gamma n})$. Then $z_{n\delta} = x_{\beta_n} - x_{\gamma n}$ is as required in the definition of $y_\delta$. Hence $f_\delta$ does not extend to an element of $A^*$, which contradicts the existence of $f$.

Let $\alpha$ be as above and define a function $\tilde{f}: H \to \mathbf{Z}$, by $\tilde{f}(h) = f(x_\beta)$ where $\alpha < \beta$ and $\theta(\beta) = h$. We now claim $\tilde{f}$ is a homomorphism. Suppose not. Choose $h_1$, $h_2$ so that $\tilde{f}(h_1) + \tilde{f}(h_2) \neq \tilde{f}(h_1 + h_2)$. Then by standard arguments we can find $\delta \in E$ so that $f \restriction \mathbf{Z}^{(\delta)} = f_\delta$ and there is an increasing sequence of ordinals $\{\alpha_n : n \in \omega\}$ with limit $\delta$ such that for each $n$, there are $\alpha_n \leq \beta_n$, $\gamma_n$, $\nu_n < \alpha_{n+1}$ such that $\theta(\beta_n) = h_1$, $\theta(\gamma_n) = h_2$ and $\theta(\nu_n) = h_1 + h_2$. Then $z_{n\delta} = x_{\beta_n} + x_{\gamma n} - x_{\nu_n}$ is as required in the definition of $y_\delta$. Hence $f_\delta$ does not extend to an element of $A^*$, which contradicts the existence of $f$.

We can now complete the proof of (5.1.3). We have shown that $\tilde{f}$ is an element of $H^* = G$. By construction, for every $\beta > \alpha$, $\psi(\tilde{f})(x_\beta) = \tilde{f}(\theta(\beta)) = f(x_\beta)$, so $f - \psi(\tilde{f}) \in B$, and hence $f = (f - \psi(\tilde{f})) + \psi(\tilde{f}) \in B + \psi[G]$. It is clear that $B \cap \psi[G] = \{0\}$.

It is easy enough to show that $B$ is a predual of $A$ by a direct argument, but we are in a position to use the machinery of section XI.1. Let $\tau_{\alpha\beta}$ denote the injection of $A_\alpha$ into $A_\beta$ and let $\pi_{\beta\alpha}$ be the restriction of $\rho_\alpha$ to $A_\beta$. Then $(A_\alpha, \tau_{\alpha\beta}, \pi_{\beta\alpha} : \alpha < \beta < \lambda)$ is an inverse-direct system. Observe that, by an application of Fodor's lemma as in the proof of 4.5, $A = \varinjlim A_\alpha = \varprojlim A_\alpha$ (see Exercise XI.14). As well, $B = \varinjlim A_\alpha^*$, so $B^* = \varprojlim A_\alpha^{**}$. But each $A_\alpha$ is reflexive, so $\varprojlim A_\alpha^{**} = \varprojlim A_\alpha = \varinjlim A_\alpha = \varinjlim A_\alpha^{**}$, and by XI.1.10, $\varphi_B$ is an isomorphism from $A$ onto $B^* = \varinjlim A_\alpha^{**}$. $\square$

One consequence of the theorem above is that, assuming V $=$ L, for every possible length-rank and non-weakly compact regular cardinal $\lambda$, there is a group of that length-rank with the additional property that each group of even index which appears in the sequence is $\lambda$-separable of cardinality $\lambda$; moreover, these groups can be very different (and thus strongly non-reflexive):

**5.2 Corollary.** (V $=$ L) *Suppose $\lambda$ is a regular non-weakly compact cardinal and $\{E_n : n \in \omega\}$ is a sequence of stationary subsets of $\lambda$ consisting of ordinals of cofinality $\omega$ such that for all $n$, $E_n \subseteq E_{n+1}$. There is a dual group $A_0$, so that for every $n$, $A_{2n}$ is a $\lambda$-separable* *group of cardinality $\lambda$ and* $\Gamma(A_{2n}) = \tilde{E}_n$. *(Here $A_{n+1} \overset{\text{def}}{=} A_n^*$.)*

PROOF. It suffices to show that for every finite $n$, there is a dual group $A_0$ of length-rank $(n, \infty)$ such that for all $k \leq n$, $A_{2k}$ is a $\lambda$-separable group of cardinality $\lambda$ and $\Gamma(A) = \tilde{E}_k$. This statement can be proved by induction on $n$, using 5.1. $\square$

These results are particularly interesting in the case that $\lambda = \aleph_1$. In this case it suffices to assume that $\diamondsuit$ holds, since $E(\aleph_1)$ is always true. If we assume that $\diamondsuit$ holds for some stationary subset of $\aleph_1$ then there is a strongly non-reflexive dual group of cardinality $\aleph_1$ which is $\aleph_1$-separable. This statement cannot be proved in ZFC:

**5.3 Corollary.** *It is independent of ZFC whether every $\aleph_1$-separable group of cardinality $\aleph_1$ is reflexive.*

PROOF. Assuming MA $+ \neg$CH, every $\aleph_1$-separable group of cardinality $\aleph_1$ is reflexive (see Corollary XI.2.9). But assuming $\diamondsuit$, 5.2 implies that there is an $\aleph_1$-separable group of cardinality $\aleph_1$ which is strongly non-reflexive. $\square$

In Corollary 5.2, we cannot hope to require that *every* $A_n$ is $\lambda$-separable, since the dual of a $\lambda$-separable group will contain a summand isomorphic to $\mathbf{Z}^\omega$, which is not even strongly $\aleph_1$-free. However, by a different proof we can obtain, assuming $V = L$, a dual group $A_0$ of cardinality $\aleph_1$ so that for every $n$, $A_n$ is strongly $\aleph_1$-free and moreover is strongly non-reflexive: see Eklof-Mekler-Shelah 1987.

We are now going to prove the analog of Theorem 5.1 using the Black Box (cf. XIII.2.1). The proof follows the proof of 5.1 in outline, but there are several differences. In the course of the proof we will not be able to assume that the groups $A_\alpha$ are free. So we will need to use a different argument to show the analog of (5.1.1), i.e., that each $A_\alpha$ is reflexive. In the proof of the following theorem, we will use the fact that tree groups are reflexive. The groups we will construct will be subgroups of $\mathbf{Z}^{\lambda \times \omega}$.

It is useful to have some notation for dealing with these groups. As usual, we view $\mathbf{Z}^{(\lambda \times \omega)}$ as the subgroup of $\mathbf{Z}^{\lambda \times \omega}$ consisting of those elements which are 0 in all but finitely many coordinates. $Z^{(\lambda \times \omega)}$ is freely generated by $\{x_{\alpha n} : \alpha < \lambda, n \in \omega\}$ where

$$x_{\alpha n}(\beta, m) = \begin{cases} 1 & \text{if } \beta = \alpha \text{ and } m = n \\ 0 & \text{otherwise.} \end{cases}$$

Similarly for $X \subseteq \lambda$ and $Y \subseteq \omega$, we identify $Z^{X \times Y}$ as the subgroup of $Z^{\lambda \times \omega}$ consisting of the those elements which are 0 on $\lambda \times \omega \setminus X \times Y$. The projection onto $Z^{\alpha \times \omega}$ is denoted by $\rho_\alpha$. For $s \subseteq \omega$, $\rho_s$ denotes the projection on $Z^{\lambda \times s}$. We let $Z^{(>\alpha n)}$ denote the subgroup generated by $\{x_{\beta m} : \beta > \alpha \text{ and } m > n\}$.

**5.4 Theorem.** *If $G = H^*$, then there is a dual group $A$ so that $A^{**}/\sigma[A] \cong G^*$, provided that there is a non-$\omega$-measurable cardinal $\lambda = \mu^+ \geq |H|$ such that $\mathrm{E}(\lambda)$ holds and $\mu^{\aleph_0} = \mu$.*

PROOF. Fix $E \subseteq \lambda$ a stationary set consisting of ordinals of cofinality $\omega$ which is not stationary in any limit ordinal less than $\lambda$. Choose $\theta : \lambda \to H$ so that for every element of $h$ of $H$, $\theta^{-1}[h]$ is cofinal in $\lambda$. Consider the free group $Z^{(\lambda \times \omega)}$. Let $\psi$ be the embedding of $G$ into $(Z^{(\lambda \times \omega)})^*$ defined by $\langle \psi(g), x_{\alpha n} \rangle = g(\theta(\alpha))$. Let the initial structure of the Black Box be $N = (\mathrm{H}(\lambda^{++}), \in, <, G, \psi)$. Let $\{(M_i, X_i) : i \in I\}$ be as in the definition of the Black Box with respect to $N$ and the stationary set $E$ (cf. XIII.2.1). Let $\{\eta_\delta : \delta \in E\}$ be a ladder system as guaranteed by clause (d) of the Black Box.

We will build $A$ as a pure subgroup of the $Z$-adic closure of $Z^{(\lambda \times \omega)}$ in $Z^{\lambda \times \omega}$ so that $A$ contains $Z^{(\lambda \times \omega)}$. Our goal in the construction of $A$ is to control $A^*$ by preventing unwanted elements of $(Z^{(\lambda \times \omega)})^*$ from being in $A^*$ while ensuring that $\psi[G] \subseteq A^*$.

We will view each $X_i$ as predicting a homomorphism from $Z^{(\lambda \times \omega)} \cap M_i$ to $Z$, which we will denote as $f_i$. The exact details of the coding are not important. For example, we can assume that the set underlying $Z^{(\lambda \times \omega)}$ is $\lambda$, $\{x_{\alpha n} : \alpha < \lambda, n \in \omega\}$ is the set of limit ordinals and both the group structure on $Z^{(\lambda \times \omega)}$ and the bijection from $\lambda \times \omega$ to $\{x_{\alpha n} : \alpha < \lambda, n \in \omega\}$ are definable. If we fix a definable bijection between $\lambda$ and $\lambda \times Z$, then any $X \subseteq \lambda$ codes a subset of

$\lambda \times \mathbf{Z}$. If the subset coded is a homomorphism, then $X$ codes that homomorphism; otherwise we can let $X$ code the homomorphism which is constantly 0.

Suppose now that $\delta \in E$. Fix some $(M_{i_0}, X_{i_0})$ with $\delta(i_0) = \delta$. Consider the following property

$(*)$     for all $\alpha < \delta$ and $n \in \omega$, there exists $z \in \mathbf{Z}^{(>\alpha n)}$ such that $f_{i_0}(z) \neq 0$, but $\langle \psi(g), z \rangle = 0$ for all $g \in G$.

If $(M_{i_0}, X_{i_0})$ does not satisfy $(*)$, then we let $y_i = 0$ for all $i$ such that $\delta = \delta(i)$.

Assume that $(M_{i_0}, X_{i_0})$ satisfies $(*)$. By part (d) of the Black Box, $M_{i_0} = \cup M_{i_0 n}$ where each $M_{i_0 n}$ is an elementary submodel of $M_{i_0}$. Since each $M_{i_0 n}$ also satisfies $(*)$ we can choose an increasing sequence of natural numbers $m_n (n \in \omega)$, and elements $z_{n i_0} \in M_{i_0 n}$ such that for all $n$: $z_{n i_0}$ is in the subgroup generated by $\{x_{\gamma k} : \eta_\delta(n) < \gamma < \eta_\delta(n+1), n < k < m_n\}$; for all $g \in G$, $\langle \psi(g), z_{n i_0} \rangle = 0$; and $f_{i_0}(z_{n i_0}) \neq 0$. Choose a sequence of integers, $\{k_n : n \in \omega\}$ so that $n!$ divides $k_n$ and no extension of $f_{i_0}$ induces a homomorphism from the pure closure of $\mathbf{Z}^{(\lambda \times \omega)} \cup \{\sum_{n \in \omega} k_n z_{n i_0}\}$ to $\mathbf{Z}$. For any $i$ such that $\delta(i) = \delta$, let $z_{ni}$ be the image of $z_{n i_0}$ under the isomorphism from $M_{i_0}$ to $M_i$. Let $y_i = \sum_{n \in \omega} k_n z_{ni}$. Since $(M_i, X_i) \cong (M_{i_0}, X_{i_0})$, if $f_i$ is the homomorphism which $X_i$ predicts then $f_i$ does not extend to the pure closure of $\mathbf{Z}^{(\lambda \times \omega)} \cup \{y_i\}$. Notice that by clause (d) of the Black Box, for all $n$, $z_{ni}$ is in the subgroup generated by $\{x_{\gamma k} : \eta_\delta(n) < \gamma < \eta_\delta(n+1), n < k < m_n\}$.

Define $A$ to be the pure closure of $\mathbf{Z}^{(\lambda \times \omega)} \cup \{y_i : i \in I\}$. We now verify that $A$ is the required group. Notice that for every $\alpha$ and finite set $s \subseteq \omega$, $\rho_\alpha[A]$ and $\rho_s[A]$ are contained in $A$. Furthermore, $\rho_s[A] = \mathbf{Z}^{(\lambda \times s)}$. Define $A_\alpha$ to be $\rho_\alpha[A]$. If we let $\tau_{\alpha\beta}$ be the inclusion map and $\pi_{\beta\alpha}$ the restriction of $\rho_\alpha$, then $(A_\alpha, \tau_{\alpha\beta}, \pi_{\beta\alpha} : \alpha < \beta)$ is an inverse-direct system. An element $f$ of $A^*$ is *bounded* if $f \upharpoonright \ker(\rho_\alpha \upharpoonright A) = 0$ for some $\alpha < \lambda$ and is *semibounded* if it is the sum of a bounded element together with an element $\psi(g)$ such that $\psi(g) \upharpoonright \ker(\rho_s \upharpoonright A) = 0$ for some finite subset $s \subseteq \omega$. Let $B$ denote the bounded elements of $A^*$ and $S$ the semibounded elements. We will show that every element of $A^*$ is the sum of a semibounded function and an element of $\psi[G]$. To finish the proof we will show that $A$ is a predual of $S$. The proof follows the same lines as that of Theorem 5.1. First we show that

(5.4.1)                  for all $\alpha$, $A_\alpha$ is a reflexive group.

The proof is by induction on $\alpha$. Note that $A_\alpha$ is the pure closure of $\mathbf{Z}^{(\alpha \times \omega)} \cup \{y_i : \delta(i) \leq \alpha\}$. The only interesting cases occur when $\alpha$ is a limit ordinal. Suppose first that either $\alpha \notin E$ or $(M_i, X_i)$ does not satisfy (*) for some $i$ such that $\delta(i) = \alpha$. Since $E$ is non-reflecting, we can choose an increasing continuous sequence $\{\beta_\nu : \nu < \kappa\}$ of ordinals outside of $E$ with limit $\alpha$. Then

$$A_\alpha = \bigcup_{\nu < \kappa} A_{\beta_\nu} = A_{\beta_0} \oplus \bigoplus_{\nu < \kappa} \ker \pi_{\beta_{\nu+1} \beta_\nu}.$$

Now $\ker \pi_{\beta_{\nu+1} \beta_\nu}$ is reflexive since it is a summand of $A_{\beta_{\nu+1}}$. Since $A_\alpha$ is a direct sum of reflexive groups, it is reflexive. (See X.1.1.)

Suppose now that $\alpha = \delta \in E$, and if $\delta(i) = \delta$ then $(M_i, X_i)$ satisfies (*). In the choice of $\{y_i : \delta(i) = \delta\}$ there is a sequence of natural numbers, $m_n$ so that each $y_i$ $(\delta(i) = \delta)$ is contained in the $\mathbf{Z}$-adic closure of $\bigoplus_{n \in \omega} \mathbf{Z}^{([\eta_\delta(n), \eta_\delta(n+1)) \times m_n)} \overset{\text{def}}{=} D$. It is easy to see that $D$ is a direct summand of $\bigcup_{\beta < \alpha} A_\beta$, with complementary summand

$$C \overset{\text{def}}{=} A_{\eta_\delta(0)} \oplus \bigoplus_{n \in \omega} (\ker(\rho_{mn} \restriction \ker \pi_{\eta_\delta(n+1)\eta_\delta(n)})).$$

By the induction hypothesis, $C$ is reflexive. Finally $A_\alpha$ is the direct sum of $C$ and the pure closure of $D \cup \{y_i : \delta(i) = \delta\}$. To prove the claim we need to see that the the pure closure of $D \cup \{y_i : \delta(i) = \delta\}$ is a reflexive group. But this group is a tree group, and so is reflexive, by 2.3.

Exactly as in the proof of (5.1.2), we can show

(5.4.2)                           $\psi[G] \subseteq A^*.$

Next we show.

(5.4.3)                           $A^* = S \oplus \psi[G].$

Suppose $f \in A^*$. Let $X$ be the subset of $\lambda$ which codes $f \upharpoonright \mathbf{Z}^{(\lambda \times \omega)}$. By the construction of $A$ and the Black Box, $(N, X)$ does not satisfy

(∗)   for all $\alpha$ and $n$, there exists $z \in \mathbf{Z}^{(>\alpha n)}$ such that $f(z) \neq 0$ but $\langle \psi(g), z \rangle = 0$ for all $g \in G$.

Choose $\alpha < \lambda$ and $n \in \omega$ so that for all $z \in \mathbf{Z}^{>\alpha n}$, if $f(z) = 0$ then $\langle \psi(g), z \rangle = 0$ for all $g \in G$. Suppose that $\alpha < \beta, \gamma$ and $n < m$, $k$. If $\theta(\beta) = \theta(\gamma)$, then $f(x_{\beta m}) = f(x_{\gamma k})$, for otherwise $x_{\beta m} - x_{\gamma k}$ contradicts the choice of $\alpha$ and $n$.

So $f$ induces a map $\tilde{f}$ on $H$ by letting $\tilde{f}(h)$ be the eventual value of $f$ on $\theta^{-1}[h] \times \omega$. Just as in the proof of 5.1, we can show that $\tilde{f}$ is a homomorphism. So $\tilde{f} \in G$ and $f - \psi(\tilde{f})$ is semibounded. Finally, we show that

(5.4.4)                   $S$ is a predual of A.

As in the proof of (5.1.4), it is easy to show that $\varinjlim A_\alpha = \varprojlim A_\alpha$ and that $B$ is the direct limit of the dual system. So $B$ is a predual of $A$. The group of semibounded functions modulo the bounded functions is isomorphic to $(\mathbf{Z}^\lambda / \mathbf{Z}^{<\lambda})^{(\omega)}$. Since by Łoś' theorem (III.3.3), $(\mathbf{Z}^\lambda / \mathbf{Z}^{<\lambda})^* = 0$, the semibounded functions also form a predual of $A$. □

**5.5 Corollary.** *Suppose that below the first measurable cardinal (if any), there are arbitrarily large cardinals $\lambda = \mu^+$ such that $\mathrm{E}(\lambda)$ holds and $\mu^{\aleph_0} = \mu$. Then for any group $H$ of non-$\omega$-measurable cardinality, there is a dual group $A$ so that $A^{**}/\sigma[A] \cong H^{**}$.* □

We can attempt to repeat the proof of 5.4 without the assumption that $\mathrm{E}(\lambda)$ holds. The only part of the proof which fails is the proof that the semibounded functions in $A^*$ form a predual of $A$ (which in turn depends on the fact that each of the $A_\alpha$ are reflexive). However in view of Theorem 3.2, we still know that $A$ is a dual group and that $G^*$ is isomorphic to a direct summand of $A^{**}/\sigma[A]$. So in the proof of Theorem 5.4, we have established the following lemma.

**5.6 Lemma.** *For every group* $H$ *whose cardinality is non-$\omega$-mea-surable and for every non-$\omega$-measurable* $\mu$ *such that* $\mu^{\aleph_0} = \mu$ *and* $|H| \leq \mu^+$ *there is a dual group* $A$ *of cardinality* $\mu^+$ *so that* $H^{**}$ *is isomorphic to a direct summand of* $A^{**}/\sigma[A]$. $\square$

It is natural to conjecture that for any dual group $G$ there is a dual group $A$ so that $G \cong A^{**}/\sigma[A]$. Since every dual is isomorphic to a direct summand of its own double dual, we have the following theorem, which is a weak form of the conjecture.

**5.7 Theorem.** *For every dual group* $G$ *of non-$\omega$-measurable car-dinality there is a dual group* $A$ *so that* $G$ *is isomorphic to a direct summand of* $A^{**}/\sigma[A]$. $\square$

Lemma 5.6 can also be used to give another proof (in ZFC) that there is a strongly non-reflexive dual group. Choose $\mu$ so that $\mu^{\aleph_0} = \mu$. Let $\lambda = \mu^+$. Let $H$ be the subgroup of $\mathbf{Z}^\lambda$ generated by the elements which are 0 on all but a countable set. So $H^* \cong \mathbf{Z}^{(\lambda)}$ and $|H| = \lambda$. By Lemma 5.6, there is a dual group $A$ of cardinality $\lambda$ so that $H^{**}$ is isomorphic to a subgroup of $A^{**}$. But $H^{**} \cong \mathbf{Z}^\lambda$. So $|A^{**}| \geq 2^\lambda$, and hence not isomorphic to $A$.

## NOTES

The material in this chapter is new. Section 1 is due to Mekler. 2.3 is due to Mekler and Shelah. 3.1 is due to Mekler, 3.2 and 3.3 are due to Shelah. Section 4 is due to Mekler, but is related to Schlitt 1986 (see XI.1 and the notes for XI). The existence of sequences of dual groups of arbitrary length was first shown consistent in Eklof-Mekler-Shelah 1987. Section 5 is due to Mekler and Shelah.

It can be shown that CH implies that there is a non-reflexive sub-group of $\mathbf{Z}^\omega$. A proof appears in the lecture notes from Eklof's 1985 course at Simon Fraser University.

The logical significance of length-rank can be understood by consid-ering sequences of dual groups, i.e., structures of the form $(A_1, A_2, \ldots, \mathbf{Z};$ $\langle\,,\,\rangle)$ where $A_1$ is a dual group and $A_{n+1} = A_n^*$ (cf. XI.3.6). Mekler has shown that the length-rank of $A_1$ characterizes the $L_{\infty\omega}$-theory of sequences of dual groups.

# OPEN PROBLEMS

## Almost free modules.

1. Is it true that for any ring which is not left-perfect, $\mathrm{Inc}'(R) = \mathrm{Inc}(\mathbf{Z})$?

2. If we give the general definition of $\kappa$-separable, are direct summands of $\kappa$-separable groups $\kappa$-separable? [Note that in the case where $|A| = \kappa$, if $A$ is a counterexample then $\Gamma(A) = 1$.]

3. Is it true that if $\kappa$-free does not imply $\kappa^+$-free, then $\kappa$-separable does not imply $\kappa^+$-separable? [Note Exercises VII.13–17.]

4. (Kaplansky test problem for $\aleph_1$-separable groups) Suppose $A$, $B$ are countable torsion-free groups so that $A$ is not isomorphic to $B$, but $A \oplus A \cong B \oplus B$. Are there $\aleph_1$-separable groups $G$, $H$ (of cardinality $\aleph_1$) so that $G \oplus G \cong H \oplus H$, the quotient type of $G$ is $A$, and the quotient type of $H$ is $B$? [cf. Thomé 1988 and 19??a and b.)

5. Find a large cardinal equiconsistent with "there is a $\kappa$, so that $\kappa$-free implies free". [Shelah has shown that the first such cardinal has some large cardinal properties.]

6. (Droste) Is it consistent that $\mathrm{Inc}(\mathbf{Z})$ is countable?

A stronger question is the following. Consider the smallest class $K$ of successor ordinals containing 1 and closed under ordinal addition and the operation $\alpha \mapsto \omega_\alpha + 1$. Magidor and Shelah 19?? (assuming GCH) and Shelah (in ZFC), have shown that $\mathrm{Inc}(\mathbf{Z})$ contains $\{\aleph_\alpha : \alpha \in K\}$. Is it consistent that $\mathrm{Inc}(\mathbf{Z})$ is exactly this set?

7. For any variety $\mathcal{V}$, is $\lambda$ in the essentially non-free incompactness spectrum if and only if there is some $n$ such that $(\mathrm{CP}_n)$ holds and there is a $\lambda$-free family of countable sets based on a $\lambda$-system of height $n$? [See VII.3A.17.]

8. Does the incompactness spectrum of groups equal that of abelian groups?

## Structure of Ext.

1. Find a combinatorial principle equivalent to the existence of a non-free W-group (of arbitrary cardinality)? [See XII.3].

2. Is it consistent that the class of W-groups of cardinality $\aleph_1$ is exactly the class of strongly $\aleph_1$-free groups of cardinality $\aleph_1$?

3. If we have, say, all the uniformization results that can be deduced from MA + ¬CH, then is every strongly $\aleph_1$-free (every Shelah) group of cardinality $\aleph_1$ a W-group?

4. If every strongly $\aleph_1$-free group of cardinality $\aleph_1$ is a W-group, are they also all $\aleph_1$-coseparable?

5. Does strongly $\aleph_1$-free plus $\aleph_1$-coseparable imply $\aleph_1$-separable (for groups of cardinality $\aleph_1$)?

6. Is it consistent that there are filtration-equivalent $\aleph_1$-separable groups of cardinality $\aleph_1$ such that one is a W-group but the other is not?

## Endomorphism rings.

1. Characterize the endomorphism rings of separable groups modulo the idea of small endomorphisms.

## Dual groups.

1. If one predual of a dual group is a dual group, must they all be dual groups?

2. Investigate the foundation rank of dual groups.

3. Suppose $A_0$, $A_1$, $A_2$, $A_3$ is a sequence of groups where $A_{i+1} = A_i^*$; if $B$ is the subgroup annihilated by ker $\rho_3$, is $B$ a maximal predual of $A_1$?

4. If $A$ is a dual group, which groups can appear as $A^{**}/\sigma[A]$? [Recall from XIV.1.7 that any such group is a dual group; cf. the results in XIV.5.]

5. Is it provable in ZFC that every $\aleph_1$-separable group of cardinality $\aleph_1$ is a dual group?

6. Is there a reflexive group of $\omega$-measurable cardinality?

7. Is it provable in ZFC that there exists $A$ such that $A^*$ and $A^{**}$ are both slender?

8. Is it provable in ZFC that there exists $A \subseteq \mathbf{Z}^\omega$ which is non-reflexive? [follows from CH: see the Notes to XIV] Even assuming CH, is there such an $A$ which is strongly non-reflexive?

9. (Huber) Is it provable in ZFC that every W-group (of cardinality $\aleph_1$) is reflexive?

10. Is there a **Z**-chain of strongly non-reflexive dual groups i.e., groups $A_n$ ($n \in \mathbf{Z}$) such that for all $n$, $A_n^* = A_{n+1}$ and $A_n$ is not isomorphic to $A_{n+2}$? (And for other partial orders.)

11. Is there a group $A$ such that $A^{**}$ is not $H^{***}$ for any $H$?

12. If $A$ is a dual group of infinite rank, is $A \cong A \oplus \mathbf{Z}$?

## Others.

1. Is $\text{Reid}_\mu$ closed under direct summands?

2. Investigat dependence/independence among the arrows in IV.2, VII.4, especially: is it consistent that every W-group is free but not every hereditarily-separable group is a W-group?

3. Does every $\aleph_1$-separable group of cardinality $\aleph_1$ have a coherent system of projections?

# BIBLIOGRAPHY

Anderson, F. W. and Fuller, K. R.
   1974   **Rings and Categories of Modules**, Springer-Verlag.
Baer, R.
   1937   *Abelian groups without elements of finite order*, Duke Math.
          J. **3**, 68–122.
Balcerzyk, S.
   1959   *On factor groups of some subgroups of a complete direct
          sum of infinite cyclic groups*, Bull. Acad. Polon. Sci., Sér.
          Sci. Math. **7**, 141–142.
   1962   *On groups of functions defined on Boolean algebras*, Fund.
          Math. **50**, 347–367.
Bass, H.
   1960   *Finitistic dimension and a homological generalization of
          semiprimary rings*, Trans. Amer. Math. Soc. **95**, 466–
          488.
Baumgartner, J. E.
   1984   *Applications of the proper forcing axiom*, in **Handbook of
          Set-Theoretic Topology**, North-Holland, 913–959.
Becker, T., Fuchs, L. and Shelah, S.
   1989   *Whitehead modules over domains*, Forum Math. **1**, 53–68.
Bell, J. L.
   1974   *On compact cardinals*, Z. Math. Logik Grundlag. Math.
          **20**, 389–393.
Bell, J. L. and Slomson, A. B.
   1969   **Models and Ultraproducts: An Introduction**, North-
          Holland.
Beller, A. and Litman, A.
   1980   *A strengthening of Jensen's □ principles*, J. Symbolic Logic
          **45**, 251–264.
Ben-David, S.
   1978   *On Shelah's compactness of cardinals*, Israel J. Math. **31**,
          34–56.
Bergman, G. and Solovay, R. M.
   1987   *Generalized Horn sentences and compact cardinals*, Ab-
          stracts Amer. Math. Soc., 832–04–13.

Blass, A.
    1987    *Near coherence of filters. II*, Trans. Amer. Math. Soc **300**, 557–581.

Blass, A. and Laflamme, C.
    1989    *Consistency results about filters and the number of inequivalent growth types*, J. Symbolic Logic **54**, 50–56.

Cartan, H. and Eilenberg, S.
    1956    **Homological Algebra**, Princeton University Press.

Chang, C. C. and Keisler, H. J.
    1973    **Model Theory**, North-Holland.

Chase, S. U.
    1960    *Direct products of modules*, Trans. Amer. Math. Soc. **97**, 457–473.
    1962a    *Locally free modules and a problem of Whitehead*, Illinois J. Math. **6**, 682–699.
    1962b    *On direct sums and products of modules*, Pacific J. Math. **12**, 847–854.
    1963a    *On group extensions and a problem of J. H. C. Whitehead*, in **Topics in Abelian Groups**, Scott, Foresman and Co., 173–197.
    1963b    *Function topologies on abelian groups*, Illinois J. Math. **7**, 593–608.

Cohn, P. M.
    1981    **Universal Algebra**, Reidel.

Comfort, W. W. and Negrepontis, S.
    1972    *On families of large oscillation*, Fund. Math. **75**, 275–290.
    1974    **The Theory of Ultrafilters**, Springer-Verlag.

Corner, A. L. S.
    1963    *Every countable reduced torsion-free ring is an endomorphism ring*, Proc. London Math. Soc.(3) **13**, 687–710.
    1969a    *On endomorphism rings of primary abelian groups*, Quart. J. Math. Oxford Ser. (2) **20**, 277–296.
    1969b    *Endomorphism algebras of large modules with distinguished submodules*, J. Algebra **11**, 155–185.

Corner, A. L. S. and Göbel, R.
  1985   *Prescribing endomorphism algebras, a unified treatment,*
          Proc. London Math. Soc. (3) **50**, 447–479.

Cutler, D.
  1987   *Abelian p-groups determined by their $p^n$-socle,* in **Abelian
          Group Theory** (ed. by R. Göbel & E. A. Walker), Gordon
          and Breach, 111–116.

Cutler, D., Mader, A. and Megibben, C.
  1989   *Essentially indecomposable abelian p-groups having a filtra-
          tion of prescribed type,* in **Abelian Group Theory**, Con-
          temporary Mathematics vol. 87, Amer. Math. Soc., 43–50.

Dales, H. G. and Woodin, W. H.
  1987   **An Introduction to Independence for Analysts,** Lon-
          don Math. Soc. Lecture Note Series No. 115, Cambridge
          University Press.

Devlin, K. J.
  1973   **Aspects of Constructibility,** Lecture Notes in Mathe-
          matics No. 354, Springer-Verlag.

  1977   **The Axiom of Constructibility: A Guide for the
          Mathematician,** Lecture Notes in Mathematics No. 617,
          Springer-Verlag.

  1978   *A note on the combinatorial principles $\Diamond(E)$,* Proc. Amer.
          Math. Soc. **72**, 163–165.

  1979   *Variations on $\Diamond$,* J. Symbolic Logic **44**, 51–58.

  1983   *The Yorkshireman's guide to proper forcing,* in **Surveys in
          Set Theory,** London Math. Soc. Lecture Note Series No.
          87, Cambridge University Press, 60–115.

  1984   **Constructibility,** Springer-Verlag.

Devlin, K. J. and Jensen, R. B.
  1975   *Marginalia to a theorem of Silver,* in **Logic Conference
          Kiel 1974,** Lecture Notes in Mathematics No.   499,
          Springer-Verlag, 115–142.

Devlin, K. J. and Johnsbråten, H.
  1974   **The Souslin Problem,** Lecture Notes in Mathematics No.
          405, Springer-Verlag.

Devlin, K. J. and Shelah, S.
   1978   *A weak version of $\Diamond$ which follows from $2^{\aleph_0} < 2^{\aleph_1}$*, Israel
          J. Math. **29**, 239–247.

Dickmann, M. A.
   1975   **Large Infinitary Languages**, North-Holland.

Dimitrić, R.
   1983   *Slender modules over domains*, Comm. in Alg. **11**, 1685–
          1700.

Drake, F. R.
   1974   **Set Theory: An Introduction to Large Cardinals**,
          North-Holland.

Dugas, M.
   1981   *Fast freie abelsche Gruppen mit Endomorphismenring Z*, J.
          Algebra **71**, 314–321.
   1985   *On reduced products of abelian groups*, Rend. Sem. Mat.
          Univ. Padova **73**, 41–47.
   1988   *On some subgroups of infinite rank Butler groups*, Rend.
          Sem. Mat. Univ. Padova **79**, 153–161.

Dugas, M. and Göbel, R.
   1979a  *Algebraisch kompakte Faktorgruppen*, J. Reine Angew.
          Math. 307/308, 341–352.
   1979b  *Die Struktur kartesischer Produkte ganzer Zahlen mod-
          ulo kartesischer Produkte ganzer Zahlen*, Math. Z. **168**,
          15–21.
   1981   *Quotients of reflexive modules*, Fund. Math. **114**, 17–28.
   1982a  *Every cotorsion-free ring is an endomorphism ring*, Proc.
          London Math. Soc.(3) **45**, 319–336.
   1982b  *Every cotorsion-free algebra is an endomorphism algebra*,
          Math. Z. **181**, 451–470.
   1982c  *On endomorphism rings of primary abelian groups*, Math.
          Ann. **261**, 359–385.
   1984   *Almost $\Sigma$-cyclic abelian p-groups in L*, in **Abelian Groups
          and Modules**, CISM Courses and Lectures No.  287,
          Springer-Verlag, 87–105.
   1985a  *On radicals and products*, Pacific J. Math. **118**, 79–104.

1985b *Endomorphism rings of separable torsion-free abelian groups*, Houston J. Math. **11**, 471–483.

Dugas, M., Hill, P. and Rangaswamy, K. M.

19?? *Butler groups of infinite rank II*, Trans. Amer. Math. Soc., to appear.

Dugas, M. and Irwin, J.

1989 *On pure subgroups of cartesian products of integers*, Results in Math. **15**, 35–52.

19?? *On basic subgroups of* $\Pi Z$, preprint.

Dugas, M., Mader, A. and Vinsonhaler, C.

1987 *Large E-rings exist*, J. Algebra **108**, 88–101.

Dugas, M. and Rangaswamy, K. M.

1988 *Infinite rank Butler groups*, Trans. Amer. Math. Soc. **305**, 129–142.

Dugas, M. and Shelah, S.

1989 *E-transitive groups in* $L$, in **Abelian Group Theory**, Contemporary Mathematics vol. 87, Amer. Math. Soc., 191–199.

Dugas, M. and Thomé, B.

19?? *The functor Bext under the negation of CH*, preprint.

Dugas, M. and Vergohsen, R.

1988 *On socles of Abelian p-groups in* $L$, Rocky Mt. J. Math. **18**, 733–752.

Dugas, M. and Zimmermann-Huisgen B.

1981 *Iterated direct sums and products of modules*, in **Abelian Group Theory**, Lecture Notes in Mathematics No. 874, Springer-Verlag, 179–193.

Dushnik, B. and Miller, E. W.

1940 *Concerning similarity transformations of linearly ordered sets*, Bull. Amer. Math. Soc. **46**, 322–326.

Eda, K.

1982 *A boolean power and a direct product of abelian groups*, Tsukuba J. Math. **6**, 187–193.

1983a *A note on subgroups of* $Z^N$, in **Abelian Group Theory**, Lecture Notes in Mathematics No. 1006, Springer-Verlag, 371–374.

1983b    *Almost-slender groups and Fuchs-44-groups*, Comment. Math. Univ. St. Paul, **32**, 131–135.

1983c    *On Z-kernel groups*, Arch. Math. (Basel) **41**, 289–293.

1983d    *On a Boolean power of a torsion free abelian group*, J. Algebra **82**, 84–93.

1983e    *A minimal flabby sheaf and an abelian group*, Tsukuba J. Math. **7**, 157–168.

1984    *Z-kernel groups of measurable cardinalities*, Tsukuba J. Math. **8**, 95–100.

1985    *A generalized direct product of abelian groups*, J. Algebra **92**, 33–43.

1987    *A characterization of $\aleph_1$-free abelian groups and its application to the Chase radical*, Israel J. Math. **60**, 22–30.

1989    *Cardinality restrictions of preradicals*, **Abelian Group Theory**, Contemporary Mathematics vol. 87, Amer. Math. Soc., 277–283.

19??a    *Boolean powers of abelian groups*, preprint.

19??b    *Slender modules, endo-slender abelian groups and large cardinals*, Fund. Math., to appear.

Eda, K. and Abe, Y.

1987    *Compact cardinals and abelian groups*, Tsukuba J. Math. **11**, 353–360.

Eda, K. and Ohta, H.

1986    *On abelian groups of integer-valued continuous functions, their Z-dual and Z-reflexivity*, in **Abelian Group Theory** (ed. by R. Göbel & E. A. Walker), Gordon and Breach, 241–257.

Ehrenfeucht, A.

1955    *On a problem of J. H. C. Whitehead concerning Abelian groups*, Bull. Acad. Polon. Sci., Cl. III, 3, 127–128.

Ehrenfeucht, A. and Łoś, J.

1954    *Sur les produits cartésiens des groupes cycliques infinis*, Bull. Acad. Polon. Sci., Sér. Sci. Math. **2**, 261–263.

Eklof, P. C.

1973    *The structure of ultraproducts of abelian groups*, Pacific J. Math. **47**, 67–79.

1974    *Infinitary equivalence of abelian groups*, Fund. Math. **81**, 305–314.

1975    *On the existence of $\kappa$-free abelian groups*, Proc. Amer. Math. Soc. **47**, 65–72.

1977a   *Homological algebra and set theory*, Trans. Amer. Math. Soc. **227**, 207–225.

1977b   *Ultraproducts for algebraists*, in **Handbook of Mathematical Logic**, North-Holland, 105–137.

1977c   *Methods of logic in abelian group theory*, in **Abelian Group Theory**, Lecture Notes in Mathematics No. 616, Springer-Verlag, 251–269.

1980    **Set Theoretic Methods in Homological Algebra and Abelian Groups**, Les Presses de L'Université de Montréal.

1982    *On singular compactness*, Alg. Univ. **14**, 310–316.

1983    *The structure of $\omega_1$-separable groups*, Trans. Amer. Math.. Soc. **279**, 497–523

Eklof P. C. and Fisher, E. R.

1972    *The elementary theory of abelian groups*, Ann. Math. Logic **4**, 115–171.

Eklof, P. C. and Fuchs, L.

1988    *Baer modules over valuation domains*, Ann. Mat. Pura Appl.(IV) **150**, 363–374.

Eklof, P. C., Fuchs, L. and Shelah, S.

19??    *Baer modules over domains*, Trans. Amer. Math. Soc, to appear.

Eklof, P. C. and Huber, M.

1979    *Abelian group extensions and the axiom of constructibility*, Comment. Math. Helv. **54**, 440–457.

1980    *On the rank of Ext*, Math. Z. **174**, 159–185.

1981    *On the p-ranks of Ext$(A, G)$, assuming CH*, in **Abelian Group Theory**, Lecture Notes in Mathematics No. 874, Springer-Verlag, 93–108.

1985    *On $\omega$-filtered vector spaces and their application to abelian p-groups:I*, Comment. Math. Helvetici **60**, 145–171.

1988    *On $\omega$-filtered vector spaces and their application to abelian p-groups:II*, Rocky Mt. J. Math. **18**, 123–136.

Eklof, P. C., Huber, M. and Mekler, A.H.

1988    *Totally Crawley groups*, J. Algebra **112**, 370–384.

Eklof, P. C. and Mekler A. H.

1977    *On constructing indecomposable groups in L*, J. Algebra **49**, 96–103.

1981    *Infinitary stationary logic and abelian groups*, Fund. Math. **112**, 1–15.

1983    *On endomorphism rings of $\omega_1$-separable primary groups*, in **Abelian Group Theory**, Lecture Notes in Mathematics No. 1006, Springer-Verlag, 320–339.

1988    *Categoricity results for $L_{\infty\kappa}$-free algebras*, Ann. Pure and Appl. Logic **37**, 81–99.

Eklof, P. C., Mekler, A. H. and Shelah, S.

1984    *Almost disjoint abelian groups*, Israel J. Math. **49**, 34–54.

1987    *On strongly non-reflexive groups*, Israel J. Math. **59**, 283–298.

Eklof, P. C. and Shelah, S.

19??    *On Whitehead modules*, preprint.

Fay, T. H., Oxford, E. P. and Walls, G. L.

1982    *Preradicals in abelian groups*, Houston J. Math. **8**, 39–52.

1983a    *Preradicals induced by homomorphisms*, in **Abelian Group Theory**, Lecture Notes in Mathematics No. 1006, Springer-Verlag, 660–670.

1983b    *Singly generated socles and radicals*, in **Abelian Group Theory**, Lecture Notes in Mathematics No. 1006, Springer-Verlag, 671–684.

Fodor, G.

1956    *Eine Bermerkung zur Theorie der regressiven Funktionen*, Acta Sci. Math. (Szeged) **17**, 139–142.

Fuchs, L.

1960    **Abelian Groups**, Pergamon Press.

1970    **Infinite Abelian Groups**, vol.I, Academic Press.

1973    **Infinite Abelian Groups**, vol.II, Academic Press.

1983    *On projective dimensions of modules over valuation domains*, **Abelian Group Theory**, Lecture Notes in Mathematics No. 1006, 589–598.

1986     *Arbitrarily large indecomposable divisible torsion modules over certain valuation domains*, Rend. Sem. Mat. Univ. Padova **76**, 247–254.

1989     *Some applications of abelian group theory to modules*, in **Abelian Group Theory**, Contemporary Mathematics vol. 87, Amer. Math. Soc., 241–248.

Fuchs, L. and Salce, L.

1985     **Modules over Valuation Domains**, Marcel Dekker.

Fuchs, L. and Shelah, S.

1989     *Kaplansky's problem on valuation rings*, Proc. Amer. Math. Soc. **105**, 25–30.

Gale, D. and Stewart, F. M.

1953     *Infinite games with perfect information*, in **Contributions to the Theory of Games**, Annals of Mathematics Studies **28**, 245–266.

Gerstner, O., Kaup, L. and Weidner, H. G.

1969     *Whitehead-Moduln abzählbaren Ranges über Hauptidealringen*, Arch. Math. (Basel) **20**, 503–514.

Göbel, R.

1975     *On stout and slender groups*, J. Algebra **35**, 39–55.

1980     *Darstellung von Ringen als Endomorphismenringe*, Arch. Math. (Basel) **35**, 338–350.

1983     *Endomorphism rings of abelian groups*, in **Abelian Group Theory**, Lecture Notes in Mathematics No. 1006, Springer-Verlag, 340–353.

1986     *Wie weit sind Moduln vom Satz von Krull-Remak-Schmidt entfernt?*, Jber. d. Dt. Math.-Verein. **88**, 11–49.

Göbel, R., Richkov, S. V. and Wald, B.

1981     *A general theory of slender groups and Fuchs 44-groups*, in **Abelian Group Theory**, Lecture Notes in Mathematics No. 874, Springer-Verlag, 194–201.

Göbel, R. and Shelah, S.

1988     *Semi-rigid classes of cotorsion-free abelian groups*, J. Algebra **93**, 136–150.

Göbel, R. and Wald, B.
  1979  *Wachstumstypen und schlanke Gruppen*, Symposia Math. **23**, 201–239.
  1980  *Martin's Axiom implies the existence of certain slender groups*, Math. Z. **172**, 107–121.

Gregory, J.
  1973  *Abelian groups infinitarily equivalent to free ones*, Notices Amer. Math. Soc. **20**, A–500.
  1976  *Higher Souslin trees and the generalized continuum hypothesis*, J. Symbolic Logic **41**, 663–671.

Griffith, P. A.
  1968  *Separability of torsion-free groups and a problem of J. H. C. Whitehead*, Illinois J. Math. **12**, 654–659.
  1969a  *A solution to the splitting mixed problem of Baer*, Trans. Amer. Math. Soc. **139**, 261–269.
  1969b  *A note on a theorem of Hill*, Pacific J. Math. **29**, 279–284.
  1970  **Infinite abelian groups**, Univ. of Chicago Press.

Grossberg, R. and Shelah, S.
  1989  *On the structure of $Ext_p(G, \mathbf{Z})$*, J. Algebra, **121**, 117–128.

Hall, M., Jr.
  1948  *Distinct representatives of subsets*, Bull. Amer. Math. Soc. **54**, 922–926.

Hall, P.
  1935  *On representatives of subsets*, J. London Math. Soc. **10**, 26–30.

Harrington, L. and Shelah, S.
  1985  *Some exact equiconsistency results in set theory*, Notre Dame J. Formal Logic **26**, 178–188.

Hausen, J.
  1981  *On generalizations of projectivity for modules over Dedekind domains*, J. Austral. Math. Soc. Ser. A **31**, 207–216.

Heinlein, G.
  1971  **Vollreflexive Ringe und schlanke Moduln**, doctoral dissertation, Erlangen.

Herrlich. H.
1967   *C-kompakte Räume*, Math Z. **96**, 228–255.

Higman, G.
1951   *Almost free groups*, Proc. London Math. Soc. (3) **1**, 284–290.
1989   *Some countably free groups*, in **Group Theory (Singapore 1987)**, deGruyter, 129–150.

Hill, P.
1969   *On the decomposition of groups*, Canad. J. Math. **21**, 762–768.
1970   *On the freeness of abelian groups: a generalization of Pontryagin's theorem*, Bull. Amer. Math. Soc. **76**, 1118–1120.
1972   *Primary groups whose subgroups of smaller cardinality are direct sums of cyclic groups*, Pacific J. Math. **42**, 63–67.
1973   *New criteria for freeness in abelian groups, I*, Trans. Amer. Math. Soc. **182**, 201–209.
1974a  *New criteria for freeness in abelian groups, II*, Trans. Amer. Math. Soc. **196**, 191–201.
1974b  *A special criterion for freeness*, Symposia Math. **13**, Academic Press, 311–314.
1974c  *On the splitting of modules and abelian groups*, Canad. J. Math. **26**, 68–77.

Hiller, H., Huber, M. and Shelah, S.
1978   *The structure of Ext(A, Z) and V = L*, Math. Z. **162**, 39–50.

Hiller, H. and Shelah, S.
1977   *Singular cohomology in L*, Israel J. Math. **26**, 313–319.

Hiremath, V. A.
1978   *Finitely projective modules over a Dedekind domain*, J. Austral. Math. Soc. Ser. A **26**, 330–336.

Hodges, W.
1981   *In singular cardinality, locally free algebras are free*, Algebra Universalis **12**, 205–220.

Huber, M.
1979   *On cartesian powers of a rational group*, Math. Z. **169**, 253–259.

1983a *On reflexive modules and abelian groups*, J. Algebra **82**, 469–487.

1983b *Methods of set theory and the abundance of separable abelian p-groups*, in **Abelian Group Theory**, Lecture Notes in Mathematics No. 1006, Springer-Verlag, 304–319.

Huber, M. and Warfield, R. B., Jr.

1979 *On the torsion subgroup of Ext(A, G)*, Arch. Math. (Basel) **32**, 5–9.

1981 *Homomorphisms between cartesian powers of an abelian group*, in **Abelian Group Theory**, Lecture Notes in Mathematics No. 874, Springer-Verlag, 202–227.

Hulanicki, A.

1957 *Algebraic characterization of abelian divisible groups which admit compact topologies*, Fund. Math. **44**, 192–197.

Hyttinen, T. and Väänänen, J.

19?? *On Scott and Karp trees of uncountable models*, J. Symbolic Logic, to appear.

Ivanov, A.V.

1978 *A problem on abelian groups*, Math. USSR Sbornik **34**, 461–474.

1979 *Direct and complete direct sums of abelian groups* (in Russian), Vestnik Mosc., Ser. 1, Matem., Mech., **37**, 96.

1981 *One class of abelian groups*, Matem. Zametki, **29**, 351–358.

Jech, T.

1978 **Set Theory**, Academic Press.

1986 **Multiple Forcing**, Cambridge University Press.

Jensen, C. U.

1972 **Les Foncteurs Dérivés de lim et leurs Applications en Théorie des Modules**, Lecture Notes in Mathematics No. 254, Springer-Verlag.

Jensen, R. B.

1972 *The fine structure of the constructible hierarchy*, Ann. Math. Logic 4, 229–308.

Kanamori, A. and Magidor, M.
  1978   *The evolution of large cardinal axioms in set theory*, in
         **Higher Set Theory**, Lecture Notes in Mathematics 669,
         Springer-Verlag, 99–275.

Kaplansky, I.
  1952   *Modules over Dedekind rings and valuation rings*, Trans.
         Amer. Math. Soc. **72**, 327–340.
  1958   *Projective modules*, Ann. of Math. (2) **68**, 372–377.
  1969   **Infinite abelian groups**, rev. ed., Univ. of Michigan
         Press.

Keef, P.
  1989   *On set theory and the balanced-projective dimension of* $C_\Omega$
         *groups*, in **Abelian Group Theory**, Contemporary Math-
         ematics vol. 87, Amer. Math. Soc., 31–41.

Keisler, H. J.
  1964a  *On cardinalities of ultraproducts*, Bull Amer. Math. Soc.
         **70**, 644–647.
  1964b  *Good ideals in fields of sets*, Ann. Math. **79**, 338–359.
  1965   *A survey of ultraproducts*, in **Logic, Methodology and
         Philosophy of Science**, North-Holland, 112–126.

König, D.
  1926   *Sur les correspondances multivoques des ensembles*, Fund.
         Math. **8**, 114–134.

Kueker, D. W.
  1981   $L_{\infty\omega_1}$-*elementarily equivalent models of power* $\omega_1$, in **Logic
         Year 1979–80**, Lecture Notes in Mathematics 859,
         Springer-Verlag, 120–131.

Kunen, K.
  1970   *Some applications of iterated ultrapowers in set theory*,
         Ann. Math. Logic **1**, 179–227.
  1972   *Ultrafilters and independent sets*, Trans. Amer. Math. Soc.
         **172**, 299–306.
  1980   **Set Theory**, North-Holland.

Lady, E. L.
  1973   *Slender rings and modules*, Pacific J. Math. **49**, 397–406.

Lawrence, J.
  1984   *Countable abelian groups with a discrete norm are free,*
          Proc. Amer. Math. Soc. **90**, 352–354.

Łoś, J.
  1955   *Quelques remarques, théorèmes et problèmes sur les classes*
          *définissables d'algèbres,* in **Mathematical Interpreta-**
          **tions of Formal Systems,** North-Holland, 98–113.

Mader, A.
  1982   *Duality and completions of linearly topologized modules,*
          Math. Z. **179**, 325–335.

  1984   *Groups and modules that are slender as modules over their*
          *endomorphism rings,* in **Abelian Groups and Modules,**
          CISM Courses and Lectures No. 287, Springer-Verlag, 315–
          327.

Magidor, M.
  1982   *Reflecting stationary sets,* J. Symbolic Logic **47**, 755–771.
Magidor, M. and Shelah, S.
  19??   *When does almost free imply free?* (*For groups, transversal*
          *etc.*), preprint.

Magidor, M. and Solovay, R.
  19??   to appear.

Martin, D A. and Solovay, R. M.
  1970   *Internal Cohen extensions,* Ann. Math. Logic **2**, 143–178.
Matlis, E.
  1972   **Torsion-free Modules,** Univ. of Chicago Press.
Megibben, C.
  1974   *Generalized pure injectivity,* Symposia Math. **13**, Academic
          Press, 257–271.

  1983   *Crawley's problem on the unique $\omega_1$-elongation of p-groups*
          *is undecidable,* Pacific J. Math. **107**, 205–212.

  1987   $\omega_1$-*separable p-groups,* in **Abelian Group Theory** (ed. by
          R. Göbel & E. A. Walker), Gordon and Breach, 117–136.

Mekler, A. H.
  1977   *The number of $\kappa$-free abelian groups and the size of Ext,* in
          **Abelian Group Theory,** Lecture Notes in Mathematics
          No. 616, Springer-Verlag, 323–331.

1980   *How to construct almost free groups*, Canad. J. Math **32**, 1206–1228.

1981   $\aleph_1$-*separable groups of mixed type*, in **Abelian Group Theory**, Lecture Notes in Mathematics No. 874, Springer-Verlag, 114–126.

1982   *Shelah's Whitehead groups and CH*, Rocky Mountain J. Math. **12**, 271–278.

1983   *Proper forcing and abelian groups*, in **Abelian Group Theory**, Lecture Notes in Mathematics No. 1006, Springer-Verlag, 285–303.

1984   *c.c.c. forcing without combinatorics*, J. Symbolic Logic **49**, 830–832.

1987   *The structure of groups that are almost the direct sum of countable abelian groups*, Trans. Amer. Math. Soc. **303**, 145–160.

19??   *Almost free groups in varieties*, J. Algebra, to appear.

Mekler, A. H. and Shelah, S.

1986a  $\omega$-*elongations and Crawley's problem*, Pacific J. Math. **121**, 121–132.

1986b  *The solution to Crawley's problem*, Pacific J. Math. **121**, 133–134.

1987   *When $\kappa$-free implies strongly $\kappa$-free*, in **Abelian Group Theory** (ed. by R. Göbel & E. A. Walker), Gordon and Breach, 137–148.

1988   *Diamond and $\lambda$-systems*, Fund. Math. **131**, 45–51.

1989   *Uniformization principles*, J. Symbolic Logic **54**, 441–459.

19??a  $L_{\infty\omega}$-*free algebras*, Algebra Universalis, **26**, 351–366.

19??b  *The p-rank of Ext*, to appear.

19??c  *The consistency strength of "every stationary set reflects"*, Israel J. Math., to appear.

Milner, E. C.

1974   *Transversal theory*, Proceedings of the International Conference of Mathematicians, Vancouver, B. C., vol. I, 155–169.

Mitchell, W. J.
  1984   *The core model for sequences of measures I*, Math. Proc.
         Camb. Phil. Soc. **95**, 229–260.

Mrówka S.
  1972   *Structures of continuous functions. VIII. Homomorphisms*
         *of groups of integer-valued continuous functions*, Bull. Acad
         Polon. Sci. Sér. Sci. Math. **20**, 563–566.

Mueller, B. J.
  1970   *On semi-perfect rings*, Illinois J. Math. **14**, 464–467.

Mycielski, J.
  1964   *Some compactifications of general algebras*, Coll. Math. **13**,
         1–9.

Neumann, H.
  1967   **Varieties of Groups**, Springer-Verlag.

Nöbeling, G.
  1968   *Verallgemeinerung eines Satzes von Herrn E. Specker*, In-
         vent. Math. **6**, 41–55.

Nunke, R. J.
  1959   *Modules of extensions over Dedekind rings*, Illinois J. Math.
         **3**, 222–241.

  1962a  *On direct products of infinite cyclic groups*, Proc. Amer.
         Math. Soc. **13**, 66–71.

  1962b  *Slender groups*, Acta Sci. Math. (Szeged) **23**, 67–73.

  1977   *Whitehead's Problem*, in **Abelian Group Theory**, Lec-
         ture Notes in Mathematics No. 616, Springer-Verlag, 240–
         250.

Pontryagin, L. S.
  1934   *The theory of topological commutative groups*, Ann.   of
         Math. **35**, 361–388.

Pope, A. L.
  1984   *Almost free groups in varieties*, J. Algebra **91**, 36–52.

Prest, M.
  1988   **Model Theory and Modules**, London Math. Soc. Lec-
         ture Notes Series No. 130, Cambridge Univ. Press.

Prikry, K. and Solovay, R. M.
  1975    *On partitions into stationary sets*, J. Symbolic Logic **40**, 75–80.

Ramsey, F. P.
  1930    *On a problem of formal logic*, Proc. London Math. Soc. (2) **30**, 264–286.

Reid, G. A.
  1967    **Almost Free Abelian Groups**, Lecture notes, Tulane University, unpublished.

Rotman, J. J.
  1961    *On a problem of Baer and a problem of Whitehead in abelian groups*, Acad. Sci. Hungar. **12**, 245–254.

  1979    **An Introduction to Homological Algebra**, Academic Press.

Rychkov, S. V.
  1980    *On cartesian products of abelian groups*, Soviet Math. Doklady **21**, 747–748.

Rychkov, S. V. and Thomé, B.
  1986    *Slender groups and related concepts*, Comm. Algebra **14(2)**, 333–387.

  1988    *Indecomposable abelian p-groups that are almost direct sums of cyclic groups*, (Russian), Mat. Zametki **43**, 705–712; translated in Math. Notes **43**, 405–408.

Sabbagh, G. and Eklof, P. C.
  1971    *Definability problems for modules and rings*, J. Symbolic Logic **36**, 623–649.

Sageev, G. and Shelah, S.
  1981    *Weak compactness and the structure of Ext(A, Z)*, in **Abelian Group Theory**, Lecture Notes in Mathematics No. 874, Springer-Verlag, 87–92.

  1985    *On the structure of Ext(A, Z) in ZFC⁺*, J. Symbolic Logic **50**, 302–315.

Schlitt, G. S.
  1986    **Some results in reflexivity**, M. Sc. Thesis, Simon Fraser Univ.

Scott, D. S.
    1961  *Measurable cardinals and constructible sets*, Bull.  Acad.
          Polon. Sci., Sér. Sci. Math. **9**, 521–524.


Shelah, S.
    1974  *Infinite abelian groups, Whitehead problem and some con-
          structions*, Israel J. Math **18**, 243–256.

    1975a *A compactness theorem for singular cardinals, free algebras,
          Whitehead problem and transversals*, Israel J. Math. **21**,
          319–349.

    1975b *Existence of rigid-like families of abelian p-groups*, in
          **Model Theory and Algebra**, Lecture notes in Mathe-
          matics No. 498 384–402.

    1977  *Whitehead groups may not be free even assuming CH, I*,
          Israel J. Math. **28**, 193–203.

    1978  **Classification Theory and the Number of Non-
          isomorphic Models**, North-Holland.

    1979a *On uncountable abelian groups*, Israel J. Math. **32**, 311–330.

    1979b *On successors of singular cardinals*, in **Logic Colloquium
          '78**, North-Holland, 357–380.

    1980  *Whitehead groups may not be free even assuming CH, II*,
          Israel J. Math. **35**, 257–285.

    1981a *On endo-rigid strongly $\aleph_1$-free abelian groups in $\aleph_1$*, Israel
          J. Math. **40**, 291–295.

    1981b *The consistency of $Ext(G, \mathbf{Z}) = \mathbf{Q}$*, Israel J. Math. **39**,
          74–82.

    1981c *Models with second order properties. III. Omitting types for
          L(Q)*, Arch. Math. Logik **21**, 1–11.

    1982  **Proper Forcing**, Lecture Notes in Mathematics 940,
          Springer-Verlag.

    1984a *A combinatorial principle and endomorphism rings I: on
          p-groups*, Israel J. Math. **49**, 239–257.

    1984b *A combinatorial theorem and endomorphism rings of
          Abelian groups II*, in **Abelian Groups and Modules**,
          CISM Courses and Lectures No. 287, Springer-Verlag, 37–
          86.

1984c *Diamonds, uniformization*, J. Symbolic Logic **49**, 1022–1033.

1985 *Incompactness in regular cardinals*, Notre Dame J. Formal Logic **26**, 195–228.

1986 *Remarks on squares*, in **Around Classification Theory of Models**, Lecture Notes in Mathematics 1182, Springer-Verlag 276–279.

1987 *On reconstructing separable reduced p-groups with a given socle*, Israel J. Math. **60**, 146–166.

19??a **Universal Classes**, *Chapter III*, manuscript.

19??b *Kaplansky test problem for R-modules*, preprint.

Solovay, R. M.
1971 *Real-valued measurable cardinals*, in **Axiomatic Set Theory**, (Part 1), Amer. Math. Soc. Proc. of Symp. in Pure Math. XIII, 397–428.

Solovay, R. M. and Tenenbaum, S.
1971 *Iterated Cohen extensions and Souslin's problem*, Ann. of Math.(2) **94**, 201–245.

Specker, E.
1950 *Additive Gruppen von Folgen ganzer Zahlen*, Portugaliae Math. **9**, 131–140.

Stein, K.
1951 *Analytische Funktionen mehrerer komplexer Veränderlichen zu vorgegebenen Periodizitätsmoduln und das zweite Cousinsche Problem*, Math. Ann. **123**, 201–222.

Steinhorn, C. I. and King, J. H.
1980 *The uniformization property for $\aleph_2$*, Israel J. Math. **36**, 248–256.

Steprāns, J.
1985 *A characterization of free abelian groups*, Proc. Amer. Math. Soc. **93**, 347–349.

Thomé, B.
1988 **Aleph-1-separable Groups, Kaplansky's Test Problems, and Endomorphism Rings**, Ph.D. Dissertation, Univ. of California, Irvine.

# BIBLIOGRAPHY

19??a  $\aleph_1$-separable groups and Kaplansky's test problems, Forum Math., to appear.

19??b  Kaplansky's second test problem for $\aleph_1$-separable groups, preprint.

Ulam, S.
1930  Zur Masstheorie in der allgemeinen Mengenlehre, Fund. Math. **16**, 140–150.

Veličković, B.
19??  Forcing axioms and stationary sets, Advances in Math., to appear.

Wald, B.
1983a  Martinaxiom und die Beschreibung gewisser Homomorphismen in der Theorie der $\aleph_1$-freien abelschen Gruppen, Manuscripta Math. **42**, 297–309.

1983b  On $\kappa$-products modulo $\mu$-products, in **Abelian Group Theory**, Lecture Notes in Mathematics No. 1006, Springer-Verlag, 362–370.

1984  The non-slender rank of an abelian group, in **Abelian Groups and Modules**, CISM Courses and Lectures No. 287, Springer-Verlag, 221–231.

1985  Integer valued functions with countable support, Arch. Math. **45**, 203–206.

1987  On the groups $Q_\kappa$, in **Abelian Group Theory** (ed. by R. Göbel & E. A. Walker), Gordon and Breach, 229–240.

Warfield, R. B., Jr.
1969  Purity and algebraic compactness for modules, Pacific J. Math. **28**, 699–719.

Wolfson, K. G.
1963  Isomorphisms of the endomorphism rings of a class of torsion-free modules, Proc. Amer. Math. Soc. **14**, 589–594.

Zeeman, E. C.
1955  On direct sums of free cycles, J. London Math. Soc. **30**, 195–212.

Zimmermann-Huisgen, B.
1979  On Fuchs' problem 76, J. Reine Angew. Math. **309**, 86–91.

Zorzitto, F.
  1985    *Discretely normed abelian groups*, Aequationes Math. **29**,
          172–174.

# INDEX